Reinforcement Learning for Sequential Decision and Optimal Control

Shengbo Eben Li

Reinforcement Learning for Sequential Decision and Optimal Control

 Springer

Shengbo Eben Li
School of Vehicle and Mobility
Tsinghua University
Beijing, China

ISBN 978-981-19-7786-2   ISBN 978-981-19-7784-8   (eBook)
https://doi.org/10.1007/978-981-19-7784-8

This Springer imprint is published by the registered company Springer Nature Singapore Pte Ltd.
The registered company address is: 152 Beach Road, #21-01/04 Gateway East, Singapore 189721, Singapore

About the Author

 Shengbo Eben Li received his M.S. and Ph.D. degrees from Tsinghua University in 2006 and 2009. He is currently a professor at Tsinghua University in the interdisciplinary field of autonomous driving and artificial intelligence. Before joining Tsinghua University, he has worked at Stanford University, University of Michigan, and UC Berkeley. His active research interests include intelligent vehicles and driver assistance, deep reinforcement learning, optimal control and estimation, etc.

He has published more than 130 peer-reviewed papers in top-tier international journals and conferences. He is the recipient of best paper awards (finalists) of IEEE ITSC, ICCAS, IEEE ICUS, IEEE IV, L4DC, etc. He has received a number of important academic honors, including National Award for Technological Invention of China (2013), National Award for Progress in Sci & Tech of China (2018), Distinguished Young Scholar of Beijing NSF (2018), Youth Sci & Tech Innovation Leader from MOST China (2020), etc. He also serves as Board of Governor of IEEE ITS Society, Senior AE of IEEE OJ ITS, and AEs of IEEE ITSM, IEEE Trans ITS, Automotive Innovation, etc.

Contact information:

Shengbo Eben Li

Ph.D., Professor

Rm 643, Lee Shau Kee Sci&Tech Building,

Tsinghua University, 100084, Beijing

Office: 86-10-62796150

Email: lishbo@tsinghua.edu.cn

Preface

Since the beginning of the 21st century, artificial intelligence (AI) has been reshaping almost all areas of human society, which has high potential to spark the fourth industrial revolution. Notable examples can be found in the sector of road transportation, where AI has drastically changed automobile design and traffic management. Many new technologies, such as driver assistance, autonomous driving, and cloud-based cooperation, are emerging at an unbelievable speed. These new technologies have the potential to significantly improve driving ability, reduce traffic accidents, and relieve urban congestion.

As one of the most important AI branches, reinforcement learning (RL) has attracted increasing attention in recent years. RL is an interdisciplinary field of trial-and-error learning and optimal control that promises to provide optimal solutions for decision-making or control in large-scale and complex dynamic processes. One of its most conspicuous successes is AlphaZero from Google DeepMind, which has beaten the highest-level professional human player. The underlying key technology is the so-called deep reinforcement learning, which equips AlphaGo with amazing self-evolution ability and high playing intelligence.

Despite a few successes, the application of RL is still in its infancy because most RL algorithms are rather difficult to comprehend and implement. RL connects deeply with statistical learning and convex optimization, and involves a wide range of new concepts and theories. As a beginner, one must undergo a long and tedious learning process to become an RL master. Without fully understanding those underlying principles, it is very difficult for new users to make proper adjustments to achieve the best application performance. This book aims to provide a systematic introduction to fundamental RL theories, mainstream RL algorithms and typical RL applications for researchers and engineers. The main topics include Markov decision processes, Monte Carlo learning, temporal difference learning, RL with function approximation, policy gradient method, approximate dynamic programming, deep reinforcement learning, etc.

The book contains 11 chapters. Chapter 1 provides an overview of RL, including its history, famous scholars, successful examples and up-to-date challenges. Chapter 2 discusses the basis of RL, including main concepts and terminologies, Bellman's optimality condition, and general problem formulation. Chapter 3 introduces Monte Carlo learning methods for model-free RL, including on-policy/off-policy methods and importance sampling technique. Chapter 4 introduces temporal difference learning methods for model-free RL, including Sarsa, Q-learning, and expected Sarsa. Chapter 5 introduces stochastic dynamic programming (DP), i.e., model-based RL with tabular representation, including value iteration DP, policy iteration DP and their convergence mechanisms. Chapter 6 introduces how to approximate policy and value functions in indirect RLs as well as the associated actor-critic architecture. Chapter 7 derives different kinds of direct policy gradients, including likelihood ratio gradient, natural policy gradient and a few advanced variants. Chapter 8 introduces infinite-horizon ADP, finite-horizon ADP and its connection with model predictive control. Chapter 9 discusses how to handle state constraints and its connection with feasibility and safety, as well as the newly proposed actor-critic-scenery learning architecture. Chapter 10 is devoted to deep reinforcement learning, including how to train artificial neural networks and typical deep RL algorithms such as DQN, DDPG, TD3, TRPO, PPO, SAC, and DSAC. Finally, Chapter 11 provides various RL relics, including robust RL, POMDP, multi-agent RL, meta-RL, inverse RL, offline RL, major RL libraries and platforms. Any comments and corrections from readers would be much appreciated. Please feel free to send your comments and suggestions to lishbo@tsinghua.edu.cn.

I wish to offer my sincere gratitude to all the members at the Intelligent Driving Laboratory (iDLab) for their great supports and contributions to the writing of this book. My understanding of reinforcement learning and dynamic programming gained significantly from continuing interactions with Jingliang Duan, Yang Guan, Haitong Ma, Yangang Ren, Jie Li, Yuhang Zhang, Yao Lyu, Wehan Cao, Ziyu Lin, Wenxuan Wang, Ziqing Gu, Baiyu Peng, Yao Mark Mu, Yuheng Lei, and many other talented students in iDLab. I want to thank Jianyu Chen, Xianyuan Zhan, Jun Zhu, Chongjie Zhang, Gao Huang from Tsinghua University, Changliu Liu from Carnegie Mellon University, Wei Zhan from UC Berkeley, Feng Gao and Guofa Li from Chongqing University, Dongbin Zhao from Chinese Academy of Sciences, Chang Liu from Peking University, Jianye Hao from Tianjin University for their priceless comments and suggestions. Special thanks should be given to my former doctoral supervisor Prof. Keqiang Li for his guidance and instruction that lead me to this attractive research field. My deep appreciation is also given to lots of other collaborators and friends who have helped improve and publish this book.

In closing, I would like to express my sincere gratitude to the entire Springer editorial staff for their remarkable support and dedication to getting this book published. Their unwavering dedication and professionalism made this book possible, and I am honored to acknowledge their hard work in the preface. Special thanks to Celine Chang, Sudha Ramachandran, Ellen Seo, Michela Castrica and Baoan Sang for their invaluable contributions throughout the editing process. Without their keen insights and meticulous attention to detail, this book would never have come to fruition. In addition, I would like to extend my heartfelt appreciation to Shuqin Li from Tsinghua University for her untiring assistance, which made the book writing and publishing process a smooth and enjoyable experience. Finally, I am grateful to my family for their love, encouragement and support while this book was being written.

The original online version of this book has been revised. The correction to the online book can be found at https://doi.org/10.1007/978-981-19-7784-8_12.

• Symbols

• General Notations

$=$	Equal to
$\approx$	Approximately equal
$\stackrel{\text{def}}{=}$	Equal by definition
$\Leftrightarrow$	If and only if
$\rightarrow$	Assign to
$\propto$	Proportional to
$\in$	Is an element of, e.g., $s \in \mathcal{S}$
$\subset$	Is a subset of, e.g., $\mathcal{S} \subset \mathbb{S}$
$\mathbb{R}$	Set of real numbers
$\Pr\{\cdot\}$	Event probability
$\Pr\{X \leq x\}$	Cumulative probability
$p(x), p_X(x)$	Probability mass function of discrete random variable X, i.e., $p_X(x) = \Pr\{X = x\}$
	Probability density function of continuous random variable X, i.e., $\int_a^b p_X(x)\mathrm{d}x = \Pr\{a \leq X \leq b\}$
$p(x,y), p_{XY}(x,y)$	Joint probability distribution of random variables X, Y
$p(x\vert y), p_{X\vert Y}(x\vert y)$	Conditional distribution of random variable X given $Y = y$
$X{\sim}\mathcal{T}(\theta)$	X obeys a certain distribution $\mathcal{T}(\theta)$
$\mathcal{N}(\mu, \sigma^2)$	Gaussian distribution with mean μ and variance σ^2
$\mathbb{E}\{X\}$	Expectation of random variable X, i.e., $\mu = \mathbb{E}\{X\} = \int xp(x)\mathrm{d}x$
$\mathbb{D}\{X\}$	Variance of random variable X, i.e., $\mathbb{D}\{X\} = \mathbb{E}\{(X - \mu)^2\} = \mathbb{E}\{X^2\} - \mu^2$
$J(\cdot)$	Scalar performance index
$\vert\cdot\vert$	Absolute value of a scalar
$\Vert\cdot\Vert_1, \Vert\cdot\Vert_2, \Vert\cdot\Vert_\infty$	1-norm, 2-norm, ∞-norm of vectors or matrices

• Markov Decision Process (MDP)

t	Time step
k	Main iteration of RL/ADP (i.e., cycle of critic and actor)
i, j	Inner iteration of RL/ADP (i.e., critic update and actor update)
s	State
s'	Next state
S	State space
$\vert S \vert$	Number of elements in finite state space
a	Action
a'	Next action
$\mathcal{A}$	Action space
$\vert \mathcal{A} \vert$	Number of elements in finite action space
π	Policy
π'	Next policy
$r, r_{ss'}^{a}$	Immediate reward from s to s' under action a
$\mathcal{P}, \mathcal{P}_{ss'}^{a}$	Transition probability of environment dynamics from state s to state s' when taking action a
γ	Discount factor
G_t	Return that starts from time t
$\pi(s)$	Deterministic policy
$\pi(a\vert s)$	Stochastic policy
$v^{\pi}(s)$	State-value function under policy π
$q^{\pi}(s,a)$	Action-value function under policy π
$\mathbb{E}_{\pi}\{\cdot\vert s\}$	Expectation under π and $\mathcal{P}$ given an initial state s; for brevity, shorten as $\mathbb{E}_{\pi}\{\cdot\}$
$d^{\mathrm{init}}(s)$	Initial state distribution
π^{*}	Optimal policy
$v_{*}(s)$	Optimal state-value function
$q_{*}(s,a)$	Optimal action-value function
π^{g}	Greedy policy
π^{ε}	ε-greedy policy
D	Set of data samples

$\mathcal{B}$ Mini-batch of data set

• Monte Carlo & Temporal Difference

$G_{t:T}$ Return that starts from time t and ends at T

π Target policy

b Behavior policy

$V^\pi(s)$ State-value estimate under policy π

$V^k(s)$ State-value estimate under policy π_k in policy iteration, i.e.,
$$V^k(s) = V^{\pi_k}(s)$$

$V_k(s)$ State-value estimate at k-th value iteration

$Q^\pi(s,a)$ Action-value estimate under policy π

$Q^k(s,a)$ Action-value estimate under policy π_k in policy iteration, i.e.,
$$Q^k(s,a) = Q^{\pi_k}(s,a)$$

$Q_k(s,a)$ Action-value estimate at k-th value iteration

π_k Policy at k-th main iteration

$V_j^\pi(s)$ Estimate of $v^\pi(s)$ at j-th inner iteration of PEV

$Q_j^\pi(s,a)$ Estimate of $q^\pi(s,a)$ at j-th inner iteration of PEV

$\rho_{t:t+h}$ Importance sampling ratio from time t to time $t+h$

$\rho_t, \rho_{t:t}$ Importance sampling ratio at time t, i.e., $\rho_t = \rho_{t:t}$

$G_{t:t+n}$ n-step return from time t to time $t+n$

G_t^λ λ-return from time t

ϵ Exploration rate

α Learning rate for critic update

β Learning rate for actor update

δ_{TD}^t One-step TD error

$\delta_{TD(n)}^t$ n-step TD error

$p_B(a)$ Boltzmann distribution

• Dynamic Programming

$s_{(i)}$ The i-th element in finite state space, $i \in \{1,2,\cdots,n\}$

$a_{(i)}$ The i-th element in finite action space, $i \in \{1,2,\cdots,m\}$

$s'_{(i)}$ Next state of i-th state element

$G_{\text{ave}}(\pi), J_{\text{ave}}(\pi)$	Average return and average cost under policy π
$J_\gamma(\pi)$	Discounted cost under policy π
π^{ave}_*	Optimal policy for average cost
$d^{\text{ave}}_*(s)$	Stationary state distribution under optimal policy π^{ave}_*
π^γ_*	Optimal policy for discounted cost
$d^\gamma_*(s)$	Stationary state distribution under optimal policy π^γ_*
$\mathcal{L}(\cdot)$	Linear operator
$\mathcal{B}(\cdot)$	Bellman operator
$G_\Delta(t)$	Differential return
$v^\pi_\nabla(s)$	Differential state-value function
$V(s,t)$	Time-dependent value function
K_{Ave}, K_γ	Feedback gain of stochastic LQ control
P_{Ave}, P_γ	Solution of Riccati equation
$\mathcal{H}(\pi)$	Entropy of a policy
$\rho(A)$	Spectral radius of matrix A
$\text{tr}(A)$	Trace of matrix A, i.e., sum of diagonal elements
● **Function Approximation**	
$F(s)$	Feature vector for linear approximation
$V(s;w)$	Parameterized state-value function with parameter w
$Q(s,a;w)$	Parameterized action-value function with parameter w
$\pi_\theta(a|s;\theta), \pi(a|s;\theta)$	Parameterized stochastic policy with parameter θ
$\pi_\theta(s;\theta), \pi(s;\theta)$	Parameterized deterministic policy with parameter θ
$\nabla J(w), \nabla_w J(w)$	Gradient of PEV loss function
$\nabla J(\theta), \nabla_\theta J(\theta)$	Gradient of PIM loss function
$\zeta_{i,j}$	One-step transition probability from $s_{(i)}$ to $s_{(j)}$
$d_\pi(s), d_b(s)$	Stationary state distribution under policy π and policy b
$d^\gamma_{\pi_\theta}(s)$	Discounted state distribution under π_θ
$\zeta(s)$	Baseline function
$\mathbb{E}_{s\sim d_\pi}$	Expectation under stationary state distribution of policy π, i.e.,

$$\sum_{s\in S} d_\pi(s)\cdot *$$

Symbol	Meaning
$\mathbb{E}_{a\sim\pi}$	Expectation under policy π given a state s, i.e., $\displaystyle\sum_{a\in A}\pi(a\mid s)\cdot *$
$\mathbb{E}_{s\sim d_\pi,a\sim\pi}$	Expectation under both $d_\pi(s)$ and $\pi(a\mid s)$, i.e., $\displaystyle\sum_{s\in S} d_\pi(s)\left\{\sum_{a\in A}\pi(a\mid s)\cdot *\right\}$
$\mathbb{E}_{s\sim d_\pi,a\sim\pi,s'\sim P}$	Expectation under initial state distribution $d_\pi(s)$, policy π and environment dynamics P
$\mathbb{E}_{s_0,a_0,s_1,a_1,\dots\sim\pi}$	Expectation under the trajectory that is generated by policy π
$\mathbb{E}_\pi$	Notation for $\mathbb{E}_\pi \overset{\text{def}}{\equiv} \mathbb{E}_{s\sim d_\pi,a\sim\pi,s'\sim P}$ or $\mathbb{E}_\pi \overset{\text{def}}{\equiv} \mathbb{E}_{s_0,a_0,s_1,a_1,\dots\sim\pi}$
$d_t(s), d(s_t)$	State distribution of state $s_t = s$ at time t
$\mathcal{D}_\pi, \mathcal{D}_b$	Set of samples from policy π or policy b
J_{Critic}	Critic loss function
J_{Actor}	Actor loss function

● **Approximate Dynamic Programming**

Symbol	Meaning
u	Action, $u \in \mathbb{R}^m$
$\mathcal{U}$	Action space, $\mathcal{U} \subset \mathbb{R}^m$
x	State, $x \in \mathbb{R}^n$
$\mathcal{X}$	State space, $\mathcal{X} \subset \mathbb{R}^n$
$V(x)$	Cost function or State-value function
$l(x,u)$	Utility function
$f(x,u)$	Deterministic model of environment dynamics
$V^*(x)$	Optimal cost function
π^*	Optimal policy
$V(x;w)$	Parameterized cost function
$\pi(x;\theta)$	Parameterized policy
$V(x,t)$	Time-dependent value function
$\pi(x,t)$	Time dependent policy
$\dfrac{\partial J}{\partial\theta}, \dfrac{\partial J}{\partial\theta^\top}$	$\dfrac{\partial J}{\partial\theta} \in \mathbb{R}^l, \dfrac{\partial J}{\partial\theta^\top} \in \mathbb{R}^{1\times l}, \text{if } \theta \in \mathbb{R}^l$

$\frac{\partial \pi^{\mathsf{T}}}{\partial \theta}$	$\frac{\partial \pi^{\mathsf{T}}}{\partial \theta} \in \mathbb{R}^{l \times m}$, if $\theta \in \mathbb{R}^{l}$ and $\pi \in \mathbb{R}^{m}$
T	Fixed terminal time
N	Length of fixed horizon
$\tau \in [t, T]$	Time interval in continuous-time domain
$i \in \{1, \cdots, N\}$	Time interval in discrete-time domain
x_{t+i}, u_{t+i}	State and action at time $t + i$ in discrete real-time domain
$x(\tau), u(\tau)$	State and action at time τ in continuous real-time domain
$x_{i|t}, u_{i|t}$	State and action at time i in virtual-time domain starting at time t
x^{R}_{t+i}	Reference at time $t + i$ in real-time domain
$x^{R}_{i|t}$	Reference at time i in virtual-time domain starting at time t
$V^{*}_{\text{Open}}(x)$	Optimal state-value function with open-loop control
J^{π}_{MPC}	MPC-like performance index with policy π
J^{ϕ}_{MPC}	MPC-like performance index with policy ϕ
$h(x_{i|t}), h(x_{t+i})$	Constraint function in virtual-time and real-time domains
$B(\cdot, \cdot)$	Barrier function
X_{Cstr}	Set of constrained states
X_{Init}	Initially Feasible Region (IFR)
X_{Edis}	Endlessly Feasible Region (EFR)
$L(\theta, \lambda)$	Lagrange function
$L_{\text{Aug}}(\theta, \lambda, \zeta)$	Augmented Lagrange function
J_{Scenery}	Scenery loss function
$\lambda(x)$	Lagrange multiplier function
$F(x)$	Feasibility function
$F^{*}(x)$	Perfect feasibility function
$F(x; \phi)$	Parameterized feasibility function
ζ	Learning rate for scenery update
X_{k}	Feasible region at k-th main iteration
u_{Safe}	Safe action

● **Deep Reinforcement Learning**

$\sigma(\cdot)$	Activation function
$\mathcal{H}(p,\hat{p})$	Cross entropy
$\text{MSE}(y,\hat{y})$	Mean squared error
$D_{\text{KL}}(p\|\hat{p})$	Kullback-leibler (KL) divergence
$\mathcal{H}(\pi)$	Policy entropy
$\mathcal{H}\big(\pi(\cdot\,|s)\big)$	Policy entropy under a given state s
J_{Reg}	Performance index with regularization
Q^{target}	Target network
$w,\overline{w}$	Parameters of value network and its target network
$\theta,\overline{\theta}$	Parameters of policy network and its delayed network
$A(s,a)$	Advantage function
$Z^{\pi}(s,a)$	Distributional return, i.e., a random variable with its mean $$\mathbb{E}_{\pi}\{Z^{\pi}(s,a)\}=q^{\pi}(s,a)$$
$\mathcal{Z}^{\pi}(\cdot\,|s,a)$	The distribution of distributional return, i.e., $$Z^{\pi}\sim\mathcal{Z}^{\pi}(\cdot\,|s,a)$$
$r_{\text{aug}}(s,a,s')$	Augmented reward signal with policy entropy
$\mathcal{D}_{\text{Replay}}$	Replay buffer
$\text{clip}(z,z_{\text{min}},z_{\text{max}})$	Clip function that bounds z between its bounds
u,w	Action and its adversarial disturbance in robust RL
T_{zw}	Transfer function from objective output to disturbance
$\min\limits_{u}\max\limits_{w} J$	Minimax optimization
$H(x,u,w,\partial V/\partial x)$	Hamiltonian in zero-sum game
(u^*,w^*)	Nash equilibrium in zero-sum game
ξ_t,ζ_t	Process noise and observation noise in HMM
h_t	History information
$V_h^{\pi}(h_t)$	State-value function with the policy $\pi(h)$ conditioned on h_t
$V_b^{\pi}(b_t)$	State-value function with the policy $\pi(b)$ conditioned on b_t
$V_h^*(h_t)$	Optimal state-value function conditioned on h_t
$V_b^*(b_t)$	Optimal state-value function conditioned on b_t

u_t, x_t, y_t	Action, state and measurement of HMM
$\hat{x}_t$	Estimate of a state
$b_t(x)$	Belief state at time t
$\mathcal{T}_i, \rho(\mathcal{T})$	Trainable task and its distribution
a_i, a_{-i}	Action of agent i; actions of agents except agent i
π_i, π_{-i}	Policy of agent i; policies of agents except agent i
$\pi_i(a_j\|s)$	Policy of agent i that maps from full observation to action of agent j
$\pi_i(a_i\|s)$	Policy of agent i that maps from full observation to its own action
$Q_i(s, a_j)$	Action-value function stored in agent i about action of agent j
$Q_i(s, a_i)$	Action-value function stored in agent i about its own action
$r_i(s, a, s')$	Reward signal of agent i
$r_\psi(s, a)$	Parameterized reward signal with parameter ψ

- Abbreviation

A2C	Advantage Actor-Critic
A3C	Asynchronous Advantage Actor-Critic
AC	Actor-Critic
ACS	Actor-Critic-Scenery
ACOE	Average Cost Optimality Equation
ADP	Approximate Dynamic Programming
ANN	Artificial Neural Network
BDQ	Bounded Double Q-functions
BP	BackPropagation
BPTT	BackPropagation Through Time
CAC	Clipped Actor Criterion
CBF	Control Barrier Function
CNN	Convolutional Neural Network
CPU	Constrained Policy Update
DDPG	Deep Deterministic Policy Gradient
DDQN	Double Deep Q-Network
DP	Dynamic Programming
DPG	Deterministic Policy Gradient
DPU	Delayed Policy Update
DQF	Double Q-Functions
DQN	Deep Q-Network
DRF	Distributional Return Function
DRL	Deep Reinforcement Learning
DSAC	Distributional Soft Actor-Critic
EFR	Endlessly Feasible Region
ELBO	Evidence-Lower-BOund
EnR	Entropy Regularization
ExR	Experience Replay
FD	Finite Difference

FDD	Feasible Descent Direction
GA	Genetic Algorithm
GPI	Generalized Policy Iteration
GPS	Guided Policy Search
GRU	Gated Recurrent Unit
HDP	Heuristic Dynamic Programming
HJB	Hamilton-Jacobi-Bellman
HJI	Hamilton-Jacobi-Isaacs
HMM	Hidden Markov Model
IFR	Initially Feasible Region
iid	independent and identically distributed
IRL	Inverse Reinforcement Learning
IS	Importance Sampling
KKT	Karush-Kuhn-Tucker
KL	Kullback-Leibler
LP	Linear Programming
LQ	Linear Quadratic
LQG	Linear-Quadratic-Gaussian
LS	Least Squares
LSTM	Long Short-Term Memory
MAP	Maximum A Posteriori
MC	Monte Carlo
MDP	Markov Decision Process
MLP	MultiLayer Perceptrons
MPC	Model Predictive Control
MSE	Mean Squared Error
NAF	Normalized Advantage Function
OCP	Optimal Control Problem
OTOI	Offline Training and Online Implementation
PDE	Partial Differential Equation

PDF	Probability Density Function
PDO	Primal-Dual Optimization
PER	Prioritized Experience Replay
PEV	Policy EValuation
PEx	Parallel Exploration
PG	Policy Gradient
PIM	Policy IMprovement
PMP	Pontryagin's Maximum Principle
POMDP	Partially Observable Markov Decision Process
PPO	Proximal Policy Optimization
QP	Quadratic Programming
RBF	Radial Basis Function
RBFN	Radial Basis Function Network
RHC	Receding Horizon Control
RMSE	Root Mean Square Error
RID	Region IDentification
RNN	Recurrent Neural Network
SAC	Soft Actor-Critic
SARSA	State-Action-Reward-State-Action
SGD	Stochastic Gradient Descent
SIS	State Initialization Sampling
SOTI	Simultaneous Online Training and Implementation
SPG	Stochastic Policy Gradient
SRF	State Rollout Forward
SSE	Sum Squared Error
SSD	Stationary State Distribution
STN	Separated Target Network
SVF	Soft Value Function
TAR	Total Average Return
TD	Temporal Difference

| TD3 | Twin Delayed Deep Deterministic policy gradient |
| TRPO | Trust Region Policy Optimization |

Contents

Chapter 1.   Introduction to Reinforcement Learning

> Out of TAO, One is born;
>
> One produces Two;
>
> Two produces Three;
>
> Out of Three, the Universe is created.
>
> -- Lao Tzu (around 6th-century BCE)

The nature of human intelligence has attracted the special interest of scientists and researchers, as humans are the most intelligent species on the planet. The way that humans adapt their behavior as they interact with surrounding environments is a crucial learning paradigm, which can be demonstrated by how a person drives a car. When learning to drive, an inexperienced driver continuously performs steering, accelerating, and braking operations according to his perception of traffic environments. Based on the built-in learning mechanisms in his brain, a driver improves his skills by choosing rewarded behaviors and avoiding punished behaviors. After continuous practice for several months, the driver eventually becomes capable of reasonably considering safety, maneuverability, and ride comfort in his daily driving manipulation. This kind of trial-and-error process is an essential way for humans to learn new skills and develop intelligence. For humans, how to learn a new skill is one of the most fundamental questions. This is also a fundamental question in artificial intelligence. Reinforcement learning (RL) is a biologically inspired learning approach that focuses on finding optimal decision or control strategies for dynamic environments. While traditional RL method focuses on learning from interactions with an unknown environment, modern RL method is more like a full-space optimizer for optimal control problem, which maximizes or minimizes a certain criterion to search for an optimal policy. The power of RL mainly lies in its strong ability to solve complex and large-scale sequential decision problems. Due to its potential in developing superhuman intelligent strategies, RL has attracted wide attention in a variety of interdisciplinary fields, including autonomous driving, game AI, robot control, and quantitative trading. In 2017, RL was listed by the *MIT Technology Review* as one of the 10 breakthrough technologies.

1.1   History of RL

The history of RL dates back to the early days of cybernetics. Its development is mainly influenced by two separated fields: (1) dynamic programming and (2) trial-and-error learning. Dynamic programming (DP) was first introduced by Richard Bellman in the 1950s. DP is one of the three pillars of optimal control, along with the calculus of variations and Pontryagin's maximum principle. The basis of DP is to divide a multistage optimization problem into a sequence of single-stage problems. As an abstraction of human and animal learning patterns, trial-and-error learning takes particular inspiration from Pavlovian conditioning and instrumental conditioning, which are rooted in the fields of psychology and neuroscience. Some historical milestones of RL development are listed as follows:

- 1977, P. Werbos: DP with function approximation [1-29][1-28]

S. E. Li, *Reinforcement Learning for Sequential Decision and Optimal Control*,
https://doi.org/10.1007/978-981-19-7784-8_1

- 1988, R. Sutton: Temporal difference learning [1-21]
- 1992, C. Watkins: Q-learning [1-26][1-27]
- 1994, G. Rummery, M. Niranjan: SARSA [1-17]
- 1996: D. Bertsekas & J. Tsitsiklis: RL convergence and stability [1-5]
- 1999, R. Sutton, D. McAllester, S. Singh: Policy gradient [1-24]
- 2014, D. Silver, G. Lever, N. Heess, et al.: Deterministic policy gradient [1-20]
- 2015, V. Mnih, K. Kavukcuoglu, D. Silver, et al.: Deep Q-Network [1-12]
- 2015 to now, successes of Atari AI, AlphaGo, MuZero, and Racing Car AI [1-12][1-19][1-18][1-33]

The terminology in RL comes from a combination of notations in several fields, including optimal control, convex optimization, and statistical learning. A few excellent literature surveys and reference books are available for researchers in this fascinating field, including Bertsekas and Tsitsiklis (1996) [1-5], Sutton and Barto (1998, 2018) [1-22][1-23], Busoniu, et al. (2009)[1-7], Powell (2011) [1-16], Zhang, et al. (2013) [1-34], Littman (2015) [1-11], Liu, et al. (2017) [1-10], Bertsekas (2017) [1-4], Dong, et al. (2020) [1-8], Meyn (2021) [1-13], etc. Thus, by no means do newcomers lack excellent references in this field.

However, what seems to be missing in the literature is a self-contained and relatively in-depth material for new learners to develop a good sense of the theoretical basis of RL principles and a broad view of mainstream algorithms. One major purpose of this book is to fill this gap. To keep things concise and easy to understand, a few compromises in material selection had to be made. The major concepts in this book are introduced in the way that a new reader can understand what the concept is and how it came about. Brief summaries of several important topics are also presented, including fixed-point iteration, average returns, stochastic optimization, receding horizon control, and safety guarantee. Naturally, we focus on the basic ideas in principles and algorithms rather than solid mathematics for theoretical completeness or coding skills for engineering applications.

1.1.1   Dynamic Programming

Optimal control addresses the problem of finding a control law for a given dynamic system to minimize a particular criterion. There are three pillars of optimal control, including calculus of variations, Pontryagin's maximum principle, and dynamic programming. The development of these theories is largely attributed to the work of Lev Pontryagin and Richard Bellman in the 1950s and that of Leonhard Euler and Joseph-Louis Lagrange in the 18th century.

Figure 1.1 Leonhard Euler

Figure 1.2 Joseph-Louis Lagrange

The calculus of variations is a subfield of mathematical analysis that uses variations to find the maxima and minima of functionals. In 1733, Euler first elaborated this subject through a partly geometric approach. In 1755, Lagrange further outlined his δ-algorithm, which led to the famous Euler-Lagrange equation. This equation is a second-order partial differential equation, of which the solutions are the extrema of a stationary functional.

Figure 1.3 Lev Pontryagin

Pontryagin's maximum principle (PMP) is an extension of the Euler–Lagrange equation in the presence of input constraints. It was first formulated by Lev Pontryagin in 1956. In general, this principle states that the Hamiltonian must take an extremum over the set of all permissible actions. When satisfied along a trajectory, the PMP is a necessary condition of the optimum.

Figure 1.4 Richard Bellman

As stated by Richard Bellman, an optimal policy has the property that whatever the initial state and the initial decision are, the remaining decisions must constitute an optimal policy with regard to the state resulting from the first decision. This statement is the well-known Bellman's principle of optimality, which allows us to divide a multistage optimization problem into some initial choices and the payoffs of their following problems [1-3]. The application of this principle in optimal control leads to two well-known equations: Bellman equation in discrete-time systems and Hamilton-Jacobi-Bellman (HJB) equation in continuous-time systems. In general, both the Bellman equation and the HJB equation are necessary and sufficient conditions of optimality.

Table 1.1 Bellman's principle of optimality

Type	Continuous-time	Discrete-time
Necessary and sufficient condition	Hamilton-Jacobi-Bellman (HJB) equation	Bellman equation

In practice, directly solving these equations is usually computationally untenable because DP suffers from the curse of dimensionality in continuous spaces and high-dimensional systems. In 1977, Paul Werbos suggested parameterizing the continuous state space and reconstructing DP in the parameter space [1-29][1-28]. This idea later evolved into today's approximate dynamic programming (ADP), which has been widely used in model-based problems. After building a parameterized policy, the optimization dimension is reduced to that of the parameter space. Thus, ADP partially avoids the need for excessive computation even though it sacrifices a certain level of policy accuracy. Most ADP algorithms fall into the family of indirect methods, whose goal is to solve Bellman equation or HJB equation to find an optimal policy. These algorithms can be further classified into two categories: policy iteration and value iteration. In the early stage, the name ADP was interchangeable with other names, for example, adaptive DP, heuristic DP, and neuro-DP. The study of Bertsekas and Tsitsiklis in the 1990s consolidated the ADP algorithm design and theoretical analysis of convergence, including both value iteration ADP and policy iteration ADP [1-25][1-5][1-6]. Their studies mainly focused on model-based optimal control problems with deterministic state-space models rather than model-free sequential decision problems in stochastic environments.

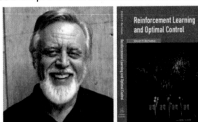

Figure 1.5 Dimitri Bertsekas and his book

1.1.2   Trial-and-Error Learning

Trial-and-error learning is one of the core mechanisms by which humans and animals acquire intelligent behaviors. By trying different responses to the same situation, humans and animals can learn what kind of behaviors are good and what kind of behaviors are bad, and thus they can repeat good behaviors and avoid bad behaviors. Eventually, the trial-and-error mechanism makes humans and animals capable of predicting future rewards and choosing the best action. In the field of artificial intelligence, this idea has inspired the development of many early RL algorithms, including Monte Carlo learning and temporal difference learning [1-14].

Figure 1.6 Claude Shannon and his maze-running mouse

Figure 1.7 Donald Michie and his MENACE system

The trial-and-error approach was one of several important thoughts in the early stage of artificial intelligence. In 1952, Claude Shannon introduced a mechanical mouse that used trial-and-error to find its path through a maze. In the early 1960s, influential work was conducted by Donald Michie, who designed a mechanical player called MENACE to play tic-tac-toe. MENACE learned in a manner that is very similar to today's Monte Carlo (MC) learner. Every time MENACE wins a game, the moves are used more in successive games. Every time it loses a game, the moves are used less in the future. As a result, MENACE becomes more likely to play moves that resulted in a previous win and less likely to play moves that resulted in a previous loss. Widrow, et al. (1973) designed a trial-and-error learning rule that could learn from success and failure signals instead of labeled training data. This algorithm is known as "learning with a critic" instead of "learning with a teacher" [1-30]. Since then, the term "critic" has appeared in the literature, and it resembles an external expert that evaluates the action's performance.

In fact, the trial-and-error learner connects responses to stimuli without any involvement of thinking, reasoning, or understanding. Rooted in neuropsychology, the connection between trial-and-error and dynamic programming was not well recognized until the late 1970s. One of the breakthrough learning approaches is called temporal difference (TD), which utilizes a bootstrapping mechanism to update a critic function on the basis of historical estimates. Witten (1977) introduced the first TD learning algorithm called TD(0) [1-32]. A broad class of TD-based algorithms emerged after researchers became aware of the inherent connection between temporal difference and dynamic programming. Two famous examples are TD(n) and TD-lambda, which were developed by Richard Sutton and Andrew Barto in the 1980s [1-1][1-2][1-21]. To date, the most famous TD algorithm might be Q-learning, which was proposed by Chris Watkins in 1989 [1-26]. In essence, Q-learning is an off-policy algorithm in which an action-value function is optimized with samples from arbitrary policies. Another important TD algorithm is SARSA [1-17], which is an on-policy algorithm whose action-value function is estimated by samples from the same policy.

Figure 1.8 Richard Sutton and his book

Both MC and TD methods focus on addressing finite-element Markov tasks and modeling the countable space with tabular functions. The tabular presentation prevents these methods from addressing large-scale and complex tasks, such as autonomous driving and playing computer games. In the 1990s, policy gradient methods began to attract particular interests of the RL community. The policy to be learned is approximated by a parameterized function whose parameters are searched along the gradient ascending direction of a long-term return. The fundamental idea is to view RL as a stochastic optimization problem, and hence its policy gradient is independent of the optimality condition. There are three kinds of popular policy gradients, including likelihood ratio gradient [1-24][1-31], natural policy gradient [1-9], and deterministic policy gradient [1-20]. The policy gradient algorithms form a new RL family called direct methods. These methods usually address high-dimensional tasks with continuous state and action spaces. Moreover, flexible policy representations can be chosen in different tasks, which allows their policy search to cooperate with mature optimization algorithms or even human domain knowledge.

The naive version of policy gradient, however, is not the salvation to practical RL problems. One main challenge is the large variance in the sample-based gradient estimation. To reduce this variance, a parameterized value function is often estimated simultaneously, which introduces additional benefits in terms of convergence and accuracy [1-22][1-23]. Methods with value function estimation can further evolve to the famous actor-critic (AC) architecture, which has become one of the most successful RL paradigms. The AC architecture learns to approximate both policy and value functions; these functions are referred to as "actor" and "critic", respectively. Barto, et al. (1983) might have introduced the first actor-critic algorithm, which takes linear functions as the critic and actor approximations [1-2]. The actor and critic affect each other in an intertwined manner: actions taken by the actor are evaluated by the critic, and the actor improves the policy according to the critic evaluation. After a sufficient number of iteration cycles, both the actor and the critic can reach their own optimum. In 1992, IBM demonstrated a program that used an actor-critic algorithm to play backgammon. This algorithm was skilled enough to rival the best human player.

Although a few successes have been achieved using the aforementioned algorithms, most RL algorithms are still limited to low-dimensional tasks. The introduction of deep neural networks is another milestone in RL history, which enables RL to solve many previously intractable problems, for example, playing video games directly from image pixels. Today, the combination of deep learning and reinforcement learning is revolutionizing the direction of artificial intelligence, which is generally believed to be a critical step toward general artificial intelligence. The inception of deep RL might be the Atari video games application, which has reached superhuman-level performance [1-12]. The standout success of deep RL is AlphaGo, which was developed by Google Deepmind for Chinese Go. In 2015, AlphaGo became the first Go program to beat a human professional player on a full-sized board [1-19].

1.2   Examples of RL Applications

One can find that RL and optimal control share a very similar problem formulation. In general, RL is more suitable to address large-scale, complex, and stochastic problems, such as computer games, Chinese Go, and autonomous driving. In contrast, optimal control often faces simple and deterministic tasks, which require the calculation of the optimal action in a real-time fashion. Today, the boundary between these two techniques is gradually diminishing due to a deep understanding of their connection in the perspective of receding horizon control. Here, we briefly present three examples to demonstrate what kinds of problems are more suitable for reinforcement learning.

1.2.1   Tic-Tac-Toe

Tic-tac-toe is a paper-and-pencil game for two players. The player who succeeds in placing three of his marks in a horizontal, vertical, or diagonal row wins the game. For example, in Figure 1.9, the first player, who is playing with the cross "×", wins the game.

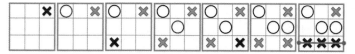

Figure 1.9 Example of tic-tac-toe game

The first AI player of this game, MENACE, was designed using 304 matchboxes and a few beads in 9 different colors. Each matchbox corresponds to a unique board layout that a player might face. The matchbox contains a few colored beads corresponding to each possible move that can be made next. When MENACE takes a turn, the operator simply selects the matchbox corresponding to the current board layout and then randomly picks one bead to move. Every time MENACE wins a game, an additional bead is added to each matchbox played, corresponding to the winning move. Every time MENACE loses, a bead corresponding to the losing move is removed. As a result, MENACE becomes more likely to play moves that have won in the past and less likely to play moves that have lost before.

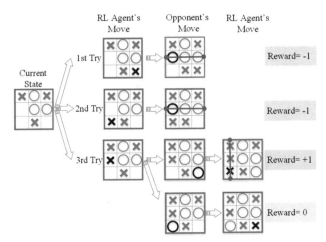

Figure 1.10 Example of possible tic-tac-toe moves

Modern RL methods have been applied to tic-tac-toe. In these methods, the game state refers to the placement of naughts "o" and crosses "×" on a 3x3 grid. Tic-tac-toe has a total of 19683 states. Many states are rotationally or reflectively identical and some states are even unreachable. They are removed from the state space to reduce the computational complexity. A value is assigned to each state to represent the winning probability starting from that state. For example, if the agent is playing with "o", the winning probability is 100% for all states with three "o" in a row. Similarly, for all states with three "×" in a row, the winning probability is zero. In either condition, the episode will terminate, and a winner will be announced. The estimate of state values, i.e., the winning possibility of each state, is learned through a trial-and-error process. The agent creates a state tree that represents all the possible successive states starting from the current state and then tries all possible moves one by one to observe the received reward (i.e., +1 for a win, -1 for a loss and 0 for a draw). In Figure 1.10, the agent has three possible moves to make. The agent would lose immediately by making the moves shown in the first and second trials, but the third trial has the potential to yield a win. Accordingly, the RL agent will make the move shown in the third trial.

1.2.2   Chinese Go

Chinese Go is a board game where two players compete to control more territories on a 19-by-19 grid board. The grid board maximally contains 361 points. Players try to border empty intersections on the board with their stones and receive points for the number of spaces they have surrounded. The player with the higher number of points wins the game. Unlike a Chess program, it is impossible to solve Go with a searching technique due to its enormous state space. It is said that the number of possible sequences of actions in Go is about 10^{170}, which is larger than that of all atoms in the universe. This game is so complicated that even the most accomplished Go players may struggle to say whether certain moves are good or bad. This is why it is difficult for a programmer to code a good Go-playing policy.

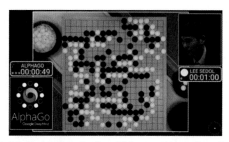

Figure 1.11 Alpha Go playing Chinese Go against Lee Sedol

Recently, RL with neural network approximation has achieved groundbreaking performance in playing Chinese Go. In 2016, Silver and his colleagues developed a deep Q-network approach for playing Chinese Go, called AlphaGo [1-19]. In this approach, the board layout, which is a 19 × 19 image, is sent to two convolutional neural networks. One convolutional neural network is called the value network, and it evaluates how good or bad a specific board layout is. The other convolutional neural network is called the policy network, which is used for sampling actions. The policy network is first initialized with the human expert's moves and is then updated along the steepest ascending direction with a reinforcement learner. Simultaneously, the value network is trained to forecast the possibility of winning or losing the game. The two networks are logically combined with Monte Carlo tree search for more efficient exploration and policy deployment. The world was shocked when AlphaGo, which used the above approaches, defeated the world-famous professional Chinese Go player Lee Sedol by winning four games to one. Later, a new version, called AlphaGo Zero, achieved a more surprising performance, winning with a ratio of 89:11 against its previous version.

1.2.3   Autonomous Vehicles

Autonomous driving is one of the most prominent applications in the field of artificial intelligence. With recent advances in deep learning algorithms and high-performance processors, driverless vehicles have grown from being seen only in science fiction materials to prototypes in the real world. An autonomous driving system can be trained with supervised learning methods but it would require a large amount of expert driving data. As an alternative, by using RL to train a self-driving policy, unknown traffic scenarios can be explored without expensive expert data. Moreover, online RL training is a promising way to develop self-evolving abilities similar to those of human drivers.

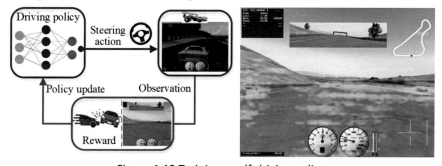

Figure 1.12 Training a self-driving policy

In the past decade, a few examples of the self-taught autonomous driving ability have been developed with virtual simulators. For example, Peng, et al. (2021) developed an end-to-end self-driving algorithm in the TORCS environment [1-15]. Their approach enabled a car to learn how to remain inside its lane without external teachers. This self-driving policy includes a deep neural network with three convolutional layers and three fully connected layers. As illustrated in Figure 1.12, the input is the image captured by a single frontal camera. When the car stays inside the lane, it continually earns positive rewards. When the car travels outside its lane, it receives high negative rewards. The training result is visualized as the saliency map, which is shown on the right-hand side of Figure 1.12. This map shows that the self-taught policy pays more attention to lane lines and the frontal road horizon. Interestingly, this self-taught behavior somehow matches the driving knowledge of human drivers. This example demonstrates that an RL agent has a kind of self-evolving ability to automatically develop high-level driving intelligence.

1.3   Key Challenges in Today's RL

In a narrow sense, reinforcement learning is a subarea of machine learning. It takes suitable action to maximize rewards in a particular situation. Generally, today's machine learning methods can be classified into three types: (1) supervised learning, (2) unsupervised learning, and (3) reinforcement learning (Figure 1.13). Supervised learning is a method of learning from a set of training data labeled by a knowledgeable external teacher, while unsupervised learning mainly involves finding hidden patterns in a collection of unlabeled data. Different from supervised learning and unsupervised learning, traditional RL is featured by trial-and-error interactions with the environment.

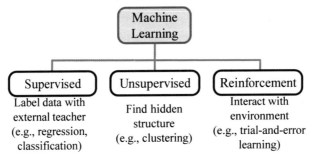

Figure 1.13 Classifications of machine learning

In a broad sense, RL has a strong connection with optimal control. Both approaches seek to optimize (either maximize or minimize) a certain performance index while being subjected to a kind of environment dynamics. The difference between these two approaches is that optimal control often requires an accurate model with some assumptions on formalism and determinism. In contrast, RL is usually model-free and requires a collection of experience from an unknown environment. Despite a few successes, RL still faces a variety of challenges when it is deployed for practical applications. For example, exploration-exploitation dilemma, uncertainty and partial observability, temporally delayed reward, infeasibility from safety constraint, nonstationary environment, and lack of generalizability are common problems that may be encountered when using RL approaches in practice.

1.3.1   Exploration-Exploitation Dilemma

Almost all RL methods face a fundamental challenge, i.e., how to balance the trade-off between exploration and exploitation. An agent tries a variety of actions to interact with the environment and has to utilize previous experience to determine a policy as quickly as possible. Meanwhile, the agent also needs to explore other actions to seek better decisions. Relying too heavily on previous experience may sacrifice the ability to explore new actions, while too much new exploration will incur more randomness in the policy search. Policy randomness can increase the exploration ability but will somewhat negatively impact the policy optimality. In some extreme cases, environmental exploration may even be forbidden, and only a pre-collected historical dataset is accessible. The RL algorithm under these conditions is called offline RL, in which only previous experience is available for policy learning.

1.3.2   Uncertainty and Partial Observability

Classic RL requires full observability and cannot be easily extended into partially observable tasks. In practice, an agent may not observe all the states; for example, a self-driving car cannot detect all surrounding vehicles in traffic environments due to vehicle-to-vehicle occlusion or sensor range limits. Partially observable Markov decision processes are more computationally intractable, and one has to estimate the hidden states as the input information to make decisions. In addition, any description of environment dynamics inevitably has a certain model mismatch. Large model errors introduce additional challenges into partially observable problems, which calls for a robust optimal policy.

1.3.3   Temporally Delayed Reward

In many episodic tasks, a sparse reward signal, e.g., either winning or losing in games, is delivered very late or even at the end of the game. Temporally delayed reward is increasingly diluted over time and it brings very little information from the distant precedent states. As a consequence, very slow convergence and poor policy accuracy often happen in this kind of episodic tasks. In this situation, a large amount of expensive environment interactions must be performed to propagate the delayed reinforcement from precedent states. One solution is to add a few intermediate signals to reduce the sparsity of rewards, but this requires a proper reward reshaping technique.

1.3.4   Infeasibility from Safety Constraint

The challenge of satisfying state constraints comes from its resultant infeasibility issue. The infeasibility behavior in constrained RL is very complicated because of two interconnected tasks, i.e., finding an optimal policy and identifying its feasible region. Obviously, a policy is meaningless if it is not defined in a feasible region. The feasibility of a region also depends on how to choose its policy. These two tasks strongly couple with each other, which is why constrained RL is hard to deal with. This infeasibility issue becomes even worse if safety is required. Safety guarantee means that the learned policy should strictly satisfy hard state constraints. To guarantee perfect safety, one must explore the environmental information from both inside and outside the feasible region and any dangerous constraint violations must not be introduced during the agent-environment interaction.

1.3.5   Non-stationary Environment

Almost all learners use stationary data to train a model under the assumption of independent and identically distributed (iid) data. Similarly, RL also needs a stationary environment to learn a policy. However, some tasks may involve time-varying structures or parameters. If the environment changes too quickly, RL is likely to fail due to the lack of iid data. This challenge cannot be easily mitigated in a real environment. Even worse, if some causes of intrinsic instability are not correctly eliminated, an RL algorithm may fail in a stationary environment. A typical unstable example comes from the deadly triad issue, i.e., the simultaneous existence of bootstrapping, off-policy, and function approximation.

1.3.6   Lack of Generalizability

The ability to adapt to unseen scenarios is a characteristic of human intelligence. However, existing RL methods often suffer from a lack of generalizability. Although today's RL methods can learn to solve complex tasks, they tend to specialize in their training domain and often breakdown when deployed to different scenarios. This poses a great challenge in applying RL to real world applications. In many tasks, an RL policy is first trained in a simulated environment and then deployed for use in a real world application. When the simulator has relatively low fidelity, a serious sim-to-real generalization issue will occur. Techniques such as domain adaptation are required to calculate transferable policies for better generalization.

1.4   References

[1-1]   Barto A, Sutton R (1982) Simulation of anticipatory responses in classical conditioning by a neuron-like adaptive element. Behav Brain Res 4(3): 221-235

[1-2]   Barto A, Sutton R, Anderson C (1983) Neuron-like adaptive elements that can solve difficult learning control problems. IEEE Trans Syst Man Cybern 13(5):834-846

[1-3]   Bellman R (1957) Dynamic programming. Princeton University Press, Princeton

[1-4]   Bertsekas D (2017) Dynamic programming and optimal control. Athena Scientific, Massachusetts

[1-5]   Bertsekas D, Tsitsiklis J (1996) Neuro-dynamic programming. Athena Scientific, Belmont

[1-6]   Bertsekas D, Tsitsiklis J, Volgenant A (2011) Neuro-dynamic programming. Encycl Optim 27(6):1687-1692

[1-7]   Busoniu L, Babuska R, De Schutter B, Ernst D (2009) Reinforcement learning and dynamic programming using function approximators. CRC, Boca Raton

[1-8]   Dong H, Ding Z, Zhang S (2020) Deep Reinforcement Learning: Fundamentals, Research and Applications. Springer, Singapore

[1-9]   Kakade S (2002) A natural policy gradient. NeurIPS, Vancouver, Canada

[1-10]   Liu D, Wei Q, Wang D, et al (2017) Adaptive dynamic programming with applications in optimal control. Springer International Publishing AG

[1-11] Littman M (2015) Reinforcement learning improves behaviour from evaluative feedback. Nature, vol 521: 445-451

[1-12] Mnih V, Kavukcuoglu K, Silver D, et al (2015) Human-level control through deep reinforcement learning. Nature 518: 529-533

[1-13] Meyn S (2021) Control systems and reinforcement learning. Cambridge University Press

[1-14] Neftci E & Averbeck B (2019) Reinforcement learning in artificial and biological systems. Nature Machine Intelligence, vol 1: 133-143

[1-15] Peng B, Sun Q, Li SE, et al (2021) End-to-end autonomous driving through dueling double deep Q-Network. Auto Innovation 4: 328–337

[1-16] Powell W (2011) Approximate dynamic programming: solve the curses of dimensionality. Wiley, New York

[1-17] Rummery G, Niranjan M (1994) On-line Q-learning using connectionist systems. Technical Report, Cambridge University

[1-18] Schrittwieser J, Antonoglou I, Hubert T, et al (2020) Mastering Atari, Go, chess and shogi by planning with a learned model. Nature, vol 588: 604-609

[1-19] Silver D, Huang A, Maddison C, et al (2016) Mastering the game of Go with deep neural networks and tree search. Nature 529:484-489

[1-20] Silver D, Lever G, Heess N, et al (2014) Deterministic policy gradient algorithms. ICML, Beijing, China

[1-21] Sutton R (1988) Learning to predict by the method of temporal differences. Mach Learn 3(1): 9-44

[1-22] Sutton R, Barto A (1998) Reinforcement learning: an introduction (First Edition). MIT Press, Cambridge

[1-23] Sutton R, Barto A (2018) Reinforcement learning: an introduction (Second Edition). MIT Press, Cambridge

[1-24] Sutton R, McAllester D, Singh S, et al (1999) Policy gradient methods for reinforcement learning with function approximation. NeurIPS, Denver, USA

[1-25] Tsitsiklis J, Roy B (1996) Feature-based methods for large scale dynamic programming. Mach Learn 22(1-3): 59-94

[1-26] Watkins C (1989) Learning from delayed rewards. Dissertation, Cambridge University

[1-27] Watkins C, Dayan P (1992) Q-learning. Mach Learn 8(3-4): 279-292

[1-28] Werbos P (1992) Approximate dynamic programming for real-time control and neural modeling. Van Nostrand Reinhold, New York

[1-29] Werbos, P (1990). Consistency of HDP applied to a simple reinforcement learning problem. Neural Netw 3(2): 179-189

[1-30] Widrow B, Gupta N, Maitra S (1973) Punish/reward: learning with a critic in adaptive threshold systems. IEEE Trans Syst Man Cybern SMC-3(5): 455-465

[1-31] Williams R (1992) Simple statistical gradient-following algorithms for connectionist reinforcement learning. Mach Learn 8(3-4):229-256

[1-32] Witten I (1977) An adaptive optimal controller for discrete-time Markov environments. Info Control 34(4): 286-295

[1-33] Wurman P, Barrett S, Kawamoto K, et al (2022) Outracing champion Gran Turismo drivers with deep reinforcement learning. Nature, vol 602: 223-228

[1-34] Zhang H, Liu D, Luo Y, Wang D (2013) Adaptive dynamic programming for control: algorithms and stability. Springer-Verlag London

Chapter 2.   Principles of RL Problems

> O how they cling and wrangle, some who claim
> For preacher and monk the honored name!
> For, quarreling, each to his view they cling.
> Such folk see only one side of a thing.
>
>   -- Gautama Buddha (563-483 BCE)

Reinforcement learning (RL) is an interdisciplinary branch of artificial intelligence and automatic control. It generally refers to a group of policy-searching algorithms that can generate optimal decisions for dynamic environments. Like a pet trained by stimuli and treats, an RL agent is penalized when an incorrect decision is made and rewarded when a correct decision is made. Thus, an RL agent, even a novice RL agent, can gradually become increasingly intelligent. From the viewpoint of artificial intelligence, RL specifies how an agent improves its policy through trial-and-error interactions with the environment (Figure 2.1). From the viewpoint of automatic control, RL is similar to a numerical optimizer defined in the whole state space to solve sequential decision problems. The former often refers to model-free RL, which is a trial-and-error learner that imitates the learning behaviors of animals. The latter usually refers to model-based RL, which shares almost the same problem formulation as that in optimal control.

In addition to the model-related classification, RL algorithms can be categorized from other perspectives, for example, continuous time or discrete time; tabular representation or function approximation representation; single agent or multiple agents; discrete space or continuous space; deterministic policy or stochastic policy; full observability or partial observability; and finite horizon or infinite horizon. For conciseness and ease of understanding, this chapter begins with model-free RL in discrete-time stochastic environments. It is assumed that we have finite state and action spaces with the Markov property. Gradually, this book will elaborate on other classes of popular RL algorithms, especially those with continuous state and action spaces, parameterized policy, deterministic environment, and finite horizon, as demanded by real-world problems.

2.1   Four Elements of RL Problems

No matter what type an RL algorithm belongs to, it generally contains four key elements: (1) state-action samples, (2) a policy, (3) reward signals, and optionally, (4) an environment model. The state is a representation of the environment dynamics, while the action is the input. The policy is a mapping from the state to the action, which defines the feedback behavior of the RL agent. Given the current state and action, an environment model can predict the resultant next state. The reward signal defines how good a policy behaves in the environment. In mathematics, the goal of RL is to optimize the total rewards it receives over the long run while being subjected to either the data distribution sampled from the environment interaction (i.e., model-free) or the environment dynamic model (i.e., model-based).

S. E. Li, *Reinforcement Learning for Sequential Decision and Optimal Control*,
https://doi.org/10.1007/978-981-19-7784-8_2

The concept of model-free RL, in which an agent interacts with the environment, is illustrated in Figure 2.1. After receiving reward signals, the agent's policy is adjusted to achieve better performance in the next round of interaction. This concept is rather abstract but flexible enough to suit a variety of sequential decision problems. For a control engineer, the term "agent" might cause some misunderstanding. Unlike the relationship between a person and his surroundings, the agent is more like a physical controller, and the environment is the system to be controlled. For example, in an autonomous driving system, the electronic control unit (ECU) is the agent, while the car itself, including sensors and actuators, is part of the environment.

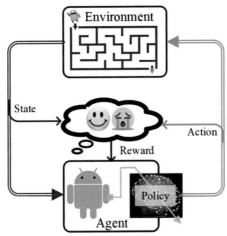

Figure 2.1 Concept of RL with an agent-environment interaction

2.1.1 Environment Model

Most stochastic and discrete-time environments can be viewed as Markov processes. The Markov process has the Markov property, i.e., the state at the current-time step depends only on the state and action at the last time step. This property is also characterized as memorylessness. In a Markov process, the current state is a sufficient representation of the environment status. A Markov chain, which has a finite state space, is the simplest Markov process. Here, we take the Markov chain as an example to show how to mathematically describe its behavior. The Markov chain has a finite number of elements in the state space $\mathcal{S}$ and the action space $\mathcal{A}$. Given any state $s \in \mathcal{S}$ and action $a \in \mathcal{A}$, the occurrence probability of a new state s' at the next-time step is

$$\mathcal{P}^a_{ss'} = p(s'|s,a) \stackrel{\text{def}}{=} \Pr\{s_{t+1} = s'|s_t = s, a_t = a\}, \qquad (2\text{-}1)$$

where t is the current time, s_t is the state at time t, a_t is the action at time t, and $\mathcal{P}^a_{ss'}$ or $p(s'|s,a)$ is the transition probability of the environment dynamics. A Markov chain can also be described by a weighted directed graph, where the weight of an edge represents the probability of shifting from one state to another. If the states are partially observable, the system can be described as a hidden Markov model (HMM), which contains the transition probability from one state to another and the observation probability from one state to its measurement. A large number of problems in various fields can be modeled as Markov processes, including chemical, electronic or mechanical systems from engineering, optimal coding, channel allocation and sensor networks from computer science, and inventory control, portfolio management and option pricing from economics.

2.1.2   State-Action Sample

Like the Markov model, the state-action sample is a representation of the environment dynamics. Formally, a sample consists of a state-action pair (s_t, a_t) and its consequent state s_{t+1}. A sequence of state-action samples ordered chronologically is called a state-action trajectory, which comes from the interaction between the agent and the environment, either physically in the real world or virtually in a computer simulator. A dataset is composed of multiple state-action samples:

$$\mathcal{D} \stackrel{\text{def}}{=} \Big\{ \underbrace{s_0, a_0, s_1}_{\text{Sample}}, a_1, s_2, a_2, \cdots \cdots, \underbrace{s_t, a_t, s_{t+1}}_{\text{Sample}}, a_{t+1}, \cdots \Big\}, \tag{2-2}$$

where $\mathcal{D}$ is the dataset. The dataset can contain either one trajectory or multiple trajectories. Obviously, each state-action pair in a trajectory is not alone, and it can have long-term consequences on subsequent states and actions.

2.1.3   Policy

The policy is a mapping from the state space to the action space. A sequential decision problem must contain a policy that describes the agent's way of behaving when interacting with the environment. The Markov environment with a stationary Markov policy is called a Markov decision process (MDP). A policy is said to be Markovian if an action is exclusively determined by the current state. A policy is said to be stationary if the policy is independent of time. The stationary Markov policy can be categorized into (1) a deterministic policy, which always selects the same action at the same state, and (2) a stochastic policy, which selects an action according to the learned probability distribution. The deterministic policy is expressed as

$$a = \pi(s), s \in \mathcal{S}. \tag{2-3}$$

For example, a linear feedback control law is a deterministic policy. A stochastic policy $\pi(a|s)$ specifies the probability of selecting action a in the current state s:

$$\pi(a|s) = \Pr\{a_t = a|s_t = s\}, s \in \mathcal{S}. \tag{2-4}$$

In general, stochastic policies are more commonly discussed than their deterministic counterparts. Specifically, a deterministic policy lacks good exploration ability because it always outputs the same action at the same state. A stochastic policy can provide better exploration because it selects actions randomly. When the states are partially observable, the stochastic policy has the potential to take the uncertainty about the hidden state into consideration [2-7].

2.1.3.1   Tabular Policy

The simplest representation of a policy is a look-up table, i.e., a tabular policy, which has been used in many early versions of RLs, such as Monte Carlo, Q-learning, and SARSA. The look-up table is suitable for low-dimensional and finite-element tasks when high-frequency decision-making or real-time control is demanded. In other words, a tabular policy is a precomputed array that holds the action to take for each state. This tabular structure provides the ability to handle each state or action element individually.

2.1.3.2   Parameterized Policy

For large-scale tasks, a tabular policy is both memory-consuming and computationally inefficient. Alternatively, a parameterized function can be selected to represent the policy to reduce the storage and computational burdens. The parameterized deterministic and stochastic policies are denoted as follows:

$$\pi_\theta(s) \overset{\text{def}}{=} \pi(s; \theta),$$
$$\pi_\theta(a|s) \overset{\text{def}}{=} \pi(a|s; \theta),$$

(2-5)

where θ is the policy parameter. The most commonly used functions include polynomial functions, Fourier functions, radial basis functions, and artificial neural networks. The parameterized policy has several advantages: (1) it has higher computational efficiency and needs less storage memory in high-dimensional or continuous space; and (2) some special parameterization choices may provide a good way of adding prior knowledge into the policy. For example, a reasonable driver model can be used as the self-driving policy, which has the potential to embed the human driver's experience into it.

2.1.4   Reward Signal

In reinforcement learning, the reward signal is the driving force to search for an optimal policy. A reward signal is a function that maps the triple of the current state, current action, and next state to a real number, which is defined as:

$$r_{ss'}^a \overset{\text{def}}{=} r_t = r(s_t, a_t, s_{t+1}).$$

(2-6)

This type of reward signal has a special separable property, i.e., starting from a given state, any reward signal depends only on the current action and the one-step transition probability. Together with the Markov environment, the separable property of reward signals allows the sequential decision problem to be broken into multistage optimization problems. This is why one can apply Bellman's principle of optimality to build the Bellman equation. In a deterministic environment, the next state is exclusively determined by the current state and the current action. Hence, (2-6) can be simplified as a function of only two variables, i.e., the current state and the current action. Instead of simply focusing on each immediate reward, RL seeks to maximize the expectation of a long-term return. The long-term return is equal to the weighted sum of the reward signals in the long run. Figure 2.2 illustrates the temporal order of the state, action, and reward signal. Each circle "O" represents a state, and each solid dot "●" represents an action.

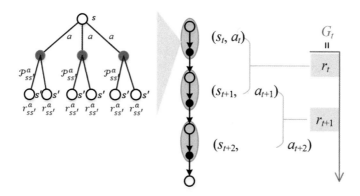

Figure 2.2 State, action and reward signal

2.1.4.1   Long-term Return

The long-term return (or return for short) is calculated by accumulating reward signals in the long run. Most control tasks, e.g., autonomous driving, are continuing, i.e., they can run without termination. We refer to them as continuing tasks or infinite-horizon tasks. In contrast, many games, such as Chess and Chinese Go, have an indefinite horizon, which means that they must terminate after a finite number of steps. This naturally yields identifiable episodes, each starting from an initial state and ending at a terminal state. These problems are called indefinite-horizon tasks, or more generally, episodic tasks. The terminology "episodic" also refers to those tasks with a finite horizon, which contain a fixed number of time steps in each episode. Here, we focus on a continuing task and define its long-term return. The most popular definition is the discounted return:

$$G_t = \sum_{i=0}^{+\infty} \gamma^i r_{t+i}, \tag{2-7}$$

where G_t represents the long-term return and $0 < \gamma < 1$ is the discount factor. This definition of long-term return emphasizes earlier rewards. If $\gamma \approx 0$, the agent is myopic and is only concerned with an immediate reward. As γ approaches 1, the agent becomes more farsighted and takes more future rewards into account. The discounted return is exactly what one prefers if he believes sayings such as "better an egg today than a hen tomorrow" or "a bird in the hand is worth two in the bush". This kind of definition also makes sense when the environment model is not perfect. The states in the long run become increasingly unpredictable over time, and the discount factor helps reduce the influence of uncertain future states. Another popular definition for continuing tasks is the average return:

$$G_t = \lim_{T \to \infty} \frac{1}{T} \sum_{i=0}^{T-1} r_{t+i}, \tag{2-8}$$

where T is the terminal time, which goes to infinity. When T is finite, (2-8) naturally reduces to a finite-horizon return for episodic tasks. The purpose of discounted and average returns is to provide a computable measure. Clearly, a return without a discount factor (or equivalently, when $\gamma = 1$) is unbounded in regard to infinite-horizon tasks. This is attributed to the stochastic behaviors of MDPs. The randomness does not allow states to settle, and the summation of stochastic reward signals becomes an infinitely large value. Unlike stochastic tasks, deterministic tasks do not have this issue since the environment will be stabilized around a known equilibrium. The accumulative sum of their reward signals is thus meaningful. In general, discounted return is more computationally friendly and has been widely used in today's RL community.

2.1.4.2 Value Function

In most stochastic tasks, the value function is defined as the expectation of the long-term return (also called the expected return), which is used to evaluate how good a policy is. There are two variants of value functions: (1) state-value function and (2) action-value function. The former is a function that only takes the state as input, defined as the expected return under a policy π starting from a certain state s:

$$v^\pi(s) \stackrel{\text{def}}{=} \mathbb{E}_\pi\{G_t|s\} = \mathbb{E}_\pi\left\{\sum_{i=0}^{+\infty}\gamma^i r_{t+i} \,|\, s_t = s\right\}$$

$$= \mathbb{E}_{a_t,s_{t+1},a_{t+1},s_{t+2},a_{t+2},\cdots}\left\{\sum_{i=0}^{+\infty}\gamma^i r_{t+i} \,|\, s_t = s\right\},$$

(2-9)

where $v^\pi(s)$ is the state-value function for policy π. In the definition of $v^\pi(s)$, the initial state s_t is enforced to be a given value, i.e., $s_t = s$. The notation $\mathbb{E}$ with the subscript π conditioned on s, i.e., $\mathbb{E}_\pi\{\cdot\,|s\}$, denotes that the random state-action variables, $a_t, s_{t+1}, a_{t+1}, s_{t+2}, a_{t+2}, \cdots$, are all generated by the policy π given an initial state $s_t = s$. Obviously, $\mathbb{E}_\pi\{\cdot\,|s\}$ is conditionally dependent on the initial state s, but it can be written as $\mathbb{E}_\pi\{\cdot\}$ for brevity. Moreover, it also implies that $v^\pi(s)$ is governed by stochastic environment dynamics, even though this fact is not explicitly stated.

The action-value function is a function of both the state and the action, defined as the expected return from a given state-action pair (s, a):

$$q^\pi(s, a) \stackrel{\text{def}}{=} \mathbb{E}_\pi\{G_t|s, a\}$$

$$= \mathbb{E}_\pi\left\{\sum_{i=0}^{+\infty}\gamma^i r_{t+i} \,|\, s_t = s, a_t = a\right\}$$

(2-10)

$$= \mathbb{E}_{s_{t+1},a_{t+1},s_{t+2},a_{t+2},\cdots}\left\{\sum_{i=0}^{+\infty}\gamma^i r_{t+i} \,|\, s_t = s, a_t = a\right\},$$

where $q^\pi(s, a)$ is the action-value function for policy π. Here, the state-action pair (s_t, a_t) at time t is enforced to be two real values (s, a), i.e., $s_t = s$ and $a_t = a$.

There are a variety of mathematical tools to represent value functions, such as tabular functions, polynomial approximations, and artificial neural networks. The tabular function is usually applied to address problems that have small and finite spaces. For problems with combinatorial and large state spaces, parameterized functions, such as polynomials and neural networks, are more suitable due to their low computational and storage burdens [2-7].

2.1.4.3 Self-consistency Condition

A fundamental property of the value function is that it naturally holds a certain recursive relationship, i.e., the self-consistency condition. Note that the self-consistency condition is an inherent property of three elements, i.e., value function, policy, and environment dynamics, and its existence does not rely on Bellman's principle of optimality. Figure 2.3 shows the backup diagram between the current state and its successor states. Looking ahead one-step from any given state to its successor states, it is easy to find the recursive relationship between $v^\pi(s)$ and $q^\pi(s, a)$:

$$v^\pi(s) = \sum_{a \in \mathcal{A}} \pi(a|s)q^\pi(s, a),$$ (2-11)

and

$$q^\pi(s, a) = \sum_{s' \in \mathcal{S}} \mathcal{P}_{ss'}^a \left(r_{ss'}^a + \gamma v^\pi(s') \right).$$ (2-12)

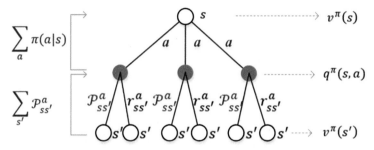

Figure 2.3 Backup diagram for $v^\pi(s)$ and $q^\pi(s, a)$

Combining (2-11) and (2-12), we have the self-consistency condition of the first kind (i.e., for the state-value function):

$$v^\pi(s) = \mathbb{E}_\pi\{r + \gamma v^\pi(s')|s\} = \sum_{a \in \mathcal{A}} \pi(a|s) \left\{ \sum_{s' \in \mathcal{S}} \mathcal{P}_{ss'}^a \left(r_{ss'}^a + \gamma v^\pi(s') \right) \right\}.$$ (2-13)

which is the recursive relationship of the state-value function. Reusing (2-12) and (2-11), we have the self-consistency condition of the second kind (i.e., for the action-value function):

$$q^\pi(s, a) = \mathbb{E}_\pi\{r + \gamma v^\pi(s')|s, a\}$$

$$= \sum_{s' \in \mathcal{S}} \mathcal{P}_{ss'}^a \left(r_{ss'}^a + \gamma \sum_{a' \in \mathcal{A}} \pi(a'|s')q^\pi(s', a') \right).$$ (2-14)

which is the recursive relationship of the action-value function. Here, the subscript π in $\mathbb{E}_\pi\{\cdot\}$ actually has a different meaning. It is easy to observe that $\mathbb{E}_\pi\{\cdot\}$ in (2-13) is equivalent to $\mathbb{E}_{a,s'}\{\cdot\}$ because the distribution of a, s' depends on the policy π. In (2-14), $\mathbb{E}_\pi\{\cdot\}$ is equivalent to $\mathbb{E}_{s'}\{\cdot\}$, and s' is governed by policy π. For brevity, most chapters in this book will not discuss the details of the expectation definition. Readers will need to distinguish their meanings from the context if necessary.

The self-consistency condition, which essentially builds a bridge between value function and policy, contains an intrinsic mechanism to evaluate how good a policy is. This evaluation is performed by calculating the true value that corresponds to a given policy.

The calculation methods can be classified into two types: (1) sample-based evaluation, in which one estimates the state values from the experience generated by the agent-environment interaction, e.g., Monte Carlo (MC) and temporal difference (TD), or (2) model-based evaluation, in which one solves the self-consistency condition with an analytical environment model, e.g., dynamic programming (DP) and approximate dynamic programming (ADP). In essence, these two methods are equivalent in the sense of finding the connection between value function and policy. The sample-based methods need to collect a sufficiently large number of samples, which is time-consuming due to the agent-environment interaction. Since environmental information is collected from interactions, the dilemma of exploration and exploitation must be addressed. Model-based methods can be more computationally efficient with the aid of the partial derivatives of analytical models. The prerequisite of using model-based methods is that an accurate environment model must be known in advance.

2.2   Classification of RL Methods

The term "reinforcement learning" refers to both a class of problems and a set of computational methods. Historically, only some computational methods that attempted to solve stochastic decision problems were known as reinforcement learning algorithms; for example, Q-learning was considered RL, but dynamic programming was often not. In fact, the definition of the RL problem covers a broad spectrum of decision-making and control problems, including discrete-time or continuous-time, linear or nonlinear, deterministic or stochastic, infinite-horizon or finite-horizon, average return or discounted return, and constrained tasks or nonconstrained tasks. The principle of RL is to search for an optimal policy for a dynamic environment that maximizes or minimizes a predefined performance index. Mathematically, it shares a problem formulation that is very similar to that of optimal control.

2.2.1   Definition of RL Problems

One might be interested in the terminological differences between objective function and expected return. The objective function, which is a scalar performance index that one wants to minimize or maximize, is a more general terminology. In contrast, the expected return is a function of the initial state, indicating a sum of reward signals in the form of expectation. Precisely speaking, most definitions of expected return cannot be regarded as a performance index because they actually are not scalars. In the literature, there are two other terminologies, i.e., cost function and loss function, which are analogous to the meaning of objective function. The term cost function is common in optimal control, and the term loss function is often used in statistical learning. Moreover, the cost function usually refers to a sum of rewards or utility functions over the time horizon, and the loss function is often associated with data points where a true value or a target is at hand.

In this book, the formal definition of RL is a class of policy-finding algorithms that seek to maximize or minimize the weighted expectation of the state-value function. Without loss of generality, we take a discrete-time and stochastic task as an example. A standard goal-oriented sequential decision or optimal control problem is mathematically defined as

$$\max_{\pi}/\min J(\pi) = \mathbb{E}_{s \sim d_{\text{init}}(s)}\{v^{\pi}(s)\},$$

subject to

$$(1)\Pr\{s_{t+1} = s'|s_t = s, a_t = a\} = \mathcal{P}_{ss'}^a,$$ (2-15)

or

$$(2)\mathcal{D} = \{s_0, a_0, s_1, a_1, s_2, a_2, \cdots\cdots, s_t, a_t, s_{t+1}, a_{t+1}, \cdots\},$$

$$a \in \mathcal{A}, s \in \mathcal{S},$$

where $J \in \mathbb{R}$ is called overall objective function (also called the performance index), π is the policy to be optimized, $d_{\text{init}}(s)$ is a predefined weighting function, and $v^{\pi}(\cdot)$ is the state-value function. Note that $d_{\text{init}}(s)$ can also be viewed as a distribution function of the initial state, which is fixed and independent of the policy. Most parts of this book use maximizers, which are popular in the machine learning community, while some chapters use minimizers, which are commonly used in the field of automatic control.

One way to classify RL is to consider whether the environment model is known, i.e., model-based RL, or unknown, i.e., model-free RL (Table 2.1). In essence, the former is an optimal control problem but requires considering the overall state space instead of only focusing on one state point at each time step like model predictive control (MPC). In the books of Powell (2011) [2-6] and Bertsekas (2017) [2-1], the family of model-based RL algorithms was discussed and referred to as neuro-dynamic programming. This type of descriptive language mainly comes from the field of optimal control. In contrast, model-free RL uses the data samples collected from environmental interactions to learn a good policy. It is easy to see that the interactive behavior of model-free RL is similar to that of a trial-and-error learner. Sutton's book mainly discusses the family of model-free RLs [2-7], including Monte Carlo RL and temporal difference RL, with descriptive language coming from the field of statistical learning.

Another way to classify RL is by checking whether the optimality condition is used to compute an optimal policy, i.e., indirect RL and direct RL (Table 2.1). The family of indirect RL methods aims to use the solution of the Bellman equation or Hamilton-Jacobi-Bellman (HJB) equation as the optimal policy. Both of these equations are based on Bellman's principle of optimality, which is the necessary and sufficient condition of optimality. Therefore, their solution is at least one optimal policy of the primal problem. The family of direct RL methods does not rely on the optimality condition. Instead, the entire policy space is searched by directly optimizing the primal problem. Combining the analysis described above, the classification of RL methods is listed in Table 2.1, and a few mainstream RL algorithms are listed in each category [2-2].

Table 2.1 Classification of RL methods [2-2]

	Model-based	Model-free
Indirect	DP, ADP, HDP, ADHDP, DHP, GDHP, ADGDHP, CDADP, DGPI	MC, SARSA, Q-learning, A2C*, A3C*, DQN, GAE, DDQN, Dueling DQN, C51, Rainbow, NAF, R2D2, HRA

Direct	PILCO, GPS, I2A, MVE, STEVE, ME-TRPO, SLBO, MBPO, DMVE, MBMF	TRPO, PPO, DPG, DDPG, Off-PAC, ACER, REACTOR, IPG, TD3, SAC, DSAC, Trust-PCL, SIL, APE-X, IMPALA, PPG

* Although A2C and A3C are classified as indirect RLs in this table, they can be derived from both direct and indirect RLs. In fact, most actor-critic algorithms, including TD3, SAC, and DSAC, can find their origins from either direct RL or indirect RL.

The names of the listed algorithms are as follows:

- DP: Dynamic Programming; ADP: Approximate Dynamic Programming; HDP: Heuristic Dynamic Programming; ADHDP: Action Dependent Heuristic Dynamic Programming; DHP: Dual Heuristic Dynamic Programming; GDHP: Globalized Dual Heuristic Dynamic Programming; ADGDHP: Action Dependent Globalized Dual Heuristic Dynamic Programming; CDADP: Constrained Deep Adaptive Dynamic Programming; DGPI: Deep Generalized Policy Iteration.
- MC: Monte Carlo; SARSA: State-Action-Reward-State-Action; A2C: Advantage Actor-Critic; A3C: Asynchronous Advantage Actor-Critic; DQN: Deep Q Networks; GAE: Generalized Advantage Estimator; DDQN: Double Deep Q Networks; Dueling DQN: Dueling Deep Q Networks; C51: Distributional Perspective on Reinforcement Learning; Rainbow: Combining Improvements in Deep Reinforcement Learning; NAF: Normalized Advantage Functions; R2D2: Recurrent Replay Distributed DQN; HRA: Hybrid Reward Architecture.
- PILCO: Model-Based and Data-Efficient Approach to Policy Search; GPS: Guided Policy Search; I2A: Imagination-Augmented Agents; MVE: Model-based Value Expansion; STEVE: Stochastic Ensemble Value Expansion; ME-TRPO: Model-Ensemble TRPO; SLBO: Stochastic Lower Bound Optimization; MBPO: Model-Based Policy Optimization; DMVE: Dynamic-horizon Model-based Value Estimation; MBMF: Model-Based Model-Free hybrid.
- TRPO: Trust Region Policy Optimization; PPO: Proximal Policy Optimization; DPG: Deterministic Policy Gradient; DDPG: Deep Deterministic Policy Gradient; Off-PAC: Off-policy Actor-Critic; ACER: Actor-Critic with Experience Replay; REACTOR: Distributional Retrace Actor-Critic; IPG: Interpolated Policy Gradient; TD3: Twin Delayed Deep Deterministic Policy Gradient; SAC: Soft Actor-Critic; DSAC: Distributional Soft Actor-Critic; Trust-PCL: Trust Path Consistency Learning; SIL: Self-Imitation Learning; APE-X: Distributed Prioritized Experience Replay; IMPALA: Importance Weighted Actor-Learner Architectures; PPG: Phasic Policy Gradient.

2.2.2   Bellman's Principle of Optimality

Bellman's principle of optimality was first proposed by Richard Bellman in the 1950s. This principle was originally stated as follows: an optimal policy has the property that whatever the initial state and the initial decision are, the remaining decisions constitute a new optimal policy with respect to the state resulting from the initial decisions. Both the Bellman equation (for a discrete-time problem) and the Hamilton-Jacobi-Bellman equation (for a continuous-time problem) are derived from Bellman's principle of optimality, which has become the bridge between the fields of optimal control and reinforcement learning. Before moving forward, we must clarify a terminological difference from the book by Sutton and Barto in 2018 [2-7]. Their book <*Reinforcement Learning: An Introduction*> uses the names "Bellman equation" and "Bellman optimality condition" to describe the two terms "self-consistency condition" and "Bellman equation" in this book. Regarding the two terminologies related to Bellman's principle of optimality, this book follows the convention of optimal control community. Readers must distinguish their names carefully to avoid misunderstanding.

2.2.2.1   Optimal Value and Optimal Policy

Here, we still use a discrete-time stochastic system as an example. For any finite MDP, there exists at least one optimal policy, i.e., π^*, which is better than or at least equal to any other policy [2-7]. An optimal policy naturally results in a corresponding optimal state-value function, $v^*(s) \stackrel{\text{def}}{=} v^{\pi^*}(s)$, and a corresponding optimal action-value function, $q^*(s, a) \stackrel{\text{def}}{=} q^{\pi^*}(s, a)$. From an engineering perspective, there may be more than one optimal policy, but their corresponding optimal values are the same. The optimal state-value function is

$$v^*(s) = \max_{\pi} v^{\pi}(s), \forall s \in \mathcal{S}. \tag{2-16}$$

The optimal action-value function is

$$q^*(s, a) = \max_{\pi} q^{\pi}(s, a), \forall s \in \mathcal{S}, \forall a \in \mathcal{A}. \tag{2-17}$$

As shown in the backup diagram for optimal value functions (Figure 2.4), the relation between $v^*(s)$ and $q^*(s, a)$ is as follows:

$$v^*(s) = \max_{a \in \mathcal{A}} q^*(s, a), \forall s \in \mathcal{S}, \tag{2-18}$$

$$q^*(s, a) = \sum_{s' \in \mathcal{S}} \mathcal{P}_{ss'}^a \left(r_{ss'}^a + \gamma v^*(s') \right), \forall s \in \mathcal{S}, \forall a \in \mathcal{A}. \tag{2-19}$$

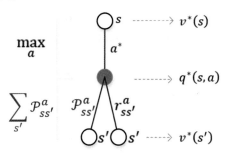

Figure 2.4 Backup diagram for $v^*(s)$ and $q^*(s, a)$

Obviously, an optimal policy can be determined by greedy search of optimal action-value function:

$$\pi^*(a|s) = \begin{cases} 1, & \text{if } a = a^* \\ 0, & \text{if } a \neq a^* \end{cases}, \tag{2-20}$$

$$a^* = \arg\max_a q^*(s, a).$$

Note that this optimal policy is deterministic, even though it is written in a stochastic form. One might question why the optimal policy must be deterministic and why it must be a stationary function of the current state. Is there any possibility that an optimal policy might depend on the historical states? In a dynamic environment with the Markov property, this possibility can be eliminated by the Bellman equation. Due to the Markov property, the Bellman equation of the first kind exactly states that the optimal action can be exclusively determined by the current state. Although multiple optimal actions may exist, they still have the same optimal value. Therefore, any of these actions may be used to construct an optimal policy. This analysis naturally leads to the fact that the optimal policy must be stationary and deterministic.

2.2.2.2   Two Kinds of Bellman Equation

Because $v^*(s)$ is still a kind of value function, it satisfies the self-consistency condition. By combining (2-18) and (2-19), we have the Bellman equation of the first kind:

$$v^*(s) = \max_{a \in \mathcal{A}} \sum_{s' \in \mathcal{S}} \mathcal{P}_{ss'}^a \left(r_{ss'}^a + \gamma v^*(s') \right), \forall s \in \mathcal{S}. \tag{2-21}$$

The benefit of this equation lies in the fact that it allows us to focus on optimizing only current action instead of the whole action sequence to be taken in the future. Therefore, the computational complexity is largely reduced due to the decreased number of variables to be optimized. Intuitively, the Bellman equation expresses the fact that the state value of a state under an optimal policy must be equal to the expected return of taking the best action from that state [2-7]. Once (2-21) is solved, an optimal policy can be found. Since $v^*(s)$ does not explicitly include any action, the environment model must be used to search for the best policy.

In addition to the state-value function, many RLs also take the action-value function as guidance to find an optimal policy. The Bellman equation of the second kind, i.e., that for $q^*(s, a)$, is derived by using (2-19) and (2-18) with the next state s' and the next action a':

$$q^*(s, a) = \sum_{s' \in \mathcal{S}} \mathcal{P}_{ss'}^a \left(r_{ss'}^a + \gamma \max_{a' \in \mathcal{A}} q^*(s', a') \right), \forall s \in \mathcal{S}, \forall a \in \mathcal{A}. \tag{2-22}$$

The information of the environment model is naturally embedded in the optimal value function, which makes choosing the optimal action even easier. The action-value function effectively stores all optimal actions. It allows optimal actions to be selected without knowing the successor states and their values beforehand. Hence, the optimal action can be found by directly maximizing the optimal action-value function $q^*(s, a)$. However, the action-value function (with a domain of $\mathcal{S} \times \mathcal{A}$) may occupy much more memory and computational resources than its state-based counterpart (with a domain of $\mathcal{S}$ only).

2.2.3   Indirect RL Methods

In most engineering tasks, the Bellman equation is the necessary and sufficient condition of optimality. Therefore, the solution of the Bellman equation is equivalent to that of the primal problem. The methods that rely on solving the Bellman equation are called indirect RL and can be further categorized into (1) policy iteration and (2) value iteration, depending on their iteration mode and convergence mechanism.

2.2.3.1   Policy Iteration

Policy iteration is a two-step method used to calculate the solution of the Bellman equation. It has two alternating steps, i.e., policy evaluation and policy improvement. The policy is evaluated by its corresponding value function in the first step, and a better policy is then searched for in the second step. Even though neither of them is optimal, they can gradually converge to their respective optima. Its convergence depends on the so-called "better policy" choice, which means that the policy will become increasingly better until reaching the optimum. This concept is illustrated in Figure 2.5.

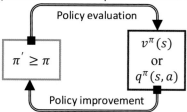

Figure 2.5 Policy iteration

2.2.3.2   Value Iteration

Different from policy iteration, value iteration views the Bellman equation as an algebraic equation. Essentially, the Bellman equation is solved by using the fixed-point iteration technique. Here, the value function is regarded as the variable to be iterated, and no intermediate policy is needed during iteration. Once the optimal value function is found, an optimal policy can be obtained simply by acting greedily with respect to that function. The convergence of value iteration is guaranteed by the fact that the Bellman operator is a contraction mapping. This concept is shown in Figure 2.6.

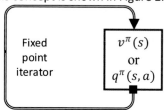

Figure 2.6 Value iteration

Although their iteration cycles are seemingly similar, policy iteration and value iteration are essentially different in their convergence mechanism. The convergence of the former is guaranteed by the "better policy" choice, which means that, at each iteration, a policy that is better than the previous policy must be found. Mathematically, this concept is equivalent to $v^{\pi_{new}}(s) \geq v^{\pi_{old}}(s)$ for any $s \in \mathcal{S}$. Thus, policy iteration produces a monotonically ascending sequence of state-value functions. The latter uses the fixed-point iterator to solve the Bellman equation. Its convergence is determined by the contractive property of the Bellman operator. The monotonicity of state-value function can be guaranteed under the condition of proper initialization, and moreover the error of any two successive values $\|v^{new}(s) - v^{old}(s)\|$ always converges to zero.

2.2.4   Direct RL Methods

Unlike indirect RL, direct RL does not explicitly employ the necessary and sufficient condition of optimality. Instead, in direct RL, a parameterized policy with respect to the scalar performance index $J(\theta)$ is "directly" searched for, in which the policy is mathematically described in the form of π_θ. The most popular direct RL learning method comes from the first-order optimization:

$$\theta^* = \arg\max_\theta J(\theta) = \arg\max_\theta \mathbb{E}_{s \sim d_{\mathrm{init}}(s)}\{v^{\pi_\theta}(s)\}, \qquad (2\text{-}23)$$

$$\theta \leftarrow \theta + \alpha \cdot \nabla_\theta J(\theta), \qquad (2\text{-}24)$$

where α is the learning rate and $\nabla_\theta J(\theta)$ is the policy gradient. The well-known first-order optimization might be the stochastic policy gradient algorithm. The underlying idea of this algorithm is to push up the probabilities of actions that lead to a higher return and to push down the probabilities of actions that lead to a lower return. This concept is illustrated in Figure 2.7. Although starting from the same initial state s_0, different policy parameters $\{\theta_0, \theta_1, \cdots, \theta_k, \cdots, \theta_\infty\}$ have different long-term returns. The policy parameter is changed along the direction that augments the overall performance. With this updating mechanism, the policy parameter is iteratively improved until an optimum is reached.

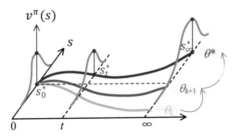

Figure 2.7 Concept of direct RL

Policy gradients can have a variety of different analytical formulas, including vanilla policy gradient, natural policy gradient, and deterministic policy gradient. The accurate estimation of policy gradients is the key to building effective direct RL algorithms. Estimating the policy gradient is a time-consuming procedure that requires many trials around the current policy. In some unfortunate cases, only a local optimum may be found. Other optimization methods, such as derivative-free optimization, can introduce some global benefits in multimodal and ill-conditioned problems, but they are very sample inefficient. Second-order optimization is believed to need fewer iterations but suffers from the high complexity of calculating the inverse Hessian matrix.

2.3   A Broad View of RL

2.3.1   Influence of Initial State Distribution

In practice, the definition of many RL algorithms does not include any weighting information provided by the initial state distribution. To explore the influence of the initial state distribution, let us define two types of optimal policies, namely, $\pi^\circ(s)$ and $\pi^*(s)$. The former is from the maximization of the scalar performance index $J(\pi)$ in the expectation form, and the latter is from the element-by-element maximization of $q^*(s, a)$ for all $s \in S$:

$$\pi^\circ(s) = \arg\max_{\pi(s)\in\Pi} J(\pi(s)),$$ (2-25)

$$\pi^*(s) = \arg\max_{a\in\mathcal{A}} \sum_{s'\in S} \mathcal{P}^a_{ss'}(r + \gamma v^*(s')), \forall s \in S,$$ (2-26)

where $\pi(s) \in \Pi$ is the admissible policy that is confined by a policy set Π. For example, a parameterized policy such as a linear function or a polynomial function actually poses a few constraints on the freedom of selecting actions, which is equivalent to forcing the parameterized policy to have a certain structure. It is worth noting that π^* in (2-26) is a true optimal policy for each state s because it directly comes from the Bellman equation without any policy constraints. Moreover, π^* receives no influence from the initial state distribution $d_{init}(s)$ because this optimal policy is elementwise calculated according to (2-26) for all $s \in S$.

2.3.1.1   When $\pi^*(s)$ is Inside Policy Set Π

First, let us discuss the case in which π^* is inside the policy set Π, i.e., $\pi^* \in \Pi$. The policy set is defined as all the potential policies that meet a predefined policy structure. For example, all linear functions constitute a policy set with structural constraints. For the performance index $J(\pi)$, π° is optimal among all policies in set Π, including $\pi^*(s)$. Therefore, we have the first inequality:

$$J(\pi^*) \le J(\pi^\circ) = \max_\pi J(\pi).$$ (2-27)

Now, let us derive the second inequality. It is clear that

$$J(\pi^\circ) = \max_\pi \mathbb{E}_{s\sim d_{init}(s)}\{v^\pi(s)\}$$

$$\le \max_\pi \mathbb{E}_{s\sim d_{init}(s)}\left\{\max_\pi v^\pi(s)\right\}$$ (2-28)

$$= \mathbb{E}_{s\sim d_{init}(s)}\left\{\max_\pi v^\pi(s)\right\}$$

$$= \mathbb{E}_{s\sim d_{init}(s)}\{v^*(s)\}$$

$$= J(\pi^*).$$

By combining (2-27) and (2-28), we have the following equality:

$$J(\pi^*) = J(\pi^\circ);$$ (2-29)

That is,

$$\mathbb{E}_{s\sim d_{init}(s)}\left\{\max_\pi v^\pi(s)\right\} = \max_\pi \mathbb{E}_{s\sim d_{init}(s)}\{v^\pi(s)\}.$$

Therefore, optimal policies π^* and π° are the same if they are both unique solutions. Hence, $\pi^\circ(s)$ is also independent of the initial state distribution $d_{init}(s)$, i.e.,

$$\max_\pi J(\pi) \iff \max_\pi v^\pi(s), \forall s \in S.$$ (2-30)

It is easy to conclude that as a weighting function, the initial state distribution does not affect the result of the optimal policy as long as it is inside the allowable policy set. This is the reason why many RL algorithms do not explicitly consider the initial state distribution when designing the performance index.

2.3.1.2   When $\pi^*(s)$ is NOT Inside Policy Set Π

When $\pi^*(s)$ does not belong to policy set Π, the first inequality (2-27) no longer holds. As a result, we only have the second inequality (2-28), i.e.,

$$\max_{\pi} \mathbb{E}_{s\sim d_{\text{init}}(s)}\{v^{\pi}(s)\} \le \mathbb{E}_{s\sim d_{\text{init}}(s)}\left\{\max_{\pi} v^{\pi}(s)\right\}, \tag{2-31}$$

A structurally constrained policy $\pi^{\circ}(s) \in \Pi$ usually gives a suboptimal policy compared with the solution of Bellman equation π^*. The left-side optimal policy cannot maximize the state-value of each state element. Therefore, the initial state distribution affects the optimal value of scalar performance index $J(\pi)$, and the learned policy π° will become dependent of $d_{\text{init}}(s)$. In many examples, the policy is usually represented by artificial neural networks with universal approximation ability. The policy set Π can be almost equal to a completely free policy space, and the discrepancy between π° and π^* will become small enough. That is, the influence of the initial state distribution is negligible.

2.3.2   Differences between RL and MPC

The definition of RL is very similar to that of model predictive control (MPC). Basically, MPC is a receding horizon control method in which actions are selected by minimizing a given cost function at each time step. Its cost function is often defined in a finite predictive horizon [2-3][2-4]. An infinite horizon is also allowable but is not common in practice. Figure 2.8 illustrates the basic concept of implementing MPC. This figure demonstrates a receding horizon controller to track a fixed reference signal.

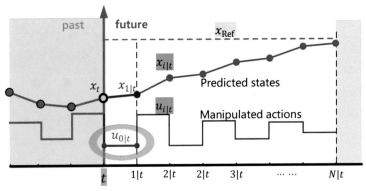

Figure 2.8 Receding horizon control in MPC

At time t, a sequence of open-loop optimal actions are obtained in the predictive horizon starting from the current time. Among all the optimal actions, only the first action is implemented as the current control input. This kind of optimization and implementation is repeated at each control instant. Based on this unique characteristic, there are several advantages of utilizing MPC, such as its closed-loop control structure, ability to pursue optimal performance, explicit handling of state constraints, and easy extension to nonlinear systems.

We can regard RL and MPC as two ways to implement optimal controllers. Despite their similarity in the problem definition, there are also some obvious differences between RL and MPC. RL has a statistical learning background, while MPC is rooted in feedback control. RL can apply to those cases where the environmental information is unknown, while in MPC, the environment model must be known beforehand [2-5]. It is easy to see that these methods have different descriptive languages, as listed in Table 2.2. Three notable differences between these methods are as follows: (1) RL maximizes the overall objective function, while MPC minimizes it; (2) RL often addresses stochastic environments, while MPC often handles deterministic systems; and (3) RL often addresses probabilistic models, while MPC usually builds state-space models.

Table 2.2 Descriptive language differences between RL and MPC

RL (Discounted cost)	MPC (Infinite horizon)
• Stochastic system (e.g., environment)	Deterministic system (e.g., plant and process)
• State & Action (s, a)	State & Control input (x, u)
• Probabilistic model $\Pr\{s_{t+1} = s' \mid s_t = s, a_t = a\} = \mathcal{P}_{ss'}^a$	State-space model $x_{t+1} = f(x_t, u_t)$
• Policy $\pi(a \mid s)$	Controller $u = \pi(x)$
• Reward signal $r(s, a, s')$	Utility function $l(x, u)$
• State-value function $v^\pi(s) = \mathbb{E}_\pi \left\{ \sum_{i=0}^{+\infty} \gamma^i r_{t+i} \mid s_t = s \right\}$	Infinite-horizon cost function $V(x) = \sum_{i=0}^{+\infty} l(x_{t+i}, u_{t+i}) \mid_{x_t = x}$
• Maximize a weighted state-value function $\max_\pi \mathbb{E}_{s \sim d(s)} \{ v^\pi(s) \}$	Minimize a cost function $\min_u V(x)$
• Self-consistency condition $v^\pi(s) = \sum_a \pi(a \mid s) \left\{ \sum_{s'} \mathcal{P}(r + \gamma v^\pi(s')) \right\}$	Self-consistency condition $V(x) = l(x, u) + V(x')$
• Bellman equation $v^*(s) = \max_a \sum_{s'} \mathcal{P}(r + \gamma v^*(s'))$	Bellman equation $V^*(x) = \min_u \{ l(x, u) + V^*(x') \}$

Another key difference between RL and MPC lies in when to find and implement optimal actions. An RL agent searches for an optimal policy defined on the whole state space.

One advantage of this design is that RL algorithms can be executed offline, and then the learned policy can be implemented online. MPC often uses a receding horizon optimization scheme, and at each moment, it only calculates the optimal action for the current state point. An exception to this process is a technique called explicit MPC, which uses multi-parametric programming to find a tabular solution. In fact, explicit MPC is more like a special model-based RL algorithm. Except for explicit MPC, most MPCs are implemented with a high online optimization burden [2-5].

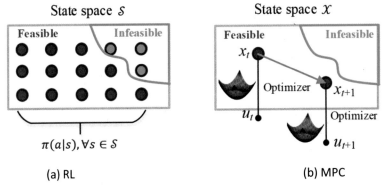

(a) RL   (b) MPC

Figure 2.9 Comparison of how RL and MPC find their optimal actions

Figure 2.9 compares how RL and MPC find their optimal actions. Since RL searches in the whole state space, its optimal policy needs to cover every state points. Hence, it is prone to fail in building an admissible policy even if only a few local states are infeasible. In contrast, MPC has better tolerance to the infeasibility issue since it does not need to be concerned with every state. The occurrence of infeasible states may be attributed to several reasons, among which the most prominent reasons are hard state constraints and the loss of closed-loop stability.

2.3.3   Various Combination of Four Elements

Almost all RL methods involve a quadruple $\{\mathcal{D}, \pi, v^{\pi}, \mathcal{P}\}$, as shown in Figure 2.10. The quadruple includes four key elements: (1) state-action samples $\mathcal{D} = \{s_0, a_0, s_1, a_1, s_2, a_2, \cdots \cdots, s_t, a_t, s_{t+1}, a_{t+1}, \cdots\}$; (2) the policy $\pi(a|s)$; (3) the reward signal r_t, or more precisely, state-value function $v^{\pi}(s)$; and (4) the environment model $\mathcal{P}_{ss'}^a = p(s'|s, a)$.

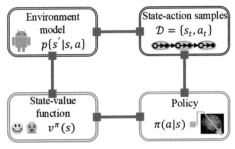

Figure 2.10 Four elements in RL

The state-action samples come from the agent-environment interaction. The collected data sequence actually contains two kinds of information: (1) $\mathcal{S} \times \mathcal{A} \mapsto \mathcal{S}$, the information of environment dynamics, and (2) $\mathcal{S} \mapsto \mathcal{A}$, the information of the currently used policy. The state-value function is the expectation of the long-term return that equals the accumulative sum of reward signals. The reward signal may not always exist in the real world. In this type of real-world tasks, reward signals need to be artificially designed.

Obviously, the information of the quadruple $\{\mathcal{D}, \pi, v^{\pi}, \mathcal{P}\}$ is redundant in finding an optimal policy. Either the environment model or the state-action samples can provide information on environment dynamics, and they correspond to model-based RL and model-free RL, respectively. Following this idea, the combination of the four elements in the quadruple can formulate various optimization-related problems, including model-free RL, model-based RL, inverse RL, parameter identification, and mixed RL (i.e., RL with both the environment model and state-action samples).

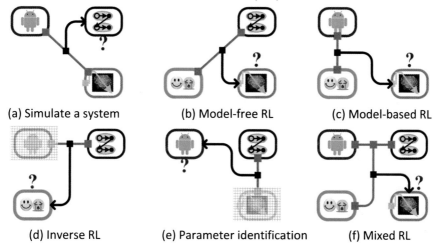

(a) Simulate a system   (b) Model-free RL   (c) Model-based RL

(d) Inverse RL   (e) Parameter identification   (f) Mixed RL

Figure 2.11 Different combinations of the four elements in RL

As demonstrated in Figure 2.11, model-free RL attempts to find π^* when $\mathcal{D}$ and $v^{\pi}(s)$ are known; model-based RL seeks to find π^* when $p(s'|s, a)$ and $v^{\pi}(s)$ are known. If both $\mathcal{D}$ and $p(s'|s, a)$ are known but neither of them is optimal, they can be combined to find a better policy. This concept defines a mixed RL problem. Inverse RL tries to find $v^{\pi}(s)$ when $\mathcal{D}$ is known (note that in Figure 2.11 (d), the shaded module $\mathcal{P}$ can either be known or unknown because $\mathcal{D}$ inherently contains the information of environment dynamics). Note that inverse RL assumes that $\mathcal{D}$ is from expert experience, which can be said to be optimal in the view angle of experts.

2.4   Measures of Learning Performance

The learning performance of an RL agent can be measured in a variety of dimensions, including policy performance, learning accuracy, learning speed, sample efficiency, and approximation accuracy. Policy performance is often used to compare the goodness of two different policies in a real environment. The learning accuracy measures how close a policy or a value function is to its optimum. The measures of learning speed mainly include wall-clock time and iteration efficiency. The sample efficiency describes how many samples are required to reach a certain policy performance. The approximation accuracy focuses on issues such as underestimation and overestimation and it measures the difference between the parameterized value and the true value.

Some deep RL studies may measure the robustness of a policy. A robust learning result is desired so that the learned policy is not sensitive to random seeds or hyperparameters. In some special conditions, when the validation environment is not available, off-policy evaluation is needed to calculate performance measures with only pre-collected data. This method allows RL to predict how well a new policy performs without deploying it into the real world.

2.4.1 Policy Performance

The total average return (TAR) is the most popular measure of policy performance. Given a policy π, its total average return is calculated by repeated Monte Carlo estimations. The basic procedure is described as follows: (1) select the initial state by a predefined state distribution $d_{init}(s)$; and (2) for each starting state, generate a few episodes and calculate the return of each episode, which is denoted as $G_i(s)$. The total average return of policy π is calculated as

$$R_{Avg} = \sum d_{init}(s)\left\{\frac{1}{N}\sum_{i=1}^{N} G_i(s)\right\},$$

$$G_i(s) = \sum_{t=0}^{L_i-1} r_t,$$

(2-32)

where $d_{init}(s)$ is the initial state distribution, N is the number of episodes, and L_i is the length of the i-th episode. The episode length can be either the same or different with each other. The initial state distribution is a critical factor that affects the calculation of policy performance. It either naturally exists in the environment model or can be selected by a human designer. For example, an autonomous driving system can be initialized using the average behavior of human drivers while a computer game simulation can be initialized with evenly distributed random states. In general, we do not include a discount factor when calculating total average returns. The purpose is to provide a fair comparison for RL algorithms trained with different discount factors.

In episodic tasks, a learned policy is directly sent into the environment for performance evaluation. Each simulation will stop automatically at terminal states. For continuing tasks, the episode length needs to be specified to terminate the simulation. Note that in some special cases, we may concern only the deterministic part of a stochastic policy during the evaluation of policy performance. This is because the optimal policy must be deterministic in almost all fully observable MDPs. The randomness in a stochastic policy is mainly for exploration but not what an engineer desires in practice. Therefore, it is reasonable to evaluate only the deterministic part of a stochastic policy if it was also used as the policy for on-line deployment.

2.4.2 Learning Accuracy

Some simple RL problems have a provable guarantee of asymptotic convergence and an analytical solution of optimal behaviors. The learning accuracy measures how far a learned policy or a learned value function is from the true optimum. When the optimal value $v^*(s)$ under an optimal policy π^* is known, the learning accuracy is measured using the root mean square error (RMSE). Here, we weight each square error with its occurrence probability instead of treating them equally. The RMSEs of $\pi(s)$ and $V^\pi(s)$ are expressed as

$$\text{RMSE}_{v\text{Opt}} = \sqrt{\sum_{s \in \mathcal{S}} d_\pi(s) \left(V^\pi(s) - v^*(s) \right)^2},$$

$$\text{RMSE}_{\pi\text{Opt}} = \sqrt{\sum_{s \in \mathcal{S}} d_\pi(s) \| \pi(s) - \pi^*(s) \|_2^2},$$

(2-33)

where $d_\pi(s)$ is the stationary state distribution, $\pi(s)$ is the learned policy, and $V^\pi(s)$ is the learned value. In practice, a stationary state distribution can be calculated by deploying the policy π into an actual environment and collecting sufficient samples for an accurate estimation. The generated samples naturally obey the stationary state distribution under the policy π. In some tasks, one can replace $d_\pi(s)$ with a uniform distribution to avoid collecting a large number of real samples.

Remark (1): The learning accuracy, which is represented by the distance from the true value or policy, is an absolute measure. In fact, the "true" optimal policy π^* and the "true" optimal value $v^*(s)$ are only accessible in some simple tasks with analytical solutions. These analytical solutions are often inaccessible in complex tasks. For these tasks, the learning accuracy cannot be calculated.

Remark (2): Obviously, $G(s)$ in TAR and $V^\pi(s)$ in RMSE are quite close in value output. Both are calculated by the data samples from environmental interactions. One crucial difference between these values is that TAR usually does not include the discount factor. In addition, $G(s)$ requires that s is the initial state of the whole trajectory, i.e., $s = s_0$, so that for one trajectory, $G(s)$ can be calculated only once. In contrast, $V^\pi(s)$ can use either the whole trajectory or a subpart that starts from the same state. Hence, for one episode, one may calculate $V^\pi(s)$ multiple times if multiple samples with $s_i = s$ are found. In practice, the difference between these values is not clearly distinguished, and a blurred use of $G(s)$ and $V^\pi(s)$ is quite popular in the literature.

2.4.3   Learning Speed

The learning speed is an important indicator of the training efficiency. Explicitly deriving theoretical results of the convergence rate is usually untenable due to the complexity of environment dynamics and function approximators. A more practical way of determining the convergence rate is to numerically measure the wall-clock time or the iteration number that is required to reach a near-optimal point, for example, 95% of the highest total average return. The wall-clock time is the time duration that the RL training process lasts on a computer and is dependent on the computer's calculation speed. In contrast, the total number of iterations from start to finish is often independent of the computer hardware.

2.4.4   Sample Efficiency

The sample efficiency describes how many samples are required to reach a certain policy performance. Here, a "sample" is defined as a triple of state, action, and next state. An indicator of sample efficiency is the total number of newly collected samples when reaching a certain percentage of the best policy performance, for example, 95% of the highest total average return. In addition, generating curves of the total average return with respect to the number of samples is an intuitive way to compare the sample efficiency of different RL algorithms. Note that sample efficiency only counts "new" samples. Repeatedly using historical samples, e.g., off-policy with experience replay, does not add any new samples, and thus it is helpful to increase the sample efficiency.

2.4.5   Approximation Accuracy

To discuss overestimation or underestimation, we need to measure the accuracy of value function approximation. This measure calculates the difference between parameterized values and true values, described by the root mean square error (RMSE):

$$
\text{RMSE}_{v\text{Appr}} = \sqrt{\sum_{s \in \mathcal{S}} d_\pi(s) \left(V^\pi(s; w) - v^\pi(s) \right)^2} ,
\tag{2-34}
$$

where $d_\pi(s)$ is the stationary state distribution, $V^\pi(s; w)$ is the parameterized value, and $v^\pi(s)$ is the true value. Similarly, knowing the true value is an inevitable component in calculating the approximation accuracy. Similar to the measure of learning accuracy, the true state value $v^\pi(s)$ is only accessible in some simple tasks with analytical solutions. In this measure, one can approximately calculate this value by sending the policy into the environment and performing sample-based estimation. To increase the approximation accuracy, an RL agent is encouraged to choose a larger parameter space. However, increasing the size of the parameter space is a double-edged sword. Since RL always uses incomplete and noisy exploration information, an excessively complicated approximation function may cause severe overfitting issue. There must be a trade-off between reducing approximation errors and preventing potential overfitting.

2.5   Two Examples of Markov Decision Processes

Since RL can gradually learn a policy, it best suits complex problems where no analytical or easily programmable solution exists. For those problems in a virtual world, such as playing Chinese Go and computer games, RL does not cause any risk in the process of environment exploration. The agent can interact with the environment by simply deploying any possible policy. In many real-world tasks, for example, autonomous driving and robot control, environmental interactions might cause unpredictable economic losses or safety issues. A common solution is to first learn a policy in a simulated environment and then deploy it to the real world after the trained policy becomes mature. Here, we introduce two simulated Markov environments with finite state and action spaces. One example is that of an indoor cleaning robot, and the other is an autonomous driving system. Both examples are assumed to have simplified dynamics and run in grid-based worlds.

2.5.1   Example: Indoor Cleaning Robot

Robotic floor cleaners alleviate the need for people to complete unpleasant daily chores and can free up their time. The tasks that a robotic cleaner needs to fulfill include obstacle avoidance, self-charging, sweeping and vacuuming. Most robotic floor cleaners use ultrasonic sensors or infrared lasers for environment perception and self-positioning. Two axial wheels, which are DC-motor driven, permit motion in the parallel direction and generate the moving torque. The single front wheel, which is servo-controlled, steers the robot by actively and promptly switching directions.

(a) Robotic floor cleaner

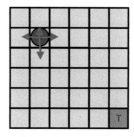
(b) Grid environment

Figure 2.12 Indoor cleaning robot

$S_{(1)}$	$S_{(2)}$	$S_{(3)}$	$S_{(4)}$	$S_{(5)}$	$S_{(6)}$
$S_{(7)}$	$S_{(8)}$	$S_{(9)}$	$S_{(10)}$	$S_{(11)}$	$S_{(12)}$
$S_{(13)}$	$S_{(14)}$	$S_{(15)}$	$S_{(16)}$	$S_{(17)}$	$S_{(18)}$
$S_{(19)}$	$S_{(20)}$	$S_{(21)}$	$S_{(22)}$	$S_{(23)}$	$S_{(24)}$
$S_{(25)}$	$S_{(26)}$	$S_{(27)}$	$S_{(28)}$	$S_{(29)}$	$S_{(30)}$
$S_{(31)}$	$S_{(32)}$	$S_{(33)}$	$S_{(34)}$	$S_{(35)}$	T

State = $\{s_{(i)}\}$, i=1, 2,..., 35, 36

Action = {North, South, West, East}

(a) State space   (b) Action space

Figure 2.13 Cleaning robot in a grid environment

Here, the cleaning robot is assumed to work in a rectangular grid environment. Only one cell is the destination (Cell T), as shaded in Figure 2.13 (a). Each cell represents a certain position and corresponds to a state in the grid environment (Cell T corresponds to $s_{(36)}$). The cleaning robot has four possible actions, namely, "North", "South", "West", and "East", shown in Figure 2.13 (b). The state space of the grid environment is

$$\mathcal{S} = \{s_{(1)}, s_{(2)}, \cdots, s_{(36)}\}. \tag{2-35}$$

The action space of the robot is

$$\mathcal{A} = \{\text{North, South, West, East}\}. \tag{2-36}$$

Each action can move the cleaning robot to its neighboring cells in a random manner. Assume that the cleaning robot is at the state s. After taking an action, its next state $s' \in \mathcal{S}$ (if available) is governed by the following robot dynamics:

$$\Pr\{s' = \text{Front Cell} \,|s, a\} = 0.8, \tag{2-37}$$

$$\Pr\{s' = \text{Left Cell} \,|s, a\} = 0.1,$$
$$\Pr\{s' = \text{Right Cell} \,|s, a\} = 0.1,$$
$$\Pr\{s' = \text{Back Cell} \,|s, a\} = 0.$$

Whenever an action is taken, the robot moves forward with a probability of 0.8. It moves leftward or rightward with a probability of 0.1. There is no possibility to move backward. In addition, when at a corner or next to a boundary, the robot must not move outside of the grid environment. If its next state is not available, the robot stays at its current position. The reward signal is designed to drive the cleaner to reach the destination as soon as possible:

$$r(s, a, s') = \begin{cases} -1, \text{if } s' \neq s_{(36)} \\ +9, \text{if } s' = s_{(36)} \end{cases}. \tag{2-38}$$

Each move results in a penalty of -1, but arriving at the destination has a reward of +9. The design of the reward signals follows the observation that each action with DC-motor and servo-motor consumes a certain amount of energy and that the destination cell will fully recharge the robot.

2.5.2   Example: Autonomous Driving System

An autonomous car is a vehicle capable of sensing its environment and operating without human involvement. The Society of Automotive Engineers (SAE) defines 6 levels of driving automation ranging from Level 0 (fully manual) to Level 5 (fully autonomous). Autonomous cars are equipped with a variety of different types of onboard sensors, such as cameras, radar and lidar, to perceive its surrounding environment. Path decision and motion control are made by the car to avoid obstacles or respond to other road users. The use of autonomous cars has the potential to increase traffic flow, provide enhanced mobility, relieve driver workload, and reduce fuel consumption.

(a) Cars operating on a curved road

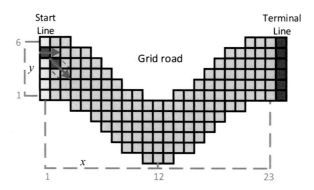

(b) Road environment grid

Figure 2.14 Autonomous cars on a curved road

Heading direction h={ } Action a={Left, Keep, Right}

Figure 2.15 Car heading direction and steering action

Here, an autonomous car is assumed to work on a curved grid road. Each car takes up two adjacent cells. The rear cell corresponds to two state dimensions, i.e., horizontal position-x and vertical position-y. The relative position between the two cells represents the third state dimension, i.e., heading direction h. The car has three steering actions, i.e., "Right", "Keep", and "Left". Note that "Left" and "Right" are defined with respect to the heading direction. The steering action can adjust the car's heading direction. For example, if the car is heading toward the right, a "Left" steering action will turn the car toward the top right side. After the steering action is executed, the car will move one step forward along the new heading direction. The terminal line is on the right side of the grid road, shaded as the self-driving destination. Once the front cell of the car occupies any cell in the terminal line, the current episode ends. The state of this car-road system is a vector with three dimensions:

$$s = [x, y, h]^\mathrm{T} \in \mathcal{S},$$
$$\mathcal{S} = \mathcal{S}_x \times \mathcal{S}_y \times \mathcal{S}_h,$$
$$\mathcal{S}_x = \left\{x_{(1)}, x_{(2)}, \cdots, x_{(23)}\right\}, \qquad (2\text{-}39)$$
$$\mathcal{S}_y = \left\{y_{(1)}, y_{(2)}, \cdots, y_{(6)}\right\},$$
$$\mathcal{S}_h = \left\{h_{(1)}, h_{(2)}, \cdots, h_{(5)}\right\}.$$

The action space is

$$\mathcal{A} = \{\text{Left}, \text{Keep}, \text{Right}\}. \qquad (2\text{-}40)$$

Each action can deterministically turn the car to a neighboring direction and move the car one step forward along the new direction. Taking the action "Right" as an example, the car dynamics is represented at state $\left[x_{(1)}, y_{(5)}, h_{(3)}\right]^{\mathrm{T}}$:

$$\Pr\left\{s' = \left[x_{(2)}, y_{(4)}, h_{(2)}\right]^{\mathrm{T}} \middle| s = \left[x_{(1)}, y_{(5)}, h_{(3)}\right]^{\mathrm{T}}, a = \text{Right}\right\} = 1. \qquad (2\text{-}41)$$

The autonomous car seeks to reach the destination as soon as possible while utilizing the least amount of steering energy. The reward signal has two sub-elements: steering reward and moving reward. Each steering action "Left" or "Right" results in a reward of -1, and such a no-steering action "Keep" has a zero reward. The moving reward is either -1 or $-\sqrt{2}$, depending on whether the travel path is horizontal (vertical) or diagonal. The car stops after reaching any terminal state:

$$r(s, a, s') = r_{\text{Steer}} + r_{\text{Move}},$$

$$r_{\text{Steer}} = \begin{cases} -1 & \text{, if } a = \text{Left} \\ 0 & \text{, if } a = \text{Keep} \\ -1 & \text{, if } a = \text{Right} \end{cases}, \qquad (2\text{-}42)$$

$$r_{\text{Move}} = \begin{cases} -1 & \text{, if } h' = 1 \text{ or } 3 \text{ or } 5 \\ -\sqrt{2} & \text{, if } h' = 2 \text{ or } 4 \end{cases}.$$

The reward r_{Steer} poses a penalty to energy-wasting behaviors in the steering actuator, while r_{Move} is equivalent to penalizing the traveling distance. In addition, collisions with the road boundaries are not allowed in the environmental exploration. This is equivalent to posing an infinitely large penalty to actions when the car runs outside the lane.

2.6   References

[2-1]   Bertsekas D (2017) Dynamic programming and optimal control. Athena Scientific, Massachusetts

[2-2]   Guan Y, Li SE, Duan J, et al (2021) Direct and indirect reinforcement learning. Intl J of Intelli Syst (36): 4439-4467

[2-3]   Li SE, Xu S, Kum D (2015) Efficient and accurate computation of model predictive control using pseudospectral discretization. NeuroComputing (172): 363-372

[2-4]   Zheng Y, Li SE, Li K et al (2017) Distributed model predictive control for heterogeneous vehicle platoons under unidirectional topologies. IEEE Tran Control Syst Tech (25) 3: 899-910

[2-5]   Maciejowski J (2000) Predictive control with constraints. Prentice Hall, Upper Saddle River

[2-6]   Powell W (2011) Approximate dynamic programming: solve the curses of dimensionality. Wiley, New York

[2-7]   Sutton R, Barto A (2018) Reinforcement learning: an introduction. MIT press, Cambridge

Chapter 3.   Model-Free Indirect RL: Monte Carlo

Progress is made by trial and failure;

the failures are generally a hundred times

more numerous than the successes;

yet they are usually left unchronicled.

-- William Ramsay (1852-1916)

Monte Carlo (MC) experiments refer to a class of computational algorithms that are used to analyze and simulate the impact of randomness or uncertainty in stochastic processes. This class of computational algorithms was named after the Casino de Monte-Carlo, the gambling hotspot in Monaco. The core idea was first developed by Stanislaw Ulam and his colleagues in the late 1940s. Immediately after Ulam's breakthrough, the importance of MC experiments was noticed by scientists and engineers in various fields. The central idea of MC experiments was quickly extended to a broad class of problems that have a probabilistic interpretation. Today, Monte Carlo techniques have become very prevalent in the fields of optimal filtering, statistical learning and stochastic control. A non-exhaustive list of relevant techniques includes mean-field MC filter, MC resampler, MC localization, and MC tree search.

In the field of reinforcement learning, Monte Carlo estimation is a central component of a large class of model-free algorithms. The MC learning algorithm is essentially an important branch of generalized policy iteration (GPI), which has two periodically alternating steps, i.e., policy evaluation (PEV) and policy improvement (PIM). In the GPI framework, each policy is first evaluated by its corresponding value function. Then, based on the evaluation result, greedy search is completed to output a better policy. The MC estimation is mainly applied to the first step, i.e., policy evaluation. The simplest idea, i.e., averaging the returns of all collected samples, is used to judge the effectiveness of the current policy. As more experience is accumulated, the estimate will converge to the true value by the law of large numbers. Hence, MC policy evaluation does not require any prior knowledge of the environment dynamics. Instead, all it needs is experience, i.e., samples of state, action, and reward, which are generated from interacting with a real environment.

Compared to dynamic programming (DP), the MC learning algorithm has several advantages. First, it can learn online from repeated interactions with the actual environment, and accordingly, an accurate model is not required. Second, in each MC policy evaluation, its environment exploration does not need to traverse the whole state space. Third, it is often less negatively impacted by the violation of the Markovian property because MC estimation does not bootstrap itself, which is a core difference from the temporal difference approach. However, MC estimation often suffers from very slow convergence due to the demand for sufficient exploration. One caveat about MC is that it can only be applied to episodic and small-scale tasks. Since MC has no bootstrapping update, it must be implemented in an episode-by-episode manner. Each

S. E. Li, *Reinforcement Learning for Sequential Decision and Optimal Control*,
https://doi.org/10.1007/978-981-19-7784-8_3

episode-level interaction must be finished before calculating any return, and therefore, one cannot say MC is a perfect online learning algorithm.

3.1   MC Policy Evaluation

MC policy evaluation is best suited for those tasks with a terminal time or an ease of termination. Here, we focus on episodic tasks with discrete state and action spaces. Under some assumptive conditions, the episodic results can extend to most continuing tasks. An obvious way to estimate the value function is to simply average the returns starting from a particular state s or a particular state-action pair (s, a). As an increasing number of returns are observed, the average will converge to the true value. This idea underlies almost all methods of MC policy evaluation. Here, we use the notation $V^\pi(s)$ to represent the estimated value of state s under a policy π and use the notation $v^\pi(s)$ to represent its true value. Note that in Figure 3.1, $s_{(i)}$ represents the i-th state element in the state space S, rather than the state at time t. The same definition applies to $a_{(i)}$ in the action space $\mathcal{A}$.

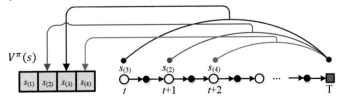

(a) Estimation of the state-value function

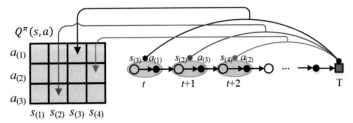

(b) Estimation of the action-value function

Figure 3.1 Monte Carlo policy evaluation

The estimate of the state-value function is the average of discounted returns starting from each state element in the whole state space:

$$V^\pi(s) = \text{Avg}\{G_{t:T}|s_t = s\},$$
$$G_{t:T} = \sum_{i=0}^{T-t} \gamma^i r_{t+i}. \tag{3-1}$$

Here, we assume that each episode starts from time t and ends at time T. The terminal time T, which determines the length of each episode, is not fixed. The average-based estimation is illustrated in Figure 3.2, in which each episode must start from the same state, i.e., $s_t = s$. Episodes may have different lengths because they may have different terminal times. Nevertheless, the returns of all episodes are summed up and divided by the number of episodes to estimate the value function. By the law of large numbers, the estimated value will converge to the true value, i.e., $V^\pi(s) \to v^\pi(s)$, as the number of episodes goes to infinity.

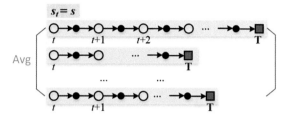

Figure 3.2 Episode-based average estimation

The MC policy evaluation in (3-1) relies only on experience. When using the state-value function to make a decision, state values alone are not sufficient to determine the best action, and an environment model is needed to calculate the return under a given action and state. If the model is not available, it is particularly useful to estimate the action-value function rather than the state-value function. To estimate the action-value function, the returns to be averaged must start at a particular state-action pair (s, a) in the state-action space:

$$Q^\pi(s, a) = \text{Avg}\{G_{t:T} | s_t = s, a_t = a\},$$

$$G_{t:T} = \sum_{i=0}^{T-t} \gamma^i r_{t+i}. \tag{3-2}$$

In the MC estimation, the returns for each state or state-action pair are independent. This property makes MC methods particularly attractive when one requires the value of only a single state or a subset of the whole state space. A common problem is that many state-action pairs may never be visited due to poor exploration. For example, when following a deterministic policy, one will observe returns only for one action from each fixed state. Without enough new returns, MC estimates may have poor accuracy for some state-action pairs. This is a serious problem because high-return action are more likely to be chosen compared to other actions at each state at the PIM step [3-3]. One way to maintain exploration is to randomly start the episodes at all available state-action pairs. This strategy, which is known as exploring starts, guarantees that all state-action pairs will be visited. However, the computational burden may be increased in large-scale tasks. Another way to maintain exploration is to consider the utilization of stochastic policies (e.g., try to train an ϵ-greedy policy). This type of stochastic policy enriches the exploration ability with a nonzero possibility of randomly selecting all possible actions at each state.

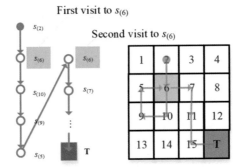

Figure 3.3 First-visit MC vs. every-visit MC

During the agent-environment interaction, a particular state s may be visited multiple times in the same episode. This phenomenon is illustrated in Figure 3.3, in which state $s_{(6)}$ is visited twice in one episode. There are two basic MC methods to handle multi-visit phenomenon: first-visit MC and every-visit MC. In the first-visit MC, only the returns from the first visit to a particular state s are averaged, whereas in the every-visit MC, all the returns from every visit to state s are averaged. In both methods, convergence to the true value occurs as the number of visits to s goes to infinity. In the first-visit MC, each return is an independent, identically distributed estimate of $v^{\pi}(s)$ with a finite variance. By the law of large numbers, the sequence of average-based estimates will converge to their expected value. Each average itself is an unbiased estimate, and the standard deviation of its error declines at the rate of $1/\sqrt{N}$, where N is the number of returns to be averaged [3-3]. The result of every-visit MC is less straightforward, but its estimate can also converge to the true value at a more satisfactory rate.

3.2   MC Policy Improvement

MC policy evaluation, which approximates the value function by averaging the returns from a collection of episodes, is the first step of MC learning algorithms. MC policy improvement, which seeks to find a better policy than the current policy, is the second step of these algorithms. The significance of selecting a better policy lies in the fact that selecting a better policy is the key to ensuring convergence, which means that the policy will continue to improve until reaching the optimal solution. A better policy can be found by searching for greedy actions with respect to the current value function. With an environment model, state values alone are sufficient to determine a better policy. One can simply look ahead one step and choose whichever action leads to the largest summation of the current reward and the value of next state. In the condition without any model, action values must be used to search for greedy actions. The values of all state-action pairs should be explicitly compared to determine a better policy.

3.2.1   Greedy Policy

The greedy search technique makes the best choice as it attempts to find the overall optimal approach to solve the entire problem. Imagine that you are going hiking and your goal is to reach the highest peak. You start hiking with a simple strategy, e.g., choose the climbing path with the largest slope. This climbing strategy is called a greedy search. A greedy policy takes the short-sighted choice at each stage in the hope of eventually finding an overall optimal solution. The greedy policy is defined as

$$\pi^g(a|s) = \begin{cases} 1, & \text{if } a = a^* \\ 0, & \text{if } a \neq a^* \end{cases},$$

subject to

(3-3)

$$a^* = \arg\max_a q^\pi(s, a).$$

Actually, a greedy search always outputs a deterministic policy, which can maximally exploit the knowledge of value functions but retains rare ability to explore the environment. To balance exploitation and exploration, an extended approach is to use the ϵ-greedy policy, which behaves greedily most of the time but occasionally selects a random action with a small probability. The ϵ-greedy policy is defined as

$$\pi^\epsilon(a|s) = \begin{cases} 1 - \epsilon + \dfrac{\epsilon}{|\mathcal{A}|}, & \text{if } a = a^* \\ \dfrac{\epsilon}{|\mathcal{A}|}, & \text{if } a \neq a^* \end{cases},$$

subject to

(3-4)

$$a^* = \arg\max_a q^\pi(s, a),$$

where ϵ is the exploration rate, i.e., a small probability of randomly selecting actions in $\mathcal{A}$, and $|\mathcal{A}|$ represents the number of actions in $\mathcal{A}$. In the ϵ-greedy policy, independent of the value functions, all non-greedy actions are assigned to a non-zero probability to maintain a certain amount of exploration.

3.2.2 Policy Improvement Theorem

The purpose of MC policy improvement is to find a better policy. When searching for a better policy, one must have a clear definition of the term "better". Generally, the superiority of a policy is measured by its corresponding state-value function. More precisely, one policy $\bar{\pi}$ is said to be better than or equal to another policy π, i.e., $\pi \leq \bar{\pi}$, if and only if $v^\pi(s) \leq v^{\bar{\pi}}(s)$ holds for each state $s \in \mathcal{S}$:

$$\pi \leq \bar{\pi} \iff v^\pi(s) \leq v^{\bar{\pi}}(s), \forall s \in \mathcal{S}. \tag{3-5}$$

- Theorem 3-1: A new policy $\bar{\pi}$ is better than or at least as good as the current policy π if the following inequality holds:

$$v^\pi(s) \leq \sum_{a \in \mathcal{A}} \bar{\pi}(a|s) q^\pi(s, a), \forall s \in \mathcal{S}. \tag{3-6}$$

Proof:

Consider the Markov property of an MDP and that $\bar{\pi}(a|s)$ is a probabilistic distribution on $\mathcal{A}$ for each state s (illustrated in Figure 3.4):

$$v^\pi(s) \quad \leq \sum_{a \in \mathcal{A}} \bar{\pi}(a|s) q^\pi(s, a)$$

$$= \mathbb{E}_{a \sim \bar{\pi}}\{q^\pi(s, a)\}$$

$$= \mathbb{E}_{a \sim \bar{\pi}} \left\{ \sum_{s' \in S} \mathcal{P}^a_{ss'} \left(r^a_{ss'} + \gamma v^{\pi}(s') \right) \right\}$$

$$= \mathbb{E}_{a,s' \sim \bar{\pi}} \{ r + \gamma v^{\pi}(s') \}$$

$$\leq \mathbb{E}_{a,s' \sim \bar{\pi}} \left\{ r + \gamma \mathbb{E}_{a',s'' \sim \bar{\pi}} \{ r' + \gamma v^{\pi}(s'') \} \right\}$$

$$= \mathbb{E}_{a,s',a',s'' \sim \bar{\pi}} \{ r + \gamma r' + \gamma^2 v^{\pi}(s'') \}.$$

Roll forward until $t \to +\infty$ to obtain

$$\leq \mathbb{E}_{a,s',a',s'',a'',s''',\cdots \sim \bar{\pi}} \{ r + \gamma r' + \gamma^2 r'' + \cdots + \gamma^{\infty} r_{\infty} \}$$

$$= v^{\bar{\pi}}(s).$$

∎

This theorem is the famous policy improvement theorem, which provides a sufficient condition for a better policy [3-3]. In (3-6), $q^{\pi}(s, a)$ is still generated by the current policy π but weighted by a new policy $\bar{\pi}$. If $\bar{\pi}$ is deterministic, the expression $\sum_{a \in \mathcal{A}} \bar{\pi}(a|s) q^{\pi}(s, a)$ can be simplified to the compact form $q^{\pi}(s, \bar{\pi}(s))$. This proof is illustrated in Figure 3.4, which implies that such a cascading inequality (3-7) holds if (3-6) holds:

$$\sum_{a \in \mathcal{A}} \bar{\pi}(a|s) q^{\pi}(s, a) \leq \sum_{a \in \mathcal{A}} \bar{\pi}(a|s) q^{\bar{\pi}}(s, a). \tag{3-7}$$

When a policy cannot be further improved, it means that the policy is already optimal. In theory, one should make two assumptions to guarantee convergence. The first assumption is that episodes have exploring starts, and the second assumption is that an infinite number of episodes are collected at each cycle. These assumptions ensure that every value in the state-action space can be touched. Therefore, the error between the estimated value and true value will converge to zero. Zero value error requires an infinite number of episodes, which is obviously unrealistic in practice. An alternative approach is to ensure that the value error is sufficiently small instead of trying to complete a perfect policy evaluation.

Figure 3.4 Cascading inequality by replacing π with $\bar{\pi}$

3.2.3   MC Policy Selection

A reasonable design of MC policy improvement is critical to the guarantee of convergence. The main task is to find a better policy $\bar{\pi}$ with respect to the estimated value under the current policy π, i.e., find a new policy $\bar{\pi}(a|s)$ that satisfies the policy improvement theorem:

$$V^\pi(s) \leq \sum_{a \in \mathcal{A}} \bar{\pi}(a|s)Q^\pi(s,a), \forall s \in \mathcal{S}. \tag{3-8}$$

Compared to (3-6), which uses true values, (3-8) replaces the true values with the estimated values for the design of practical algorithms. Two popular choices are the greedy policy $\pi^g(a|s)$ and the ϵ-greedy policy $\pi^\epsilon(a|s)$, which are defined in (3-3) and (3-4), respectively. When the environment model is known, the optimal action can be calculated by reshaping the greedy search with the state-value estimate:

$$a^* = \arg\max_a \sum_{s' \in \mathcal{S}} \mathcal{P}_{ss'}^a \left(r_{ss'}^a + \gamma V^\pi(s') \right). \tag{3-9}$$

When the environment model is unknown, the optimal action can be calculated by directly maximizing the action-value estimate:

$$a^* = \arg\max_a Q^\pi(s,a). \tag{3-10}$$

Note that (3-9) requires to know the knowledge of the environment model, but it is not always available in the real world. This is why most MC learning algorithms only evaluate the action-value function. However, it is easy to see from Figure 3.1 that the estimation of the action-value function occupies much larger memory than that of the state-value function, which definitely requires more computational resources and more samples from environment interactions to reach the same satisfactory accuracy.

3.3   On-Policy Strategy vs. Off-Policy Strategy

Model-free RL always faces a dilemma; i.e., it aims to learn an optimal policy but needs to behave non-optimally to explore all potential actions. This is the famous exploration-exploitation dilemma. In the on-policy strategy, a compromise must be made between exploitation and exploration. A soft-greedy policy adds some randomness to greedy actions to provide a certain exploration ability, but the optimality of the finally learned policy must be sacrificed. In the off-policy strategy, the solution is to use two different policies: the first is the target policy (denoted as π), which is a greedy policy that is used to pursue optimality, and the second is the behavior policy (denoted as b), which must be stochatistic for the sake of sufficient environment interactions. The target policy can better exploit information from the collected experience to find the optimal result, while the behavior policy can better explore the environment to collect sufficient information.

3.3.1   On-Policy Strategy

An on-policy strategy evaluates or improves the same policy that is used to explore the environment. The first choice that comes to mind is the greedy policy, which always chooses the action with the highest value. One shortcoming of this choice is that the greedy policy itself is deterministic and lacks exploration, i.e., only the locally best action is selected, which deprives the true optimal action of sufficient data. In stochastic environments, the lack of exploration can be partially compensated for by the random behavior of the environment dynamics. A greedy policy still works even though it lacks the exploration ability. However, in deterministic environments, the lack of exploration can be destructive and may result in poor convergence and large errors in the final result.

- Theorem 3-2: The greedy policy $\pi^g(a|s)$ satisfies the policy improvement theorem:

$$v^{\pi^g}(s) \le \sum_{a\in\mathcal{A}} \pi_{\text{new}}^g(a|s)q^{\pi^g}(s,a), \forall s \in \mathcal{S}, \tag{3-11}$$

where π_{new}^g is the new greedy policy after policy improvement:

$$a^* = \arg\max_a q^{\pi^g}(s,a),$$

$$\pi_{\text{new}}^g(a|s) = \begin{cases} 1, & \text{if } a = a^* \\ 0, & \text{if } a \ne a^* \end{cases}.$$

Proof:

Because π^g is actually a deterministic policy, we have

$$v^{\pi^g}(s) = q^{\pi^g}\left(s, \pi^g(s)\right) \le \max_a q^{\pi^g}(s,a) = q^{\pi^g}\left(s, \arg\max_a q^{\pi^g}(s,a)\right)$$
$$= q^{\pi^g}\left(s, \pi_{\text{new}}^g(a|s)\right).$$

∎

The inequality above holds because a^* is the best action among all the actions. This means that choosing a greedy policy can guarantee the monotonic improvement property of policy updates. In the on-policy strategy, one simple solution to enhance exploration is to choose actions with the best value most of the time while also choosing actions at random with a small probability. This kind of design is called a soft-greedy policy. The ϵ-greedy search is one representative type of soft-greedy policy. The ϵ-greedy policy can be regarded as a combination of two parts: (1) a random part, which selects uniformly among all the actions with a probability of ϵ (that is, each action is selected with a probability of $\epsilon/|\mathcal{A}|$), and (2) a greedy part, which selects the best action with a probability of $1-\epsilon$. Figure 3.5 illustrates the structure of the ϵ-greedy policy that separates the random part from the greedy part.

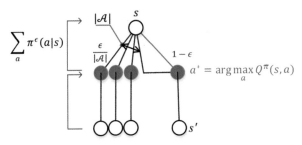

Figure 3.5 Two types of actions in an ϵ-greedy policy

- Theorem 3-3: The ϵ-greedy policy $\pi^\epsilon(a|s)$ satisfies the policy improvement theorem:

$$v^{\pi^\epsilon}(s) \leq \sum_{a \in \mathcal{A}} \pi^\epsilon_{new}(a|s) q^{\pi^\epsilon}(s, a), \forall s \in \mathcal{S}, \tag{3-12}$$

where $\pi^\epsilon_{new}(a|s)$ is the next ϵ-greedy policy after policy improvement.

Proof:

$$v^{\pi^\epsilon}(s) = \sum_a \pi^\epsilon(a|s) q^{\pi^\epsilon}(s, a)$$

$$= \frac{\epsilon}{|\mathcal{A}|} \sum_a q^{\pi^\epsilon}(s, a) + \sum_a \left(\pi^\epsilon(a|s) - \frac{\epsilon}{|\mathcal{A}|} \right) q^{\pi^\epsilon}(s, a).$$

Note that a^* is from the maximization of action-value $q(s, a)$ among all actions. Thus, we have

$$\leq \frac{\epsilon}{|\mathcal{A}|} \sum_a q^{\pi^\epsilon}(s, a) + \sum_a \left(\pi^\epsilon(a|s) - \frac{\epsilon}{|\mathcal{A}|} \right) \max_a q^{\pi^\epsilon}(s, a)$$

$$= \frac{\epsilon}{|\mathcal{A}|} \sum_a q^{\pi^\epsilon}(s, a) + (1 - \epsilon) \max_a q^{\pi^\epsilon}(s, a)$$

$$= \sum_a \pi^\epsilon_{new}(s, a) q^{\pi^\epsilon}(s, a).$$

∎

An intuitive explanation of the proof in Theorem 3-3 is that the ϵ-greedy policy has two completely decoupled parts, of which the random part always stays the same while the greedy part is gradually improved. To balance the needs for exploration and exploitation at different stages, a common approach in the ϵ-greedy policy is to decrease the value of the exploration rate ϵ over time. Initially, the policy has a large probability of selecting random actions to increase its exploration ability, but it will gradually shift to a deterministic policy to obtain better exploitation performance.

One issue of an ϵ-greedy search is that it treats all actions equally in the random part. Each action is chosen uniformly, even though some actions may be better than others. A straightforward solution is to choose actions with an unequal distribution, for example, with probabilities proportional to their evaluation values. A typical example is the well-known Boltzmann exploration, which builds a softmax policy to treat actions unequally. Specifically, in the Boltzmann exploration, an action a is selected with a probability that depends on its action-value function $Q(s, a)$:

$$\pi^B(a|s) = \frac{\exp(\tau^{-1} Q(s, a))}{\sum_{a \in \mathcal{A}} \exp(\tau^{-1} Q(s, a))}, \forall a \in \mathcal{A}, s \in \mathcal{S}, \tag{3-13}$$

where π^B is the Boltzmann policy and τ is the temperature coefficient. This policy spans a spectrum that has the "exploit-only" greedy policy at one end ($\tau \to 0$) and the "explore-only" uniform policy at the other ($\tau \to \infty$). A properly designed Boltzmann exploration often works better than most soft-greedy policies if the best action is deterministic and has no correlation with non-optimal actions. To achieve these conditions, the temperature coefficient needs to be adjusted with a regret perspective, and its decreasing rate should be manually tuned with great care [3-2]. Other advanced policy choices can be seen in the literature, including upper-confidence-bound action, Thompson sampling strategy, count-based exploration, and entropy regularization.

3.3.2   Off-Policy Strategy

Compared with its on-policy counterpart, an off-policy strategy can balance exploitation and exploration in a more elegant way. Usually, the off-policy strategy requires the assumption of full coverage. With this assumption, every action that is possibly selected by the target policy must be taken at least occasionally under a behavior policy. The target policy can be either deterministic or stochastic, but the behavior policy must be stochastic to cover all possible actions. An off-policy RL algorithm generates samples by a stochastic behavior policy but calculates value functions under the target policy. In fact, the two policies do not match each other and one cannot directly calculate the value function of the target policy from behavior samples. This mismatch problem must be addressed by using the importance sampling (IS) technique.

3.3.2.1   Importance Sampling Transformation

The IS technique is a statistical method that is used to estimate the expectation of a random variable under a particular distribution while only having samples from a different distribution [3-1][3-3]. In statistics, IS is usually applied to predict the probability of a rare event. Rare events can be found on the tails of some probability distributions, which makes it extremely difficult to collect enough samples. Along with MC simulation, the IS technique is used to evaluate the expectation of target return by collecting the samples from the behavior distribution. The mathematical logic underlying this transformation lies in a simple rearrangement of terms in the target integral [3-1]:

$$
\begin{aligned}
\mathbb{E}_\pi\{h(x)\} &= \int d_\pi(x)h(x)\mathrm{d}x \\
&= \int d_b(x)\frac{d_\pi(x)}{d_b(x)}h(x)\mathrm{d}x \\
&= \int d_b(x)\rho(x)h(x)\mathrm{d}x \\
&= \mathbb{E}_b\{\rho(x)h(x)\},
\end{aligned}
\tag{3-14}
$$

where x is the random variable, $d_\pi(x)$ is the target distribution, $d_b(x)$ is the behavior distribution, and $\rho(x)$ is the IS ratio, which is defined as

$$
\rho(x) = \frac{d_\pi(x)}{d_b(x)}.
$$

The main goal of the IS transformation is to maintain the same expectation under the data samples from two different distributions. Since the two expectations are equal, the samples from one distribution can be used to estimate the other's expectation. This kind of equivalence does not mean that the variances of the two samples are identical.

Therefore, one can intentionally select a proper behavior distribution to reduce the target variance.

- Theorem 3-4: $\mu_\pi = \mu_b$, where $\mu_\pi = \mathbb{E}_\pi\{h(x)\}$ and $\mu_b = \mathbb{E}_b\{\rho(x)h(x)\}$.
- Corollary 3-5: $\sigma_b^2 \neq \sigma_\pi^2$.

Proof:

$$\sigma_\pi^2 \stackrel{\text{def}}{=} \mathbb{D}_\pi\{h(x)\}$$
$$= \mathbb{E}_\pi\{(h(x) - \mu_\pi)^2\}$$
$$= \mathbb{E}_b\{\rho(x)(h(x) - \mu_\pi)^2\}$$
$$= \mathbb{E}_b\{\rho(x)h^2(x)\} - \mu_\pi^2,$$

while

$$\sigma_b^2 \stackrel{\text{def}}{=} \mathbb{D}_b\{\rho(x)h(x)\}$$
$$= \mathbb{E}_b\{(\rho(x)h(x) - \mu_b)^2\}$$
$$= \mathbb{E}_b\{\rho^2(x)h^2(x)\} - \mu_b^2.$$

∎

Let us look at how to estimate the expectation and variance of a function of a random variable. Given a trajectory of samples $x_0, x_1, x_2, \cdots, x_{T-1}$ from the behavior distribution $d_b(x)$, the estimate of μ_b is

$$\hat{\mu}_b = \frac{1}{T} \sum_{t=0}^{T-1} h(x_t)\rho(x_t), \; x_t \sim d_b(x). \tag{3-15}$$

The estimate $\hat{\mu}_b$ is unbiased, i.e., $\mathbb{E}\{\hat{\mu}_b\} = \mu_b = \mu_\pi$. Meanwhile, the estimate of variance with the behavior samples is

$$\hat{\sigma}_b^2 = \frac{1}{T-1} \sum_{t=0}^{T-1} (h(x_t)\rho(x_t) - \hat{\mu}_b)^2. \tag{3-16}$$

Lowest-variance estimation is desired because it means the highest precision. Let us revisit Corollary 3-5: one interesting observation is that σ_b^2 becomes zero if we select a behavior distribution like $d_b(x) = h(x)d_\pi(x)/\mu_\pi$. However, this option is not realistic since it requires knowing the true expectation. Even worse, it is not legitimate either because $h(x)$ can be negative at some state points but a distribution must always be positive. One alternative is to choose the following condition:

$$\widetilde{d}_b(x) \propto |h(x)|d_\pi(x). \tag{3-17}$$

- Theorem 3-6: Choosing (3-17) leads to the lowest variance among all legitimate behavior distributions.

Proof: Self-normalize $\widetilde{d}_b(x)$ by its integration with respect to x:

$$\widetilde{d}_b(x) = \frac{|h(x)|d_\pi(x)}{\int |h(x)|d_\pi(x)dx}.$$

Then,

$$\begin{aligned}
\widetilde{\sigma}_b^2 &= \int \frac{h(x)^2 d_\pi(x)^2}{\widetilde{d}_b(x)^2} \widetilde{d}_b(x)dx - \mu_b^2 \\
&= \left(\int |h(x)|d_\pi(x)dx \right)^2 - \mu_b^2 \\
&= \left(\int |h(x)|\rho(x)d_b(x)dx \right)^2 - \mu_b^2.
\end{aligned}$$

By using the Cauchy-Schwarz inequality, we have the following derivation:

$$\begin{aligned}
&\leq \int h(x)^2 \rho(x)^2 d_b(x)dx \int d_b(x)dx - \mu_b^2 \\
&= \int h(x)^2 \rho(x)^2 d_b(x)dx - \mu_b^2 \\
&= \sigma_b^2.
\end{aligned}$$

∎

Obviously, the IS transformation is able to reduce the estimation variance if the behavior distribution is properly selected. The underlying mechanism is that certain values of a random variable have more impact on its statistical estimation. If these important values are emphasized by sampling more frequently, the variance can be effectively reduced. That is why this technique is referred to as importance sampling. In RL, the IS transformation provides the foundation for off-policy strategy as it tells us how to reuse samples from a behavior policy.

3.3.2.2   Importance Sampling Ratio for State-Values

The IS ratio works as a weighting coefficient in estimating value functions. While the IS ratio is consistent and unbiased, its associated transformation may result in a high variance in the updates of value functions. The high variance mainly comes from the quotient structure of target action distribution and behavior action distribution. Any small uncertainty in the two distributions will cause a large error in the calculation of IS ratio. Assume that we have already recorded a trajectory from t-step to T-step. For any state-value function, the current state s_t is known, and the function to be estimated is

$$h(X) \overset{\text{def}}{=} G_{t:T}(a_t, s_{t+1}, \cdots, a_{T-1}, s_T).$$

Here, we must view $X = \{a_t, s_{t+1}, \cdots, a_{T-1}, s_T\}$ as a multivariate random variable, and the probability of generating such a trajectory is actually the multivariate distribution of multiple random variables $a_t, s_{t+1}, \cdots, a_{T-1}, s_T$, i.e., a joint probability distribution. The probability mass functions under target policy π and behavior policy b are

$$d_\pi(a_t, s_{t+1}, \cdots, a_{T-1}, s_T) = \Pr\{a_t, s_{t+1}, \cdots, s_T | \pi, s_t\} = \prod_{i=t}^{T-1} \pi(a_i|s_i)p(s_{i+1}|s_i, a_i),$$

$$d_b(a_t, s_{t+1}, \cdots, a_{T-1}, s_T) = \Pr\{a_t, s_{t+1}, \cdots, s_T | b, s_t\} = \prod_{i=t}^{T-1} b(a_i|s_i)p(s_{i+1}|s_i, a_i).$$

Their quotient becomes a general multistep IS ratio:

$$\rho_{t:T-1} = \frac{d_\pi(a_t, s_{t+1}, \cdots, a_{T-1}, s_T)}{d_b(a_t, s_{t+1}, \cdots, a_{T-1}, s_T)} = \prod_{i=t}^{T-1} \frac{\pi(a_i|s_i)}{b(a_i|s_i)}. \tag{3-18}$$

Although the occurrence probability of one trajectory depends on the environment dynamics, their effects on the IS ratio are coincidentally eliminated by the cancellation of the denominator and numerator. Accordingly, the IS ratio depends only on the two policies, irrelevant to the environment dynamics. Depending on the length of the episodic horizon, the IS ratio $\rho_{t:T-1}$ may have only one or multiple factors in the continued multiplication. The expectation of $G_{t:T}$ using the samples from the behavior policy can be easily calculated:

$$v^\pi(s)$$
$$= \mathbb{E}_\pi\{G_{t:T}|s\}$$
$$= \sum_{(a_t, s_{t+1}, \cdots, a_{T-1}, s_T)} d_\pi(a_t, s_{t+1}, \cdots, a_{T-1}, s_T)G_{t:T}(a_t, s_{t+1}, \cdots, a_{T-1}, s_T)|_{s_t=s}$$
$$= \sum_{(a_t, s_{t+1}, \cdots, a_{T-1}, s_T)} \frac{d_\pi(*)}{d_b(*)} d_b(*)G_{t:T}(*)|_{s_t=s}$$
$$= \sum_{(a_t, s_{t+1}, \cdots, a_{T-1}, s_T)} d_b(*)\rho_{t:T-1}G_{t:T}(*)|_{s_t=s}$$
$$= \mathbb{E}_b\{\rho_{t:T-1}G_{t:T}|s\},$$

where $G_{t:T}$ is regarded as the function to be estimated, which depends on the random variables $a_t, s_{t+1}, \cdots, a_{T-1}, s_T$ from t to T, and $d(*)$ is the probability mass function of their joint distribution. If the time length is only from t to $t+1$, (3-18) is reduced to a one-step IS ratio:

$$\rho_t \stackrel{\text{def}}{=} \rho_{t:t} = \frac{\pi(a_t|s_t)}{b(a_t|s_t)}. \tag{3-19}$$

The one-step IS ratio for the state-value function is not as trivial as that for the action-value function, and the two policies must be accurately known as the numerator and denominator of (3-19). Usually, a large variance inevitably occurs if the two policies move away from each other, especially when there are some rare actions with a small probability. A small perturbation in rare actions may lead to large errors in the calculation of the one-step IS ratio. As a result, the estimate errors may be remarkably amplified in the calculation of returns.

3.3.2.3   Importance Sampling Ratio for Action-Values

Different from the state-value function, both current state and current action (s_t, a_t) are known in the action-value function. Here, we must view their following states and actions $s_{t+1}, \cdots, a_{T-1}, s_T$ as random variables. One can compute the probability mass functions of $s_{t+1}, \cdots, a_{T-1}, s_T$ under the target policy π and the behavior policy b. The multistep IS ratio for the action-value function becomes

$$\rho_{t+1:T-1} = \frac{d_\pi(s_{t+1}, \cdots, a_{T-1}, s_T)}{d_b(s_{t+1}, \cdots, a_{T-1}, s_T)} = \frac{\Pr\{s_{t+1}, \cdots, s_T | \pi, s_t, a_t\}}{\Pr\{s_{t+1}, \cdots, s_T | b, s_t, a_t\}} = \prod_{i=t+1}^{T-1} \frac{\pi(a_i|s_i)}{b(a_i|s_i)}.$$

Then, one can estimate the action-value function using the samples from the behavior policy b:

$q^\pi(s,a)$

$= \mathbb{E}_\pi\{G_{t:T}|s,a\}$

$$= \sum_{(s_{t+1},\cdots,a_{T-1},s_T)} d_\pi(s_{t+1}, \cdots, a_{T-1}, s_T) G_{t:T}(s_{t+1}, \cdots, a_{T-1}, s_T)|_{s_t=s,a_t=a}$$

$$= \sum_{(s_{t+1},\cdots,a_{T-1},s_T)} \frac{d_\pi(\#)}{d_b(\#)} d_b(\#) G_{t:T}(\#)|_{s_t=s,a_t=a}$$

$$= \sum_{(s_{t+1},\cdots,a_{T-1},s_T)} d_b(\#)\rho_{t+1:T-1}G_{t:T}(\#)|_{s_t=s,a_t=a}$$

$= \mathbb{E}_b\{\rho_{t+1:T-1}G_{t:T}|s,a\},$

where # represents proper random variables. Since both s_t and a_t are known, the multistep IS ratio $\rho_{t+1:T-1}$ starts at time $t+1$, instead of time t. The calculation of this ratio ends at time T. Clearly, the one-step IS ratio for the action-value function is reduced to

$$\rho_{t+1:t} = \frac{d_\pi(s_{t+1})}{d_b(s_{t+1})} = \frac{\Pr\{s_{t+1}|\pi, s_t, a_t\}}{\Pr\{s_{t+1}|b, s_t, a_t\}} = \frac{p(s_{t+1}|s_t, a_t)}{p(s_{t+1}|s_t, a_t)} = 1. \qquad (3\text{-}20)$$

Interestingly, this formula leads to a special feature of off-policy action-value estimation. That is to say that in some special cases, off-policy estimation does not contain any explicit IS ratio. For example, when applying one-step bootstrapping into the estimation of action-value function, one does not need to distinguish the behavior policy from the target policy. In this case, the IS transformation becomes unnecessary, and the samples from an arbitrary policy can be seamlessly used in the one-step temporal difference (TD) estimation. This feature has become a built-in property in Q-learning, which utilizes off-policy strategy without any explicit IS ratio. It also forms the basis for applying the trick of experience replay in many deep RL algorithms, including deep Q-network (DQN), double deep Q-network (DDQN), and deep deterministic policy gradient (DDPG).

3.4   Understanding Monte Carlo RL from a Broad Viewpoint

It is easy to see that Monte Carlo learning algorithms belong to the aforementioned generalized policy iteration (GPI) framework, which is a mainstream branch of the indirect RL family. Almost all MC learning algorithms need to estimate the action-value function since the environment model is unknown. The learning procedure is to alternately perform PEV and PIM until convergence to an optimum occurs (Figure 3.6). The standard MC learning algorithm is implemented on an episode-by-episode basis. In each cycle, the action-value function is estimated to approximate its true quantity, and the policy is then improved with respect to the action-value estimate. Intuitively, PEV and PIM operate against each other because each creates a renewed target for the other. Their iteration will gradually shift both policy and value function to their optima.

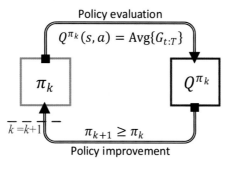

$$\pi_0 \xrightarrow{\text{PEV}} Q^{\pi_0} \xrightarrow{\text{PIM}} \pi_1 \xrightarrow{\text{PEV}} Q^{\pi_1} \xrightarrow{\text{PIM}} \cdots \xrightarrow{\text{PIM}} \pi^* \xrightarrow{\text{PEV}} q^*$$

Figure 3.6 Cycle of PEV and PIM in Monte Carlo RL

3.4.1   On-Policy MC Learning Algorithm

MC learning algorithms can be designed with either on-policy or off-policy strategies. A typical on-policy MC algorithm is illustrated in Figure 3.7, in which k represents the number of the main cycle. It contains an on-policy sampler and a Monte Carlo trainer. At each cycle, the on-policy sampler can generate a batch of episodes (i.e., batch episodes) with the current policy. The PEV and PIM together constitute a Monte Carlo trainer, which aims to output a new policy for the next round of interaction. During each cycle, all episodes come from the same policy to facilitate the average-based value estimation.

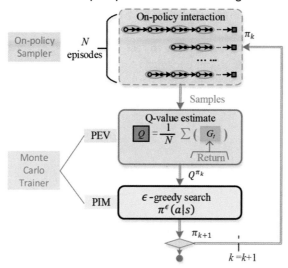

Figure 3.7 On-policy MC algorithm

The sampler randomly selects initial state-action pairs to cover the whole state-action space. The episodes that start from the same state-action pair are extracted to calculate the return of that pair. The returns that start from (s, a) and terminate at time T are denoted as

$$G_{t_1:T}^{(1)}, G_{t_2:T}^{(2)}, \cdots, G_{t_i:T}^{(i)}, \cdots G_{t_N:T}^{(N)},$$

where N is the number of episodes. Note that all the returns above are required to start from the same state-action pair. The subscript t_i represents the starting time, while the superscript (i) represents the episode index. And it should be noticed that the terminal time T varies in different episodes. Hence, the batch average of all these episodes is

$$Q^\pi(s, a) = \frac{1}{N} \sum_{i=1}^{N} G_{t_i:T}^{(i)},$$

$$\forall s_{t_i} = s, a_{t_i} = a.$$

(3-21)

This formula, which is called batch MC estimation, reduces to an episode-by-episode estimation if only one episode is left, i.e., $N = 1$. On-policy MC only has one policy, which not only serves as the target to be optimized but also decides the agent behavior for environment interaction. Either greedy policy or ϵ-greedy policy is an available option for on-policy MC. The greedy policy can reach the true optimal solution but lacks good exploration ability. In contrast, the ϵ-greedy policy has better exploration ability, but cannot converge to the true optimum because its random term always exists to maintain good exploration.

3.4.2   Off-Policy MC Learning Algorithm

In theory, the off-policy strategy can achieve the perfect balance between exploitation and exploration. Together with Monte Carlo estimation, this strategy leads to various off-policy MC learning algorithms. As illustrated in Figure 3.8, a standard off-policy algorithm contains an off-policy sampler and a Monte Carlo trainer. The former uses the behavior policy to interact with the environment, and the latter tries to calculate an optimal target policy with reliance on the importance sampling technique. In this algorithm, the greedy search often serves as the target policy to reach the optimal performance. In contrast, the ϵ-greedy policy is taken as the behavior policy for better exploration.

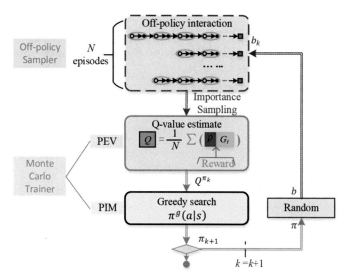

Figure 3.8 Off-policy MC algorithm

One additional benefit of off-policy MC is that it allows the use of historical samples from the last policy or even older policies to calculate the value function. The reuse of past experience can significantly increase the sample efficiency, even though it might introduce unpredictable training instability. Accordingly, there is a high probability that the returns that start from (s, a) are from different behavior policies:

$$G_{t_1:T}^{(1)}, G_{t_2:T}^{(2)}, \cdots, G_{t_i:T}^{(i)}, \cdots G_{t_N:T}^{(N)},$$

and each return has its own IS ratio:

$$\rho_{t_1+1:T-1}^{(1)}, \rho_{t_2+1:T-1}^{(2)}, \cdots, \rho_{t_i+1:T-1}^{(i)}, \cdots \rho_{t_N+1:T-1}^{(N)}.$$

To estimate the action-value function, we simply scale returns by their IS ratios and average all the returns that are from different policies. This method is called the ordinary batch average and is represented as:

$$Q^\pi(s, a) = \frac{1}{N} \sum_{i=1}^{N} \rho_{t_i+1:T-1}^{(i)} G_{t_i:T}^{(i)},$$

$$\forall s_{t_i} = s, a_{t_i} = a.$$

An alternative method is called the weighted batch average, which uses weighted IS ratios to perform the weighted averaging operation:

$$Q^\pi(s, a) = \frac{1}{\sum_{i=1}^{N} \rho_{t_i+1:T-1}^{(i)}} \sum_{i=1}^{N} \rho_{t_i+1:T-1}^{(i)} G_{t_i:T}^{(i)},$$

$$\forall s_{t_i} = s, a_{t_i} = a.$$

The two kinds of batch averages look similar but differ slightly in their biases and variances. The ordinary batch average is unbiased, whereas the weighted version is biased (even though the bias is able to converge asymptotically to zero). The variance of the former is often higher because some IS ratios may have very large errors. In contrast, in the latter, the largest weight is limited to one, and its variance caused by IS ratios is always bounded. In batch MC estimation, returns can have very different IS ratios even if they share the same behavior policy. This is because the IS ratio depends on not only the behavior policy but also the length of each episode. If episodes are sufficiently long, we can simplify the IS ratio to only a function of policies. This simplification builds a new ordinary batch average:

$$Q^\pi(s, a) = \frac{1}{N} \rho_{t+1:T-1} \sum_{i=1}^N G_{t_i:T}^{(i)}.$$

Clearly, the batch version of off-policy MC must be implemented on a batch-by-batch basis [3-3]. The large computational burden of off-policy MC may be its biggest bottleneck when applied to real-world problems.

3.4.3   Incremental Estimation of Value Function

Monte Carlo learning algorithms are often used in episodic tasks without knowing their dynamic models. The sampler is responsible for gathering samples, and the trainer is responsible for averaging the collected samples and calculating a better policy. As more experience is collected, the learned policy will gradually shift to its optimum. Figure 3.9 compares the flow charts of on-policy MC and off-policy MC.

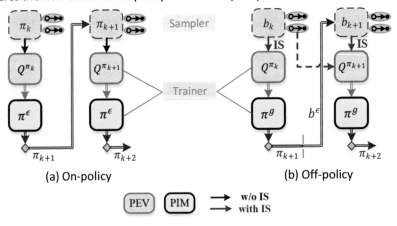

(a) On-policy   (b) Off-policy

Figure 3.9 Monte Carlo RL without initialization

In theory, MC estimation does not need any initialization value, and its action-value calculation starts from random scratch at each cycle. This requires a large number of samples to maintain satisfactory accuracy, which is a definite waste of computing resources. In practice, one could initialize the action-value function using the last step estimate even though it somewhat lacks strong theoretical support as an MC technique. Value initialization can effectively reduce the number of used samples and is helpful to achieve high-level estimation accuracy. Figure 3.10 demonstrates an MC learning algorithm with value initialization. Its value estimation is equal to a weighted sum of the value of the last cycle and the sample average of the current cycle:

$$Q^{\pi_{k+1}}(s, a) = (1 - \lambda) Q^{\pi_k}(s, a) + \lambda \cdot \text{Avg}\{G | s, a; \pi_{k+1}\},$$

where $\lambda \in (0,1]$ is the incremental rate. This technique is also called incremental MC estimation. The higher the incremental rate becomes, the more emphasis is placed on newly generated samples, and vice versa. When $\lambda = 1$, the incremental MC estimation reduces to an average-based estimation. A proper value initialization in incremental MC estimation also accelerates the learning speed and improves the learning stability. Algorithm 3-1 is the pseudocode of Monte Carlo RL with first-visit MC and incremental MC estimation. Here, the initial action-value function is set to zero, and the initial stochastic policy is chosen to be the ϵ-greedy policy with evenly distributed random actions.

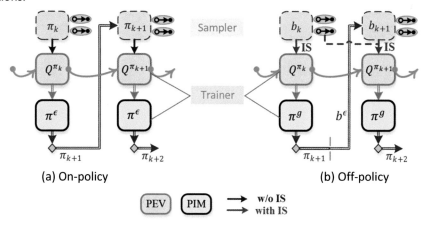

(a) On-policy   (b) Off-policy

Figure 3.10 Monte Carlo RL with initialization

Algorithm 3-1: On-policy Monte Carlo RL with value initialization

Hyperparameters: Discount factor γ, Exploration rate ϵ, Incremental rate λ, Episode number per iteration M
Initialization: $Q(s,a) \leftarrow 0$, $\pi^\epsilon(a\|s) \leftarrow 1/\|\mathcal{A}\|$
Repeat (indexed with k)
Initialize visit counter $\mathcal{N}(s,a) \leftarrow 0$
For j in $1,2,\dots,M$
$s_0 \sim d_{\text{init}}(s)$
//Observe one trajectory
$s_0, a_0, r_0, s_1, \dots, s_{T-1}, a_{T-1}, r_{T-1}, s_T$
//Calculate returns in one trajectory
For t in $0,1,2,\dots,T-1$
If (s_t, a_t) is first visited in the trajectory
$$G_j(s_t, a_t) \leftarrow \sum_{i=0}^{T-t} \gamma^i r_{t+i}$$
$$\mathcal{N}(s_t, a_t) \leftarrow \mathcal{N}(s_t, a_t) + 1$$
End

> **End**
>
> > **End**
> >
> > **Sweep** (s, a) in $S \times \mathcal{A}$
> >
> > > //Calculate average return for each visited state-action pair
> > >
> > > $$\bar{G}(s, a) \leftarrow \frac{1}{N(s, a)} \sum_{j=1}^{M} G_j(s, a), \forall(s, a) \in \{N(s, a) > 0\}$$
> > >
> > > //Incremental MC estimation
> > >
> > > $$Q(s, a) \leftarrow (1 - \lambda)Q(s, a) + \lambda\bar{G}(s, a), \forall(s, a) \in \{N(s, a) > 0\}$$
> >
> > **End**
> >
> > //Calculate greedy actions and update policy
> >
> > $$a^* \leftarrow \arg\max_a Q(s, a)$$
> >
> > $$\pi^\epsilon(a|s) = \begin{cases} 1 - \epsilon + \epsilon/|\mathcal{A}|, & \text{if } a = a^* \\ \epsilon/|\mathcal{A}|, & \text{if } a \neq a^* \end{cases}$$
>
> **End**

∎

3.5   Example of Monte Carlo RL

An episodic example is trained to demonstrate how to apply an MC learning algorithm. The environment is a grid world, and a cleaning robot seeks to reach its destination as quickly as possible to charge itself. Along with a random initial policy and a zero initial value function (Figure 3.11), an on-policy MC algorithm is applied to train a tabular policy. This algorithm is designed with incremental MC estimation, ϵ-greedy policy, and first-visit estimation. A few key hyperparameters are repeatedly simulated to examine their influences on the training performance.

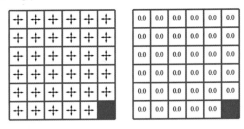

(a) Tabular policy   (b) Tabular value function

Figure 3.11 Tabular representation

3.5.1   Cleaning Robot in a Grid World

The cleaning robot works in a rectangular environment that is a 6×6 grid. Except for the destination cell at the bottom-right corner, each cell in the grid represents one state. The environment has 36 states and 4 possible actions. The state and action spaces are

$$S = \{s_{(1)}, s_{(2)}, \cdots, s_{(36)}\},$$
$$\mathcal{A} = \{\text{North, South, East, West}\}.$$

$$(3\text{-}22)$$

The goal of RL is to find an optimal policy to move the robot to the destination cell as soon as possible. Each action moves the agent to its neighboring cells in a random manner. The robot moves forward with a probability of 80% and moves leftward or rightward with a respective probability of 10%. However, the robot cannot move backward. Figure 3.12 shows a few examples of generated episodes, which start with different initial states. It is easy to see that some states are visited more than once in the same episode due to the randomness of robot dynamics. The first-visit MC technique is utilized to average returns without considering the repeated visits in each episode. The occurrence of each state in a certain number of episodes is shown in Figure 3.13, in which the height of each bar represents the visiting frequency of that state. As more episodes are collected, the states that are closer to the destination cell are more frequently visited, while those states farther away from the destination cell are less frequently visited.

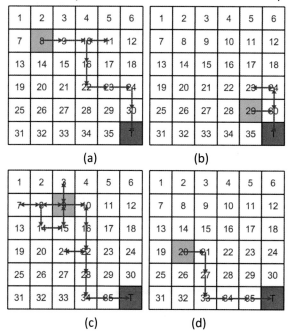

Figure 3.12 Episodes in the MC simulation

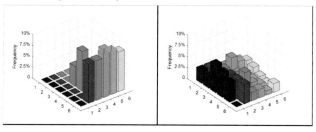

(a) in 1 episode   (b) in 10 episodes

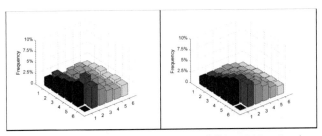

(c) in 50 episodes   (d) in 1000 episodes

Figure 3.13 Average occurrences of states in the given number of episodes

3.5.2   MC with Action-Value Function

When the robot arrives at the destination, the environment interaction stops at once, and one episode is generated. After a certain number of episodes, the values of all state grids are computed with average-based estimation. Each element in the tabular value function is updated individually, and it eventually converges to the true value. In Figure 3.14, the arrows represent optimal actions in the learned policy, which is deterministic due to the greedy search mechanism. For the ease of demonstration, state-value functions at some cycles are drawn in Figure 3.15 instead of action-value functions. Figure 3.16 shows the final value and the final policy. As more cycles are completed, the learned policy becomes an optimal policy, in which all arrows tend to point to the destination cell.

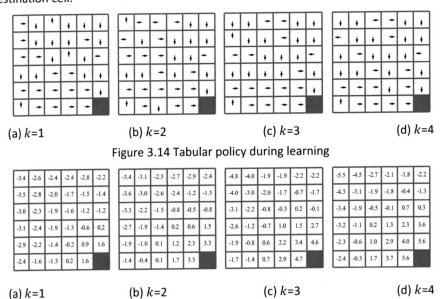

(a) $k=1$ (b) $k=2$ (c) $k=3$ (d) $k=4$

Figure 3.14 Tabular policy during learning

(a) $k=1$ (b) $k=2$ (c) $k=3$ (d) $k=4$

Figure 3.15 Tabular state-value during learning

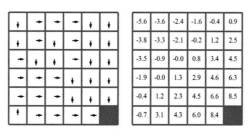

(a) Policy   (b) State-value function

Figure 3.16 Final results

3.5.3   Influences of Key Parameters

In the MC learning algorithm, the three key hyperparameters are Ep/PEV, ϵ and λ, where Ep/PEV represents the number of episodes used in each PEV, ϵ is the exploration rate, and λ is the incremental rate. The episode number is a key index that reflects the episode efficiency, i.e., how many episodes are needed to perform a useful policy evaluation. The exploration rate, which reflects the ability to balance exploration and exploitation, is one critical parameter in the ϵ-greedy policy. The incremental rate is a key parameter to determine the effectiveness of value initialization.

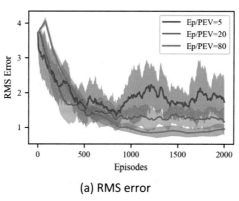

(a) RMS error

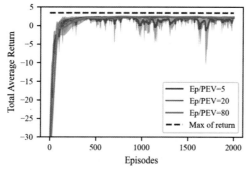

(b) Total average return

Figure 3.17 Influence of the episode number

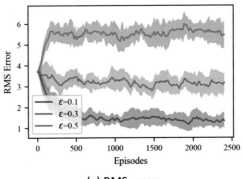

(a) RMS error

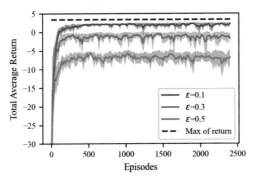

(b) Total average return

Figure 3.18 Influence of the exploration rate

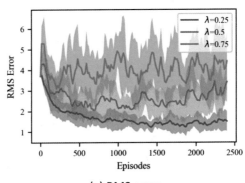

(a) RMS error

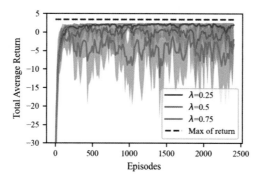

(b) Total average return

Figure 3.19 Influence of the incremental rate

In this example, the influences of the three hyperparameters are tested with repeated MC simulations. In these simulations, one parameter is varied using three options and the other two parameters are fixed at their medium values. Figure 3.17, Figure 3.18 and Figure 3.19 illustrate the learning curves of the root mean square (RMS) error and the total average return. The solid lines correspond to the mean and the shaded regions correspond to 95% confidence interval over multiple runs. It is observed that a large Ep/PEV can reduce the RMS error and improve the learning stability. In this example, a large Ep/PEV does not change the convergence rate very much even though the frequency of policy improvement becomes sparse in the same number of total episodes. A large exploration rate has a negative impact on the policy optimality, leading to severe sacrifices in both RMS error and total average return. This is because for any finite MDP, its optimal policy should be deterministic, and a stochastic policy with a large exploration rate must sacrifice the ability to achieve the true optimum. Incremental MC estimation is an effective way to enhance the accuracy of action-value estimation. A small incremental rate implies more reliance on historical experience, which is helpful in stabilizing the training process and improving the policy performance.

3.6 References

[3-1] Doucet A, De Freitas N, Gordon N (2001) Sequential Monte Carlo methods in practice. Springer, Berlin Heidelberg New York

[3-2] Kaelbling L, Littman M, Moore A (1996) Reinforcement learning: a survey. J Artif Intell, 4(1): 237-285

[3-3] Sutton R, Barto A (2018). Reinforcement learning: an introduction. MIT press, Cambridge

Chapter 4. Model-Free Indirect RL: Temporal Difference

If I have seen further,

it is by standing on the shoulders of giants.

-- Sir Isaac Newton (1643-1727)

The estimation of value function is critical for model-free RL algorithms. Unlike Monte Carlo (MC) methods, temporal difference (TD) methods learn the value function by reusing existing value estimates. The underlying mechanism in TD is bootstrapping. The word "bootstrapping" originated in the early 19th century with the expression "pulling oneself up by one's own bootstraps". Initially, this expression implied an obviously impossible feat. Later, it became a metaphor for achieving success with self-assistance. In statistical learning, bootstrapping can be interpreted as a sample reuse technique that uses historical estimates in the update step for the same kind of estimated value. In temporal difference, bootstrapping is a mechanism to reuse historical value estimates to update current value function. Similar to MC, TD only uses experience to estimate the value function without knowing any prior knowledge of the environment dynamics. The advantage of TD lies in the fact that it can update the value function based on its current estimate. Therefore, TD learning algorithms can learn from incomplete episodes or continuing tasks in a step-by-step manner, while MC must be implemented in an episode-by-episode fashion.

As stated by Andrew Barto and Richard Sutton, if one had to identify one idea as central and novel to reinforcement learning, it would undoubtedly be temporal difference. TD has the ability to learn from an incomplete sequence of events without waiting for the final outcome. Unlike MC estimation, TD has the ability to approximate the future return as a function of the current state [4-8]. The checkers-playing program developed in 1959 might have been the earliest TD attempt [4-6]. A similar idea can be seen in the adaptive critic element in the 1970s, which has evolved into today's one-step TD [4-2]. In 1988, Sutton formally extended the idea of one-step TD to TD-lambda using eligibility traces. In 1995, Tesauro applied the TD-lambda algorithm to some real-world games and created TD-Gammon, which is a famous program that learned to play the backgammon game at the level of expert human players [4-10]. In 2004, Bertsekas and his colleagues introduced an infinite-horizon TD algorithm with a discounted return, which has high speed and reliability [4-3]. When combined with function approximation and off-policy training, many TD algorithms tend to become unstable and easily diverge, even in simple tasks [4-1][4-11]. Some special examples, such as off-policy TD-lambda, have been proven to be convergent when using the linear function approximation [4-4]. Nevertheless, there are a large number of counterexamples in which theoretical convergence cannot be mathematically proven under off-policy settings.

Thus far, temporal difference learning has received extra attention in the interdisciplinary fields of neuroscience and psychology. The term reinforcement is used to describe the mechanism of instrumental conditioning and the self-learning property of first predicting upcoming stimuli and then changing behavior regardless of the contingency on previous behavior. A few physiological studies have found similar

S. E. Li, *Reinforcement Learning for Sequential Decision and Optimal Control*,
https://doi.org/10.1007/978-981-19-7784-8_4

learning mechanism in human brains to TD learning. For example, the firing rate of dopamine neurons in the brain appears to be proportional to a reward error. The prediction mechanism in the brain reports the difference between the estimated reward and the actual reward. The larger the reward difference is, the more a new behavior is reinforced. When this process is paired with a stimulus that reflects a reward in the future, current behavior can be changed to receive more positive rewards.

4.1   TD Policy Evaluation

The central concept of temporal difference (TD) is bootstrapping, in which reinforcement occurs through recursive correction toward a more accurate prediction. From a humorous perspective, bootstrapping is very similar to the quote "turtles all the way down". This saying alludes to the mythological idea that the earth rests on the back of a large turtle. This turtle, in turn, stands on the back of another turtle. There is no end, as there is always another turtle standing on top of a column of turtles. More formally, TD is an approach to estimating a quantity that depends on its historical values. The value estimate is recursively updated to bring the value closer to the true quantity. The name TD originates from its usage in differences over successive time steps to drive the learning process. By looking one or more steps ahead, a more accurate prediction of the value function can be obtained. To better illustrate the bootstrapping mechanism, let us take the state-value function as an example. Unlike MC policy evaluation, TD policy evaluation estimates the state value of a certain state as a weighted sum of the existing estimates of this state value and the currently observed returns starting from the same state. At time t, there is one sample $(s_t = s, a_t = a, s_{t+1} = s')$, and its reward signal $r_t = r$. The TD estimation starts from a weighted sum of two elements:

$$V(s) \leftarrow (1 - \alpha)V(s) + \alpha \cdot G_t = V(s) + \alpha(G_t - V(s)), \qquad (4\text{-}1)$$

where $V(s)$ is the state-value function for the estimation, G_t is the return starting from time t, and α is the learning rate. Using the self-consistency condition of the first kind to replace the return G_t, we have the simplest TD policy evaluation, known as TD(0) or one-step TD:

$$V(s) \leftarrow V(s) + \alpha(r + \gamma V(s') - V(s)). \qquad (4\text{-}2)$$

At each step, the value function is updated with the value of the next state and the reward signal received. This bootstrapping mechanism allows us to utilize existing estimates to accelerate convergence. The observed reward plays a key role in driving the evaluation process to eventually converge to the true quantity. Figure 4.1 illustrates the TD estimation for the state-value function, in which one single sample (s_t, a_t, s_{t+1}) can update the state-value $V(s)$ at a particular state s. In theory, perfect convergence requires a sufficient number of samples to completely explore the environment dynamics. In practice, iterations can always be stopped after a finite number of updates, as long as the estimation accuracy is sufficiently good.

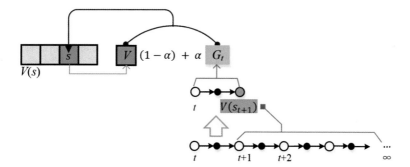

Figure 4.1 One-step TD evaluation with a state-value function

We refer to both TD and MC policy evaluations as sample updates (i.e., updates with experience), while DP requires expected updates (i.e., updates with a model). The sample update process in TD involves looking ahead to a sample successor state, using the state value of the successor and the reward to compute a backed-up value. In an expected update, each iteration updates all the values of states in the whole space to produce a new evaluation. Obviously, the sample update differs greatly from the expected update in that the former only deals with a single sample successor rather than a complete distribution of all possible successors.

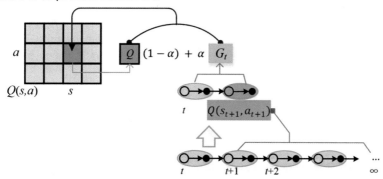

Figure 4.2 One-step TD evaluation for the action-value function

Figure 4.2 illustrates the TD estimation for the action-value function. Although TD learns a guess from other guesses, it can still guarantee convergence to the correct estimate. For an admissible policy, TD(0) has been proven to converge to the true value with a probability of 1 if the step-size parameter decreases to zero with certain conditions. Compared to MC, TD(0) can be implemented in an online and fully incremental manner. Batch MC estimation minimizes the mean-squared error without reliance on the Markov property. In contrast, batch TD(0) is based on the Markov process. Suppose that the state measurement is correct; then, TD(0) can have the certainty-equivalence estimate of almost any value function [4-9].

4.2 TD Policy Improvement

Similar to that in MC, the main goal of TD policy improvement is to find a better policy, which is critical to convergence guarantee. A common approach to finding a better policy is to search for a policy that satisfies the policy improvement theorem. The greedy search and the ϵ-greedy search are both popular choices. Both on-policy strategy and off-policy strategy can be applied to design TD learning algorithms. The former only has a single policy, while the latter must handle two different policies, i.e., a target policy and a behavior policy.

4.2.1   On-Policy Strategy

On-policy RL improves the same policy that is used to explore the environment. A natural choice is the greedy policy. However, a greedy search at an early stage often leads to a local minimum because the greedy policy is deterministic and lacks a good exploration ability. The exploration-exploitation dilemma requires an RL agent to choose random actions. Random actions are less optimal but can drive the RL agent to enter unexplored zones. The ϵ-greedy policy is stochastic and can better balance exploration and exploitation. Using this policy, the best action is selected most of the time, and other actions are chosen randomly with a small probability. This policy can become a deterministic policy by reducing the exploration rate. For example, an exploration rate can be chosen that is reversely proportional to the cycle number.

4.2.2   Off-Policy Strategy

Off-policy RL provides a better compromise between exploitation and exploration. The IS transformation allows the estimation of the value function under target policy π from the data samples of behavior policy b. For the state-value function, the current state is known, i.e., $s_t = s$, and its following state and action are random variables. Suppose that we have already recorded a sample trajectory from time t to time $t + 1$. The probabilities of generating such a trajectory under π and b are

$$d_\pi(a, s') = \Pr\{a, s' | \pi, s\} = \pi(a|s)p(s'|s, a),$$
$$d_b(a, s') = \Pr\{a, s' | b, s\} = b(a|s)p(s'|s, a),$$

(4-3)

where s' and a are two random variables, and $d(a, s')$ is the probability mass function of their bivariate distribution. The quotient of the aforementioned two distributions is

$$\rho_{t:t} = \frac{d_\pi(a, s')}{d_b(a, s')} = \frac{\pi(a|s)}{b(a|s)},$$

(4-4)

where $\rho_{t:t}$ is the one-step IS ratio. The expectations of three key functions, including the reward signal and value function under the target policy π and the behavior policy b are

$$\mathbb{E}_\pi\{r(s, a, s')\} = \mathbb{E}_b\left\{\frac{d_\pi(a, s')}{d_b(a, s')}r(s, a, s')\right\} = \mathbb{E}_b\{\rho_{t:t}r(s, a, s')\},$$

$$\mathbb{E}_\pi\{V(s')\} = \mathbb{E}_b\left\{\frac{d_\pi(a, s')}{d_b(a, s')}V(s')\right\} = \mathbb{E}_b\{\rho_{t:t}V(s')\},$$

(4-5)

$$\mathbb{E}_\pi\{V(s)\} = \mathbb{E}_b\{V(s)\}.$$

Note that the equivalence of the left-right expectation holds under the condition that (a, s') are random variables. As mentioned before, $\mathbb{E}_\pi\{\cdot\}$ is an abbreviation for $\mathbb{E}_\pi\{\cdot | s\}$, which conditionally depends on the given state. Moreover, their distribution is governed by the behavior policy. Therefore, the one-step TD evaluation with off-policy becomes

$$V(s) \leftarrow V(s) + \alpha \left(\rho_{t:t}(r + \gamma V(s')) - 1 \cdot V(s) \right), \forall (a, s') \sim b. \qquad (4\text{-}6)$$

Here, only the item $r + \gamma V(s')$ is reshaped with the IS transformation because it is a function of two random variables, s' and a. In contrast, $V(s)$ is not affected by the IS transformation because the current state s is a known variable. In practice, (4-6) often induces a large variance. An enhanced version of (4-6) was proposed by Sutton and his colleagues that contains a trick that replaces (4-6) by multiplying the last item by the IS ratio:

$$V(s) \leftarrow V(s) + \alpha \rho_{t:t}(r + \gamma V(s') - V(s)), \forall (a, s') \sim b. \qquad (4\text{-}7)$$

This formula can effectively improve the quality of value estimation in TD updates. The quantity $\delta_t \overset{\text{def}}{=} r + \gamma V(s') - V(s)$ has a special name, TD error or more precisely, one-step TD error. The rationality of performing such a replacement is that (4-7) does not change the expectation of (4-6) under the behavior policy, which is formally stated in Theorem 4-1.

• Theorem 4-1: $\mathbb{E}_b\{V(s)\} = \mathbb{E}_b\{\rho_{t:t}V(s)\}$.

Proof:

For the two random variables $(a, s') \sim b$, the function $V(s)$ in (4-7) can be viewed as a constant because the current state s is known. Therefore, $V(s)$ is independent of any function of (a, s'). Thus we have the following equality:

$$
\begin{aligned}
\mathbb{E}_{a \sim b}\{\rho_{t:t}V(s)\} &= \mathbb{E}_{a \sim b}\left\{\frac{\pi(a|s)}{b(a|s)}\right\} \cdot \mathbb{E}_{a \sim b}\{V(s)\} \\
&= \sum_a \left\{ b(a|s) \frac{\pi(a|s)}{b(a|s)} \right\} \cdot \mathbb{E}_{a \sim b}\{V(s)\} \qquad (4\text{-}8)\\
&= \sum_a \pi(a|s) \cdot \mathbb{E}_{a \sim b}\{V(s)\} \\
&= 1 \cdot \mathbb{E}_{a \sim b}\{V(s)\}.
\end{aligned}
$$

∎

Therefore, the two expectations above are exactly the same. Replacing (4-6) with (4-7) does not change the expectation of bootstrapping estimation but can reduce the variance caused by measurement noise and approximation errors. A concise explanation about variance reduction is that in (4-7), the TD error is multiplied by the IS ratio and the variance of their errors is less amplified than in (4-6). In the first update formula (4-6), the two items $r + \gamma V(s')$ and $V(s)$ coexist in a separate manner, and their errors cannot be easily cancelled out because one has the IS ratio and the other does not.

4.3   Typical TD Learning Algorithms

Temporal difference (TD) is the most critical and distinctive idea in model-free RL. It learns a guess from its historical guesses, i.e., bootstrapping. The advantages of TD include (1) better training efficiency, i.e., learning much faster on Markov problems; (2) lower cost of implementation, i.e., reduced demands on storage memory and peak computing power; and (3) learning from incomplete sequences. Early TD learning algorithms, including SARSA, Q-learning and expected SARSA, are often designed with tabular representation, where every state is distinct and treated individually. In essence, almost all TD learning algorithms are implemented under the generalized policy iteration (GPI) framework, which involves alternating processes of PEV and PIM. In particular, experience replay can be implicitly embedded into some off-policy TDs, such as Q-learning and expected SARSA.

4.3.1   On-Policy TD: SARSA

SARSA is an on-policy TD learning algorithm that was first proposed by Rummery and Niranjan (1994) [4-5]. The same policy is used to interact with the environment and perform monotonic policy updates. The RL agent learns the action-value function according to the experience that is collected from environment interactions. The action-value function at each cycle is estimated with a bootstrapping mechanism. At the k-th cycle, the update formula of PEV is

Loop N times

$$Q^\pi(s_t, a_t) \leftarrow Q^\pi(s_t, a_t) + \alpha\big(r_t + \gamma Q^\pi(s_{t+1}, a_{t+1}) - Q^\pi(s_t, a_t)\big). \tag{4-9}$$

End

Here, N is the number of samples used at each PEV. An update can be taken every time we encounter a 5-tuple $(s_t, a_t, r_t, s_{t+1}, a_{t+1})$. The 5-tuple, i.e., State-Action-Reward-State-Action, is the exact process that SARSA is named after. Of course, PEV updates can be performed either once or arbitrary times, with a range from $N = 1$ to $N \to \infty$. A large number of updates (or samples) are helpful to improve the accuracy of value estimates. However, too many samples also increase the computational burden. In the PIM step, we search for a new ϵ-greedy policy based on the current estimate of the action-value function:

$$\pi'(a|s) \leftarrow \begin{cases} 1 - \epsilon + \epsilon/|\mathcal{A}|, & \text{if } a = a^* \\ \epsilon/|\mathcal{A}|, & \text{if } a \neq a^* \end{cases},$$

subject to

$$a^* = \arg\max_a Q^\pi(s, a), \forall s \in \mathcal{S}. \tag{4-10}$$

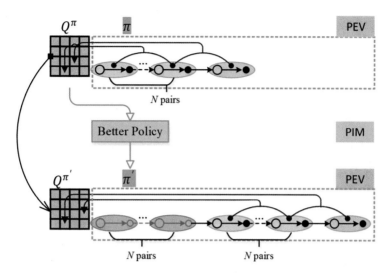

Figure 4.3 Flow chart of SARSA

It is easy to see that SARSA has three main hyperparameters: learning rate α, discount factor γ, and initial value guess Q_{init}. An initial condition must be assigned to value function estimation before each PEV update occurs for SARSA. Proper initialization can introduce some specific advantages into SARSA. For instance, a better choice of the initial value function would be that for the previous policy. As a result, the experience of old policies is implicitly inherited, which is beneficial to achieve high sample efficiency and convergence speed. The pseudocode of SARSA is shown in Algorithm 4-1, in which N PEV updates are performed after each policy update.

Algorithm 4-1: SARSA

Hyperparameters: Discount factor γ, exploration rate ϵ, learning rate α, pairs per PEV N

Initialization: $Q(s,a) \leftarrow 0, \pi^\epsilon(a|s) \leftarrow 1/|\mathcal{A}|$

Repeat (indexed with k)

    // Environment initialization

    $s_0 \sim d_{init}(s)$

    $a_0 \sim \pi^\epsilon(a|s_0)$

    $s \leftarrow s_0, a \leftarrow a_0$

    Repeat until each episode terminates

        Rollout one step with action a, observe next state s' and reward r

        Sample action a' under next state s': $a' \sim \pi^\epsilon(a|s')$

        //Do action-value update

        $Q(s,a) \leftarrow Q(s,a) + \alpha\big(r + \gamma Q(s',a') - Q(s,a)\big)$

        If N steps since last policy update, update $\pi^\epsilon(a|s)$ with $Q(s,a)$

        $s \leftarrow s', a \leftarrow a'$

End
End

■

4.3.2   Off-Policy TD: Q-Learning

One of the most important breakthroughs in TD was the development of Q-learning (Watkins, 1989) [4-12]. Its convergence proof was presented by Watkins and Dayan in 1992 [4-13]. In 2014, Google's DeepMind introduced a successful variant of Q-learning with a deep neural network, namely, the deep Q-network. This algorithm is able to play Atari 2600 games at a level comparable to a professional human player. The simplest form of Q-learning is

$$Q(s_t, a_t) \leftarrow Q(s_t, a_t) + \alpha \left(r_t + \gamma \max_a Q(s_{t+1}, a) - Q(s_t, a_t) \right). \tag{4-11}$$

In theory, Q-learning is a special off-policy one-step TD algorithm. It combines PEV and PIM into one single formula. In addition, with a form similar to that of SARSA, Q-learning can be broken down into two separate steps: (a) a one-step TD policy evaluation and (b) a greedy policy search:

Loop only once

$$Q^\pi(s_t, a_t) \leftarrow Q^\pi(s_t, a_t) + \alpha \left(r_t + \gamma Q^\pi(s_{t+1}, \pi(s_{t+1})) - Q^\pi(s_t, a_t) \right).$$

End

$$\pi'(s) = \arg \max_a Q^\pi(s, a), \forall s \in \mathcal{S}.$$

One benefit of Q-learning is that its behavior policy can take arbitrary form because it does not require explicit IS transformation, as shown in Theorem 4-2. In fact, Q-learning can be more flexible in behavior policy selection, and even an unknown policy can be used as the behavior policy. This special property allows Q-learning to be executed with precollected samples. The experience replay trick is applied to reuse historical samples to accelerate convergence and improve sample efficiency. In this trick, one does not need to know the structure and parameters of the behavior policies. This kind of flexibility is a very large advantage of Q-learning and most of its variants, such as deep Q-networks (DQN).

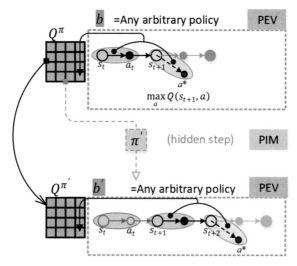

Figure 4.4 Flow chart of Q-learning

- Theorem 4-2: Q-learning is a one-step off-policy TD without explicitly containing the IS transformation.

Proof:

The behavior policy b can be any arbitrary policy. The current state and action (s, a) are known, and (s', a') are two random variables governed by the behavior policy b. Note that we only need to consider importance sampling for random variables. Then, the off-policy version of TD(0) policy evaluation is

$$Q(s, a) \leftarrow Q(s, a) + \alpha\big(\rho_{t+1:t}r + \gamma\rho_{t+1:t+1}Q(s', a') - Q(s, a)\big), \qquad (4\text{-}12)$$

where

$$\mathbb{E}_\pi\{r(s, a, s')\} = \mathbb{E}_b\{\rho_{t+1:t}r(s, a, s')\},$$
$$\rho_{t+1:t} = \frac{\Pr\{s'|\pi, s, a\}}{\Pr\{s'|b, s, a\}} = \frac{p(s'|s, a)}{p(s'|s, a)} = 1, \qquad (4\text{-}13)$$

and

$$\mathbb{E}_\pi\{Q(s', a')\} = \mathbb{E}_b\{\rho_{t+1:t+1}Q(s', a')\},$$
$$\rho_{t+1:t+1} = \frac{p(s'|s, a)\pi(a'|s')}{p(s'|s, a)b(a'|s')} = \frac{\pi(a'|s')}{b(a'|s')}. \qquad (4\text{-}14)$$

Here, we must note that the equivalence above holds under the condition that only (s', a') are random variables. Since one only has samples from the behavior policy, the IS transformation is required to address the issue of policy mismatch. The expectations of TD errors are equal under the target policy π and the behavior policy b. As a result, the TD error sampled by the behavior policy can be used to update the target policy. In Q-learning, the main goal is to prove the equivalence of the following two expectations:

$$\mathbb{E}_b\{\rho_{t+1:t}r + \gamma\rho_{t+1:t+1}Q(s', a')\} = \mathbb{E}_b\Big\{r + \gamma \max_a Q(s', a)\Big\}.$$

Using the fact that the IS ratio depends on random variables, we obtain the following equality:

$$\mathbb{E}_b\{\rho_{t+1:t}r + \gamma\rho_{t+1:t+1}Q(s',a')\} = \mathbb{E}_{s'\sim\mathcal{P}}\left\{1\cdot r + \gamma\mathbb{E}_{a'\sim b}\left\{\frac{\pi(a'|s')}{b(a'|s')}Q(s',a')\right\}\right\}.$$

Because the next action a' is actually from the greedy search, it can be replaced with the maximization operation $a' = \arg\max_a Q(s',a)$. Therefore, only one random variable remains, i.e., the next state s'. Hence, we have

$$= \mathbb{E}_{s'\sim\mathcal{P}}\left\{r + \gamma\sum_{a'} b(a'|s')\frac{\pi(a'|s')}{b(a'|s')}Q(s',a')\right\}$$

$$= \mathbb{E}_{s'\sim\mathcal{P}}\left\{r + \gamma\sum_{a'} \pi(a'|s')Q(s',a')\right\}$$

$$= \mathbb{E}_{s'\sim\mathcal{P}}\{r + \gamma\mathbb{E}_{a'\sim\pi}\{Q(s',a')\}\}$$

$$= \mathbb{E}_b\left\{r + \gamma\max_a Q(s',a)\right\}.$$

∎

In theory, Q-learning is able to output an optimal policy that maximizes the expected value of the total reward. The main operation of Q-learning is to maximize the action-value function over all possible actions in the next state. This operation eliminates the deployment of the importance sampling (IS) technique. The updating process of Q-learning is surprisingly efficient and can quickly converge, even though the environment interaction is insufficient. The letter "Q" in Q-learning stands for "quality", which originates from the fact that this method returns the quality of a given state-action pair. In its simplest form, tabular Q-learning quantizes the state and action spaces and stores them in tables. The tabular representation often fails to handle large-scale tasks due to the curse of dimensionality. One solution is to represent the action-value function with an approximate function. This technique makes Q-learning suitable for complex problems with continuous state spaces. The pseudocode of Q-learning is illustrated in Algorithm 4-2, in which any random behavior policy can be selected to explore the environment. For example, one can even select $b(a|s)$ to be a stochastic policy with a uniform action distribution.

Algorithm 4-2: Q-learning

Hyperparameters: Discount factor γ, Learning rate α

Initialization: $Q(s,a) \leftarrow 0$, Behavior policy $b(a|s) \leftarrow 1/|\mathcal{A}|$

// Environment initialization

$s_0 \sim d_{\text{init}}(s)$

$s \leftarrow s_0$

Repeat until episode terminates (indexed with k)

        //Sample action a under state s from behavior policy

$a \sim b(a|s)$

Rollout one step with action a, observing s' and r

//Do action-value update

$$Q(s, a) \leftarrow Q(s, a) + \alpha \left(r + \gamma \max_a Q(s', a) - Q(s, a) \right)$$

$s \leftarrow s'$

End

Calculate greedy policy $\pi(a|s)$ from $Q(s, a)$

■

4.3.3   Off-Policy TD: Expected SARSA

Expected SARSA is another example of a one-step TD algorithm. Its structure is very similar to that of SARSA and Q-learning, except that it uses the expected value of the next state-action pair instead of the maximum value. We know that SARSA is on-policy and Q-learning is off-policy. Expected SARSA is an off-policy TD algorithm. The formula for expected SARSA is

Loop only once

$$Q(s_t, a_t) \leftarrow Q(s_t, a_t) + \alpha \left(r_t + \gamma \sum_{a \in \mathcal{A}} \pi(a|s_{t+1}) Q(s_{t+1}, a) - Q(s_t, a_t) \right).$$

End

Expected SARSA takes the weighted sum of the values of all possible next actions with respect to the probability of taking that action. Given the next state, expected SARSA moves deterministically in the same direction as SARSA moves in expectation. In general, the average operation makes expected SARSA more computationally complex than SARSA. In return, expected SARSA can reduce the large variance caused by occasionally bad samples and increase the training stability compared to SARSA.

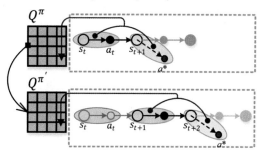

(a) Q-learning

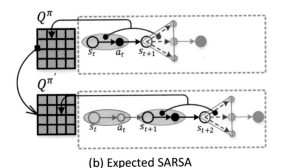

(b) Expected SARSA

Figure 4.5 Backup diagrams of Q-learning and expected SARSA

If the target policy π is from a greedy search, expected SARSA exactly reduces to Q-learning. From this perspective, expected SARSA generalizes Q-learning while being more efficient than Q-learning. Except for additional computational cost, expected SARSA usually performs better than common TD algorithms because its expectation operation provides more stable bootstrapping updates. Both Q-learning and expected SARSA are off-policy TD algorithms without explicit IS ratio. The common reason is that their update formulae do not contain the next action. Q-learning eliminates the next action by replacing it with the greedy action. Expected SARSA eliminates this action with the expectation operation under the current policy.

4.3.4   Recursive Value Initialization

TD often faces continuing tasks without termination. As a typical TD algorithm, SARSA uses multiple samples in each cycle. The number of samples can range from one to infinity. The larger the sample length is, the more accurate its PEV becomes. However, a long sample length decreases the frequency of policy updates. As a result, the convergence speed of the overall algorithm is sacrificed to a certain extent. In contrast, Q-learning is a one-step off-policy TD without any IS transformation and uses only one sample in each policy update. Both SARSA and Q-learning can inherit the final value of the last cycle, which is known as recursive value initialization. In the MC learning algorithm, recursive value initialization is equivalent to incremental MC estimation.

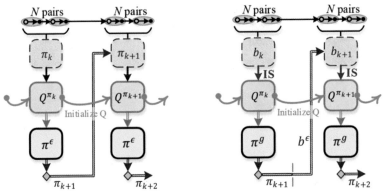

(a) On-policy TD (i.e., SARSA)   (b) Off-policy TD (incl. Q-learning)

Figure 4.6 On-policy TD vs. Off-policy TD

Figure 4.6 compares on-policy TD and off-policy TD. With recursive value initialization, both on-policy TD and off-policy TD maintain the records of historical experience in the form of an initial action-value function. On-policy TD generates samples with the target policy to be optimized, while off-policy TD uses the behavior policy to perform the same task. In off-policy TD, historical samples may come from a variety of different policies. To harmonize with action-value estimation, most TD learning algorithms (except for a few special methods such as Q-learning and expected SARSA) should go through the IS transformation to maintain identical expectations between the target policy and the behavior policy. The IS transformation is able to maximize the capacity of reusing historical samples but often suffers from large variance caused by the cascading multiplication in the IS ratio.

4.4   Unified View of TD and MC

Temporal-difference (TD) and Monte Carlo (MC) are common approaches for sample-based policy evaluation. One might ask what kind of gap lies between TD and MC. The answer can be disclosed through n-step TD or TD-lambda. Both of them were proposed to address the bias-variance tradeoff regarding the reliance on either existing estimates or new samples. They also unify one-step TD and MC, which are actually two extremes of TD learning algorithms.

4.4.1   n-Step TD Policy Evaluation

The update of one-step TD looks only one step ahead and bootstraps by taking the state value of the previous step as a proxy for the true value. In contrast, MC performs updates for each state based on the entire sequence of observed rewards. What locates between MC and one-step TD is the idea of n-step TD, which performs updates based on a limited number of samples, i.e., more than one sample but less than the entire sequence. This idea is also called TD(n) policy evaluation. Figure 4.7 illustrates the backup diagram of n-step TD, where one extreme is one-step TD and the other is Monte Carlo.

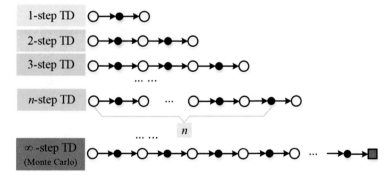

Figure 4.7 Block diagram of n-step TD

More formally, let us consider the current state s_t and its following reward sequence, $r_t, r_{t+1}, \cdots, r_{t+n-1}, \cdots$, until episode termination. Any long-term return has an inherent self-consistency relation, which provides an explicit way to divide it into n-step samples and another long-term return:

$$G_t = r_t + \gamma r_{t+1} + \gamma^2 r_{t+2} + \cdots + \gamma^{n-1} r_{t+n-1} + \gamma^n G_{t+n}, \qquad (4\text{-}15)$$

where n represents the number of samples used in TD updates. Taking the expectation of two sides at a given state or a given state-action pair, we have the following n-step self-consistency conditions:

$$v^\pi(s) = \mathbb{E}_\pi\{G_{t:t+n-1} + \gamma^n v^\pi(s_{t+n})|s_t = s\},$$

$$q^\pi(s,a) = \mathbb{E}_\pi\{G_{t:t+n-1} + \gamma^n q^\pi(s_{t+n}, a_{t+n})|s_t = s, a_t = a\},$$

where $G_{t:t+n-1}$ is called the n-step return from t to $t+n-1$ and is defined as

$$G_{t:t+n-1} \stackrel{\text{def}}{=} \sum_{i=0}^{n-1} \gamma^i r_{t+i}.$$

Let us take the state-value function as an example to discuss its properties. A better estimate of the state value is selected as the update target. The bootstrapping update of the state-value estimation $V^\pi(s_t)$ becomes

$$V^\pi(s_t) \leftarrow V^\pi(s_t) + \alpha\big(v^\pi(s) - V^\pi(s_t)\big)$$

$$\qquad\qquad (4\text{-}16)$$

$$\cong V^\pi(s_t) + \alpha\big(G_{t:t+n-1} + \gamma^n V^\pi(s_{t+n}) - V^\pi(s_t)\big),$$

where α is the learning rate. This formula is still called the temporal difference (TD) update because it changes an earlier estimate based on how it differs from a later estimate. In this formula, the following quantity is called the "n-step TD error", which is the difference between the n-step TD estimate and the current state-value estimate:

$$\delta_t^{\text{TD}(n)} \stackrel{\text{def}}{=} G_{t:t+n-1} + \gamma^n V^\pi(s_{t+n}) - V^\pi(s_t).$$

The backup diagram of the n-step TD policy evaluation is shown in Figure 4.8.

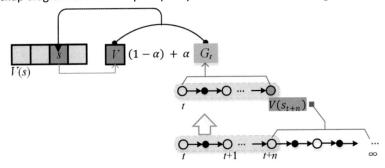

Figure 4.8 Backup diagram of the n-step TD

The n-step TD uses the estimated value function of s_{t+n} to correct the missing rewards that are beyond time $t+n$. In this method, at least the samples until time $t+n$ are available, which means that the estimated value function cannot be performed immediately after one sample; instead, there is an n-step delay. An important property of n-step return is that its expectation is often a better estimate of the state value. More specifically, the worst-case error of an expected n-step return is less than or at least equal to γ^n times that of an estimated state value:

$$\max_s |\mathbb{E}_\pi\{G_{t:t+n-1} + \gamma^n V^\pi(s_{t+n})|s_t = s\} - v^\pi(s)| \qquad\qquad (4\text{-}17)$$

$$\leq \gamma^n \max_{s}|V^\pi(s) - v^\pi(s)|.$$

Obviously, this inequality describes an error reduction property of discounted returns [4-9]. The more samples the value estimation uses, the more accurate it will become. Because of this error reduction property, one can believe that the n-step TD is able to converge to a correct evaluation under some proper initial conditions.

4.4.2   TD-Lambda Policy Evaluation

The TD-lambda is another strategy for creating an advanced bootstrapping technique. It is a generalization of the n-step TD. In the TD-lambda, multiple n-step TDs are summed together to produce a family of more reliable policy evaluations. This method was first proposed by Singh and Sutton (1996) in their eligibility trace studies [4-7]. In addition to n-step TD, TD-lambda is also a bridge between MC and one-step TD. Figure 4.9 illustrates the backup diagram of truncated TD-lambda. A truncated TD-lambda $G^\lambda_{t:t+T}$ can be understood as the weighted sum of a group of n-step TDs. The number of samples used in each TD is selected from the set $\{1,2,\cdots,T\}$. When the maximal sample length $T \to \infty$, the TD-lambda return is also called the λ-return, which is defined as

$$G^\lambda_t = (1-\lambda)\sum_{n=1}^{\infty} \lambda^{n-1}\left(G_{t:t+n-1} + \gamma^n V^\pi(s_{t+n})\right), \qquad (4\text{-}18)$$

where $\lambda \in [0,1]$ is the trace-decay parameter. When $\lambda = 0$, TD-lambda degenerates into a one-step TD estimation, and when $\lambda = 1$, TD-lambda becomes a Monte Carlo estimation. From the perspective of forward view, the TD-lambda return is often computationally inefficient. The eligibility trace offers an elegant algorithmic mechanism from a backward view with some computational advantages. The eligibility trace allows us to compute TD-lambda in a backward recursive manner. Its effectiveness was demonstrated by TD-Gammon, a famous program that learned to play backgammon at the level of expert players [4-10].

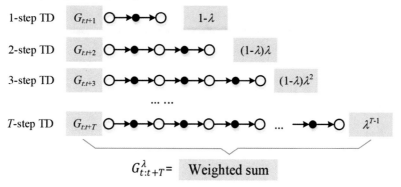

Figure 4.9 Backup diagram for truncated TD-lambda

As stated by Sutton and Barto (2018), there are two basic ways to understand the mechanism of eligibility traces [4-9]. When TD methods are augmented with eligibility traces, they produce a family of error self-correction methods spanning a spectrum that has MC at one end and one-step TD at the other. From this viewpoint, the eligibility trace unifies TD and MC in a revealing way. The other way to understand eligibility traces is much more mechanistic. The eligibility trace is a temporary record of the occurrence of an event, such as the visiting of a state or the taking of an action. The trace marks the memory parameters associated with the event as eligible for undergoing learning changes. When a TD error occurs, only the eligible states or actions are assigned credit or blame for the error. Hence, the eligibility trace is a basic mechanism of temporal credit assignment.

4.5   Examples of Temporal Difference

TD learning algorithms can handle both continuing and episodic tasks. Using these methods, the learning procedure continuously generates samples to extract information from environment dynamics. As more samples are accumulated, the learned policy will gradually converge to its optimum. To provide a fair comparison in the cleaning robot example, the robot agent is trained and validated with typical TD algorithms, including Q-learning and SARSA. The cleaning robot works in a rectangular grid environment with 36 states and 4 actions. The goal is to find an optimal policy to move the robot to the destination cell as quickly as possible. The state and action spaces are

$$S = \{s_{(1)}, s_{(2)}, \cdots, s_{(36)}\},$$
$$A = \{\text{North}, \text{South}, \text{West}, \text{East}\}.$$

(4-19)

For the cleaning robot, each action can move the agent randomly to its neighboring cells. The robot moves forward with a probability of 80% and moves leftward or rightward with a respective probability of 10%, but it can never move backward. Tabular SARSA and Q-learning are applied to train the optimal policy and corresponding value function.

4.5.1   Results of SARSA

A standard tabular SARSA algorithm is implemented in the cleaning robot setting, in which a multistep PEV and an ϵ-greedy policy are utilized. The experimental result of SARSA is shown in Figure 4.10, including the learned policy and the state-value function. The shaded grid at the bottom-right corner is the destination. Figure 4.10 shows the trained policy, whose arrows tend to drive the cleaning robot to its destination. The three hyperparameters, i.e., learning rate α, exploration rate ϵ, and number of state-action pairs per cycle (pairs/PEV), are repeatedly simulated to examine their influences on the policy performance. Policy performance is measured by the RMS error and total average return. It is easily seen from Figure 4.11 that when using SARSA, a high learning rate increases the convergence speed but often leads to a large variance. Similar to MC and other TDs, SARSA is significantly affected by the exploration rate in the ϵ-greedy policy. Figure 4.12 shows that while seemingly having no significant effect on the convergence speed, the reduction in the exploration rate helps decrease the RMS error. In this example, the pair number per cycle has very little influence on both RMS error and total average return (Figure 4.13). Mostly, the total number of collected samples determines the final policy performance. Fewer steps in policy improvement is helpful to reduce the high computational burden that is caused by the maximization operation.

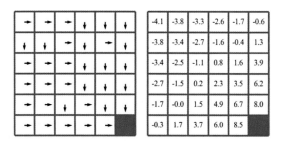

(a) Final policy   (b) Final value function

Figure 4.10 Learned policy and value function

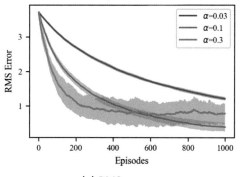

(a) RMS error

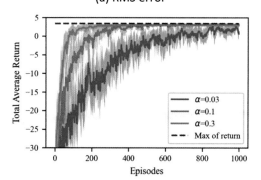

(b) Total average return

Figure 4.11 SARSA with different learning rates

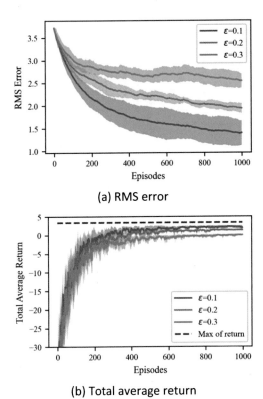

(a) RMS error

(b) Total average return

Figure 4.12 SARSA with different exploration rates

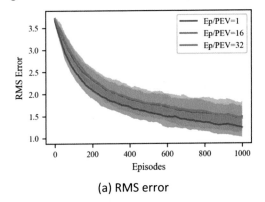

(a) RMS error

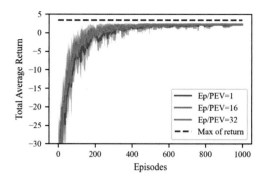

(b) Total average return

Figure 4.13 SARSA with different pairs per PEV

4.5.2   Results of Q-Learning

A tabular Q-learning algorithm is applied to train the cleaning robot. The experimental result of Q-learning is shown in Figure 4.14, including the learned policy and the state-value function. Q-learning seems to output a more reasonable policy than SARSA, which means its action arrows are closer to the optimum. The learning rate α is the exclusive hyperparameter in Q-learning. Figure 4.15 shows that a high learning rate can effectively increase the convergence speed. The total average return can almost reach the optimum since Q-learning is an off-policy algorithm and a greedy target policy is pursued.

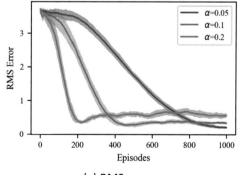

(a) Final policy   (b) Final value function

Figure 4.14 Learned policy and value function

(a) RMS error

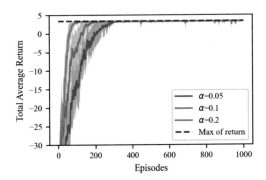

(b) Total average return

Figure 4.15 Q-learning with different alpha

4.5.3   Comparison of MC, SARSA, and Q-Learning

Figure 4.16 compares the results of three model-free algorithms, MC, SARSA and Q-learning, under analogous settings. Q-learning and SARSA share the same learning rate: $\alpha = 0.1$. Both MC and SARSA use the ϵ-greedy policy with an exploration rate of $\epsilon = 0.1$, while Q-learning uses a random behavior policy. The discount factor γ is set to 0.95, and each episode is limited to at most 64 steps. Other algorithm-specific hyperparameters are set to sensible values to ensure a fair comparison. In general, Q-learning performs the best, with the lowest RMS error, highest convergence speed, and largest total average return. SARSA ranks second, as its performance is worse than that of Q-learning but better than that of MC. Even though MC appears to be the worst algorithm, the adoption of incremental MC estimation effectively improves its training performance. A properly designed MC may achieve a performance close to that of SARSA because its incremental estimation actually introduces a long-term updating mechanism that is very similar to bootstrapping.

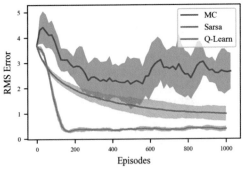

(a) RMS error

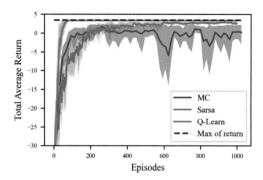

(b) Total average return

Figure 4.16 Comparison of MC, SARSA and Q-learning

4.6   References

[4-1]   Baird L (1995) Residual algorithms: reinforcement learning with function approximation. ICML, Tahoe City, USA

[4-2]   Barto A, Sutton R, Anderson C (1983) Neuronlike adaptive elements that can solve difficult learning control problems. IEEE Trans Syst Man Cybern 13(5):834-846

[4-3]   Bertsekas D, Borkar V, Nedic A (2004) Improved temporal difference methods with linear function approximation. In: Handbook of Learning and ADP, Wiley, New York

[4-4]   Precup D, Sutton R, Dasgupta S (2001) Off-policy temporal-difference learning with function approximation. ICML, Williamstown

[4-5]   Rummery G, Niranjan M (1994) On-line Q-learning using connectionist systems. Technical Report, Cambridge University

[4-6]   Samuel A (1959). Some studies in machine learning using the game of checkers. IBM J Res Dev 3(3): 210-229

[4-7]   Singh S, Sutton R (1996) Reinforcement learning with replacing eligibility traces. Mach Learn 22(1-3):123-158

[4-8]   Sutton R (1988) Learning to predict by the method of temporal differences. Mach Learn 3(1): 9-44

[4-9]   Sutton R, Barto A (2018) Reinforcement learning: an introduction. MIT press, Cambridge

[4-10]   Tesauro G (1995) Temporal difference learning and TD-Gammon. Communication ACM 38(3): 58-68

[4-11]   Tsitsiklis J, Van R (1997) An analysis of temporal-difference learning with function approximation. IEEE Trans Automatic Control 42(5):674-690

[4-12]   Watkins C (1989) Learning from delayed rewards. Dissertation, Cambridge University

[4-13]   Watkins C, Dayan P (1992) Q-learning. Mach Learn 8(3-4): 279-292

Chapter 5. Model-Based Indirect RL: Dynamic Programming

> Nothing in life is to be feared,
> it is only to be understood.
> Now is the time to understand more,
> so that we may fear less.
> -- Marie Curie (1867-1934)

Many distinct communities, including physics, economics, finance, transportation, energy, and the internet, have actively worked on finding optimal decisions or control inputs for stochastic systems. In the sequential decision process, which is one type of mathematical abstraction, multistage decisions are made, and certain rewards are received at each stage. Each current decision may affect which future decisions will be made as well as the total return. Hence, the desire to optimize the rewards induced by the present action and future actions needs to be balanced. Bellman's principle of optimality has served as a leading method for solving goal-oriented sequential decision problems. This principle is essentially a dimensionality reduction scheme that uses recursive optimization. In other words, a multistage problem is broken down into a series of overlapping subproblems, and each optimal decision is solved recursively. As may be inferred by the name, this principle was first developed by Richard Bellman in the 1950s.

The goal-oriented sequential decision has a very close connection with stochastic optimal control in dynamic environments. The aim of stochastic optimal control is to find an admissible action sequence that minimizes the expectation of a particular criterion. In the continuous-time domain, the optimal control law obeys a partial differential condition derived from Bellman's principle of optimality, i.e., the Hamilton–Jacobi–Bellman (HJB) equation. In the discrete-time domain, the discretization of the HJB equation can lead to a recursive formula, which is known as the Bellman equation. In the RL community, dynamic programming (DP) refers to a collection of model-based planning methods that compute the optimal policy through the optimality condition. The term "dynamic" denotes how to handle a temporally sequential system with transition behaviors, while the term "programming" denotes how to optimize the accumulative rewards to find the best policy. Due to its reliance on the optimality condition, DP belongs to the class of indirect RL methods. Different from direct RL, indirect RL takes the solution of the Bellman equation or the HJB equation as the optimal policy. The success of this replacement is attributed to the fact that either equation is the necessary and sufficient optimality condition in most engineering tasks.

To the best of our knowledge, the connection between DP and RL was first discussed by Minsky in 1961 in his comments on Samuel's checker player. In 1989, Watkins explicitly stated the connection between DP and RL and characterized a class of RL algorithms as incremental DP. DP's main benefits are its wide flexibility and excellent compatibility with various kinds of environment dynamics, including linear and nonlinear, discrete-time and continuous-time, deterministic and stochastic, and differentiable and nondifferentiable dynamics. A classic way of using DP is to quantize the state and action spaces as countable grids and apply backward recursion from the terminal time to the initial time. In practice, classic DP with a tabular representation is of limited usefulness because of

© The Author(s), under exclusive license to Springer Nature Singapore Pte Ltd. 2023, Corrected Publication 2024
S. E. Li, *Reinforcement Learning for Sequential Decision and Optimal Control*,
https://doi.org/10.1007/978-981-19-7784-8_5

the curse of dimensionality and the lack of generalization. Even though it is not quite efficient for continuous-space and large-scale problems, classic DP still plays an important role in the theoretical analysis of optimum existence and convergence speed.

5.1   Stochastic Sequential Decision

The design of a DP algorithm involves two prerequisites: (1) modeling the transition dynamics, including how to reasonably select state variables and action variables, and (2) designing a reward function and selecting a proper policy structure. Given the diversity of engineering problems, it is not surprising that DPs may look very different in various fields. Today, there is a growing trend toward understanding different variants of DP from a unified viewpoint. This chapter attempts to bridge two major fields, i.e., reinforcement learning and optimal control, and identify opportunities for cross-fertilization.

5.1.1   Model for Stochastic Environment

A stochastic environment is mathematically described by a stochastic model, whose counterpart is a deterministic model. A deterministic model does not contain any uncertainty and outputs the same states for a particular set of inputs. Unfortunately, the natural world is filled with stochasticity. The stochastic model possesses some inherent randomness either in its structure or parameters. Hence, the same set of initial conditions will lead to an ensemble of different outputs. Here, we consider a stochastic state-space model in the discrete-time domain:

$$s_{t+1} = f(s_t, a_t, \xi_t), \tag{5-1}$$

where $f(\cdot)$ is the environment model, and ξ_t is the random noise with a known distribution. Here, ξ_t is assumed to be independent of its historical counterparts $\{\xi_0, \xi_1, \xi_2 \ldots, \xi_{t-1}\}$ and the initial state s_0. Each action a_t is arbitrary and independent of historical states and random noises. This assumption can make the transition dynamics Markovian. The Markov property of a general state-space model is proved as follows:

$$p(s_{t+1}|s_t, \ldots, s_2, s_1, s_0) = p(f(s_t, a_t, \xi_t)|s_t, \ldots, s_2, s_1, s_0).$$

Because $\{\xi_0, \xi_1, \xi_2 \ldots, \xi_t\}$ is an independent sequence:

$$= p(f(s_t, a_t, \xi_t)|s_t, \ldots, s_2, s_1)$$
$$= p(f(s_t, a_t, \xi_t)|s_t, \ldots, s_2, f(s_0, a_0, \xi_0))$$
$$= p(f(s_t, a_t, \xi_t)|s_t, \ldots, s_3, s_2)$$
$$= p(f(s_t, a_t, \xi_t)|s_t, \ldots, s_3, f(f(s_0, a_0, \xi_0), a_1, \xi_1))$$
$$= p(f(s_t, a_t, \xi_t)|s_t, \ldots, s_3).$$

Then, roll forward until s_t at time t:

$$= p(f(s_t, a_t, \xi_t)|s_t)$$
$$= p(s_{t+1}|s_t).$$

Clearly, the next state s_{t+1} is exclusively determined by the current-time state s_t, the current-time action a_t, and the current-time noise ξ_t. It is observed that the existence of the Markov property relies on two conditions: (1) the current-time noise ξ_t is independent of its historical information and the initial state s_0, and (2) actions $a_0, \cdots, a_t$ are open-loop sequences that are independent of historical states and random noises. Hence, such a stochastic state-space model can be exactly converted into a Markovian probabilistic model. Without loss of generality, we suppose that ξ_t is an additive noise with the probability density function (PDF) $p_\xi(\xi_t; \theta)$; i.e., $s_{t+1} = f(s_t, a_t) + \xi_t$. As a result, the transition dynamics from the current state s_t to the next state s_{t+1} become

$$\mathcal{P}_{ss'}^a \overset{\text{def}}{=} p_\xi(\xi_t; \theta) = p_\xi(s_{t+1} - f(s_t, a_t); \theta), \tag{5-2}$$

where $\mathcal{P}_{ss'}^a$ is the one-step transition probability and p_ξ is the function of random noise distribution with θ as its parameter. This probabilistic model has exact equivalence with the aforementioned state-space model. Let us take a linear Gaussian system to demonstrate their equivalency. Consider a linear state-space model with additive Gaussian noise:

$$s_{t+1} = As_t + Ba_t + \xi_t, \tag{5-3}$$

where (A, B) is controllable, $A \in \mathbb{R}^{n \times n}$, $B \in \mathbb{R}^{n \times m}$, and $\xi_t \sim \mathcal{N}(\mu, \mathcal{K})$ is the additive noise that follows a Gaussian distribution:

$$p_\xi(\xi_t) = \frac{1}{(2\pi)^{n/2}|\mathcal{K}|^{1/2}} \exp\left(-\frac{1}{2}(\xi_t - \mu)^{\mathrm{T}}\mathcal{K}^{-1}(\xi_t - \mu)\right),$$

where $\mu \in \mathbb{R}^n$ is the mean vector and $\mathcal{K} \in \mathbb{R}^{n \times n}$ is the covariance matrix. The conditional probability of the next state s_{t+1} depends on the current state s_t and the current action a_t:

$$p(s_{t+1}|s_t, a_t) = \frac{1}{(2\pi)^{n/2}|\mathcal{K}|^{1/2}} \exp\left(-\frac{1}{2}z^{\mathrm{T}}\mathcal{K}^{-1}z\right),$$
$$z = s_{t+1} - (As_t + Ba_t + \mu),$$

where s_{t+1} also follows the form of Gaussian distribution, i.e.,

$$s_{t+1} \sim \mathcal{N}\left((As_t + Ba_t + \mu), \mathcal{K}\right).$$

This kind of environment description is also called a probabilistic model. This model is a one-step transition probability from the current state to its next state. The covariance matrix of this new state distribution is identical to that of additive random noise. If $s \in \mathcal{S}$ and $a \in \mathcal{A}$ are in finite element spaces, (5-2) can be reduced to a Markov chain. If some state variables are not observable, a hidden Markov model that contains additional observation dynamics can be obtained. Similar to any state-space model, a probabilistic model can cover different kinds of uncertainties, including exogenous disturbances, structural errors, parametric errors, and even unmodelled high-order dynamics. Instead of being explicitly stated in a state-space model, these uncertainties are buried inside the one-step transition probability.

In summary, both state-space models and probabilistic models are powerful tools to describe stochastic environments. The built-in Markov property of these models is critical to applying Bellman's principle of optimality. The probabilistic model with one-step transition probability naturally has the Markov property, while a few assumptions on the structure and uncertainty of the stochastic state-space model are often needed to maintain the Markov property.

5.1.2   Average Cost vs. Discounted Cost

The mathematical formulation of sequential decision problems can be very diverse. One reason may be the variety of overall objective functions, which are sometimes referred to as cost functions to follow the convention used in the area of feedback control. For the infinite-horizon problem, the cost function must be defined carefully to ensure that it is always finite-valued, even though the number of time stages goes to infinity. The discounted cost approach, in which returns rapidly decrease in future stages, is a simple way to achieve this requirement. An alternative approach is to formulate the average cost, in which the cumulative returns are divided by the total number of stages.

5.1.2.1   Average Cost

An average return can be directly taken as a cost function since it is a scalar independent of the initial state. This kind of performance measure has been widely used in real-world applications, such as supply chain management, network control, and route planning. A policy that is associated with the average return implies that the agent cares just as much about delayed rewards as the immediate reward. Because the average return does not concern the behaviors at the transient stage, it is more suitable for cases focusing on steady-state performance. The average cost (or average return) is defined as

$$G_{\text{avg}}(\pi) = \lim_{T \to \infty} \frac{1}{T} \mathbb{E}\left\{ \sum_{i=0}^{T-1} r_{t+i} \right\}, \tag{5-4}$$

where T is the length of time horizon. Usually, the average cost is only associated with a stationary policy. Finding an optimal policy of (2-15) remains a challenging task for general stochastic control problems. In mathematics, this kind of average-cost problem is equivalent to optimizing an overall objective function weighted by its stationary state distribution:

$$J_{\text{avg}}(\pi) = \sum d_{\pi}(s) v_{\gamma}^{\pi}(s), \tag{5-5}$$

where $J_{\text{avg}}(\pi)$ is the equivalent cost, $v_{\gamma}^{\pi}(s)$ is the state-value function with a discount factor, and $d_{\pi}(s)$ is the stationary state distribution governed by policy π. Here, $d_{\pi}(s)$ is also the weighting function. The stationary state distribution varies with the learned policy, which poses severe difficulties in searching for an optimal policy. Note that the weighting choice in (5-5) differs from the commonly used discounted cost. In previous chapters, a fixed initial state distribution, which does not change with the policy, was used in the RL problem definition. As illustrated in Theorem 5-1, the benefit of such an equivalent cost (5-5) lies in its connection with the average cost for arbitrary policies.

- Theorem 5-1: For an arbitrary policy π, the following equality always holds:

$$J_{\text{avg}}(\pi) = \frac{1}{1-\gamma} G_{\text{avg}}(\pi).$$

Proof:

For brevity, we assume that $s \in \mathcal{S}$ and $a \in \mathcal{A}$ are finite element spaces. Hence, we have the following derivation:

$$J_{\text{avg}}(\pi) = \sum d_\pi(s) v_\gamma^\pi(s)$$

$$= \sum d_\pi(s) \sum_{a \in \mathcal{A}} \pi(a|s) \left\{ \sum_{s' \in \mathcal{S}} \mathcal{P}_{ss'}^a \left(r_{ss'}^a + \gamma v_\gamma^\pi(s') \right) \right\}$$

The first term is equal to the average return because all the rewards obey the stationary state distribution:

$$= G_{\text{avg}}(\pi) + \gamma \sum d_\pi(s) \sum_{a \in \mathcal{A}} \pi(a|s) \sum_{s' \in \mathcal{S}} \mathcal{P}_{ss'}^a v_\gamma^\pi(s')$$

The stationary state distribution will remain the same after one-step transition:

$$= G_{\text{avg}}(\pi) + \gamma \sum_{s' \in \mathcal{S}} d_\pi(s') v_\gamma^\pi(s')$$

$$= G_{\text{avg}}(\pi) + \gamma J_{\text{avg}}(\pi)$$

$$= G_{\text{avg}}(\pi) + \gamma G_{\text{avg}}(\pi) + \gamma^2 J_{\text{avg}}(\pi)$$

$$= G_{\text{avg}}(\pi) + \gamma G_{\text{avg}}(\pi) + \gamma^2 G_{\text{avg}}(\pi) + \cdots$$

$$= \frac{1}{1-\gamma} G_{\text{avg}}(\pi).$$

∎

Clearly, $J_{\text{avg}}(\pi)$ can be separated into two components that are multiplied: (1) a γ-related constant and (2) an average return. Furthermore, we derive an equivalence condition between the maximization of $J_{\text{avg}}(\pi)$ and that of $G_{\text{avg}}(\pi)$:

$$\max_\pi J_{\text{avg}}(\pi) \iff \max_\pi G_{\text{avg}}(\pi).$$

Therefore, their optimization must have exactly the same optimal policy:

$$\pi_{\text{avg}}^* = \arg\max_\pi J_{\text{avg}}(\pi) = \arg\max_\pi G_{\text{avg}}(\pi),$$

where π_{avg}^* is the optimal policy of both the average return and its equivalent cost. This kind of equivalence discloses an interesting property: even though $J_{\text{avg}}(\pi)$ has the discounting information, its optimal policy is essentially independent of the discount factor. Obviously, their corresponding stationary state distributions are also identical, which is denoted as the same notation $d_{\text{avg}}^*(s)$. Substituting π_{avg}^* and $d_{\text{avg}}^*(s)$ into (2-15) and (5-5), we have the following equality:

$$\frac{1}{1-\gamma} G_{\text{avg}}(\pi_{\text{avg}}^*) = J_{\text{avg}}(\pi_{\text{avg}}^*) = \sum d_{\text{avg}}^*(s) v_\gamma^{\pi_{\text{avg}}^*}(s). \tag{5-6}$$

This equality condition strengthens the fact that the two kinds of cost functions have completely identical optimal policy and stationary state distributions. If one cost function is difficult to optimize, the other cost function can replace it. This idea has been applied in some average-cost RL approaches, such as the vanishing discount factor technique.

5.1.2.2   Discounted Cost

A discounted cost is composed of two key elements: (1) an initial state distribution as the weighting coefficient and (2) a discounted return as a performance measure. Here, the initial state distribution must be fixed, which means that it is not allowed to change with the policy to be optimized. The discounted return is mathematically expressed in the form of expectation under a certain initial state distribution. As indicated in Chapter 2, such a selection can result in an optimal policy that is decoupled from the initial state distribution. Specifically, we can choose the optimal state distribution $d^*_{\text{avg}}(s)$ from the average return as the weighting function for fair comparison:

$$J_\gamma(\pi) = \sum d^*_{\text{avg}}(s)v^\pi_\gamma(s), \tag{5-7}$$

where $J_\gamma(\pi)$ is the discounted cost. Its corresponding optimal policy is obtained as the overall criterion reaches the maximum value:

$$\pi^*_\gamma = \arg\max_\pi J_\gamma(\pi),$$

where π^*_γ is the optimal policy from the discounted cost. One might wonder which of the following costs is higher in a given environment: the average cost under its corresponding optimal policy, i.e., $J_{\text{avg}}(\pi^*_{\text{avg}})$, or the discounted cost under its corresponding optimal policy, i.e., $J_\gamma(\pi^*_\gamma)$. One straightforward but incorrect idea is to think that J_{avg} is higher than J_γ because the former seems to have more freedom in its policy optimization. In fact, the correct result is just the opposite. This is because an average cost has an equivalent discounted cost, which is composed of a policy-varying weighting function and a state-value function. These two functions are strongly intertwined with each other, which reversely reduces the freedom of policy optimization. The solid proof is stated as follows.

- Theorem 5-2: The performances for the two optimal policies π^*_{avg} and π^*_γ have the following inequality:

$$J_{\text{avg}}(\pi^*_{\text{avg}}) \le J_\gamma(\pi^*_\gamma). \tag{5-8}$$

Proof:

The maximization of the discounted cost is independent of the weighting function if the weighting function does not change with the policy to be trained. This concept is equivalent to maximizing the state-value function $v^\pi_\gamma(s)$ at each individual state $s \in \mathcal{S}$. Hence, $\pi^*_\gamma = \arg\max v^\pi_\gamma(s)$ must be the best among all the feasible policies:

$$v^{\pi^*_{\text{avg}}}_\gamma(s) \le v^{\pi^*_\gamma}_\gamma(s), \text{for all } s \in \mathcal{S}.$$

Thus,

$$J_{\text{avg}}(\pi^*_{\text{avg}}) \quad = \sum d_{\pi^*_{\text{avg}}}(s)v^{\pi^*_{\text{avg}}}_\gamma(s)$$

$$\le \sum d_{\pi^*_{\text{avg}}}(s)v^{\pi^*_\gamma}_\gamma(s)$$

$$= J_\gamma(\pi_\gamma^*).$$

∎

This proof implies the fact that the maximization of an average cost is quite different from that of a discounted cost. To examine the difference between these two costs, we can cross the two cost functions and their optimal policies to construct four kinds of performance measures. The two overall objective functions are $G_{avg}(\cdot)$ and $J_\gamma(\cdot)$, and their optimal policies are π_{avg}^* and π_γ^*, respectively. Let us examine the two optimal policies under the measures of two cost functions. Obviously, they are the best under their own criteria:

$$G_{avg}(\pi_{avg}^*) \geq G_{avg}(\pi_\gamma^*),$$
$$J_\gamma(\pi_{avg}^*) \leq J_\gamma(\pi_\gamma^*).$$

Therefore, combining (5-6) and (5-8), we have

$$\frac{1}{1-\gamma} G_{avg}(\pi_\gamma^*) \leq \frac{1}{1-\gamma} G_{avg}(\pi_{avg}^*) = J_\gamma(\pi_{avg}^*) \leq J_\gamma(\pi_\gamma^*). \qquad (5\text{-}9)$$

In engineering, the average cost is often the most desirable performance measure. However, since its stationary state distribution $d_\pi(s)$ changes with the policy to be optimized, it is not easy to find its optimal solution. This is mainly because one cannot easily take the derivative of $d_\pi(s)$ with respect to the varying policy. This difficulty becomes the bottleneck of numerically optimizing general average cost problems. One popular alternative is to replace the average cost with the discounted cost, which uses a predefined and fixed initial state distribution, to eliminate the coupling issue of the stationary state distribution and the policy to be optimized.

5.1.3   Policy Iteration vs. Value Iteration

The core of DP is to numerically solve the Bellman equation with a known environment model. DP can be classified into two algorithmic methods: (1) policy iteration and (2) value iteration. They differ greatly in the mechanism that is used to ensure convergence. The policy iteration method repeats the following two steps: (1) calculate the value function given the current policy and (2) find a better new policy according to the current value function. These two steps are repeated until the optimum is reached. In this method, convergence is ensured by the mechanism of searching for a better policy. The value iteration method first calculates the optimal value function using the fixed-point iteration technique and then calculates the optimal policy with greedy search. In this method, convergence is ensured by the contractive property of the Bellman operator.

5.1.3.1   Concept of Policy Iteration

The policy iteration method consists of two cyclic steps: policy evaluation (PEV) and policy improvement (PIM). The idea of the policy iteration method is illustrated in Figure 5.1. Neither PEV nor PIM is optimal, but they can alternately roll forward and gradually shift to the optimum. During RL iteration, the next policy is enforced to be better than the current policy, which is critical to convergence guarantee. It has been proven that policy iteration relates to the mechanism of the Newton-Raphson iteration.

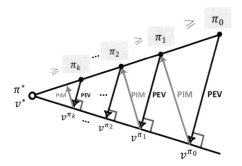

Figure 5.1 Concept of the policy iteration mechanism

- Policy evaluation (PEV):

The policy evaluation step exists in both model-free RL, such as MC and TD, and model-based RL, such as DP. In PEV, the goal is to find the true value function $v^\pi(s)$ for a given policy π. The value function serves as an evaluation index for whether the policy is good or bad. This task can be achieved by either sample-based estimation when the model is unknown (e.g., model-free PEV) or solving the self-consistency condition when the model is known (e.g., model-based PEV). Although seemingly different on the surface, PEVs in model-free RL and model-based RL essentially have the same goal, that is to find an accurate evaluation of the current policy and send it to PIM as the basis of searching for a better policy.

- Policy improvement (PIM):

The policy improvement step is used to calculate a better policy π' according to the true evaluation function $v^\pi(s)$. In other words, $\pi' \geq \pi$ is a mandatory requirement. The superiority of a better policy is critical to ensuring the convergence of policy iteration. In most RL methods, this type of new policy can be strictly obtained by greedy search, which satisfies the sufficient condition of better policy. This sufficient condition can also be described as the policy improvement theorem [5-8].

5.1.3.2   Concept of Value Iteration

The value iteration method takes the Bellman equation as an operator and uses the fixed-point iteration technique to calculate the optimal solution. Its concept is illustrated in Figure 5.2. In numerical analysis, fixed-point iteration is a method of iteratively computing the solution of an equation. The Bellman equation defines a contractive operator, which is a nonlinear mapping from one point to another in the Banach space. A unique fixed point exists if the Bellman operator satisfies the Banach fixed point theorem. The convergence of value iteration depends on the property that the Bellman operator is contractive.

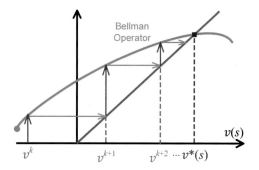

Figure 5.2 Concept of the value iteration mechanism

5.2   Policy Iteration Algorithm

Policy iteration is a two-step cyclic algorithm that alternates between policy evaluation (PEV) and policy improvement (PIM). Initially, an admissible policy must be carefully selected; otherwise, there is no way to build a stable policy iteration. Special initialization of value function is unnecessary, and normal zero-value initialization can work very well in many cases. It is generally believed that policy iteration requires fewer cycles to converge and possesses better stability than value iteration.

5.2.1   Policy Evaluation (PEV)

The self-consistency condition has no connection with optimality. This condition is an inherent property embedded inside the Markovian structure of MDPs and defines the relationship of three key elements, including value function, current policy, and environment model. Although this condition does not appear to be very complex, it plays a central role in performing policy evaluation. The self-consistency condition of the first kind is

$$v^\pi(s) = \sum_{a \in \mathcal{A}} \pi(a|s) \left\{ \sum_{s' \in \mathcal{S}} \mathcal{P}^a_{ss'} \left(r^a_{ss'} + \gamma v^\pi(s') \right) \right\}, \forall s \in \mathcal{S},$$

where $v^\pi(s)$ represents the true value under a given policy π. This equation holds for all $s \in \mathcal{S}$ because all the state elements must be considered when utilizing the expectation of the current action and the next state. In model-based RL, such as DP, PEV is used to iteratively solve the self-consistency condition, and its result is a natural measure of policy quality. In model-free RL, such as MC and TD, PEV carries out a similar task as in model-based RL with sample-based estimation. However, its result still obeys the self-consistency condition.

5.2.1.1   Iterative PEV Algorithm

In finite MDPs, the self-consistency condition forms a system of linear equations. Its analytical solution is not readily accessible due to high-dimensional state and action spaces. Instead, DPs use a simple iterative PEV algorithm to numerically calculate the fixed point of self-consistency condition:

Repeat j until infinity   (5-10)

$$V_{j+1}^{\pi}(s) \leftarrow \sum_{a \in \mathcal{A}} \pi(a|s) \left\{ \sum_{s' \in \mathcal{S}} \mathcal{P}\left(r + \gamma V_{j}^{\pi}(s')\right) \right\}, \forall s \in \mathcal{S}$$

End

In this iterative algorithm, $V_0^{\pi}(s)$, $V_1^{\pi}(s)$, $\cdots$, and $V_{\infty}^{\pi}(s)$ is the sequence of intermediate state values, and j denotes the inner step of the PEV iteration. The PEV iteration stops as j approaches infinity. Eventually, $V^{\pi}(s)$ converges to its true value, that is, $V_{\infty}^{\pi}(s) = v^{\pi}(s)$. In addition, the PEV iteration is needed to perform expected updates, i.e., sweeping the whole state space. This sweeping behavior requires equally dealing with every state element in the state space. This kind of operation is called the expected update because it is actually an expectation operation over all the possible next states [5-8]. Eventually, PEV outputs a high-quality evaluation of the current policy, which is also called state-value function.

5.2.1.2   Fixed-Point Explanation

In numerical analysis, fixed-point iterators are used to calculate approximate solutions of linear equations [5-7]. The mechanism of iterative PEV can be well explained by the fixed-point theory. Advanced PEV iterative algorithms may be built with other fixed-point schemes. The challenge in the construction of these algorithms is that (5-10) is not in the standard form of fixed-point iteration because its two sides have different variables, $V(s)$ and $V(s')$. A standard fixed-point iteration scheme must deal with the same variable defined in the Banach space, for example, the Euclidean space. Here, we take finite MDPs as an example to explain the essence of the iterative PEV algorithm. The state space in a finite MDP has a finite number of state elements $\mathcal{S} = \{s_{(1)}, s_{(2)}, \cdots, s_{(n)}\}$. When $\mathcal{P}_{ss'}^a$, $\pi(a|s)$ and $r_{ss'}^a$ are known, a system of linear equations can be derived:

$$V^{\pi}(s_{(1)}) = \sum \pi \sum \mathcal{P}_{s_{(1)}s'_{(1)}}^a \left(r + \gamma V^{\pi}(s'_{(1)})\right),$$

$$V^{\pi}(s_{(2)}) = \sum \pi \sum \mathcal{P}_{s_{(2)}s'_{(2)}}^a \left(r + \gamma V^{\pi}(s'_{(2)})\right),$$

$$\cdots \cdots$$

$$V^{\pi}(s_{(n)}) = \sum \pi \sum \mathcal{P}_{s_{(n)}s'_{(n)}}^a \left(r + \gamma V^{\pi}(s'_{(n)})\right),$$

where $V^{\pi}(s_{(i)})$ represents the variables to be solved, $s_{(1)}, s_{(2)}, \cdots, s_{(n)}$ are the state elements in $\mathcal{S}$, and $s'_{(1)}, s'_{(2)}, \cdots, s'_{(n)}$ are their corresponding next states. Obviously, all the next states are always in the state space $\mathcal{S}$ after each one-step transition. Since s' is still a state element in $\mathcal{S}$, we can rearrange the order of $V^{\pi}(s'_{(j)}), j = 1, \cdots, n$ to be consistent with that of $V^{\pi}(s_{(i)}), i = 1, \cdots, n$. Therefore, one can obtain a compact form of the linear equations above:

$$X = \gamma BX + b,$$

$$X = \left[V^{\pi}(s_{(1)}), V^{\pi}(s_{(2)}), \cdots, V^{\pi}(s_{(n)})\right]^{\mathrm{T}} \in \mathbb{R}^n,$$

$$B = \{B_{ij}\}_{n \times n} \in \mathbb{R}^{n \times n},$$

$$b = \{b_i\}_{n \times 1} \in \mathbb{R}^n,$$

(5-11)

$$B_{ij} = \sum_{a \in \mathcal{A}} \pi\big(a|s_{(i)}\big)p\big(s_{(j)}|s_{(i)}, a\big),$$

$$b_i = \sum_{a \in \mathcal{A}} \pi\big(a|s_{(i)}\big) \sum_j p\big(s_{(j)}|s_{(i)}, a\big)r^a_{s_{(i)}s_{(j)}}.$$

Figure 5.3 illustrates the compact form of the self-consistency condition in finite MDPs. We can easily see that (5-11) becomes a standard form that suits the theory of fixed-point iteration, as its two sides have the same variable in the Euclidean space $\mathbb{R}^n$, and its corresponding operator is a self-mapping from $\mathbb{R}^n$ to $\mathbb{R}^n$.

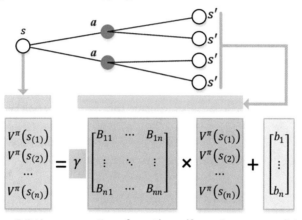

Figure 5.3 Linear equations from the self-consistency condition

An iterative PEV algorithm such as (5-10) is a multivariable fixed-point iteration, which must synchronously update all state elements. This is the reason we also refer to (5-10) as the full-width update (its original name was the expected update). At each iteration, one must sweep the whole state space to update the state-value function. The iteration stops when the state-values of every state element are sufficiently close to their true solutions. With this sweeping behavior, the convergence mechanism of the Banach fixed-point theorem can be explained very well. In addition to convergence examination, one can also check whether (5-11) has a unique fixed point. First, define a new matrix:

$$A \overset{\text{def}}{=} I_{n \times n} - \gamma B. \tag{5-12}$$

It is known that a unique solution exists if $A \in \mathbb{R}^{n \times n}$ is fully ranked. If $\gamma = 1$, we know that A is not always fully ranked. A counterexample is $A = 0$ when B is an identity matrix. If $\gamma \in (0,1)$, i.e., a discounted cost is adopted, A is invertible, and its mathematical proof is shown as follows. The invertibility of A determines the solution uniqueness of the self-consistency condition.

- Theorem 5-3: If $0 < \gamma < 1$, $\text{rank}(A) = n$.

Proof:

Recall the fact that $0 \le B_{ij} \le 1$ and that the sum of each row is equal to 1:

$$\sum_{j=1}^{n} B_{ij} = 1.$$

We know that $\|B\|_{\infty} = \max_i |\sum_{j=1}^n B_{ij}| = 1$, and then,

$$\|\gamma B\|_{\infty} = \gamma \|B\|_{\infty} = \gamma < 1.$$

The spectral radius of γB has the following inequality:

$$\rho(\gamma B) \leq \|\gamma B\|_{\infty} < 1,$$

where $\rho(\cdot)$ is the spectral radius. The spectral radius is the largest absolute value of all the eigenvalues. In an identity matrix, all eigenvalues equal one. Thus all eigenvalues of $A = I - \gamma B$ are positive and A is invertible.

∎

5.2.2   Policy Improvement (PIM)

To ensure the convergence of policy iteration, a better policy must be found according to the previous policy evaluation. Each better update can gradually shift the intermediate policy to the optimum. The term "better" must be defined with respect to a specific measure; that is, we need a quantitative definition about what a better policy is. A popular measure of "better" is the state-value function. A policy π' is said to be better than another policy π if $v^{\pi}(s) \leq v^{\pi'}(s), \forall s \in \mathcal{S}$. Clearly, this definition is an element-by-element condition that compares the values of every state element. With this definition, one classic choice in PIM is to compute greedy actions:

$$\pi'(a|s) = \arg\max_{\pi'} \left\{ \sum_{a \in \mathcal{A}} \pi'(a|s) q^{\pi}(s, a) \right\}, \forall s \in \mathcal{S},$$

(5-13)

or

$$\pi'(s) = \arg\max_{\pi'} q^{\pi}(s, \pi'(s)), \forall s \in \mathcal{S}.$$

The first equation is appropriate for a stochastic policy, and the second equation is appropriate for a deterministic policy. The rationality of this choice has been proven by Sutton & Barto (2018) [5-8]. A greedy search is actually a natural extension of the policy improvement theorem. It is easy to prove that any policy that maximizes the right side of the following policy improvement condition is a better policy:

$$v^{\pi}(s) \leq \sum_{a \in \mathcal{A}} \pi'(a|s) q^{\pi}(s, a), \forall s \in \mathcal{S},$$

(5-14)

or

$$v^{\pi}(s) \leq q^{\pi}(s, \pi'(s)), \forall s \in \mathcal{S}.$$

This policy improvement theorem provides a sufficient condition for a better policy. In DP, (5-13) cannot be used directly since its action value is unknown. In this case, $q^{\pi}(s, a)$ can be expanded with a model $\mathcal{P}$ and a state-value function with the next state $v^{\pi}(s')$. Therefore, PIM maximizes the following criterion:

(a) For stochastic policy

$$\pi'(a|s) \leftarrow \arg\max_{\pi'} \left\{ \sum_{a \in \mathcal{A}} \pi'(a|s) \sum_{s' \in \mathcal{S}} \mathcal{P}(r + \gamma V_{\infty}^{\pi}(s')) \right\},$$

(5-15)

(b) For deterministic policy

$$\pi'(s) \leftarrow \arg\max_{\pi'} \left\{ \sum_{s' \in \mathcal{S}} \mathcal{P}\left(r + \gamma V_\infty^\pi(f(s, \pi'(s), \xi))\right) \right\}.$$

Here, $v^\pi(\cdot)$ is replaced with its estimate $V^\pi(\cdot)$ for practical computation. In theory, a greedy search with a tabular representation always yields a deterministic policy. The first two equations in (5-13) and (5-15) actually do not complement this analysis well, and their existence must connect with stochastic policies, in which a parameterized policy is often used with gradient-based optimization. With a parameterized policy, each update does not output a perfect greedy policy, and the stochastic property is always maintained.

5.2.3   Proof of Convergence

The overall convergence of policy iteration is dominated by two key conditions: (a) the self-consistency condition in each PEV is solvable and (b) each PIM outputs a new policy that is better than its previous step. The former condition requires that for an admissible policy, its corresponding state value is always finite. The latter condition means that the intermediate policy can improve and eventually converge to the solution of the Bellman equation. From the control perspective, the finiteness of state-value function is equivalent to the property that its associated closed-loop system never diverges. More concisely, each intermediate policy is feasible (or simply called "stable") and can successfully interact with the environment. It is suggested from this statement that policy iteration needs the property of recursive stability. The stability of every policy depends on that of its previous policy, and we must choose an initially admissible policy to start the training process.

5.2.3.1   Convergence of Iterative PEV

Define a linear operator $\mathcal{L}$ that equals the self-consistency condition:

$$\mathcal{L}(X) = \gamma BX + b, \tag{5-16}$$

where $B \in \mathbb{R}^{m \times m}$ is the iteration matrix, $X \in \mathbb{R}^m$ is the unknown variable, and γ is the discount factor. This operator has the property of γ-contraction, which makes any pair of adjacent value functions increasingly closer to each other. The error of adjacent value functions is at least bounded by the following inequality:

$$\left\| \mathcal{L}(X_{j+1}) - \mathcal{L}(X_j) \right\|_\infty = \gamma \left\| B(X_{j+1} - X_j) \right\|_\infty.$$

Because all elements in B are nonnegative:

$$\leq \gamma \left\| B \max_{i \in \{1,2,\cdots,n\}} (|X_{j+1}^i - X_j^i|) \right\|_\infty$$

$$= \gamma \left\| B \|X_{j+1} - X_j\|_\infty \right\|_\infty.$$

Since $\|X_{j+1} - X_j\|_\infty$ is a scalar,

$$= \gamma \|B\|_\infty \|X_{j+1} - X_j\|_\infty.$$

Because $\|B\|_\infty = \max_i \sum_{j=1}^n B_{i,j} = 1$:

$$= \gamma \|X_{j+1} - X_j\|_\infty.$$

This is a typical contraction mapping. The iteration procedure of $\mathcal{L}(X)$ converges to a unique fixed point at the linear convergence rate of γ. Note that the linear operator not only handles a single element but also needs to sweep the whole value space. As an analogy to DP, a model-free RL agent should traverse the whole environment dynamics to collect sufficient data for every state element. Clearly, this requirement is not easily satisfied, and limited local exploration might cause severe instability in MC or TD learning algorithms.

- Theorem 5-4: Using (5-10), $V_j^\pi(s)$ converges to $v^\pi(s)$ as the PEV iteration goes to infinity.

Proof:

The linear operator has a unique fixed point, namely, $v^\pi(s)$. By the contraction mapping theorem, $V_j^\pi(s)$ should converge to the true value $v^\pi(s)$.

∎

Theorem 5-4 only means that $V_j^\pi(s)$ converges to a certain fixed value. The finite-valued property of PEV is a necessary condition of DP convergence. Following the cyclic alternation between PEV and PIM, a DP policy iteration algorithm will eventually converge to the true value $v^*(s)$ when PIM finally stops improvement.

5.2.3.2   Convergence of DP Policy Iteration

The cyclic flow of policy iteration is demonstrated in Figure 5.4. The core of the convergence proof lies in two important steps: (1) $\{V_\infty^{\pi_k}(s)\}$ is monotonically increasing, and (2) $\{\pi_k\}$ eventually converges to the optimum π^*. The former step is called the monotonically increasing property, which is mainly determined by selecting a new policy at each cycle. The latter step will hold when PIM stops, which indicates that no better policy can be found.

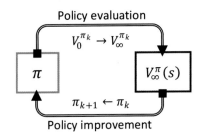

Figure 5.4 Iteration cycle of PEV and PIM

- Theorem 5-5: In DP policy iteration, $\{V_\infty^{\pi_k}(s)\}, k = 0,1,2,\cdots,\infty$, are monotonically increasing. More clearly,

$$V_\infty^{\pi_0}(s) \le V_\infty^{\pi_1}(s) \le V_\infty^{\pi_2}(s) \le \cdots \le V_\infty^{\pi_k}(s) \le V_\infty^{\pi_{k+1}}(s) \le \cdots \le v^*.$$

Proof:

In each PIM, a better policy π_{k+1} is found; i.e., $\pi_k \le \pi_{k+1}$, and we have

$$v^{\pi_k}(s) \le v^{\pi_{k+1}}(s), \text{ for } \forall s \in \mathcal{S}.$$

From Theorem 5-4, we have

$$V_\infty^{\pi_k}(s) = v^{\pi_k}(s).$$

Further, replace v^{π_k} and $v^{\pi_{k+1}}$ with $V_\infty^{\pi_k}$ and $V_\infty^{\pi_{k+1}}$, and it is easy to know the following inequality:

$$V_\infty^{\pi_k}(s) \le V_\infty^{\pi_{k+1}}(s), \forall s \in \mathcal{S}.$$

■

The bounded property of $\{V_\infty^{\pi_k}(s)\}$ is achieved by the bounded property of reward signals. One must distinguish the following two notations: the first notation is π_∞, which is generated from the DP iteration, and the second notation is π^*, which comes from the Bellman equation. Even though the sequence $\{\pi_0, \pi_1, \cdots, \pi_k, \cdots, \pi_\infty\}$ can converge, it does not necessarily converge to an optimal policy. We need to prove that $\pi_\infty = \pi^*$ when PIM stops improvement.

- Theorem 5-6: When PIM stops improvement, i.e., $\pi_\infty = \pi^*$ if π^* is unique.

Proof:

Since the value sequence is monotonically increasing and bounded, PIM will finally reach the stationary status. When PIM stops the improvement process, $\pi_\infty = \pi_{\infty+1}$. Therefore,

$$v^{\pi_\infty}(s) = v^{\pi_{\infty+1}}(s) = \max_a \sum_{s' \in \mathcal{S}} P(r + \gamma v^{\pi_\infty}(s')), \forall s \in \mathcal{S}, \tag{5-17}$$

and the Bellman equation is satisfied. If π^* is the unique solution of the Bellman equation, $v^{\pi_\infty}(s) = v^{\pi_{\infty+1}}(s) = v^*(s)$ for all $s \in \mathcal{S}$.

■

5.2.4 Explanation with Newton-Raphson Mechanism

One might be interested in the intrinsic mechanism behind policy iteration. In theory, a policy iteration algorithm can be viewed as a special variant of the Newton-Raphson iteration. In 1979, M. Puterman and S. Brumelle proved their mathematical equivalence in stationary problems with discounted costs [5-6]. The case for average cost was later proven by M. Ohnishi in 1992 with a fractional programming approach [5-4]. This equivalence proof allows us to obtain the convergence rate and error bounds for policy iteration. This session provides a simplified explanation with a finite MDP and a deterministic policy. The classic Newton-Raphson iteration solves a nonlinear equation $g(X) = 0$ by using

$$X_{k+1} = X_k - [\nabla g(X_k)]^{-1} g(X_k),$$

where $\nabla g(X_k)$ is the Jacobian matrix. Take the finite MDP with countable states and actions as an example. The state space of the finite MDP is assumed to have only two elements: $\mathcal{S} = \{s_{(1)}, s_{(2)}\}$, and its action space is assumed to have two elements: $\mathcal{A} = \{a_{(1)}, a_{(2)}\}$. The two-state and two-action Bellman equation is

$$V^\pi(s_{(1)}) = \max_a \sum_{s'} P^a_{s_{(1)}s'}(r + \gamma V^\pi(s')), \tag{5-18}$$

$$V^\pi\left(s_{(2)}\right) = \max_a \sum_{s'} \mathcal{P}^a_{s_{(2)}s'}\left(r + \gamma V^\pi(s')\right),$$

$$a \in \left\{a_{(1)}, a_{(2)}\right\}.$$

Such an equation can be abstracted into the Bellman operator $X = \mathcal{B}(X)$, in which $X^{\mathrm{T}} \overset{\text{def}}{=} \left[V^\pi\left(s_{(1)}\right), V^\pi\left(s_{(2)}\right)\right]$ is the variable to be solved. The Bellman operator is a nonlinear mapping function due to the existence of maximum operation. Let us define a new function $g(X): \mathbb{R}^2 \to \mathbb{R}^2$ as

$$g(X) = \mathcal{B}(X) - X,$$
$$g_i(X) = \max_j \left(\gamma \beta_{ij}^{\mathrm{T}} X + \lambda_{ij}\right) - X_i , i,j \in \{1,2\}, \tag{5-19}$$

where β_{ij}, λ_{ij} are the corresponding coefficients when taking action $a_{(j)}$ at state $s_{(i)}$:

$$\beta_{ij} = \left[\mathcal{P}^{a_{(j)}}_{s_{(i)}s_{(1)}}, \mathcal{P}^{a_{(j)}}_{s_{(i)}s_{(2)}}\right]^{\mathrm{T}}, \lambda_{ij} = \sum_{s'} \mathcal{P}^{a_{(j)}}_{s_{(i)}s'} r^{a_{(j)}}_{s_{(i)}s'}.$$

The i-th component of $g(X)$ can be written as a piecewise linear function:

$$g_i(X) = \begin{cases} \gamma \beta_{i1}^{\mathrm{T}} X + \lambda_{i1} - X_i, & \text{for } X \in \mathbb{I}_{i1} \\ \gamma \beta_{i2}^{\mathrm{T}} X + \lambda_{i2} - X_i, & \text{for } X \in \mathbb{I}_{i2} \end{cases} \tag{5-20}$$

where $\mathbb{I}_{i1}$ and $\mathbb{I}_{i2}$ are its two subintervals. Figure 5.5 demonstrates the first component of $g(X)$. The maximization in (5-19) yields a nonlinear mapping from $\mathbb{R}^2$ to $\mathbb{R}^2$, which contains two piecewise linear components, and each component has two subintervals.

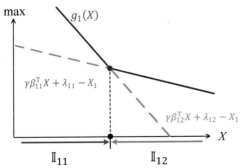

Figure 5.5 Piecewise linear function $g_1(X)$

Therefore, $g(X)$ can also be written as a multistage piecewise linear function:

$$g(X) = \gamma \beta_\sigma X + \lambda_\sigma - X, \text{for } X \in \mathbb{U}_\sigma, \sigma \in \{1,2,3,4\}$$

$$\beta_\sigma = \begin{bmatrix} \mathcal{P}^{a^*}_{s_{(1)}s_{(1)}} & \mathcal{P}^{a^*}_{s_{(1)}s_{(2)}} \\ \mathcal{P}^{a^*}_{s_{(2)}s_{(1)}} & \mathcal{P}^{a^*}_{s_{(2)}s_{(2)}} \end{bmatrix}, \lambda_\sigma = \begin{bmatrix} \sum_{s'} \mathcal{P}^{a^*}_{s_{(1)}s'} r^{a^*}_{s_{(1)}s'} \\ \sum_{s'} \mathcal{P}^{a^*}_{s_{(2)}s'} r^{a^*}_{s_{(2)}s'} \end{bmatrix}, \tag{5-21}$$

where $a = \pi^*(s)$ is the optimal action, and $\mathbb{U}_\sigma$ is the subregion. Figure 5.6 shows the relationship among four subregions $\mathbb{U}_\sigma, \sigma \in \{1,2,3,4\}$ and four subintervals $\mathbb{I}_{11}, \mathbb{I}_{12}, \mathbb{I}_{21}, \mathbb{I}_{22}$. In this figure, the dashed line is the boundary of adjacent subintervals, and the two dashed lines in $g_1(X)$ and $g_2(X)$ divide the overall region into four subregions. At each subregion, the coefficients β_σ and λ_σ are fixed. Each subregion has a separated linear function, whose optimal action remains unchanged. Figure 5.7 illustrates the multistage piecewise linear function along the section line in Figure 5.6. Clearly, $g_1(X)$ and $g_2(X)$ are two piecewise linear functions, and their intersection with the X-axis is the fixed point of Bellman equation.

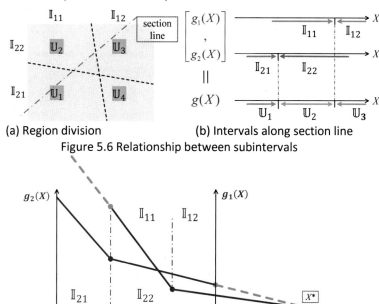

(a) Region division   (b) Intervals along section line

Figure 5.6 Relationship between subintervals

Figure 5.7 Functions along the section line

It is easy to see that $g(X)$ is differentiable except for the inflection points between two adjacent subintervals. For these inflection points, one can use either the right- or left-side derivative to build a substitute for the Jacobian matrix. Fortunately, this kind of substitution does not change the convergence property of the Newton-Raphson iteration [5-5]. The Jacobian matrix becomes $\nabla g(X) = \gamma \beta_\sigma - I$, for $X \in \mathbb{U}_\sigma, \sigma = 1,2,3,4$. The Newton iteration of solving the Bellman equation is written as

$$
\begin{aligned}
X_{k+1} &= X_k - [\nabla g(X)]^{-1} g(X_k) \\
&= X_k - (\gamma \beta_\sigma - I)^{-1}(\gamma \beta_\sigma X_k + \lambda_\sigma - X_k) \\
&= -(\gamma \beta_\sigma - I)^{-1} \lambda_\sigma .
\end{aligned}
\tag{5-22}
$$

Thus far, we have derived an update formula from the Newton-Raphson iteration. This is one side of mathematical proof. Next, we need to prove the other side of the mathematical proof. That is, the DP policy iteration has an identical update formula. A policy iteration algorithm always has two cyclic steps, i.e., policy evaluation (PEV) and policy improvement (PIM). Given the current state value X_k at the k-th iteration, PIM seeks to calculate a greedy policy:

$$
\pi_{k+1} = \arg \max_{\pi} \{\gamma B_\pi X_k + b_\pi\}.
\tag{5-23}
$$

According to the definition in (5-11), we have

$$B_{\pi(i,j)} = \sum_{a \in \mathcal{A}} \pi(a|s_{(i)}) \mathcal{P}(s_{(j)}|s_{(i)}, a),$$

$$b_{\pi(i)} = \sum_{a \in \mathcal{A}} \pi(a|s_{(i)}) \sum_{j} \mathcal{P}(s_{(j)}|s_{(i)}, a) r^a_{s(i)s(j)}.$$

(5-24)

Since PEV is performed immediately after PIM at the previous step, the next state X_{k+1} satisfies the self-consistency condition under the greedy policy π_{k+1}:

$$X_{k+1} = \gamma B_{\pi_{k+1}} X_{k+1} + b_{\pi_{k+1}},$$

(5-25)

and its solution is

$$X_{k+1} = \left(I - \gamma B_{\pi_{k+1}}\right)^{-1} b_{\pi_{k+1}}.$$

(5-26)

When adopting a deterministic policy, $\pi(a|s_{(i)}) = 1$ only for $a = a^*$ and $\pi(a|s_{(i)}) = 0$ for any other actions. By rewriting $B_{\pi_{k+1}}, b_{\pi_{k+1}}$ as the coefficients with greedy actions, we have

$$B^*_{ij} = \sum_{a \in \mathcal{A}} \pi^*(a|s_{(i)}) \mathcal{P}(s_{(j)}|s_{(i)}, a) = \mathcal{P}^{a^*}_{s(i)s(j)},$$

$$b^*_i = \sum_{a \in \mathcal{A}} \pi^*(a|s_{(i)}) \sum_{j} \mathcal{P}(s_{(j)}|s_{(i)}, a) r = \sum_{s' \in \mathcal{S}} \mathcal{P}^{a^*}_{s(i)s'} r^{a^*}_{s(i)s'}.$$

(5-27)

Comparing (5-27) with (5-21), one can easily find the following equivalence:

$$B_{\pi_{k+1}} = \beta_\sigma, b_{\pi_{k+1}} = \lambda_\sigma.$$

(5-28)

Obviously, (5-25) is a linear equation, and its solution exactly becomes (5-22). Therefore, the cycle of policy iteration is equal to one step of the Newton-Raphson iteration. Now, we can conclude that policy iteration is a special case of the Newton-Raphson iteration for solving the Bellman equation. Even though the analysis above is limited to the two-element state and two-element action, its result can be extended to general policy iteration algorithms. The Newton-Raphson method is known for having a stable quadratic convergence rate for well-conditioned problems. This also explains why the policy iteration usually requires fewer cycles to converge compared to the value iteration and moreover it has highly robust behavior.

5.3   Value Iteration Algorithm

One disadvantage of policy iteration is that its cycle involves an infinite-step policy evaluation, which requires lengthy computation due to repeated sweeps across the state space. Value iteration explicitly solves the Bellman equation by a fixed-point iterator, providing an alternative way to calculate the optimal value function. Moreover, when using value iteration, any intermediate policy is not explicitly generated, which is one central difference from policy iteration. The benefit of not having an intermediate policy lies in the elimination of specifying an initially feasible policy. Another difference is that convergence for value iteration is ensured by the contraction property of the Bellman operator rather than the process of "always finding a better policy".

5.3.1   Explanation with Fixed-point Iteration Mechanism

The Bellman operator is a nonlinear function that is used to map from one point to another within the vector space of state values. Viewing the Bellman equation as a contractive operator is useful when building new DP algorithms. The properties of a value iteration algorithm can be easily analyzed with existing theorems of contractive operator. The Bellman equation of the first kind is

$$v^*(s) = \max_a \sum_{s' \in \mathcal{S}} \mathcal{P}_{ss'}^a \left(r_{ss'}^a + \gamma v^*(s') \right).$$

Its associated Bellman operator satisfies the contraction mapping theorem. The simplest value iteration algorithm can be derived by the Picard fixed-point iteration. This scheme simply turns the Bellman equation into an update formula:

Repeat k until infinity

$$V_{k+1}(s) \leftarrow \max_a \sum_{s' \in \mathcal{S}} \mathcal{P}_{ss'}^a \left(r_{ss'}^a + \gamma V_k(s') \right), \forall s \in \mathcal{S}. \tag{5-29}$$

End

Here, the value function $V(s)$ does not have the superscript π because its value iteration does not correspond to any intermediate policy. Similar to PEV in policy evaluation, value iteration requires an infinite number of iterations to converge accurately. Under the fixed-point theory, we actually have two kinds of viewpoints to explain this formula. From the perspective of continuous state space, one can directly select $V(s)$ as a functional variable, which belongs to a special Banach space, i.e., $V(s) \in \mathbb{C}(\mathcal{S})$. Since $V(s') \in \mathbb{C}(\mathcal{S})$, the Bellman operator is self-mapped from $\mathbb{C}(\mathcal{S})$ to $\mathbb{C}(\mathcal{S})$; thus, its fixed-point iteration converges. From the perspective of discrete state space, a more straightforward understanding is to construct a vector variable with finite elements in the Euclidean space $\mathbb{R}^n$:

$$X = V(s) = \left[V\left(s_{(1)}\right), V\left(s_{(2)}\right), \cdots, V\left(s_{(n)}\right) \right]^{\mathrm{T}} \in \mathbb{R}^n.$$

In this situation, the Bellman operator is reduced to a self-mapping from $\mathbb{R}^n$ to $\mathbb{R}^n$. By gridding the state space, we have a system of nonlinear equations such as (5-18), whose fixed point is the solution of the Bellman equation. Such an equation can be abstracted into a nonlinear Bellman operator. The existence and uniqueness of its fixed point depend on whether the Bellman operator is contractive.

5.3.2   Convergence of DP Value Iteration

The nonlinearity of Bellman operator poses a few difficulties in convergence proof. For discounted problems, it has been proven that the Bellman operator is contractive if the discount factor is less than one. First, define a Bellman operator $\mathcal{B}$:

$$\mathcal{B}\left(V(s)\right) \overset{\text{def}}{=} \max_a \sum_{s' \in \mathcal{S}} \mathcal{P}_{ss'}^a \left(r_{ss'}^a + \gamma V(s') \right). \tag{5-30}$$

This operator has a property called the γ-contraction [5-2]. For an arbitrary state $s_{(i)} \in \mathcal{S}$, we have the following inequality:

$$\left| \mathcal{B}\left(V_{k+1}\left(s_{(i)}\right)\right) - \mathcal{B}\left(V_k\left(s_{(i)}\right)\right) \right|$$

$$= \left| \max_a \sum_{s' \in \mathcal{S}} \mathcal{P}\left(r + \gamma V_{k+1}(s')\right) - \max_a \sum_{s' \in \mathcal{S}} \mathcal{P}\left(r + \gamma V_k(s')\right) \right|$$

$$\leq \max_a \left| \sum_{s' \in \mathcal{S}} \mathcal{P}\left(r + \gamma V_{k+1}(s')\right) - \sum_{s' \in \mathcal{S}} \mathcal{P}\left(r + \gamma V_k(s')\right) \right|$$

$$= \gamma \max_a \left| \sum_{s' \in \mathcal{S}} \mathcal{P}V_{k+1}(s') - \sum_{s' \in \mathcal{S}} \mathcal{P}V_k(s') \right|$$

$$\leq \gamma \max_a \sum_{s' \in \mathcal{S}} \mathcal{P}\left|V_{k+1}(s') - V_k(s')\right|$$

$$\leq \gamma \max_a \sum_{s' \in \mathcal{S}} \mathcal{P}\max_{s \in \mathcal{S}}\left|V_{k+1}(s) - V_k(s)\right|$$

$$= \gamma \max_a \left\{ 1 \times \max_{s \in \mathcal{S}}\left|V_{k+1}(s) - V_k(s)\right| \right\}$$

$$= \gamma \max_{s \in \mathcal{S}}\left|V_{k+1}(s) - V_k(s)\right|$$

$$= \gamma \|V_{k+1}(s) - V_k(s)\|_\infty .$$

Considering all the state elements, and taking the infinity norm of two sides, we have

$$\left\| \mathcal{B}\left(V_{k+1}(s)\right) - \mathcal{B}\left(V_k(s)\right) \right\|_\infty = \max_s \left| \mathcal{B}\left(V_{k+1}(s)\right) - \mathcal{B}\left(V_k(s)\right) \right|$$

$$\leq \gamma \max_s \left|V_{k+1}(s) - V_k(s)\right|$$

$$= \gamma \|V_{k+1}(s) - V_k(s)\|_\infty .$$

Based on the Banach contraction theorem, $\mathcal{B}(V(s))$ converges to a unique fixed point, at least in the linear convergence rate of the discount factor. Since the discounted Bellman equation has only one unique solution, this unique fixed point unquestionably becomes the optimal solution.

5.3.3   Value Iteration for Problems with Average Costs

Unlike a discounted cost problem, various technical challenges arise when solving an average cost problem. The average cost problem often takes the average return, which is independent of the initial state, as the scalar performance measure. One challenge of maximizing the average return is that its optimum may not even exist in general nonlinear environments. Fortunately, previous studies have shown that if a finite MDP is unichain, its optimum exists. Nonetheless, searching for the optimal policy through conventional Bellman equation is still a very challenging task. Instead of dealing with average return, we define a differential return as the sum of errors between reward signals and their average return:

$$G_\Delta(s_t) = \sum_{i=0}^{\infty} \left(r_{t+i} - G_{\text{avg}}(\pi) \right),\tag{5-31}$$

where $G_\Delta(s_t)$ is the differential return and $G_{\text{avg}}(\pi)$ is the average return. It is easy to see that the differential return can be used to define a new state-value function:

$$v_\Delta^\pi(s) = \mathbb{E}_\pi\{G_\Delta(s_t)|s_t = s\} = \mathbb{E}_\pi\left\{\sum_{i=0}^{\infty}\left(r_{t+i} - G_{\text{avg}}(\pi)\right)\right\},$$

where $v_\Delta^\pi(s)$ is called the differential state-value function. Its corresponding differential Bellman equation is

$$v_\Delta^*(s) = \max_{a \in \mathcal{A}} \sum_{s' \in \mathcal{S}} \mathcal{P}_{ss'}^a \left(r_{ss'}^a - G_{\text{avg}}^*(\pi) + v_\Delta^*(s') \right).\tag{5-32}$$

Because $G_{\text{avg}}^*(\pi)$ is independent of the state variable, we have another popular equation:

$$v_\Delta^*(s) + G_{\text{avg}}^*(\pi) = \max_{a \in \mathcal{A}} \sum_{s' \in \mathcal{S}} \mathcal{P}_{ss'}^a \left(r_{ss'}^a + v_\Delta^*(s') \right).$$

In the literature, this equation is often referred to as the average cost optimality equation (ACOE) [5-8]. The stationary policy is an appealing choice for the average cost. A policy is said to be stationary if the action it chooses depends only on the current state and has no connection with the historical information. A general average cost problem may need a few regularity conditions to ensure the existence of optimal policy. In particular, the validity is tied to certain recurrence or ergodicity properties with some compactness and continuity conditions.

5.3.3.1   Finite-Horizon Value Iteration

There are several approaches that can be used to solve the differential Bellman equation, such as finite-horizon value iteration, relative value iteration, and vanishing discount factor approach. The basic idea of finite-horizon value iteration is to simply reduce the infinite horizon to a finite horizon [5-2]. Selecting actions that optimize the finite-horizon cost becomes very close to that of the average cost as the number of horizon stages becomes large enough. However, this may cause the overall costs to increase quickly toward an infinitely large value. Only a finite number of iterations can be executed in practical algorithms. Consider a multistage accumulated return:

$$V(s, N) = \sum_{t=0}^{N-1} r_t,\tag{5-33}$$

where $V(s, N)$ is the multistage value function, and N is the horizon length of accumulated return. Clearly, such an average cost problem with finite horizon satisfies a time-dependent Bellman equation. This Bellman equation can be calculated in a step-by-step manner due to its backward recursion structure. After finding its numerical solution, the optimal cost can be approximated by the accumulated return divided by the horizon length. That is,

$$G(\pi^*) \approx \frac{V^*(s,N)}{N}.$$

Often, a horizon length that is long enough is selected to obtain an accurate value approximation. In theory, $V^*(s,N)/N$ will converge to the optimal average cost if the stage length goes to infinity. However, a too long horizon length may have very high computational complexity and thus reduce the usability of the finite-horizon approximation.

5.3.3.2   Relative Value Iteration

The aforementioned value iteration technique is simple and straightforward. In some cases, a few components in the multistage value function $V(s,N)$ may easily diverge to infinity, and the calculation of the limit of $V(s,N)/N$ becomes numerically impractical. An improved version of value iteration is called relative value iteration and is designed to subtract a fixed constant from all the components of $V(s,t)$. Choosing a fixed state ζ as the baseline, the value difference between an arbitrary state and this fixed state is

$$h(s) = V(s,N) - V(\zeta,N), N \to \infty,$$

where $h(s)$ is called the difference value function and s is an arbitrary state. Note that ζ must be selected to be a fixed state, which does not change with the iteration cycle. After applying the fixed-point iteration to the difference value function, we have the following update formula:

$$
\begin{aligned}
& h_{k+1}(s) \\
&= \max_a \sum \mathcal{P}_{ss'}^a \left(r_{ss'}^a + V_k(s',N) \right) - \max_a \sum \mathcal{P}_{\zeta\zeta'}^a \left(r_{\zeta\zeta'}^a + V_k(\zeta',N) \right) \\
&= \max_a \sum \mathcal{P}_{ss'}^a \left(r_{ss'}^a + h_k(s') + V_k(\zeta,N) \right) - \max_a \sum \mathcal{P}_{\zeta\zeta'}^a \left(r_{\zeta\zeta'}^a + h_k(\zeta') + V_k(\zeta,N) \right) \\
&= \max_a \sum_{s' \in \mathcal{S}} \mathcal{P}_{ss'}^a \left(r_{ss'}^a + h_k(s') \right) - \max_a \sum \mathcal{P}_{\zeta\zeta'}^a \left(r_{\zeta\zeta'}^a + h_k(\zeta') \right).
\end{aligned}
$$

If the relative value iteration finally converges, i.e., $h^*(s) = \lim_{k \to \infty} h_k(s)$, we have a new equality condition that replaces the difference value function:

$$h^*(s) = -\rho^* + \max_a \sum_{s' \in \mathcal{S}} \mathcal{P}_{ss'}^a \left(r_{ss'}^a + h^*(s') \right),$$

$$\rho^* \stackrel{\text{def}}{=} \max_a \sum \mathcal{P}_{\zeta\zeta'}^a \left(r_{\zeta\zeta'}^a + h^*(\zeta') \right),$$

(5-34)

where ρ^* must be a constant because ζ is a fixed state. If one can initialize ρ and $h(s)$, this equality condition naturally builds a fixed-point iteration algorithm. The optimal pair $[\rho^*, h^*(s)]$ is the solution of the differential Bellman equation (5-32). Interestingly, if looking deeply into (5-32) and (5-34), $h^*(s)$ is a function of only the state variable, and moreover, we have the following relation $h^*(s) = v_\Delta^*(s)$ and $\rho^* = G_{\text{avg}}^*(\pi)$. A straightforward understanding is that (5-34) in nature contains the self-consistency condition except for the information of differential optimality condition. Different from the condition in the discounted problems, this self-consistency condition is still related to the maximum operator, which essentially comes from the definition of a differential return.

5.3.3.3   Vanishing Discount Factor Approach

Generally, many practical average-cost problems are ill-conditioned, and finding their optimal policies usually requires some strong regularity conditions. The vanishing discount factor approach is an alternate method to address these ill-conditioned problems. The main idea of the vanishing discount factor approach is to replace the average cost with a discounted cost and enforce the discount factor to be as close to one as possible. In theory, such a replacement is justified because of the following property:

$$
\begin{aligned}
G_{\mathrm{avg}}(\pi) &= \lim_{N\to\infty}\lim_{\gamma\to 1}\frac{\mathbb{E}\{\sum_{i=0}^{N-1}\gamma^i r_{t+i}\}}{\sum_{i=0}^{N-1}\gamma^i}\\[2mm]
&= \lim_{\gamma\to 1}\frac{\lim_{N\to\infty}\mathbb{E}\{\sum_{i=0}^{N-1}\gamma^i r_{t+i}\}}{\lim_{N\to\infty}\sum_{i=0}^{N-1}\gamma^i}\\[2mm]
&= \lim_{\gamma\to 1}(1-\gamma)\mathbb{E}\left\{\sum_{i=0}^{\infty}\gamma^i r_{t+i}\right\}\\[2mm]
&= \lim_{\gamma\to 1}(1-\gamma)v_{\gamma}(s).
\end{aligned}
$$

This property builds the connection between the average return $G_{\mathrm{avg}}(\pi)$ and the discounted cost $v_{\gamma}(s)$. If the discount factor γ is very close to one, the discounted cost is approximately linear with the average cost. This approach is preferred in practice since discounted cost algorithms hold a few satisfactory properties. For instance, one does not need to worry about unichain and nonstationary environments, and the reward signal does not need to be positive definite to its inputs, such as state and action. Moreover, both value iteration DP and policy iteration DP have well-behaved stability and convergence properties.

5.4   Stochastic Linear Quadratic Control

The stochastic control problem can be found in a variety of areas, including supply chain management, advertising, dynamic resource allocation, and automatic control. We have previously demonstrated that in both stochastic and deterministic problems, the discounted cost has a strong association with the average cost. Here, we will demonstrate this relationship with stochastic linear quadratic (LQ) controllers. Considering a stochastic linear model (5-3) with Gaussian random noise, we can define the reward signal as a quadratic function of a given state and action:

$$
r(s_t, a_t) = s_t^{\mathrm{T}} Q s_t + a_t^{\mathrm{T}} R a_t, \tag{5-35}
$$

where $Q \leq 0$ is negative semidefinite and $R < 0$ is negative definite. The random noise is a Gaussian function with known mean $\mathbb{E}\{\xi_t\} = 0$ and known variance $\mathbb{E}\{\xi_t \xi_t^{\mathrm{T}}\} = \sigma^2 I$. Our approach is to use DP to obtain an optimal yet analytical policy for those control problems with a discounted cost and an average cost. Traditionally, in the RL community, the negative cost function is utilized with a maximization operator. In contrast, a minimization operator is widely used for the positive cost function in the optimal control community.

5.4.1   Average Cost LQ Control

When the discount factor is used, its Bellman operator exhibits excellent contractive properties, which always guarantees the existence of optimal policy. However, the same result cannot be said for an average cost problem. Fortunately, stochastic LQ control problems always possess a kind of optimal control law under both the average and discounted criteria because of their linear-Gaussian structure. According to the analysis in (5-32), the differential Bellman equation for an average cost LQ regulator is

$$h^*(s) + \rho^* = \max_a \left\{ s^{\mathrm{T}} Q s + a^{\mathrm{T}} R a + \mathbb{E}_\xi \{ h^*(A s + B a + \xi) \} \right\}, \tag{5-36}$$

where $h(s)$ is the differential value function and ρ^* is the optimal average cost. The pair $\{h^*(s), \rho^*\}$ is said to be optimal if they are the solution of a differential Bellman equation. Fortunately, for such a quadratic cost, $h(s)$ is a quadratic function of the state variable. One can assume that $h(s) = s^{\mathrm{T}} P_{\mathrm{Avg}} s$ is an optimal value function, where P_{Avg} is the Riccati coefficient with proper dimension. The optimal coefficient is easily deduced by plugging it into the LQ-based ACOE. Eventually, its optimal average cost becomes

$$G_{\mathrm{avg}}(\pi^*_{\mathrm{avg}}) = \sigma^2 \mathrm{tr}(P_{\mathrm{Avg}}),$$

where $\tag{5-37}$

$$P_{\mathrm{Avg}} = A^{\mathrm{T}} P_{\mathrm{Avg}} A - A^{\mathrm{T}} P_{\mathrm{Avg}} B \left(B^{\mathrm{T}} P_{\mathrm{Avg}} B + R \right)^{-1} B^{\mathrm{T}} P_{\mathrm{Avg}} A + Q.$$

The optimal policy is a linear control law, whose feedback gain is dependent on the optimal Riccati coefficient:

$$\pi^*_{\mathrm{avg}}(s) = -K_{\mathrm{Avg}} s,$$

where $\tag{5-38}$

$$K_{\mathrm{Avg}} = \left(B^{\mathrm{T}} P_{\mathrm{Avg}} B + R \right)^{-1} B^{\mathrm{T}} P_{\mathrm{Avg}} A.$$

One interesting phenomenon is that the optimal feedback gain is independent of the variance of random noise. An intuitive explanation is that a linear system has the superposition property, and the maximization of expected return is decoupled with additive Gaussian noise. Thus, such an optimal policy only suppresses the mean behavior but not the variance of the closed-loop system. Although having been studied for a long time, general MDPs with average cost are still far from being fully understood. The linear quadratic controller is a rare case that has an explicit optimal structure and an analytical solution.

5.4.2   Discounted Cost LQ Control

The stochastic LQ controller for discounted cost is somewhat more complicated than the average cost in terms of their optimal structures. The Bellman equation for discounted cost is expressed in (5-39) for the whole state space:

$$v^*_\gamma(s) = \max_a \mathbb{E}_\xi \left\{ s^{\mathrm{T}} Q s + a^{\mathrm{T}} R a + \gamma v^*_\gamma (A s + B a + \xi) \right\}, \tag{5-39}$$

where $v^*_\gamma(x)$ is the optimal value for the discounted cost and $0 < \gamma < 1$ is the discount factor. Before continuing, a reasonable guess should be given to the structure of the optimal value function. We assume that $v_\gamma^*(x) = s^T P_\gamma s + M$ is the quadratic structure of optimal state-value function, in which $P_\gamma \leq 0$ is called discounted Riccati coefficient and M is a constant. Plugging this function into (5-39), its optimal value function becomes

$$v_\gamma^*(s) = s^T P_\gamma s + M,$$

where

$$P_\gamma = \gamma A^T P_\gamma A - \gamma^2 A^T P_\gamma B (\gamma B^T P_\gamma B + R)^{-1} B^T P_\gamma A + Q, \tag{5-40}$$

$$M = \frac{\gamma}{1-\gamma} \sigma^2 \mathrm{tr}(P_\gamma).$$

It is easy to see that the optimal policy is still a linear feedback law whose feedback gain depends on the discounted Riccati coefficient:

$$\pi_\gamma^*(s) = -K_\gamma s,$$

where

$$K_\gamma = \gamma (\gamma B^T P_\gamma B + R)^{-1} B^T P_\gamma A. \tag{5-41}$$

As seen from (5-40) and (5-41), both $v_\gamma^*(s)$ and $\pi_\gamma^*(s)$ depend on the discount factor. Clearly, this factor significantly affects the behavior of the optimal solution. The benefit of linear quadratic control lies in an interesting property, i.e., optimal solutions are dominated by the Riccati equations. Depending on the types of cost functions, the Riccati equations are slightly different from each other but inherently share a very similar common structure. If $\gamma \to 1$, it is obvious that $P_\gamma \to P_{\mathrm{Avg}}$ and $K_\gamma \to K_{\mathrm{Avg}}$. In addition, the average cost will be approached by a weighted average of discounted cost as $\gamma \to 1$; i.e.,

$$\lim_{\gamma \to 1} \frac{1}{\sum_{i=0}^{\infty} \gamma^i} v_\gamma^*(s) = \lim_{\gamma \to 1} \frac{\sum_{i=0}^{\infty} \gamma^i (s_{t+i}^T Q s_{t+i} + a_{t+i}^T R a_{t+i})}{\sum_{i=0}^{\infty} \gamma^i}$$

$$= \lim_{\gamma \to 1} \frac{s^T P_\gamma s + \frac{\gamma}{1-\gamma} \sigma^2 \mathrm{tr}(P_\gamma)}{\sum_{i=0}^{\infty} \gamma^i}$$

$$= \lim_{\gamma \to 1} \{(1-\gamma) s^T P_\gamma s + \gamma \sigma^2 \mathrm{tr}(P_\gamma)\}$$

$$= \sigma^2 \mathrm{tr}(P_\gamma)$$

$$= G_{\mathrm{avg}}(\pi_{\mathrm{avg}}^*).$$

The average cost function is often unsolvable in general nonlinear systems. Instead, a discounted cost problem with a sufficiently large discount factor can be optimized, which provably introduces a fairly good approximation to the average cost problem.

5.4.3 Performance Comparison with Simulations

Let us simulate a simple example to compare the performance under two kinds of performance measures. A one-dimensional linear model, in which $A = 2$ and $B = 1$, is used for this example. The additive uncertainty is Gaussian noise with mean $\mu = 0$ and variance $\sigma^2 = 1$. The coefficients of quadratic cost function are selected to be $Q = -1$ and $R = -2$. In the discounted cost function, the discount factor is selected to be $\gamma = 0.7$. The feedback gains for the average cost and discounted cost are numerically computed by (5-38) and (5-41), respectively. In the following simulations, the initial state is enforced to start with zero. The two feedback controllers are simulated multiple times to statistically compute their stationary state distributions. Figure 5.8 shows the simulated and real results of optimal costs, $G_{\mathrm{avg}}(\pi_{\mathrm{avg}}^{*})$ and $v_{\gamma}^{*}(s)$. It is easy to see that the abovementioned theoretical models have rather accurate prediction abilities.

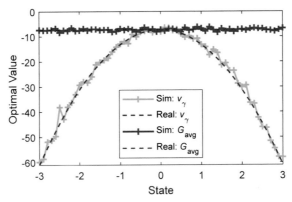

Figure 5.8 Comparison of optimal costs

The policy from the average cost is denoted as π_{avg}^{*}, and the policy from the discounted cost is denoted as π_{γ}^{*}. Now, we can compare their stationary state distributions. The closed-loop control system with an optimal linear controller is

$$s_{t+1} = (A - BK)s_t + \xi_t .$$

The uncertainty is Gaussian noise with zero mean, i.e., $\xi_t \sim \mathcal{N}(0, \sigma^2)$. It is easy to know the following property after reaching stationarity:

$$s \sim \mathcal{N}\left(0, \frac{\sigma^2}{1 - (A - BK)^2}\right).$$

Substituting π_{avg}^{*} and π_{γ}^{*} into the equation above, two stationary state distributions for average cost and discounted cost are derived, respectively:

$$d_{\pi_{\mathrm{avg}}^{*}}(s) = \sqrt{1 - (A - BK_{\mathrm{Avg}})^2} \frac{1}{\sqrt{2\pi}\,\sigma} \exp\left(-\frac{s^2}{2\sigma^2}\left(1 - (A - BK_{\mathrm{Avg}})^2\right)\right), \quad (5\text{-}42)$$

$$d_{\pi_{\gamma}^{*}}(s) = \sqrt{1 - (A - BK_{\gamma})^2} \frac{1}{\sqrt{2\pi}\,\sigma} \exp\left(-\frac{s^2}{2\sigma^2}\left(1 - (A - BK_{\gamma})^2\right)\right). \quad (5\text{-}43)$$

To examine the correctness of these two equations, simulated and real stationary state distributions are compared in Figure 5.9. Obviously, the average cost and discounted cost do not have the same stationary state distribution. Even though their means remain zero, their variances are very different from each other. Compared with theoretical prediction, statistical estimation has some small discrepancies caused by sampling randomness. This result hints us that the environment model can provide more accurate information to reinforcement learners and is helpful to accelerate convergence and ensure stability.

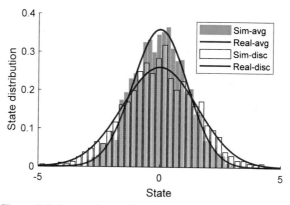

Figure 5.9 Comparison of stationary state distributions

Thus far, we have known two kinds of cost functions, i.e., J_γ and G_{avg}. The maximization of these cost functions leads to two different optimal policies, i.e., π^*_{avg} and π^*_γ, which are optimal to their respective cost functions. One might be interested in the value of each cost under the other policy. By crossing their evaluation, we have four kinds of performance measures, i.e., $J_\gamma(\pi^*_{avg})$, $J_\gamma(\pi^*_\gamma)$, $G_{avg}(\pi^*_{avg})$ and $G_{avg}(\pi^*_\gamma)$. Let us use the stationary state distribution to calculate their theoretical predictions ($e \stackrel{def}{=} A - BK_{Avg}$):

$$J_\gamma(\pi^*_\gamma) = \sum d_{\pi^*_{avg}}(s)v^*_\gamma(s) = \sigma^2 P_\gamma \left(\frac{\gamma}{1-\gamma} + \frac{1}{1-e^2}\right),$$

$$G_{avg}(\pi^*_\gamma) = (1-\gamma)\sum d_{\pi^*_{avg}}(s)v^*_\gamma(s) = \sigma^2 P_\gamma \left(\gamma + \frac{1-\gamma}{1-e^2}\right),$$

$$J_\gamma(\pi^*_{avg}) = \sigma^2 P_{Avg}\frac{1}{1-\gamma},$$

$$G_{avg}(\pi^*_{avg}) = \sigma^2 P_{Avg}.$$

From Theorem 5-1, the discounted cost has a policy-dependent weighting distribution and a linear relationship with the average cost. The linear coefficient $1/(1-\gamma)$ can be used to convert an average cost into a discounted cost if one wants to build a fair comparison between them. Table 5.1 compares the four performance measures, in which the error of real and simulated performance is less than 2.0%.

Table 5.1 Performance measure verification

Performance measure	Real	Simulation	Error
$J_\gamma(\pi^*_\gamma)$	-14.64	-14.38	1.9%
$J_\gamma(\pi^*_{avg})$	-16.17	-16.00	1.1%
$(1-\gamma)^{-1}G_{avg}(\pi^*_{avg})$	-16.17	-16.41	1.5%
$(1-\gamma)^{-1}G_{avg}(\pi^*_\gamma)$	-21.28	-21.22	0.2%

As seen from Figure 5.10, the simulation results can validate the correctness of Theorem 5-2 and (5-9). In a physical world, the average cost is what one wants to pursue. However, the existence of its optimal solution is not always guaranteed, and it is also difficult both theoretically and empirically to obtain the optimal policy. Instead, we generally solve an approximate discounted cost problem because it has better convergence and stability properties. As the discount factor $\gamma \to 1$, $G_{\text{avg}}(\pi_\gamma^*)$ will become increasingly closer to $G_{\text{avg}}(\pi_{\text{avg}}^*)$. This approximation property is the central idea of the vanishing discount factor technique.

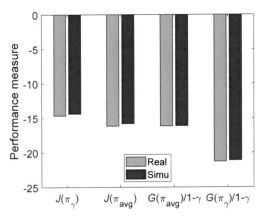

Figure 5.10 Comparison of four performance measures

5.5   Additional Viewpoints about DP

In summary, DP uses full-width synchronous backups, i.e., sweeping the whole state space at each iteration. Every successor state and action must be considered when updating state-value and policy. The environment model is able to fulfill this requirement because the knowledge contained by the model covers every corner of the environment dynamics. Generally, DP is very effective for small-sized problems (e.g., fewer than ten states). However, for medium-sized and large-scale problems (e.g., hundreds, thousands, or even millions of states), DP severely suffers from the curse of dimensionality. This is because the number of elements in the tabular representation grows exponentially with the dimensions of state and action. One solution is to generalize the policy and value with proper approximate functions, yielding so-called approximate dynamic programming (ADP). The utilization of generalization technique can reduce the tabular space of a state or an action to the parameter space, which is beneficial to reduce the computational complexity.

5.5.1   Unification of Policy Iteration and Value Iteration

Although not explicitly stated, the value iteration step, as described in (5-29), can be decomposed into a one-step PEV and its successive PIM. In the policy iteration step, an infinite number of sweeping updates are completed through PEV iterations. If PEV can stop after just one sweep, policy iteration will become value iteration. In practical DP algorithms, PEV does not actually converge to the true value, and the stopping condition often terminates its iteration at an early stage. For example, PEV can stop when the error of two successive values is sufficiently small:

$$\max_{s \in \mathcal{S}} |V_{j+1}^\pi(s) - V_j^\pi(s)| < \Delta.$$

In many other cases, PEV stops after a fixed number of iterations. This stopping condition yields what is referred to as truncated PEV, which fortunately does not result in a loss of convergence in DP. As a result, policy iteration and value iteration can be integrated into a unified framework, that is, DP with an n-step PEV $(n = 1, \cdots, \infty)$ and a complete PIM. The notation n represents the number of PEV iterations. The term "complete" means that the original PIM is employed without any modification. The two extremes of n-step PEV are $n = 1$ for DP value iteration and $n = \infty$ for DP policy iteration. The DP with an ∞-step PEV and a one-step PEV are shown in Figure 5.11 and Figure 5.12, respectively. This framework is formally referred to as generalized policy iteration (GPI), in which repeated cycles of complete PIMs and truncated PEVs are performed. The pseudocode of synchronous DP is shown in Algorithm 5-1. Here, one must select a proper number of PEV iterations to balance the training efficiency and computational burden.

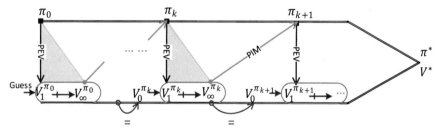

Figure 5.11 DP with an infinite-step PEV (i.e., policy iteration)

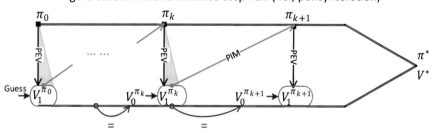

Figure 5.12 DP with a one-step PEV (i.e., value iteration)

Algorithm 5-1: Synchronous DP with n-step PEV

Hyperparameters: Discount factor γ, PEV iteration number n

Initialization: $V(s) \leftarrow 0$, $\pi(s) \leftarrow$ arbitrary action

Repeat (indexed with k)

    (1) Evaluate policy

   Repeat n times

       Sweep $s \in \mathcal{S}$

$$V^{\pi_k}(s) \leftarrow \sum_{s'} p(s'|s, \pi_k(s))(r + \gamma V^{\pi_k}(s'))$$

    End

  End

  (2) Greedy search

  Sweep $s \in \mathcal{S}$

$$\pi_{k+1}(s) \leftarrow \arg\max_a \sum_{s'} p(s'|s, a)(r + \gamma V^{\pi_k}(s'))$$

  End

End

∎

Thus far, DP backups are synchronous. Each cycle must update $V^\pi(s)$ from $V^\pi(s')$ for all $s \in \mathcal{S}$. In synchronous DP, two copies of value functions need to be stored: the old copy will be replaced by a new copy after all the states $s \in \mathcal{S}$ are updated. The complexity per iteration is approximately $\mathcal{O}(mn^2)$ for m actions and n states if using the state-value function and $\mathcal{O}(m^2n^2)$ if using the action-value function. DP can also be asynchronously implemented, and its algorithmic variants include in-place DP, prioritized sweeping, and knowledge-based DP [5-8]. When utilizing in-place DP, only one copy of value function is stored, and the updated value is immediately stored in the same memory. In prioritized sweeping, the magnitude of the Bellman error is used to guide the state selection and perform updates with the largest remaining Bellman error. This process can be implemented efficiently by maintaining a priority queue. In knowledge-based DP, states only update according to human experience. In other words, the designer's prior knowledge is used to determine which state should be updated first and which state should be delayed.

5.5.2   Unification of Model-Based and Model-Free RLs

Most theoretical results in DP can be applied to model-free RLs, including MC and TD. Therefore, we can unify model-based RL and model-free RL in the same GPI framework. The unified framework is shown in Figure 5.13 (model-based RL) and Figure 5.14 (model-free RL). This framework still has two cyclic steps, i.e., policy evaluation (PEV) and policy improvement (PIM). The goal in the PEV step is to calculate a new value function. In either model-based PEV or model-free PEV, an intermediate policy and its corresponding value function should satisfy the self-consistency condition. The goal of the PIM step is to calculate greedy actions according to the new value function.

Either a data sample or a dynamic model is a kind of description of environment dynamics. Model-based RL uses the dynamic model to calculate the optimal policy. The dynamic model is able to describe the full-width distribution of all the next states. With a model, the state-value function is enough to calculate greedy actions. Without the model, one needs the action-value function to calculate greedy actions. Note that each sample can only describe a part of the environment dynamics. To output a perfect optimal policy, an infinite number of samples is required to cover every corner of the environment.

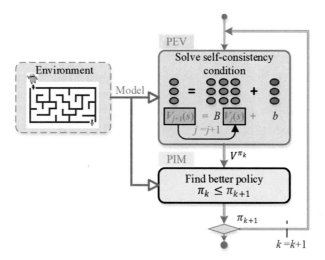

Figure 5.13 Unified framework for model-based RL

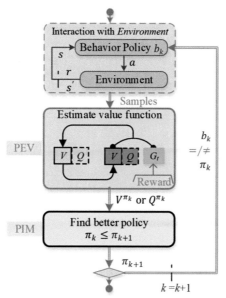

Figure 5.14 Unified framework for model-free RL

In general, environment interaction is very inefficient. Therefore, model-free RL is often slower than model-based RL. A model can be seen as the prior knowledge of environment dynamics in the whole state-action space, but a data sample is only a description of local environment dynamics. However, building an accurate model is not an easy task, and there must exist some errors between the actual environment and its analytical model. In contrast, data samples are collected from sensor measurements, and the sensing accuracy is often better than that of the analytical model. Here, the shortcoming is that sensor measurements occur in a local region and do not easily cover the whole state space. Therefore, it is very difficult to say which type of RL is better: model-based RL or model-free RL. Their performance comparison must depend on the specific task to be solved and the hyperparameters that are chosen.

5.5.3   PEV with Other Fixed-point Iterations

The PEV step, either in model-free or model-based RL, uses an iterative procedure, which can be viewed as a certain fixed-point iteration. There are five classic fixed-point iteration schemes: Picard iteration, Krasnoselskij iteration, Mann iteration, Ishikawa iteration, and Kirk iteration (see Appendix). As listed in Table 5.2, PEV in policy iteration DP is similar to the Picard iteration, and model-free PEVs, such as those in SARSA and expected SARSA, can find their similarity with the Krasnoselskij iteration or Mann iteration. In addition, n-step TD and TD-lambda are comparable to the Kirk iteration.

Table 5.2 View of fixed-point iterations for PEV in policy iteration DP

	Model-based	Model-free
Picard	PEV in DP policy iteration	--
Krasnoselskij	--	SARSA, Expected SARSA (with constant learning rate)
Mann	--	SARSA, Expected SARSA (with varying learning rate)
Ishikawa	--	--
Kirk	--	TD(n), TD-lambda

5.5.3.1   Model-Based Policy Evaluation

With an accurate model, PEV in policy iteration DP exactly follows the form of the Picard iteration. Its operator is a linear mapping function defined in a finite state space:

$$f(X) \overset{\text{def}}{=} \gamma BX + b. \tag{5-44}$$

According to the Banach fixed-point theorem, the Picard iteration converges to its fixed point if $f(X)$ is a contraction mapping in the Euclidean space. Because $\rho(\gamma B) < 1$, the linear operator (5-44) is contractive. Moreover, its fixed point is unique, which means that the self-consistency condition has a unique solution.

5.5.3.2   Model-Free Policy Evaluation

The fixed-point explanation becomes complicated when model-free cases are encountered. A model-free PEV algorithm updates in a sample-by-sample manner as the environment interactions continue. It is difficult to find a contractive operator to describe the mapping relationship of two adjacent random samples. This problem must be handled considering the expectation version of the linear operator. Let us take one-step SARSA as an example. The PEV update rule of one-step SARSA with a varying learning rate is

$$Q_{j+1}(s, a) = Q_j(s, a) + \alpha_j \left(r + \gamma Q_j(s', a') - Q_j(s, a) \right),$$

where $\{\alpha_j\}$ is a sequence of learning rates. Different from expected updates in DP, sample updates in SARSA introduce some randomness into each value estimate. Therefore, such an update formula is essentially stochastic. To remove the estimate randomness, we take the expectation of the two sides with one-step transition probability:

$$\mathbb{E}_\pi\{Q_{j+1}(s,a)\} = (1 - \alpha_j)\mathbb{E}_\pi\{Q_j(s,a)\} + \alpha_j\mathbb{E}_\pi\{r + \gamma Q_j(s',a')\}.$$

Here, $\mathbb{E}_\pi\{\#\}$ is equivalent to $\sum_{s'\in\mathcal{S}}\{\mathcal{P}_{ss'}^a \cdot \#\}$, which denotes taking expectation under the randomness of the environment dynamics. Taking expectation means that the whole environment dynamics will be considered when updating action values. Thus, we obtain a new linear operator in the expectation form:

$$f\big(Q(s,a)\big) \stackrel{\text{def}}{=} \mathbb{E}_\pi\{r + \gamma Q(s',a')\} = \sum_{s'\in\mathcal{S}} \mathcal{P}_{ss'}^a\big(r + \gamma Q(s',a')\big). \tag{5-45}$$

This operator comes from the self-consistency condition of the second kind. Its contraction property determines whether its fixed-point iteration algorithm converges. Since this type of operator is contractive, we have a Mann-like iteration scheme:

$$Q_{j+1}(s,a) = (1 - \alpha_j)Q_j(s,a) + \alpha_j f\big(Q_j(s,a)\big).$$

First, we define a functional variable in a special Banach space, i.e., $X = Q(s,a) \in \mathbb{C}(\mathcal{S},\mathcal{A})$, and a Mann-like iteration is naturally derived, i.e., $X_{j+1} = (1 - \alpha_j)X_j + \alpha_j f(X_j)$. The above analysis shows that one-step SARSA is exactly in the class of the Mann iteration. The Mann iteration can converge to a fixed point when its operator is contractive. Here, the operator is a self-mapping from $\mathbb{C}(\mathcal{S},\mathcal{A})$ to $\mathbb{C}(\mathcal{S},\mathcal{A})$. As long as every state-action pair is visited infinitely, one-step SARSA converges almost surely to the fixed point. This result also shows the necessity of sufficient exploration in model-free RL. Other model-free RLs, such as expected SARSA, n-step TD, and TD-lambda, can also find an analogy to some fixed-point iteration schemes.

5.5.4   Value Iteration with Other Fixed-Point Iterations

Value iteration bears a strong similarity with the fixed-point iteration. As listed in Table 5.3, DP value iteration is exactly in the class of Picard iteration, and Q-learning can find its origin from Krasnoselskij iteration or Mann iteration. The convergence of Krasnoselskij iteration or Mann iteration also obeys the Banach fixed-point theorem, which means that the contractive Bellman operator is the key to convergence guarantee.

Table 5.3 Viewpoint of fixed-point iterations for value iteration

	Model-based	Model-free
Picard	DP value iteration	--
Krasnoselskij	--	Q-learning with a constant learning rate
Mann	--	Q-learning with a varying learning rate
Ishikawa	--	--
Kirk	--	--

5.5.4.1   Model-Based Value Iteration

DP value iteration is naturally a kind of Picard iteration. The mapping function is the Bellman operator $\mathcal{B}$ (for countable state space in finite MDPs):

$$f(X) \stackrel{\text{def}}{=} \mathcal{B}(X).$$

Since f is a contraction mapping, the Picard iteration eventually reaches its fixed point. One might ask whether it is feasible to design new value iteration algorithms from other fixed-point iteration schemes. The answer is definitely yes, although one might still prefer to choose the simplest form, i.e., the Picard iteration, due to its concise structure and satisfactory convergence.

5.5.4.2   Model-Free Value Iteration

Q-learning is an off-policy algorithm, and its optimal value function is computed without considering how each sample is generated. Therefore, Q-learning is preferred due to its fast convergence and wide adaptability to various tasks. Its sample-based update introduces some randomness that needs to be converted to expected update. Taking Q-learning with varying learning rates as an example, its sample-based update formula is

$$Q_{k+1}(s,a) = Q_k(s,a) + \alpha_k \left(r + \gamma \max_{a'} Q_k(s',a') - Q_k(s,a) \right),$$

We rewrite this formula into an expectation-based update formula:

$$Q_{k+1}(s,a) = (1-\alpha_k)Q_k(s,a) + \alpha_k \mathbb{E}_\pi \left\{ r + \gamma \max_{a'} Q_k(s',a') \right\}.$$

The Bellman operator of the second kind is chosen as the mapping function:

$$f(Q(s,a)) = \mathbb{E}_\pi \left\{ r + \gamma \max_{a'} Q(s',a') \right\}. \tag{5-46}$$

Then, Q-learning becomes a perfect Mann iteration, i.e., $X_{k+1} = (1-\alpha_k)X_k + \alpha_k f(X_k)$. Here, $f: \mathbb{C}(\mathcal{S},\mathcal{A}) \to \mathbb{C}(\mathcal{S},\mathcal{A})$ is a contractive operator, and $X = Q(s,a) \in \mathbb{C}(\mathcal{S},\mathcal{A})$ is the functional variable defined in a special Banach space. It is easy to conclude that Q-learning belongs to the Mann iteration. In fact, Q-learning works asynchronously due to its sample-by-sample updating behavior. By the law of large numbers, its value estimate will eventually converge to the true value as long as enough samples are collected. Knowing the fixed-point mechanism of Q-learning, a variety of new value iteration algorithms can be designed. For example, a Picard Q-learning algorithm is given as

$$Q_{k+1}(s,a) = r + \gamma \max_{a'} Q_k(s',a'). \tag{5-47}$$

Obviously, this algorithm is updated with bootstrapping. Inspired by the Ishikawa iteration, a two-step Q-learning algorithm can be developed by adding a temporary value function:

$$Q^{\text{temp}}(s,a) = (1-\alpha_k)Q_k(s,a) + \alpha_k \left\{ r + \gamma \max_{a'} Q_k(s',a') \right\},$$

$$Q_{k+1}(s,a) = (1-\beta_k)Q_k(s,a) + \beta_k \left\{ r + \gamma \max_{a'} Q^{\text{temp}}(s',a') \right\}. \tag{5-48}$$

where $Q^{\text{temp}}(s,a)$ is a temporary action-value function. This two-step value iteration algorithm is able to accelerate the learning speed because it bootstraps the action value twice in one update. Its convergence, however, needs to satisfy the following conditions:

$$0 \le \alpha_k, \beta_k < 1,$$

$$\lim_{k \to \infty} \beta_k = 0, \lim_{k \to \infty} \alpha_k = 0,$$

$$\sum_{k=1}^{\infty} \beta_k = \infty, \sum_{k=1}^{\infty} \alpha_k = \infty.$$

When $\alpha_k = 0$, this algorithm reduces to traditional Q-learning, in which only one parameter β_k remains. In addition, the Ishikawa iteration may have stronger convergence for a weakly contractive operator. One can consider this kind of two-step Q-learning algorithm if the Bellman operator is ill-conditioned.

5.5.5 Exact DP with Backward Recursion

Classic DP is computed with gridded state and action spaces. This type of DP is often referred to as "exact DP". One significant issue when using exact DP is the curse of dimensionality. In the literature, both forward and backward recursions have been seen, in which the former starts from an initial time and ends at the final time, while the latter computes in the reverse order. Backward recursion, which appears to be more logical, is preferred in the optimal control field. This section discusses the connection of exact DP and value iteration DP to reveal a deep relationship between reinforcement learning and optimal control. Let us consider a discrete-time problem with fixed terminal time, as shown in Figure 5.15. Obviously, its optimal value function is dependent on the time stage:

$$V^*(s,t) = \max_{\pi} \left\{ \sum_{i=0}^{T-t} r_{t+i} \mid s_t = s \right\}. \tag{5-49}$$

Note that $V(s,t)$ in (5-49) has different augments from $V(s,N)$ in (5-33). The notation t in (5-49) represents the starting time (its terminal state is denoted as s_{T+1}), and the notation N in (5-33) represents the horizon length.

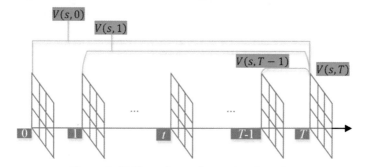

Figure 5.15 Time-dependent value function

The time-dependent value function does not allow us to build a single Bellman equation. Therefore, its corresponding Bellman equation becomes a time-dependent recursive form, which is referred to as the multistage Bellman equation:

$$V^*(s, 0) = \max\{r + V^*(s', 1)\},$$
$$V^*(s, 1) = \max\{r + V^*(s', 2)\},$$

$$\dots \dots$$

$$V^*(s, t) = \max\{r + V^*(s', t+1)\}, \qquad (5\text{-}50)$$

$$\dots \dots$$

$$V^*(s, T-1) = \max\{r + V^*(s', T)\},$$
$$V^*(s, T) = \max\{r\}.$$

The advantage of (5-50) is its natural ability to accommodate a stage-by-stage backward computation [5-3]. Such an exact DP algorithm can be thought of as a kind of temporal recursion in the discrete-time domain. The optimal policy of each stage is computed in temporal order by considering the optimal value of all the following stages. Figure 5.16 illustrates how exact DP is implemented in a gridded state space. The optimal action at each stage is searched in the finite state space. The grid mismatch issue may occur when one state point does not exactly transfer to the grid point at its next stage through a certain discrete action. This limitation imposes large challenges when using exact DP in real-world applications.

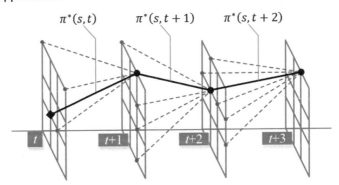

Figure 5.16 Exact DP with backward recursion

Obviously, exact DP can only handle finite-horizon problems because a starting "final time" in backward recursion must be specified. Therefore, the optimal value function is linked not only to the state s but also to the time t:

$$V^*(s, t) = \max_a \sum_{s' \in \mathcal{S}} \mathcal{P}_{ss'}^a \left(r + \gamma V^*(s_{t+1}, t+1) \right). \qquad (5\text{-}51)$$

One straightforward idea to use the fixed-point iteration for the time-dependent Bellman equation is to rewrite (5-50) into a collective form:

$$V_{\text{aug}}^*(s) = [V^*(s, 0) \quad \cdots \quad V^*(s, T)]^{\mathrm{T}}$$

$$V_{\text{aug}}^*(s) = \max \left\{ \begin{bmatrix} r \\ r \\ \vdots \\ r \end{bmatrix} + \begin{bmatrix} 0 & 1 & 0 & \cdots & 0 \\ 0 & 0 & 1 & \ddots & \vdots \\ 0 & 0 & 0 & \ddots & 0 \\ \vdots & \ddots & \ddots & \ddots & 1 \\ 0 & \cdots & 0 & 0 & 0 \end{bmatrix} V_{\text{aug}}^*(s') \right\}. \qquad (5\text{-}52)$$

One can solve (5-52) by taking $V_{\text{aug}}^*(s)$ as the new variable to be iterated. However, this idea is very inefficient since its searching dimension is extremely high. A better approach to understand the connection between value iteration DP and exact DP is to extend (5-51) from a finite horizon to an infinite horizon. When extending to infinity, the value function of each stage becomes independent of time:

$$V(s) \stackrel{\text{def}}{=} V^*(s, 0) = V^*(s, 1) = \cdots = V^*(s, t) = \cdots = V^*(s, \infty). \tag{5-53}$$

In particular, if all the value functions are the same in structure, each equation in (5-50) can degenerate into one iterative formula, i.e., a value iteration algorithm such as (5-29):

$$\left.\begin{aligned}
V(s) &\leftarrow \max\{r + V(s')\}, t = 0 \\
V(s) &\leftarrow \max\{r + V(s')\}, t = 1 \\
&\cdots \\
V(s) &\leftarrow \max\{r + V(s')\}, t = \infty - 1 \\
V(s) &= \max\{r\}, t = \infty
\end{aligned}\right\} \Rightarrow V(s) \leftarrow \max\{r + V(s')\}. \tag{5-54}$$

Therefore, value iteration DP can be regarded as a limiting case of recursive DP when the problem horizon goes to infinity. Although value iteration DP is literally a limiting case, its connotation and origin are very different from exact DP. Value iteration DP essentially comes from the Picard fixed-point iteration with a contractive Bellman operator, and its recursive formula has a variety of other fixed-point iteration schemes. In exact DP, the direct computation of backward recursion or forward recursion occurs in the temporal domain.

5.6   More Definitions of Better Policy and Its Convergence

The behavior of policy improvement plays a central role in convergence guarantee. The prerequisite to search for a better policy depends on what a better policy is defined as. Previous chapters have introduced a standard definition of better policy, which is suitable for discrete state space. This definition requires checking the value superiority of each state element individually. Therefore, policy improvement must be conducted in an element-by-element manner, which is referred to as the so-called greedy search. Previously, MC, TD and DP have used an element-by-element greedy search as the guarantee of policy improvement. One might ask whether greedy search is the exclusive choice. The answer is definitely no. In this chapter, an extended definition of better policy in the form of weighted expectation is developed to design a convergent algorithm, wherein all state elements are summed up as an integration. We will prove that this kind of weighted greedy search is also a sufficient condition to guarantee convergence.

5.6.1   Penalizing a Greedy Search with Policy Entropy

The greedy search outputs a deterministic policy, which is a classic choice of convergence guarantee. One well-known exception is the ϵ-greedy policy, which does not belong to deterministic greedy search. While the final solution is nonoptimal, convergence to the neighborhood of an optimum can still be guaranteed, and the convergence property still holds. The principle of ϵ-greedy search can be mimicked to design new rules to maintain the policy improvement ability. To achieve this goal, we need to introduce the definition of policy entropy. The entropy of a policy with discrete actions is defined as

$$\mathcal{H}(\pi) = \sum_{a \in \mathcal{A}} -p(a) \log p(a),$$

where a is the action in a discrete space $a \in \mathcal{A}$ and $p(a)$ is short for $\pi(a|s)$. Obviously, the more stochastic a policy is, the higher its policy entropy becomes. One can maximize an entropy-based PIM criterion to output a stochastic policy. The benefit of entropy maximization is that its resultant action distribution depends on what kind of constraint is added. This property allows us to generate random actions with the desired distribution. Typical action distributions and their corresponding constraints are listed in Table 5.4. The simplest case is unconstrained entropy optimization: if there is no constraint, a uniform action distribution is optimal.

Table 5.4 Action distribution in entropy maximization

Type	Probability density	Constraints	Action space
Uniform (discrete)	$p(a) = 1/|\mathcal{A}|$	--	$a \in \{a_1, a_2, \dots, a_{|\mathcal{A}|}\}$
Uniform (cont.)	$p(a) = 1/(a_{\text{high}} - a_{\text{low}})$	--	$a \in [a_{\text{low}}, a_{\text{high}}]$
Bernoulli	$p(a) = \mu^a(1 - \mu)^{1-a}$	$\mathbb{E}(a) = \mu$	$a \in \{0, 1\}$
Normal	$p(a) = \dfrac{1}{\sqrt{2\pi\sigma^2}} \exp\left(-\dfrac{(a - \mu)^2}{2\sigma^2}\right)$	$\mathbb{E}(a) = \mu$ $\mathbb{D}(a) = \sigma^2$	$a \in (-\infty, +\infty)$

One might ask what kind of PIM criterion corresponds to the ϵ-greedy policy. The answer is the greedy criterion with special entropy regularization. Let us construct the following PIM optimization problem:

$$\pi_{k+1} = \arg\max_{\pi} \{p(a^*) + \kappa \mathcal{H}(\pi)\},$$

subject to

$$a^* = \max_a q^{\pi_k}(s, a), \forall s \in \mathcal{S}, \tag{5-55}$$

$$p(a^*) + \sum_{a \in \mathcal{A} \setminus a^*} p(a) = 1,$$

where κ is the weighting coefficient. This entropy-based greedy criterion has the ability to achieve a balance between maintaining optimal actions from greedy search and enlarging policy entropy for better exploration. One can easily prove that its optimum exactly equals the ϵ-greedy policy if choosing the following condition:

$$\kappa = \frac{1}{\ln(1 - |\mathcal{A}|(\epsilon - 1)/\epsilon)}. \tag{5-56}$$

The method of Lagrange multipliers is used to solve this PIM optimization problem:

$$L(\pi, \lambda) = p(a^*) + \kappa \sum_{a \in \mathcal{A}} -p(a) \log p(a) + \lambda \left(p(a^*) + \sum_{a \in \mathcal{A} \setminus a^*} p(a) - 1 \right).$$

Consider its first-order optimality condition:

$$\frac{\partial L(\pi, \lambda)}{\partial p(a)} = 0,$$

and we obtain the probabilities of optimal action and non-optimal action:

$$p(a^*) = \exp \left(\frac{\lambda + 1}{\kappa} - 1 \right), p(a)|_{a \neq a^*} = \exp \left(\frac{\lambda - \kappa}{\kappa} \right).$$

According to the fact that $\sum_{a \in \mathcal{A}} p(a) = 1$, the following equality naturally holds:

$$\left(e^{1/\kappa} + |\mathcal{A}| - 1 \right) p(a)|_{a \neq a^*} = 1.$$

Combined with (5-56), we have the following result:

$$p(a^*) = 1 - \epsilon + \frac{\epsilon}{|\mathcal{A}|}, p(a)|_{a \neq a^*} = \frac{\epsilon}{|\mathcal{A}|}.$$

Such an entropy-based PIM loss function in (5-55) can serve as a new design of policy improvement, and its optimization outputs an ϵ-greedy policy. One can easily conclude that the conventional greedy search is not the only choice for policy improvement. This result inspires us to rethink other definitions of better policy and more general way to guarantee convergence.

5.6.2 Expectation-Based Definition for Better Policy

In previous chapters, the element-by-element definition is used to describe what a better policy is. This definition allows us to update the action of each state individually. One extended definition is to sum up the evaluation of all state elements in the whole state space and compare the overall sum instead of each element. In this section, the expectation-based definition is proposed to describe what a better policy is. With this new definition, a policy $\bar{\pi}$ is said to be better than π if

$$\mathbb{E}_{s \sim d(s)} \{ v^{\pi}(s) \} \leq \mathbb{E}_{s \sim d(s)} \{ v^{\bar{\pi}}(s) \},$$

where $d(s)$ is a certain state distribution for weighting purposes. Figure 5.17 compares the two definitions. Obviously, the element-by-element definition is a sufficient condition of the expectation-based definition.

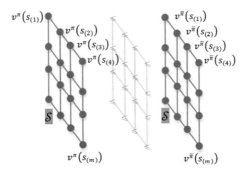

(a) Element-by-element definition

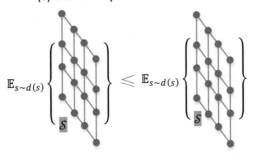

(b) Expectation-based definition

Figure 5.17 Comparison of the definition of a better policy

In essence, the aim of the PIM step is to search for a new policy that satisfies a kind of better policy definition. The aforementioned two definitions are both suitable to design the PIM updating rules. Considering the expectation-based definition, PIM can be formulated as a constrained optimization problem:

$$\pi_{k+1} = \arg \max_{\pi}\{\delta - \rho(\pi, \pi_k)\},$$

subject to   (5-57)

$$\mathbb{E}_{s \sim d(s)}\{v^{\pi}(s)\} = \delta + \mathbb{E}_{s \sim d(s)}\{v^{\pi_k}(s)\},$$

$$\delta \geq 0,$$

where $\rho(\cdot, \cdot) \geq 0$ is a distance measure of two policies and $\delta \geq 0$ is the value margin. The distance measure is zero if the two policies are identical. This new PIM optimization problem can guarantee convergence of the overall RL algorithm at least to a local optimum. The key is to first prove that $\mathbb{E}_{s \sim d(s)}\{v^{\pi_k}(s)\}$ is monotonically increasing and then prove that $v^{\pi_\infty}(s)$ is equal to the optimal value $v^*(s)$ for any $s \in \mathcal{S}$. Let us recall the following RL performance measure:

$$J(\pi) \stackrel{\text{def}}{=} \mathbb{E}_{s \sim d(s)}\{v^{\pi}(s)\}.$$

- Theorem 5-7: Sequence $\{J(\pi_0), J(\pi_1), \cdots, J(\pi_\infty)\}$ is monotonically increasing. More specifically,

$$J(\pi_0) \leq J(\pi_1) \leq \cdots \leq J(\pi_k) \leq J(\pi_{k+1}) \leq \cdots \leq J^*,$$

where $J^* = J(\pi^*)$ is the performance measure for the optimal policy.

Proof:

This monotonically increasing property is guaranteed by maximizing (5-57) with the inequality condition $\delta \geq 0$. Obviously, π^* is the best among all policies. For any policy π, we always have the fact that its performance is no better than that of the optimal policy:

$$J(\pi) \leq J^* = \mathbb{E}_{s \sim d(s)}\{v^{\pi^*}(s)\}.$$

∎

A key question is whether the maximization of (5-57) has a solution even if an equality constraint is added. The good news is that this problem always has a solution; at least, one can select the last-step policy, i.e., $\pi_{k+1} = \pi_k$, which means that the value margin becomes zero. Moreover, PIM stops when two successive policies are the same. At this moment, an optimal policy is found.

- Theorem 5-8: Assume that $J(\cdot)$ is strictly convex. When PIM stops at $k \to \infty$, we have

$$\pi_{\infty}(s) = \pi^*(s), \forall s \in \mathcal{S}.$$

Proof:

Since $J(\pi_k)$ is a constant at the $(k+1)$-th step PIM, (5-57) is equivalent to the following optimization problem:

$$\pi_{k+1} = \arg\max_{\pi}\{J(\pi) - \rho(\pi, \pi_k)\},$$

subject to

$$J(\pi) - J(\pi_k) \geq 0.$$

Let us define a surrogate function:

$$g(\pi, \pi_k) \stackrel{\text{def}}{=} J(\pi) - \rho(\pi, \pi_k).$$

This surrogate function satisfies the following inequality:

$$g(\pi, \pi_k) \leq J(\pi), \forall \pi. \tag{5-58}$$

This inequality condition actually builds a minorize-maximization (MM) optimization. According to the analysis in [5-9], the limiting case of π_k as $k \to \infty$ will stop improvement. Moreover, its final policy satisfies the first-order optimality condition, i.e., $\nabla J(\pi_{\infty}) = 0$. Since $J(\cdot)$ is strictly convex, π^* is the unique optimal result. That is to say π_{∞} is equal to π^*.

∎

Figure 5.18 illustrates the mechanism of using (5-57) to perform the PIM updates. From the perspective of the MM optimization, its surrogate function is a lower bound of the original RL objective function. The advantage of using surrogate function is that there is no special requirement on the types and properties of the distance measure $\rho(\pi, \pi_k)$. The monotonic improvement property exists as long as $\rho(\pi, \pi_k)$ is greater than zero. In addition, monotonic convergence is always guaranteed due to the nonnegative definiteness of the value margin.

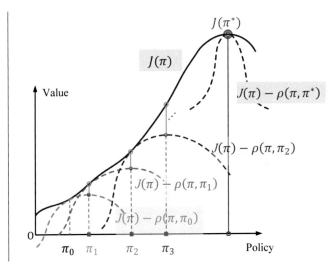

Figure 5.18 Policy improvement under expectation conditions

Thus far, the expectation-based definition is able to provide an alternative design for policy improvement. Obviously, the element-by-element definition is a special case of this expectation-based definition. If the state distribution is selected to be that of collected samples, the following new observation can be obtained. RL does not need to improve each element in the state space, and the weighted policy improvement is strong enough to guarantee convergence. Interestingly, in the final policy, every state element still has its corresponding optimal action.

5.6.3   Another Form of Expectation-Based Condition

The expectation-based definition of better policy has an alternative version, which is the natural extension of classic policy improvement theorem. Such a policy $\bar{\pi}$ that satisfies the following expectation-based condition in (5-59) is a better policy:

$$\mathbb{E}_{s\sim d(s)}\{v^{\pi}(s)\} \le \mathbb{E}_{s\sim d(s)}\left\{\sum_{a\in\mathcal{A}} \bar{\pi}(a|s)q^{\pi}(s,a)\right\}. \tag{5-59}$$

This is the extended version of classic policy improvement theorem, which provides a sufficient condition to guarantee convergence. We assume that there exists a unique optimal policy in the Markov chain. The uniqueness means that each state has a unique optimal action.

- Theorem 5-9: The maximization of (5-59) is equivalent to a greedy search:

$$\bar{\pi}^* \Leftrightarrow \tilde{\pi}^*,$$

where

$$\bar{\pi}^* = \arg\max_{\bar{\pi}} \mathbb{E}_{s\sim d(s)}\left\{\sum \bar{\pi}(a|s)q^{\pi}(s,a)\right\},$$

$$\tilde{\pi}^* = \arg\max_{\tilde{\pi}}\left\{\sum \tilde{\pi}(a|s)q^{\pi}(s,a)\right\}, \forall s \in \mathcal{S}.$$

Proof:

Suppose that $\tilde{\pi}^*$ results from the maximization of (5-59), but it fails to maximize $q^\pi(s, a)$ for $s \in S$. If $\tilde{\pi}^*$ is the optimal policy from the greedy search, it maximizes the weighted action-value function; i.e., $\mathbb{E}_{s \sim d(s)}\{\sum \tilde{\pi}(a|s)q^\pi(s, a)\}$. Because $\tilde{\pi}^*$ maximizes $q^\pi(s, a)$ for all state-action pairs, we have

$$\mathbb{E}_{s \sim d(s)}\left\{\sum \tilde{\pi}^*(a|s)q^\pi(s, a)\right\} \geq \mathbb{E}_{s \sim d(s)}\left\{\sum \pi^*(a|s)q^\pi(s, a)\right\},$$

which conflicts with our assumption. Therefore, π^* and $\tilde{\pi}^*$ must be identical.

∎

Under this kind of equivalence property, one can design another useful updating mechanism that is able to improve the policy monotonically. Here, the PIM step is formulated as a constrained optimization problem:

$$\pi_{k+1} = \arg\max_\pi\{\delta - \rho(\pi, \pi_k)\},$$

subject to

$$\mathbb{E}_{s \sim d(s)}\left\{\sum \pi(a|s)q^{\pi_k}(s, a)\right\} = \delta + \mathbb{E}_{s \sim d(s)}\left\{\sum \pi_k(a|s)q^{\pi_k}(s, a)\right\}, \qquad (5\text{-}60)$$

$$\delta \geq 0.$$

Such a policy π_{k+1} from (5-60) satisfies the classic policy improvement theorem. The key is to examine whether π_{k+1} is a better policy under the element-by-element definition. We suppose that there exists one state element $s_{(i)} \in S$ for which such an examination does not hold:

$$v^{\pi_k}(s_{(i)}) \not\leq \sum \pi_{k+1}(a|s_{(i)})q^{\pi_k}(s_{(i)}, a).$$

Let us construct a new policy $\tilde{\pi}$:

$$\tilde{\pi}(a|s) = \begin{cases} \pi_{k+1}(a|s), s \in S \setminus \{s_{(i)}\} \\ \pi_k(a|s), s \in \{s_{(i)}\} \end{cases},$$

which obviously satisfies $v^{\pi_k}(s) \leq \sum \tilde{\pi}(a|s)q^{\pi_k}(s, a), \forall s \in S$. Its corresponding value margin is

$$\tilde{\delta} = \mathbb{E}_{s \sim d(s)}\left\{\sum \tilde{\pi}(a|s)q^{\pi_k}(s, a)\right\} - \mathbb{E}_{s \sim d(s)}\{v^{\pi_k}(s)\}.$$

The corresponding value margin of policy π_{k+1} is

$$\delta_{k+1} = \mathbb{E}_{s \sim d(s)}\left\{\sum \pi_{k+1}(a|s)q^{\pi_k}(s, a)\right\} - \mathbb{E}_{s \sim d(s)}\{v^{\pi_k}(s)\}.$$

Now, let us examine the sign of the error $\Delta = \tilde{\delta} - \delta_{k+1}$:

$$\Delta$$

$$= \mathbb{E}_{s \sim d(s)}\left\{\sum \tilde{\pi}(a|s)q^{\pi_k}(s, a)\right\} - \mathbb{E}_{s \sim d(s)}\left\{\sum \pi_{k+1}(a|s)q^{\pi_k}(s, a)\right\}$$

$$= \mathbb{E}_{s \sim d(s)}\left\{v^{\pi_k}(s_{(i)}) - \sum \pi_{k+1}(a|s_{(i)})q^{\pi_k}(s_{(i)}, a)\right\}$$

$$> 0.$$

Since $\tilde{\pi}$ is exactly the same as π_k for $s \in \{s_{(i)}\}$ and equals π_{k+1} for $s \in S \setminus \{s_{(i)}\}$, it is easy to know that

$$\rho(\tilde{\pi}, \pi_k) \leq \rho(\pi_{k+1}, \pi_k).$$

Combining the analysis above, one has the following inequality:

$$\left(\tilde{\delta} - \rho(\tilde{\pi}, \pi_k)\right) > \left(\delta_{k+1} - \rho(\pi_{k+1}, \pi_k)\right),$$

which means that $\tilde{\pi}$ is better than π_{k+1}. Obviously, this result conflicts with the fact that π_{k+1} is optimal in (5-60). Therefore, there is no such a state point $s_{(i)}$ that does not satisfy the classic policy improvement theorem. This PIM optimization problem provides a good balance between the goodness of a better policy and the penalty of obtaining a new policy from the last policy. Its maximization always has a solution; at least, we can select $\pi_{k+1} = \pi_k$, of which both   value margin and policy penalty become zero. In addition, when $\rho = 0$ is selected, this kind of optimization problem reduces to the classic greedy search.

5.7   Example: Autonomous Car on a Grid Road

In this example, an autonomous car is assumed to run on a grid road, seeking to reach its destination as soon as possible. The car has three-dimensional discrete states, i.e., longitudinal position x, lateral position y, and heading direction h. The autonomous car itself always occupies two adjacent grids and has a one-dimensional action with three discrete elements, i.e., "right", "keep", and "left". Note that "left" and "right" are defined with respect to the current heading direction, and the steering action is used to adjust the car's heading direction. For instance, if the car is heading straight, a "left" steering action will turn the car into the leftward direction. After each action is executed, the car will move one step forward along its new heading direction. The destination grids are located on the rightmost side of this road, and each episode will end if the car reaches any of these grids.

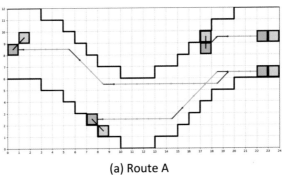

(a) Route A

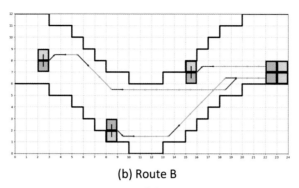

(b) Route B

Figure 5.19 Routes of the autonomous car

Figure 5.19 shows several routes of the car with different actions and initial conditions. The car is represented by two adjacent squares with different colors, in which the yellow square represents its head, and the green square represents its tail. The arrow along the route indicates the heading direction toward which the car will move. The color of the arrow indicates the steering action taken by the car in a step. More precisely, the blue arrow denotes the action "right", the cyan arrow denotes the action "keep", and the red arrow denotes the action "left".

5.7.1   Results of DP

Figure 5.20 shows the learned policy and value function in DP (when PEV iteration number $n = 3$). The arrow indicates the new heading direction of the car guided by the learned policy. The digital numbers in the grid road represent the value of each state. If there is no arrow in a cell, it means that there is no admissible policy that is able to drive the car to reach its destination from that state. Figure 5.20(a) shows that the car will turn right unquestionably in these states with $h = 5$, i.e., the previous heading direction is upward. Similar results but with the opposite direction can be seen in Figure 5.20(e), in which the previous heading direction is downward, i.e., $h = 1$. This kind of policy is easy to understand. Since these two heading directions are not toward the destination, it is better to immediately adjust its direction to run forward. The policy becomes slightly lazy in the other three cases, i.e., $h = 4,3,2$ (Figure 5.20(b), (c) and (d), respectively). One can easily see that the car does not like to change its direction if a turning action is not necessary. As demonstrated in Figure 5.20(c), the car will keep moving forward unless it hits either of the two road boundaries. We can observe that the closer to the destination the car is, the greater its cell value will become. Moreover, these cells close to the road boundaries often have smaller values than the middle cells. Obviously, the maximization of such a state-value function will take action to either let the car reach its destination or avoid boundary collision.

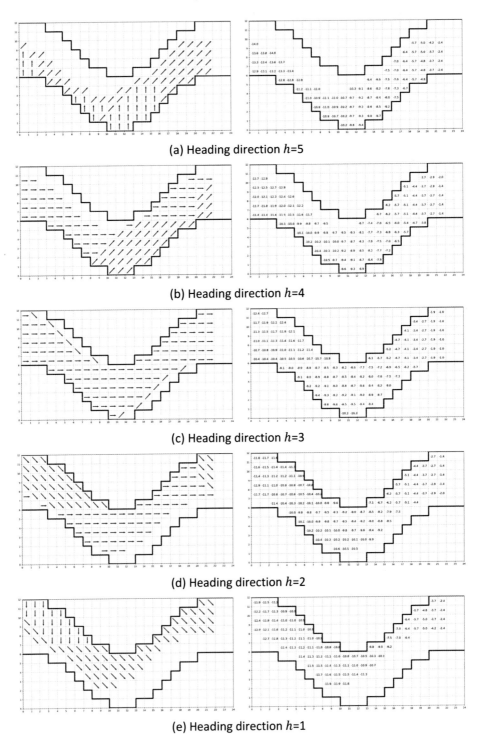

(a) Heading direction $h=5$

(b) Heading direction $h=4$

(c) Heading direction $h=3$

(d) Heading direction $h=2$

(e) Heading direction $h=1$

Figure 5.20 Learned policy and value functions

5.7.2   Influences of Key Parameters on DP

The influence of the PEV iteration number (denoted as Step/PEV) is studied by a series of simulations. As shown in Figure 5.21, the policy evaluation will be more accurate with a larger number of iterations, and its following PIM is more likely to find a close-to-optimal policy. As a result, DP will achieve a higher total average return with fewer main cycles. It should be clarified that more iteration steps in PEV also require more computation time, and the overall training efficiency may be somewhat sacrificed.

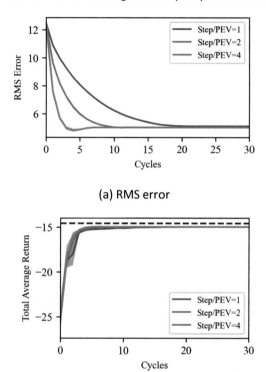

(a) RMS error

(b) Total average return

Figure 5.21 DP with different numbers of PEV iterations

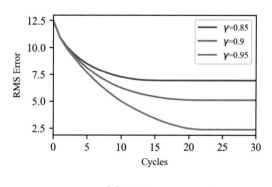

(a) RMS error

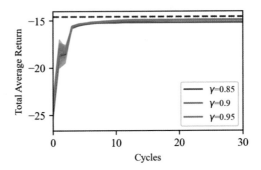

(b) Total average return

Figure 5.22 DP with different discount factors

Figure 5.22 shows the influence of the discount factor. One can conclude that the closer to 1 the discount factor is, the higher the total average return an RL agent achieves. A reasonable explanation is that a higher discount factor pushes the agent to consider more long-term rewards, and consequently, it will take a far-sighted view to explore how to obtain higher average rewards.

5.7.3   Influences of Key Parameters on SARSA and Q-learning

We also tested model-free RL algorithms, including SARSA and Q-learning, using the same example and studied their performance differences under different hyperparameter settings. Figure 5.23 shows how many state-action pairs are used in each PEV (Pair/PEV) for SARSA. It turns out that Pair/PEV poses almost no impact on both the RMS error and the total average return, and it has little influence on the policy performance of SARSA. One short explanation is that the self-driving environment is deterministic, and accordingly more state-action pairs at each cycle do not reduce the variance of value estimation, thus providing almost no support for the accuracy improvement of policy updates. Figure 5.24 shows the influence of the learning rate α in Q-learning. The learning rate has a significant impact on the convergence speed. Generally, the greater the learning rate is, the faster the Q-learning algorithm is. However, the selection of the learning rate does not affect the final policy, as shown by the curve of the total average return.

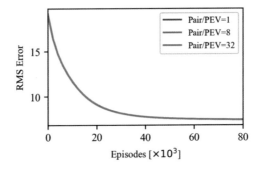

(a) RMS error

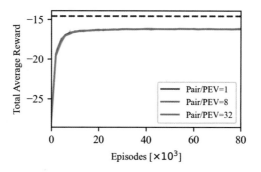

(b) Total average return

Figure 5.23 SARSA with different PEV pairs

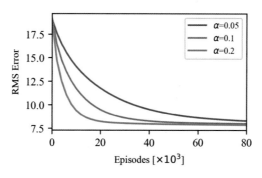

(a) RMS error

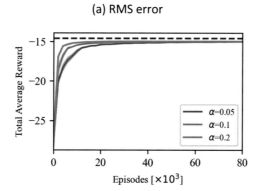

(b) Total average return

Figure 5.24 Q-learning with different learning rates

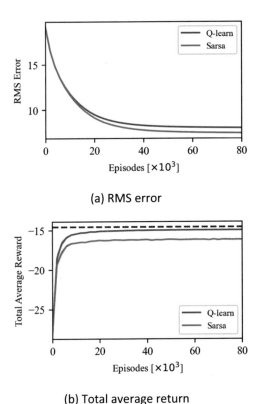

(a) RMS error

(b) Total average return

Figure 5.25 Comparison of Q-learning and SARSA

Figure 5.25 compares Q-learning and SARSA with comparable hyperparameter settings. In this example, SARSA performs as well as Q-learning in terms of convergence speed, but Q-learning can achieve a higher total average return. This is because Q-learning is off-policy while SARSA is on-policy. As an on-policy algorithm, a stochastic ϵ-greedy policy must be learned in SARSA, but the true optimal policy is deterministic. A stochastic policy can never become optimal due to its random actions. In contrast, in Q-learning, a deterministic policy is actually taken as the target policy, and the goal is to learn a policy that is optimal. In SARSA, if the exploration rate does not gradually shrink to zero, Q-learning should perform better than SARSA in terms of the final policy optimality.

5.8   Appendix: Fixed-point Iteration Theory

In numerical analysis, fixed point is defined as the solution at which an iterated function's input and output are equal. Existence and convergence are two important properties of fixed points. The study of fixed-point theory might have been initiated by Jules Poincare in the 19th century. This theory was largely extended by many famous mathematicians in the 20th century. In numerical analysis, fixed-point iteration is the basic method of computing fixed points of iterated functions. As a common equation solving mechanism, fixed-point iteration is one of the most powerful tools in numerical analysis and has been widely used in solving complex nonlinear equations, including Bellman equation and HJB equation.

5.8.1   Procedures of Fixed-Point Iteration

The Banach space is defined as a complete space with a normed vector. One famous result related to Banach spaces is the contraction mapping theorem. Let $\mathbb{X}$ be the Banach space and $f: \mathbb{X} \to \mathbb{X}$ be an operator in the Banach space. One can say that $X^* \in \mathbb{X}$ is a fixed point of f if $X^* = f(X^*)$. The operator f is said to be a contraction mapping if $\|f(X) - f(Y)\| \leq \gamma \cdot \|X - Y\|, \forall X, Y \in \mathbb{X}$ holds, where $0 < \gamma < 1$ is a Lipschitz coefficient.

- Theorem 5-10 [5-1]: If $f: \mathbb{X} \to \mathbb{X}$ is a contraction mapping, it has a unique fixed point X^*, and its Cauchy sequence converges to the fixed point:

$$X_n \to X^*, \text{as } n \to \infty,$$

where X_n is from the Picard iteration with an arbitrary initial value X_0.

∎

This is the well-known Banach fixed-point theorem, which states both the uniqueness of a fixed point and the convergence of its Cauchy sequence. It is also known as the contraction mapping theorem. This theorem contains a constructive method of generating the convergent sequence, which is called the Picard iteration. In the Cauchy sequence, which is named after Augustin-Louis Cauchy, the sequence elements become arbitrarily close to each other as the sequence progresses. In addition to the Picard iteration, there are four other types of common iteration schemes, including the Krasnoselskij iteration, the Mann iteration, the Ishikawa iteration, and the Kirk iteration.

(1) The Picard iteration is defined by

$$X_n = f(X_{n-1}).$$

As stated by the Banach fixed-point theorem, the Picard iteration builds a simple Cauchy sequence, which is able to converge to the fixed point if f is contractive.

(2) The Krasnoselskij iteration is defined by

$$X_n = (1 - \lambda)X_{n-1} + \lambda f(X_{n-1}),$$

where $\lambda \in (0,1)$ is the constant. The Krasnoselskij iteration is exactly given by the Picard iteration when introducing an averaged operator, such as $f_\lambda = (1 - \lambda)I + \lambda f$, where I is an identity operator. This iteration scheme is very similar to bootstrapping, which updates itself based on the weighted sum of a known value and new information.

(3) The Mann iteration is defined by

$$X_n = (1 - \alpha_n)X_{n-1} + \alpha_n f(X_{n-1}),$$

where $\alpha_n \in [0,1]$ are real numbers. If $\alpha_n = \lambda$, the Mann iteration reduces to the Krasnoselskij iteration.

(4) The Ishikawa iteration is defined by

$$X_{n+1} = (1 - \beta_n)X_n + \beta_n f\big((1 - \alpha_n)X_n + \alpha_n f(X_n)\big),$$

where $\alpha_n, \beta_n \in [0,1]$ are real numbers. Its convergence requires the coefficients to satisfy the following conditions:

$$0 \le \alpha_n, \beta_n < 1, \lim_{n \to \infty} \beta_n = 0, \lim_{n \to \infty} \alpha_n = 0$$

$$\sum_{n=1}^{\infty} \alpha_n = \infty, \sum_{n=1}^{\infty} \beta_n = \infty.$$

When $\alpha_n = 0$, the Ishikawa iteration reduces to the Mann iteration. Despite their structural connection, the Mann iteration and the Ishikawa iteration have different convergence properties. The Ishikawa iteration often establishes a stronger convergence for a pseudo-contractive self-map in a convex compact space.

(5) The Kirk iteration is defined by

$$X_{n+1} = c_0 X_n + c_1 f(X_n) + c_2 f(f(X_n)) + \cdots + c_k f^k(X_n),$$

where k is a fixed integer, $k \ge 1, c_i \ge 0, c_1 > 0$, and $\sum_i c_i = 1$. This iteration scheme reduces to the Krasnoselskij iteration when $k = 1$.

All five iteration schemes can form their own Cauchy sequences. The convergence speed of a Cauchy sequence $\{X_n\}$ is measured by the distance of each iteration X_n to the true fixed point X^*. In the Picard iteration, the distance is at least bounded by two key factors: (a) the Lipschitz coefficient γ and (b) the error distance of the first iteration:

$$\|X_n - X^*\| \le \frac{\gamma^n}{1 - \gamma} \|X_1 - X_0\|,$$

where X_0 is an initial point. One interesting fact is that, similar to the Picard iteration, the other four iteration schemes have the same order of convergence speed for most well-defined problems [5-1].

5.8.2   Fixed-Point Iteration for Linear Equations

Fixed-point iteration has wide applications in solving large-scale linear equations. Its computational cost is on the order of m^2 operations for each iteration (m is the number of linear equations), while the overall cost of direct linear equation solvers (for example, Gauss–Jordan elimination) is on the order of m^3 operations. Besides, fixed-point iteration is computationally competitive since the iteration number required to converge is often independent of the number of equations [5-7]. Given a large-scale linear equation $AX = b$, with A being square and fully ranked, its associated Cauchy sequence X_n is generated by

$$X_{n+1} = PX_n + (I - P)A^{-1}b,$$
$$P = I - A,$$

where P is called the iteration matrix. The iterative algorithm is convergent if and only if $\rho(P) < 1$, where $\rho(P)$ is the spectral radius. Because $\rho(P)$ is the largest absolute value of the eigenvalues of P, it is reasonable to believe that X_n converges faster when $\rho(P)$ is smaller. That is to say the spectral radius provides a sound indicator of convergence speed. To obtain better convergence, a popular trick is to split matrix A into the form of $A = M - N$, where M and N are two proper matrices. Such a splitting technique can lead to a series of fast linear iteration algorithms, including the Jacobi, Gauss-Seidel and relaxation methods. The Gauss-Seidel method has an asymptotic convergence rate that is twice as fast as that of the Jacobi method [5-7]. The relaxation method is a generalization of the Gauss-Seidel method, and its convergence speed depends on the choice of relaxation coefficient.

5.9 References

[5-1]   Berinde V (2007) Iterative approximation of fixed points. Springer Berlin, Heidelberg

[5-2]   Bertsekas D (2017) Dynamic programming and optimal control (4th edition). Athena Scientific, Nashua NH

[5-3]   Luus R (2019) Iterative dynamic programming. CRC Press, New York

[5-4]   Ohnishi M (1992) Policy iteration and Newton-Raphson methods for Markov decision processes under average cost criterion. Comp & Math with Appl 24(1–2): 147-155

[5-5]   Pollatschek M, Avi-Itzhak B (1969) Algorithms for stochastic games with geometrical interpretation. Management Science 15 (7):399-415

[5-6]   Puterman ML, Brumelle SL (1979) On the convergence of policy iteration in stationary dynamic programming. Math of Operations Res 4(1): 60-69

[5-7]   Quarteroni A, Sacco R, Saleri F (2010) Numerical mathematics. Springer Berlin, Heidelberg

[5-8]   Sutton RS, Barto AG (2018) Reinforcement learning: an introduction. MIT Press, Cambridge

[5-9]   Jeff W (1983) On the convergence properties of the EM algorithm. The Annals of Statistics 11(1): 95–103

Chapter 6.   Indirect RL with Function Approximation

> Knowing is not enough, we must apply.
>
> Willing is not enough, we must do.
>
> -- Johann Wolfgang von Goethe (1749-1832)

Solving sequential decision problems with a large state space requires large storage memory and time-consuming computation. For example, there are over 10^{20} possible states in Backgammon and approximately 10^{170} possible states in Chinese Go. Moreover, many real-world tasks like autonomous driving has continuous state spaces. These large-scale problems not only quickly deplete memory but also require a tremendous amount of expensive computation. When accurately learning all the tabular value and policy entries, this problem, which is known as the curse of dimensionality, becomes even worse. The curse of dimensionality refers to the exponential growth of a computational burden with respect to the dimension of state space and action space. The urgent need for solving large-scale problems has drawn great attention to the generalization technique.

In RL, one of the popular generalization techniques is called function approximation, in which value function and policy are approximated with proper parameterized functions. Some representation accuracy is sacrificed, but the demand for storage memory and computational resources can be effectively reduced. Sacrificing accuracy in policy representation is acceptable in engineering practice, where the goal of an agent is to find a sufficiently good solution that is not necessarily an exact optimum. The function approximation technique can be applied in either value function or policy, or both of them. Hence, according to which component is parameterized, there are three categories of RL with function approximation: (1) value approximation-only, (2) policy approximation-only, and (3) actor-critic approximation. The value approximation-only algorithm relies on the fixed-point iteration to update the parameterized value function. After successful convergence, the agent selects actions by greedily consulting the optimal value function without explicitly outputting any policy. Typical algorithms in this class include deep Q-network (DQN) and its variants. Policy approximation-only algorithms generally refer to some naive policy gradient algorithms, in which an optimal solution is searched for in a policy parameter space. A typical example is REINFORCE, which uses the Monte Carlo estimation to calculate the value function. A few finite-horizon neuro-DPs also belong to the class of policy approximation-only algorithms, in which the value function can be computed by the sum of rewards in the finite horizon.

The actor-critic (AC) approximation, which is the most popular RL architecture, is used to generalize both policy and value function. As discussed in Konda and Tsitsiklis (2004) [6-7], actor-critic methods combine the strengths of policy approximation and value approximation. In this architecture, the critic updates a parameterized value function to provide a more accurate evaluation, and the actor guides the policy parameter toward better performance. These actor-critic algorithms, as long as they are based on gradient-based optimization, often have desirable convergence properties. Moreover, a few variance reduction methods, such as the baseline technique, can be utilized to achieve

© The Author(s), under exclusive license to Springer Nature Singapore Pte Ltd. 2023, Corrected Publication 2024

S. E. Li, *Reinforcement Learning for Sequential Decision and Optimal Control*, https://doi.org/10.1007/978-981-19-7784-8_6

more stable convergence. Typical actor-critic algorithms include A3C (Asynchronous Advantage Actor-Critic), TD3 (Twin Delayed Deep Deterministic policy gradient), SAC (Soft Actor-Critic), DSAC (Distributional Soft Actor-Critic), etc. In fact, the architecture of actor-critic algorithms can be understood from the viewpoint of either indirect RL or direct RL. In the former, both policy evaluation and policy improvement are replaced with gradient-based optimization, resulting in critic updates and actor updates, respectively. In the latter, actor-critic algorithms are natural extensions of policy gradient algorithms, whose value functions are recursively estimated to reduce gradient variance. Even though a few successful applications have been achieved, practical actor-critic methods suffer from several issues, such as the lack of general convergence theory, the unpredictable deadly triad, and the tendency to remain in local optima.

6.1   Linear Approximation and Basis Functions

The function approximation seeks to select a parameterized function that closely matches a target function. In RL methods, the explicit form of the target function is often unknown, and only some equality conditions or collected data samples about the target function are accessible. In addition, the conditions and data samples are imperfect and can be extremely inaccurate or noisy. Therefore, selecting a proper function approximation method is critical to generalization. Perhaps the simplest choice is linear approximation, which is used to approximate the target function as a linear combination of basis functions (also called features). Thus far, most theoretical results on convergence and stability properties have been restricted to the linear approximation. In addition to linear approximation, nonlinear approximation, such as artificial neural network (ANN), has been widely used in practical RL algorithms. ANN is considered an "almost-almighty" approximation with many good properties, such as automatic feature extraction, generalization under incomplete information, and parallel processing ability.

In the linear approximation, an approximate function $g(\cdot; w) \in \mathbb{R}$ is the dot product of the weight vector $w \in \mathbb{R}^l$ and the feature vector $F(s) \in \mathbb{R}^l$:

$$g(\cdot; w) = w^{\mathrm{T}} \cdot F(s)$$
$$w = [w_1, \cdots, w_l]^{\mathrm{T}} \in \mathbb{R}^l \tag{6-1}$$
$$F(s) = [f_1(s), f_2(s), \cdots, f_l(s)]^{\mathrm{T}} \in \mathbb{R}^l,$$

where l is the dimension of the feature vector. Each feature $f_i(s)$ is a function of the state variable, which is the basis of the linear approximation. Feature engineering is a continuous task to successfully apply large-scale RL. Proper feature selection results in an increased searching flexibility, high approximation accuracy, and fast training speed. Given access to the state $s \in \mathbb{R}^n$, features can be formulated by different types of basis functions, including binary function, polynomial function, Fourier function, and radial basis function.

6.1.1   Binary Basis Function

The binary basis function is a popular feature choice in computer science. The feature vector is represented as $F(s) \in \{0,1\}^l$, where l is the dimension of the binary feature. One simple design method is state aggregation, which is first used to partition the state space into a number of disjoint subsets. Then, one-hot encoding is used to build binary features. A more popular design method is tile coding, which is suitable for multidimensional continuous spaces. The range of the state space is grouped into a few partitions. Each partition is called a tiling, and is divided into a few tiles. Each element of binary features is built by indicating whether a state point is located inside a certain tile [6-12]. Both state aggregation and tile coding are computationally efficient if the state is sparse. However, as the state dimension increases, the number of required tiles grows exponentially. When there is a high-dimensional state space, its binary approximation may quickly become computationally intractable.

6.1.2   Polynomial Basis Function

In mathematics, the polynomial basis function is the building block of a polynomial ring. The most common polynomial basis is the monomial basis. The monomials form a basis because every polynomial can be uniquely written as a finite linear combination of monomials. Suppose $s = [s_1, s_2, \cdots, s_n]^T \in \mathbb{R}^n$ has n components. For an d-order polynomial approximation, each monomial feature $f_i, i \in \{1,2, \cdots, l\}$ has the following form:

$$f_i(s) = \prod_{j=1}^{n} s_j^{c_{i,j}},$$
$$\sum_j c_{i,j} \le d, c_{i,j} \in \{0,1, \dots, n\},$$

(6-2)

where $c_{i,j}$ is the order of state s_j in feature f_i, and d is the highest order of polynomials. Obviously, increasing the polynomial order enhances the fitting ability, but it also increases the sensitivity to data noise. Other useful polynomial bases include the Bernstein basis and orthogonal polynomials. The orthogonal polynomials are a family of real functions in which any two different polynomials are orthogonal to each other under a certain inner product. These polynomials have the advantage of removing some unnecessary items for a coarse but still accurate approximation. Popular orthogonal families include Hermite polynomials, Chebyshev polynomials, and Legendre polynomials.

6.1.3   Fourier Basis Function

The Fourier basis function is a class of trigonometric functions with different frequencies. This class of basis functions is particularly useful to handle periodic data or data with known boundaries. Suppose $s = [s_1, s_2, \cdots, s_n]^T \in \mathbb{R}^n$ has n components and that each component is bounded in the interval $0 \le s_j \le 1, j \in \{1,2, \cdots, n\}$. For a d-order Fourier cosine approximation, each feature $f_i, i \in \{1,2, \cdots, l\}$ has the following form:

$$f_i(s) = \cos(\pi c_i^T s),$$
$$c_i = [c_{i,1}, c_{i,2}, \cdots, c_{i,n}]^T, c_{i,j} \in \{0,1, \cdots, d\},$$

(6-3)

where $c_{i,j}$ is the order of state s_j in feature f_i, and d is the highest order of the Fourier series. The Fourier series can make full use of the orthogonal property of trigonometric functions, which is known as harmonic analysis and is extremely useful as an approach to break up an arbitrary periodic function into a set of simple sine and cosine items. In some special cases where the Fourier series can be summed into a closed form, this technique can even yield analytical approximate solutions.

6.1.4   Radial Basis Function

A radial basis function (RBF) is defined by its distance from a center. This method can be easily extended to the approximation of multivariate functions with a linear combination of univariate radial basis functions. The Gaussian radial basis function, which is a popular RBF, is used to measure the Euclidean distance between the state s and its feature center $\mu_i \in \mathbb{R}^n$ and is weighted by the feature width $\sigma_i \in \mathbb{R}$. The multidimensional Gaussian RBF is defined as

$$f_i(s) = \exp\left(-\frac{\|s - \mu_i\|^2}{2\sigma_i^2}\right), \tag{6-4}$$

where $s = [s_1, s_2, \cdots, s_n]^{\mathrm{T}} \in \mathbb{R}^n$ has n components. The Gaussian RBF approximation is actually a collocation process in a set of scattered nodes, i.e., mean and variance. The computational cost of this approximation increases nonlinearly with the dimensionality of unknown nodes but linearly with the number of data samples.

To better approximate the data distribution, the sum of multiple RBFs is used instead of a single RBF. This sum is also interpreted as the radial basis function network. In addition to its function approximation ability, the RBF network works very well in many other fields, such as interpolation, classification and time series prediction. An RBF network in its most basic form consists of three layers, namely, an input layer, a hidden layer and an output layer. The hidden layer consists of a few hidden neurons with Gaussian activation functions. Since it has only one hidden layer, RBF network has a much faster convergence rate than multilayer perceptron (MLP). For low-dimensional data, where deep feature extraction is not needed, an RBF network is preferred. However, RBF network is often less useful than MLP when the underlying characteristic feature is embedded deeply inside high-dimensional data.

6.2   Parameterization of Value Function and Policy

In previous chapters, where MC, TD, and DP were described, both value function and policy are represented by low-dimensional look-up tables. Each state or action element corresponds to an entry of tabular features. For high-dimensional or continuous-time problems, there are a large or even infinite number of states and actions to be handled, and accordingly the computation of tabular representation quickly becomes intractable. The solution for large-scale RL is to parameterize both policy and value function, which generalizes them from seen states (i.e., tabular representations) to unseen states (i.e., parameterized representations). Only the parameters need to be stored and updated for better approximation of value function and policy. This process can significantly reduce the memory demand and computational burden.

6.2.1   Parameterized Value Function

In numerical analysis, function approximation is a technique of estimating an unknown function using the observed data. In reinforcement learning, the value function approximation more closely resembles a "functional" approximation, which learns to approach a target function in the whole state space. Both state-value and action-value can be approximated by some kinds of parameterized functions, including linear approximation and nonlinear approximation. Usually, such a nonlinear approximation like neural network is more powerful than a linear approximation, even though its approximation ability has no solid theoretical guarantee. In total, there are three types of value function approximation methods:

State-value approximation:

$$V(s; w) \approx v^{\pi}(s), \forall s \in \mathcal{S}.$$

Action-value approximation:

- Type-I: Suitable for both continuous and discrete action spaces   (6-5)

$$Q(s, a; w) \approx q^{\pi}(s, a), \forall s \in \mathcal{S}, \forall a \in \mathcal{A}.$$

- Type-II: Only suitable for discrete action spaces

$$Q(s, a_i; w) \approx q^{\pi}(s, a_i), \forall s \in \mathcal{S}, i = 1, \cdots, m.$$

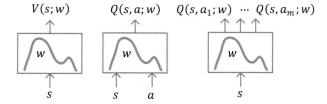

$$V(s; w) \qquad Q(s, a; w) \quad Q(s, a_1; w) \quad \cdots \quad Q(s, a_m; w)$$

(1) State-value   (2) Action-value: Type-I   (3) Action-value: Type-II

Figure 6.1 Three types of parameterized value functions

Here, $w \in \mathbb{R}^{l_w}$ is referred to as the value parameter, and l_w is the dimension of value parameter. Once the parameterized value function has been obtained, its greedy action can be calculated using one of the following formulae:

$$a^* = \arg\max_a \left\{ \sum_{s'} \mathcal{P}_{ss'}^a \left(r_{ss'}^a + \gamma V(s'; w) \right) \right\},$$ (6-6)

and

$$a^* = \arg\max_a \{Q(s, a; w)\},$$

$$a^* = \arg\max_{\{a_1, a_2, \cdots, a_m\}} \{Q(s, a_1; w), \cdots, Q(s, a_m; w)\}.$$ (6-7)

Compared with action-value function $Q(s, a)$ in $\mathcal{S} \times \mathcal{A}$ space, state-value function $V(s)$ has a lower input space, that is the state space $\mathcal{S}$. Therefore, to reach the same level of accuracy, the approximation of state-value function needs less data than that of the action-value function. The former has a computational benefit compared to dealing with the latter. In contrast, an action-value function (including both type-I and type-II functions) not only evaluates a learned policy but also contains the information about environment dynamics. This characteristic is particularly useful to compute an optimal policy when the environment model is unknown. The action-value function of type-I can be applied to those problems with both continuous and discrete actions, while type-II is more suitable for discrete actions. The benefit of type-II is that it requires only one feedforward operation to calculate the Q-value of each discrete action, which is helpful to improve the computational efficiency.

6.2.2   Parameterized Policy

In finite-element MDPs, a tabular policy is often used if both state and action spaces have low dimensions. However, when either continuous state or continuous action emerges, the parameterized policy becomes necessary to build a mapping from state to action. Similar to their tabular counterparts, parameterized policies can be categorized into two classes: (1) deterministic policy, which outputs a deterministic action for the same state; and (2) stochastic policy, which outputs a probability distribution in the action space. A parameterized stochastic policy is further divided into two types, according to whether the action space is continuous or discrete. In total, there are three types of parameterized policies:

Deterministic policy:

$$a = \pi(s; \theta) \approx \pi(s), \forall s \in S$$

Stochastic policy:

$$a \sim \pi(a|s; \theta) \approx \pi(a|s), \forall s \in S$$

(6-8)

- Type-I: Continuous action space

$$p(a; \psi(s, \theta)) \approx \pi(a|s), \forall s \in S, \forall a \in \mathcal{A}$$

- Type-II: Discrete action space

$$p(a_i|s; \theta) \approx \pi(a_i|s), \forall s \in S, i = 1, \cdots, m,$$

where $\theta \in \mathbb{R}^{l_\theta}$ is referred to as the policy parameter, l_θ is the dimension of policy parameter, $p(.;.)$ is the parameterized distribution, and $\psi(s, \theta)$ is an approximate coefficient function. The parameterization of a deterministic policy is easy to understand, which is equivalent to find an approximate function with similar input-and-output. When encountering a stochastic policy, we usually choose a parameterized action distribution and train its coefficient as the output of an approximate function. In the case of continuous action space, the commonly used stochastic policies are Gaussian functions, whose mean μ and standard deviation σ are two outputs of the approximate function, i.e., $[\mu, \sigma] = \psi(s, \theta)$ (as shown in Figure 6.2 (a)).

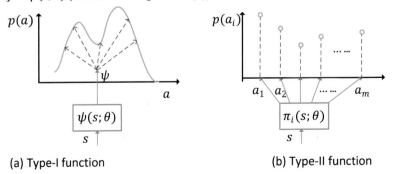

(a) Type-I function   (b) Type-II function

Figure 6.2 Stochastic policy approximations

In the case of discrete action space, we can directly output the probability of each action, instead of choosing a predefined action distribution. In other words, each discrete action is individually parameterized, as shown in Figure 6.2 (b). In this case, the Boltzmann policy is a popular choice, in which a softmax function is utilized to normalize the regression value of each action. The softmax function transforms all the real values into a closed interval between 0 and 1, whose summation must be equal to 1. Accordingly, its outputs can be interpreted as action probabilities. This normalization is also helpful to distinguish the optimal action from other non-optimal actions by raising its probability by a large margin.

The high exploration ability is one essential advantage of a stochastic policy. At each roll-out step during the training process, instead of choosing the deterministic greedy action, actions are randomly sampled from their distribution to interact with the environment. This randomly selection is equipped with stochastic behavior and can help better explore the whole space. Different from the training process, a deterministic action should be chosen when evaluating the trained policy. This is because finding an optimal policy is our desire, but this kind of policy is always deterministic for a fully observable system. A brief explanation is shown as follows. Given that $q^*(s, a)$ is at hand, assume that its optimal policy is stochastic. For an arbitrary policy $\pi(a|s)$, one always has the following inequality:

$$\int_a \pi(a|s)q^*(s, a)\mathrm{d}a \leq q^*(s, a^*) = \int_a \delta(a - a^*)q^*(s, a)\mathrm{d}a, \qquad (6\text{-}9)$$

where $\delta(\cdot)$ is the Dirac distribution, and a^* is assumed to be the unique optimal action that maximizes the optimal action-value function $q^*(s, a)$. The satisfaction of this inequality means that the optimal stochastic policy must be in the form of $\delta(a - a^*)$. It is easy to see that $\delta(a - a^*)$ forms a deterministic policy. Since the optimal policy to be pursued must be deterministic, it is highly recommended to use deterministic actions to evaluate the learned policy. There are two common approaches to find the deterministic actions from a stochastic policy. The first approach is to select the mode of the action distribution, which is similar to choosing the greedy action in a tabular policy. The other approach is to select actions with average possibilities. For example, for a Gaussian policy, only the mean of this policy is kept and its standard deviation is discarded. Fortunately, the mean of the Gaussian policy is equal to the peak probability action, which gives the most favorite action in most cases. Note that the inequality (6-9) holds only on the condition that the reward signal receives no influence from the action distribution. An RL agent with entropy regularization will have stochastic actions, and its optimal policy is not completely deterministic.

6.2.3   Choice of State and Action Representation

Thus far, we have two basic ways to represent value function and policy, namely, tabular function or approximate function. The tabular function is suitable for small-scale tasks with discrete state and action spaces. The approximate function can handle complex and large-scale tasks but inevitably introduce undesirable approximation errors. Which one should be selected depends on the types of state space and action space, as well as the problem scale. As listed in Table 6.1, RL tasks are divided into four categories according to whether their action and state spaces are discrete (marked with subscript D) or continuous (marked with subscript C).

Table 6.1 Choices of representation for RL tasks

(a) Discrete state space + discrete action space

$\mathcal{S}$ size	$\mathcal{A}$ size	V/Q	Policy
Small	Small	Tab.	Tab.
Large	Large	Appr.	Appr.

(b) Continuous state space + discrete action space

$\mathcal{S}$ Dim.	$\mathcal{A}$ size	V/Q	Policy
Low	Small	Appr.	Appr.
High	Large	Appr.	Appr.

(c) Discrete state space + continuous action space

$\mathcal{S}$ size	$\mathcal{A}$ Dim.	V/Q	Policy
Small	Low	V-Tab. Q-Appr.	Tab.
Large	High	Appr.	Appr.

(d) Continuous state space + continuous action space

$\mathcal{S}$ Dim.	$\mathcal{A}$ Dim.	V/Q	Policy
Low	Low	Appr.	Appr.
High	High	Appr.	Appr.

One example of $\mathcal{S}_D + \mathcal{A}_D$ is the indoor cleaning robot in a grid environment. Tabular methods are very suitable for this kind of task, even though the sizes of $\mathcal{S}$ and $\mathcal{A}$ are tens or even hundreds. When the space size becomes much larger, the curse of dimensionality may cause a large computational burden; therefore, proper approximate function is preferred. A typical task with $\mathcal{S}_C + \mathcal{A}_D$ in the real world is gear-shifting control in a passenger car with mechanical transmission. Its states, such as engine speed and vehicle speed, are continuous, but its action is discrete because a mechanical transmission has finite gear positions. In this task, the approximate function becomes the only choice. The task with $\mathcal{S}_D + \mathcal{A}_C$ is not very common in control practice but may appear in games. An example is the coin toss problem, in which the tossing force is continuous, while its outcome is discrete, i.e., the coin lands on either heads or tails. If this task has low-dimensional state and action, a tabular representation can still be used; however, action-value function must be approximated since its action is continuous. In the task with $\mathcal{S}_C + \mathcal{A}_C$, both state and action are continuous and the two elements, i.e., policy and value function, must be represented by their approximate functions.

6.3   Value Function Approximation

The generalization of the value function can be formulated as an optimization problem, which aims to minimize the difference between an approximated function and its target function according to a certain criterion. Let us take the state-value function as an example. The target is the true value function $v^\pi(s)$, which is approximated by a parameterized function $V(s; w)$. The purpose of the value function approximation is to find the best parameter w that minimizes a loss function $J(w)$, which measures the error between the target function and approximate function. The PEV loss function to be minimized is

$$\min_w J(w) = \mathbb{E}_{s \sim d(s)}\{\phi(v^\pi(s), V(s; w))\}, \tag{6-10}$$

where $\phi(\cdot, \cdot)$ is the measure function, and $d(s)$ is a particular weighting function. As demonstrated in Figure 6.3, the weighting function $d(s)$ strongly affects the result of the function approximation. It is easy to see that a normal weighting distribution outputs a very different approximation result from a uniform distribution. In the former, more samples are located in the high-value region, and its linear approximate function is higher than that in the latter. Accordingly, the selection of proper weighting distribution truly matters because it determines how to sample from the target function.

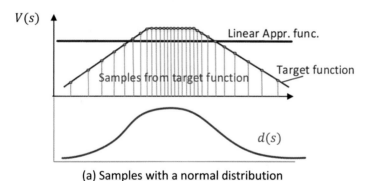

(a) Samples with a normal distribution

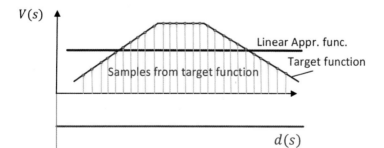

(b) Samples with a uniform distribution

Figure 6.3 Importance of the weighting function

In model-free RL, the value function approximation depends on sample-based estimation. The distribution density of collected samples determines which section of the state space is more emphasized and which section is less emphasized. For on-policy value approximation, the stationary state distribution (SSD) under the target policy π must be chosen as the weighting function $d(s)$. In theory, the same SSD should be made for an off-policy approximation because our goal is to evaluate the quality of the target policy rather than any other behavior policies. In reality, selecting target SSD in off-policy approximation might result in high variance due to large uncertainties in the calculation of IS ratios. Instead, the SSD under the behavior policy b is often selected to prevent the occurrence of high variance in off-policy value approximation.

One popular technique used to solve (6-10) is the gradient descent algorithm, which is a first-order optimization method. The stochastic gradient represents the slope of a cost function in the average sense. Starting from an initial guess, the gradient descending iteration is performed repeatedly until the minimum possible value is found. The updating formula is

$$w \leftarrow w - \alpha \nabla_w J(w), \qquad (6\text{-}11)$$

where $\alpha > 0$ is the learning rate, and $\nabla_w J(w) = \partial J(w)/\partial w$ is called the value gradient. The iteration stops when a local or the global minimum is found. The function approximation for large-scale problems raises the issue of low data efficiency, especially when considering the necessity to manage the exploration-exploitation trade-off. The value gradient needs to be accurately calculated, and its estimation variance needs to be subtly reduced.

6.3.1   On-Policy Value Approximation

The selection of the weighting function has a strong influence on the accuracy of the value approximation. The value approximation is on-policy if the collected samples are exactly from the target policy. The associated optimization problem is to minimize the mean squared error (MSE) under the target state distribution:

$$\min_w J(w) = \mathbb{E}_{s \sim d_\pi}\left\{\left(v^\pi(s) - V(s; w)\right)^2\right\}, \qquad (6\text{-}12)$$

where $d_\pi(s)$ is the SSD under target policy π. In essence, $d_\pi(s)$ represents the fraction of time that the agent stays in a certain state of the environment. The benefit of using $d_\pi(s)$ is that its associated data samples are compatible with the stationary behaviors of an RL agent as long as the environment interaction is long enough. The mean squared error is a measure of how well the approximate function fits the true value. The derivative of $J(w)$ with respect to the value parameter w is

$$\nabla_w J(w) \propto -\mathbb{E}_{s \sim d_\pi}\left\{\left(v^\pi(s) - V(s; w)\right)\frac{\partial V(s; w)}{\partial w}\right\}. \qquad (6\text{-}13)$$

The notation $\nabla_w J(w)$ in (6-13) is the value gradient. The quality of its estimation has a strong influence on the value approximation. An accurate value gradient can effectively increase the training efficiency and stability. Usually, more consistent data samples yield more accurate gradient estimation, but longer interaction time is also required. A balance must be reached between maintaining good estimation quality and reducing high computational burden to achieve a satisfactory overall runtime.

6.3.1.1   True-gradient in MC and Semi-gradient in TD

One issue in computing the value gradient is that the true value of $v^\pi(s)$ in (6-13) is actually unknown. The idea in RL is to substitute the true value with a manually selected value target. For example, MC uses the episodic return $G_{t:T}$ as the value target, and TD(0) uses one-step TD as the value target:

$$v^\pi(s) \cong \mathbb{E}_\pi\{R_t|s\},$$
$$R_t = \begin{cases} G_{t:T} & \text{for} \quad \text{MC} \\ r + \gamma V(s'; w) & \text{for} \quad \text{TD}(0) \end{cases}. \tag{6-14}$$

The selection of a proper value target is critical to efficiency and stability. Here, we take Monte Carlo RL as an example to rewrite the value gradient:

$$\nabla J_{\text{MC}}(w) = -\mathbb{E}_{s\sim d_\pi}\left\{\left(\mathbb{E}_\pi\{G_{t:T}\} - V(s; w)\right)\frac{\partial V(s; w)}{\partial w}\right\}. \tag{6-15}$$

Unlike (6-16) with one-step TD, (6-15) with MC is said to be a true value gradient. In theory, the MC estimation is unbiased for episodic tasks, which is guaranteed to at least converge to a local optimum. Similarly, we can take one-step TD to rewrite the value gradient:

$$\nabla J_{\text{TD}}(w) = -\mathbb{E}_{s\sim d_\pi}\left\{\left(\mathbb{E}_{a\sim\pi,s'\sim\mathcal{P}}\{r + \gamma V(s'; w)\} - V(s; w)\right)\frac{\partial V(s; w)}{\partial w}\right\}. \tag{6-16}$$

This TD-based formula is referred to as semi-gradient. The notation "semi-gradient" is used because the target of TD(0) contains the value parameter w during bootstrapping, but it is not considered when taking derivatives. Although the derivative of the TD target also affects the parameter update, it is often neglected in practice to avoid potential instability. This kind of compromise exists in many bootstrapping-based RLs [6-9]. Usually, TD-based gradient estimation is biased because of the truncated sample usage in calculating the value target. Nevertheless, its corresponding gradient descent algorithm can still converge to the bounded neighborhood of a minimum with stable behavior [6-13].

6.3.1.2   Sample-based estimation of value gradient

It has been noted that $\nabla_w J(w)$ in (6-13) is in the form of expectation. The value gradient needs to be estimated using the data samples from environmental interactions. In typical stochastic optimization, a small batch is taken from the whole dataset at each iteration. Although using the whole dataset is less noisy, high computational burden arises when the dataset is very large. Two practical algorithms to estimate value gradients are stochastic gradient descent and incremental least-squares.

(1) Stochastic gradient descent (SGD) algorithm: The standard version of SGD uses only a single sample to perform each parameter update, in which each sample is randomly selected from the dataset. When applying the SGD estimation, there is no need to wait until all the samples are collected and implementation can be achieved in a sample-by-sample manner. The enhanced version of SGD uses a minibatch to reduce the randomness of a single sample. Considering the fact that $\mathbb{E}_{s\sim d_\pi}\{h(s)\} \approx 1/N \sum_{i=1}^{N} h(s_i)$, if s_i is sampled from the target SSD, we have the minibatch-based gradient estimate:

$$\nabla_w J(w) \propto -\frac{1}{N} \sum_{\mathcal{D}_{\text{SGD}}} (R_i - V(s_i; w)) \frac{\partial V(s_i; w)}{\partial w}$$

$$(6\text{-}17)$$

$$\mathcal{D}_{\text{SGD}} = \{(s_i, R_i)\}, i \in \{1: N\},$$

where $\mathcal{D}_{\text{SGD}}$ is the minibatch data and N denotes its size. The advantage of SGD comes from the "stochastic" part of its name. In other words, samples are shuffled and chosen randomly from the whole dataset. In comparison, sequential samples appear in the order as they jump out from a dynamic environment. The sequential data has a strong temporal connection in two successive samples, which often leads to very poor estimation quality.

(2) Incremental least squares (ILS) algorithm: The least squares (LS) regression extends value approximation with a few advantages, such as easy recursive implementation and fairly transparent behaviors. Particularly, when a linear approximate function is adopted, it is relatively easy to gain some insight into why failure or success occurs. Batch LS regression is used to minimize the sum of squared errors (SSE) with a set of training data:

$$\min_w J(w) = \sum_{\mathcal{D}} (R_t - w^{\mathsf{T}} \cdot F(s_t))^2,$$

subject to

$$(6\text{-}18)$$

$$\mathcal{D} = \{(s_t, R_t)\}, t \in \{1: T\}$$

$$R_t = \begin{cases} G_{t:T} & \text{for} \quad \text{MC} \\ r + \gamma w^{\mathsf{T}} F(s_{t+1}) & \text{for} \quad \text{TD}(0). \end{cases}$$

where $F(s)$ is the feature vector in linear approximation. The batch LS regression builds an overly constrained algebraic equation, and its solution is a fixed point in the value function space. Therefore, the solution of optimal problem can be analytically derived without any help of gradient descent. This idea is also the training mechanism behind least-squares MC, least-squares TD, and least-squares TD(λ) [6-4]. The LS technique is highly stable and accurate. However, it requires high computational complexity. Using the Sherman-Morrison formula, an incremental LS algorithm can be derived to perform parameter updates, which is analogous to the recursive LS estimator. In the incremental LS algorithm, the computational complexity of each update is reduced to $\mathcal{O}(d^2)$, where d is the dimension of the parameter space.

6.3.1.3   Other designs for value function approximation

One can easily extend (6-10) to a general form, which guides us to select more options of value function approximation. The selection of $d(s)$ reflects how to emphasize different initial states, and $\phi(\cdot, \cdot)$ can be any measure between the true value and the parameterized function. In model-free RL, another common measure is the multistep TD error:

$$\delta_t^n(s) = r + \gamma r' + \cdots + \gamma^{n-1} r^{(n-1)} + \gamma^n V(s^{(n)}; w) - V(s; w). \tag{6-19}$$

Hence, the value function approximation changes to minimize the following mean squared TD error:

$$J(w) = \mathbb{E}_{s \sim d_\pi(s)} \left\{ \left(\delta_t^n(s) \right)^2 \right\}.$$

Its true gradient with the on-policy strategy becomes

$$\nabla_w J(w) \propto \mathbb{E}_\pi \left\{ \delta_t^n(s) \left(\gamma^n \nabla V\left(s^{(n)}; w\right) - \nabla V(s; w) \right) \right\}.$$

This formula is the true gradient of on-policy value approximation (6-12). Its one-step version is called the naive residual-gradient algorithm (Baird, 1995) [6-1]. In theory, multistep TD error has a solid convergence property. In reality, it does not converge as robustly as a semi-gradient method and may not converge to a desirable optimum. The minimization of multistep TD error may have multiple minimums since its target varies with the value parameter. In contrast, the semi-TD gradient has a fixed target, and its updating process can be more robustly stable.

6.3.2   Off-Policy Value Approximation

The off-policy approach is an effective way to balance exploitation and exploration. Different from on-policy value approximation, off-policy value approximation uses data samples that are from single behavior policy or multiple historical policies. In the previous tabular representation, tabular off-policy algorithm could be easily extended from its on-policy version. However, the extension from on-policy value approximation to off-policy value approximation is much more challenging, which will cause a few additional issues, such as the "deadly triad". The basis of off-policy value approximation is to perform gradient-based search using the dataset from the behavior policies. In theory, the same criterion as the on-policy version (6-12), which is weighted by the target state distribution, should be selected as off-policy PEV criterion. This selection is perfect in theory because our desire is to evaluate the quality of the target policy. However, severe policy mismatch issue will occur since the available samples are not from the target policy. With the technique of IS transformation, the off-policy version of semi-gradient is derived as follows:

$$\nabla_w J(w) \propto -\mathbb{E}_{s \sim d_b} \left\{ \frac{d_\pi(s)}{d_b(s)} \left(v^\pi(s) - V(s; w) \right) \nabla V(s; w) \right\}. \tag{6-20}$$

where $d_\pi(s)/d_b(s)$ is called the stationary distribution ratio, which is the quotient of the target SSD and the behavior SSD. Note that the stationary distribution ratio is a special but more complex importance sampling (IS) ratio, and its existence depends on the randomness of the state variable. Figure 6.4 compares the basic procedures of handling on-policy value approximation and off-policy value approximation. In practice, the estimation of (6-20) is computationally intractable because the SSDs of both the target policy π and the behavior policy b are not easily accessible. The quotient of these two SSDs $d_\pi(s)/d_b(s)$ is an unknown coefficient. An intuitive idea is to use sufficiently long target and behavior samples to estimate this ratio. However, the estimation accuracy can be very poor due to the lack of enough samples.

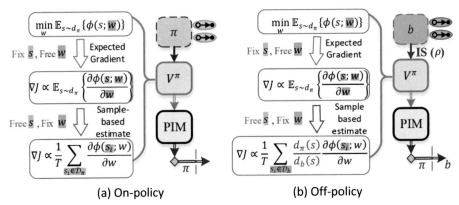

(a) On-policy   (b) Off-policy

Figure 6.4 Comparison of two value approximations

As discussed before, the selection of the weighting function reflects which state is more important and which state is less important. In theory, it is desirable to emphasize the states that appear more frequently under the target policy, but its associated stationary distribution ratio is not easily accessible. A more practical design is to change the criterion to be weighted by the behavior state distribution:

$$\min_{w} J(w) = \mathbb{E}_{s \sim d_b}\left\{\left(v^{\pi}(s) - V(s; w)\right)^2\right\}. \tag{6-21}$$

This alternative criterion allows estimating gradients without needing to explicitly know the stationary distribution ratio:

$$\nabla_w J(w) \propto -\mathbb{E}_{s \sim d_b}\left\{\left(v^{\pi}(s) - V(s; w)\right)\frac{\partial V(s; w)}{\partial w}\right\}. \tag{6-22}$$

The computation of the new value gradient (6-22) becomes easier since it can utilize the samples that are collected under behavior policies. Even though selecting a behavior state distribution as the weighting function is not theoretically perfect, it introduces huge benefit to the fast computation of off-policy value gradients. The slight sacrifice on theoretical solidness is generally acceptable in the off-policy version of policy evaluation. In the rest of this book, we will exclusively use (6-21) as the standard criterion of off-policy value approximation.

6.3.2.1   Estimation of expected gradient in MC and TD

Similar to its on-policy version, off-policy MC needs to replace $v^{\pi}(s)$ with its estimate from the samples of behavior policies. In this replacement, the issue of policy mismatch still exists because $v^{\pi}(s)$ is defined with a target policy while its samples are collected from behavior policies. Using the IS transformation, we have an off-policy MC replacement of $v^{\pi}(s)$:

$$v^{\pi}(s) \cong \mathbb{E}_{\pi}\{G_{t:T}\} = \mathbb{E}_b\{\rho_{t:T-1}G_{t:T}\}$$

$$\rho_{t:T-1} = \prod_{k=t}^{T-1}\frac{\pi(a_k|s_k)}{b(a_k|s_k)}. \tag{6-23}$$

Note that the IS ratio $\rho_{t:T-1}$ is very different from the stationary distribution ratio in (6-20). The previous stationary distribution ratio comes from the SSDs of target policy and behavior policy, whose accurate estimation should use a large amount of sufficiently long trajectories. In contrast, the IS ratio in (6-23) comes from samples with finite length, whose estimation is more computationally tractable. The basic way is to call the known target and behavior policies and calculate their ratio with a short trajectory. Accordingly, the off-policy value gradient with MC estimation becomes

$$\nabla_w J(w) \propto -\mathbb{E}_{s \sim d_b} \left\{ \left(\rho_{t:T-1} G_{t:T} - V(s;w) \right) \frac{\partial V(s;w)}{\partial w} \right\}. \tag{6-24}$$

Similarly, we have one-step TD replacement of $v^\pi(s)$:

$$v^\pi(s) \cong \mathbb{E}_{a \sim \pi, s' \sim \mathcal{P}} \{ r + \gamma V(s';w) \} = \mathbb{E}_{a \sim b, s' \sim \mathcal{P}} \{ \rho_{t:t}(r + \gamma V(s';w)) \},$$
$$\rho_{t:t} = \frac{\pi(a_t|s_t)}{b(a_t|s_t)}. \tag{6-25}$$

After this TD replacement, the new IS ratio $\rho_{t:t}$ is much simpler than the previous stationary distribution ratio. This IS ratio equals the quotient of target policy and behavior policy only at the current state-action pair. The off-policy value gradient with one-step TD is shown as follows:

$$\nabla_w J(w) \propto -\mathbb{E}_{s \sim d_b} \left\{ \left(\mathbb{E}_{a \sim b, s' \sim \mathcal{P}} \{ \rho_{t:t}(r + \gamma V(s';w)) \} - V(s;w) \right) \frac{\partial V(s;w)}{\partial w} \right\}$$
$$= -\mathbb{E}_{s \sim d_b, a \sim b, s' \sim \mathcal{P}} \left\{ \left(\rho_{t:t}(r + \gamma V(s';w)) - V(s;w) \right) \frac{\partial V(s;w)}{\partial w} \right\} \tag{6-26}$$
$$= -\mathbb{E}_b \left\{ \left(\rho_{t:t}(r + \gamma V(s';w)) - V(s;w) \right) \frac{\partial V(s;w)}{\partial w} \right\}.$$

In the value approximation, both MC-based and TD-based value gradients should contain the IS ratio in the off-policy setting. With the IS technique, we can reuse the historical data, as well as newly collected data, to calculate value gradients more accurately. If enough data are used, the quality of gradient estimation can be considerably improved in the off-policy strategy. In some special cases, off-policy value gradients may only have unknown behavior policies. For example, in autonomous driving, pre-collected data with expert drivers implicitly constitute a set of behavior policies, but these policies are not exactly known because their principles originate from human beings. In such cases, accurate estimation of IS ratios becomes difficult, and large gradient variance will occur. This in turn negatively impacts the training stability, leading to the so-called deadly triad issue.

6.3.2.2 Variance Reduction in Off-policy TD

It has been demonstrated that off-policy value gradient has higher variance than its on-policy counterpart. This result is not surprising because a behavior policy can stay far away from the target policy, and its data distribution deviates greatly from the target state distribution. A large departure in the data distribution will result in high uncertainties in the calculation of IS ratios, which quickly deteriorates the quality of value gradient estimation. In the TD-based approximation, it is useful to use the following equality to reduce the variance:

$$
\mathbb{E}_b\left\{(1-\rho_{t:t})V(s;w)\frac{\partial V(s;w)}{\partial w}\right\}
$$

$$
= \mathbb{E}_{s\sim d_b}\left\{\mathbb{E}_{a\sim b}\left\{(1-\rho_{t:t})V(s;w)\frac{\partial V(s;w)}{\partial w}\right\}\right\}
$$

$$
= \mathbb{E}_{s\sim d_b}\left\{V(s;w)\frac{\partial V(s;w)}{\partial w}\mathbb{E}_{a\sim b}\{(1-\rho_{t:t})\}\right\}
$$

$$
= \mathbb{E}_{s\sim d_b}\left\{V(s;w)\frac{\partial V(s;w)}{\partial w}\sum_a\left[b(a|s)\left(1-\frac{\pi(a|s)}{b(a|s)}\right)\right]\right\} \qquad (6\text{-}27)
$$

$$
= \mathbb{E}_{s\sim d_b}\left\{V(s;w)\frac{\partial V(s;w)}{\partial w}\left(\sum_a b(a|s)-\sum_a\pi(a|s)\right)\right\}
$$

$$
= \mathbb{E}_{s\sim d_b}\left\{V(s;w)\frac{\partial V(s;w)}{\partial w}\cdot 0\right\}
$$

$$
= 0.
$$

Adding (6-27) into (6-26), we have the modified form of the off-policy value gradient:

$$
\nabla_w J(w) \propto -\mathbb{E}_b\left\{\rho_{t:t}\big(r+\gamma V(s';w)-V(s;w)\big)\frac{\partial V(s;w)}{\partial w}\right\}. \qquad (6\text{-}28)
$$

This modified gradient can help polish the quality of the value function approximation. One interpretation is that the IS ratio often has high uncertainty due to overly small probabilities of some rare actions in the behavior policy. When multiplying the IS ratio by the TD error in (6-28), its uncertainty has a relatively weak amplification effect compared with the TD target in (6-26). This is because the TD error is much smaller than the TD target. To date, off-policy value approximation is still a tempting topic in large-scale tasks, which requires excellent skills and expert knowledge to guarantee the training stability.

6.3.3   Deadly Triad Issue

In economics, the deadly triad of poverty, unemployment and economic inequality is a catalyst for social tensions. Reinforcement learning also has a dangerous deadly triad, because of which an off-policy TD learner often fails under mild conditions. As pointed out by Richard Sutton and Andrew Barto (2018), a few examples have demonstrated that the danger of training instability arises when an RL agent combines the following three elements: (1) function approximation; (2) bootstrapping; and (3) off-policy [6-11]. Function approximation is a powerful, scalable way of representing the state and action spaces with minimal computational and memory resources. In bootstrapping, the update formula uses the historical estimate as the target (as in TD) and does not exclusively rely on complete returns (as in MC). An off-policy trainer reuses stored data and updates values weighted by the behavior policy. When all three elements come together, the convergence of successive value approximation is not always guaranteed, and severe instability may occur. This is why these three elements are considered as "deadly triad". A well-known example can be found in the book by Bertsekas and Tsitsiklis [6-2]. In this example, the environment is formulated as a two-state Markov chain, in which each state is described by a scalar feature $f(s) \in \{1,2\}$, such that

$$f\big(s_{(1)}\big) = 1, f\big(s_{(2)}\big) = 2.$$

The two states are denoted as $s_{(1)}$ and $s_{(2)}$. The transition possibility is defined as

$$H = \begin{bmatrix} 0 & \varepsilon \\ 1 & 1 - \varepsilon \end{bmatrix},$$

where ε is a sufficiently small parameter. Assume that all the transitions carry zero rewards. Let $0 < \gamma < 1$ be the discount factor. Obviously, the true value function is always zero. The stationary state distribution (SSD) is

$$d(s) = \left[\frac{\varepsilon}{1 + \varepsilon}, \frac{1}{1 + \varepsilon}\right]^{\mathrm{T}}.$$

This two-state Markov chain is illustrated below.

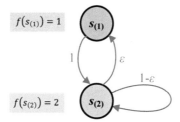

Figure 6.5 A two-state Markov chain

Consider a linear approximate function $V(s; w) = w \cdot f(s), w \in \mathbb{R}$. The basic bootstrapping update rule of the value parameter w is

$$\begin{aligned} w_{t+1} &= w_t + \alpha \mathbb{E}\{f(s_t)(0 + \gamma f(s_{t+1})w_t - f(s_t)w_t)\} \\ &= w_t + \alpha \Pr\big(s_t = s_{(1)}\big)(2\gamma - 1)w_t \\ &\quad + \alpha \Pr\big(s_t = s_{(2)}\big)(\varepsilon(2\gamma - 4)w_t + (1 - \varepsilon)(4\gamma - 4)w_t), \end{aligned} \tag{6-29}$$

where $\alpha = 0.1$ is manually chosen. Taking a specified behavior state distribution as an example, i.e., $\Pr\big(s_t = s_{(1)}\big) = \Pr\big(s_t = s_{(2)}\big) = 0.5$, we can derive

$$\frac{w_{t+1}}{w_t} = 1 + \frac{1}{20}(6\gamma - 5 + O(\varepsilon)),$$

The deadly triad issue occurs when $\gamma > 5/6$ and ε are small enough. The updating process diverges because the parameter w is multiplied by a constant factor larger than 1. If one of the three elements in the deadly triad is removed, it will converge whatever the discounted factor γ is chosen. For example:

(1) Without off-policy:

$$\frac{w_{t+1}}{w_t} = 1 + \frac{2}{5}(\gamma - 1) + O(\varepsilon) < 1$$

(2) Without bootstrapping:

$$\frac{w_{t+1}}{w_t} = 1 - \frac{1}{20}(5 + O(\varepsilon)) < 1$$

(3) Without function approximation:

$$\begin{bmatrix} V_{t+1}(s_{(1)}) \\ V_{t+1}(s_{(2)}) \end{bmatrix} = B \cdot \begin{bmatrix} V_t(s_{(1)}) \\ V_t(s_{(2)}) \end{bmatrix},$$

$$B = \begin{bmatrix} 1 - \dfrac{1}{20} & \dfrac{\gamma}{20} \\ \dfrac{\gamma}{20}\varepsilon & 1 + \dfrac{\gamma(1-\varepsilon)-1}{20} \end{bmatrix}, \rho(B) < 1.$$

Based on the analysis above, instability can easily occur even for simple tasks with linear approximate function and sufficiently accurate approximation. For more examples, see [6-2], [6-3] and [6-14].

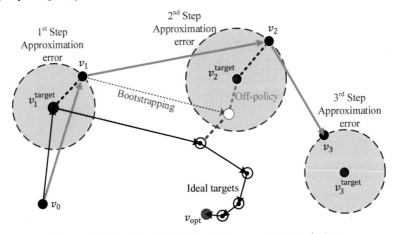

Figure 6.6 The deadly triad leads to error accumulation

A general explanation of the deadly triad is illustrated in Figure 6.6. The training instability is attributed to the mechanism of error accumulation in the target value estimation. Each element in the deadly triad contributes to a certain part of the error accumulation, which has an intrinsic amplification effect and quickly worsens. This error accumulation can be explained by the structure of a TD target value in the off-policy value approximation:

$$v^{\text{target}} = \frac{\pi(a|s)}{b(a|s)}(r + \gamma V(s'; w)). \tag{6-30}$$

First, the function approximation has an initial error due to value function parameterization. The approximation process cannot be perfect. As a result, the parameterized function is unable to reach its target accurately, especially for out-of-distribution data. Due to the lack of generalizability, the approximation error may become extremely large in less explored areas (Figure 6.7). Second, bootstrapping uses the parameterized value function as a part of its target and thus induces the last-step approximation error into the new target. Third, the off-policy approach further enlarges the error because the IS ratio is generally very inaccurate. This becomes particularly problematic when the current policy goes far from the behavior policy. The mismatch of these two policies has a strong amplification effect on the accumulated error. Thus, the deadly triad results in a vicious circle in which the target error will gradually accumulate and eventually lead to the divergence of training algorithms.

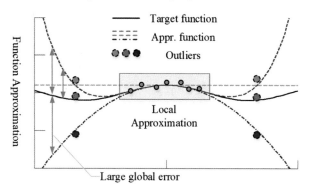

Figure 6.7 Function approximation error

In advanced RL algorithms, many tricks have been proposed to reduce the dangerous instability caused by the deadly triad. These tricks can be roughly divided into three categories. First, a better balance of approximation accuracy and generalization ability helps alleviate the error accumulation. Hasselt et al. (2018) pointed out that unbounded divergence disappears when adopting approximate functions with better generalizability, such as artificial neural networks [6-14]. Second, more "true" rewards in bootstrapping are beneficial to accurately estimate the target. For example, compared to a one-step TD target, a multistep TD target can be used to reduce the propagation of accumulated errors from historical steps. Third, restricting the IS ratio to a reasonable interval (i.e., clipping technique) can limit the variance increase of target estimation. Off-policy RLs with ratio clipping techniques, such as IMPALA [6-5] and Retrace [6-8], show better stability than their non-clipping counterparts.

6.4   Policy Approximation

Indirect RL and direct RL are two families of RL algorithms. Both of these families need function approximation to search for a quasi-optimal policy, but there is an essential difference in their policy gradient derivation. Indirect RL contains two cyclic steps: policy evaluation (PEV) and policy improvement (PIM). The policy search occurs only in the PIM step, in which its value function is already known from the PEV step. The known value function is fixed and does not change with the policy parameter. In contrast, direct RL needs to regard the weighted value function (if any) as a function of the policy parameter, which introduces a nonstationary state distribution into the policy gradient derivation. In this section, only the policy gradient derivation in the indirect RL family is discussed. With the tabular representation, PIM seeks to find a greedy policy, which gives the best action for each state element:

$$\pi^g(s) = \arg\max_a \{Q(s, a)\}, \forall s \in \mathcal{S}. \tag{6-31}$$

When extended from tabular representation to function approximation, it becomes impossible to perform the greedy search for each state element. Instead, we convert the greedy search into a stochastic optimization problem. For simplicity, we begin the PIM problem formulation with a deterministic policy:

$$\theta = \arg\max_{\theta}\{J(\theta)\}$$

with

(6-32)

$$J(\theta) = \mathbb{E}_{s \sim d(s)}\{Q(s, \pi_\theta(s))\},$$

where $\pi_\theta(s)$ is the parameterized policy, and $d(s)$ is a particular weighting function. The notation $J(\theta)$ represents a PIM loss function that guides the searching direction of the policy parameter. Clearly, the weighting function $d(s)$ has a noticeable influence on the optimal solution. Its selection, however, is not completely free and usually comes from that of the collected data. Such a PIM optimization problem can be solved by various derivative-free or derivative-based numerical optimization algorithms. A popular version is the SGD algorithm, which is given as:

$$\theta \leftarrow \theta + \beta \nabla_\theta J(\theta),$$

(6-33)

where β is the learning rate, and $\nabla_\theta J(\theta) = \partial J(\theta)/\partial \theta$ is called "indirect policy gradient". The term "indirect" is intentionally used to distinguish it from policy gradients in direct RL. One advantage of indirect policy gradient lies in its structural simplicity. Its derivation starts from (6-32), and $Q(s, a)$ does not change with the policy parameter. The structural separation of value updates and policy updates remarkably reduces the complexity of deriving an analytical policy gradient in the PIM step. In contrast, direct RL must handle the complexity of policy-dependent value function. Its gradient derivation must go through a complicated cascading recursion with a few assumptive conditions on the initial state distribution.

6.4.1   Indirect On-Policy Gradient

In fact, this type of PIM problem formulation for policy approximation can address both deterministic policy and stochastic policy. Considering the popularity of stochastic policy, we mainly discuss the derivation of stochastic policy gradients. The PIM loss function can be written into (6-34) if combined with a parameterized stochastic policy:

$$J(\theta) = \mathbb{E}_{s \sim d(s)}\left\{\sum_a \pi_\theta(a|s)Q(s, a)\right\}.$$

(6-34)

In the on-policy strategy, the target state distribution $d_\pi(s)$ replaces the predefined weighting function $d(s)$ for accurate gradient estimation. Unlike the value function approximation, the policy approximation must take care of the order of "taking derivation" and "weight replacement". The correct order is to first take the derivative under a fixed weighting function $d(s)$ and then to replace the weight $d(s)$ with a target SSD $d_\pi(s)$. The merit of keeping this order comes from the fact that $d_\pi(s)$ is a function of the policy parameter while the weighting function $d(s)$ is not. If the order is reversed, the policy parameter will affect the SSD, and one cannot easily derive the policy gradient.

6.4.1.1   Policy gradient with action-value function

The action-value function $Q(s,a)$ is fixed, i.e., independent of the policy parameter, since its estimate is already determined in the PEV step. The weighting function $d(s)$ is a predefined fixed coefficient and is independent of the policy parameter. Based on this point of view, a large simplification is introduced when taking derivatives with respect to the policy parameter. Accordingly, we only need to handle the derivative of $\pi_\theta(a|s)$ in (6-34) but not that of the action-value function and its weighting function. The primitive version of the indirect policy gradient is easily derived:

$$\nabla_\theta J(\theta) = \sum_s d(s) \sum_a \nabla_\theta \pi_\theta(a|s) Q(s,a). \tag{6-35}$$

Now, let us change our viewpoint about what $d(s)$ truly stands for. One can further regard this weighting function as an SSD to facilitate sample-based estimation. Since an RL agent is more concerned about the stationary performance under the target policy, $d(s) = d_\pi(s)$ is reasonably chosen as the weighting function:

$$
\begin{aligned}
\nabla_\theta J(\theta) &= \sum_s d_\pi(s) \sum_a \nabla_\theta \pi_\theta(a|s) Q(s,a) \\
&= \sum_s d_\pi(s) \sum_a \pi_\theta(a|s) \nabla_\theta \log \pi_\theta(a|s) Q(s,a) \\
&= \mathbb{E}_{s \sim d_\pi, a \sim \pi}\{\nabla_\theta \log \pi_\theta(a|s) Q(s,a)\},
\end{aligned}
\tag{6-36}
$$

where $\nabla_\theta J(\theta)$ is the on-policy version of indirect policy gradient. Actually, the first formula in (6-36) cannot be directly used because it is not in the form of an expectation. In the second formula, a popular statistical technique called "log-derivative trick" is used to construct its log-probability form, i.e., $\nabla_\theta \pi_\theta(a|s) = \pi_\theta(a|s) \nabla_\theta \log \pi_\theta(a|s)$. This log-derivative reformulation finally generates the third formula, whose expectation depends on the state and action distribution of the same policy. Such an expectation formula is computable with sample-based estimation, which allows us to estimate indirect policy gradients with on-policy collected samples.

6.4.1.2   Policy gradient with state-value function

If only the state-value function is available, a new policy gradient can be derived by expanding the action-value to be the sum of the current reward and the state-value of the next state. Recall the self-consistency condition between $Q(s,a)$ and $V(s)$, and we have the following relationship:

$$Q(s,a) \cong \mathbb{E}_{s' \sim \mathcal{P}}\{r + \gamma V(s')\}. \tag{6-37}$$

Substituting (6-37) into (6-36) and rearranging the positions of three random variables (s, a, s'), we have a policy gradient with the state-value function:

$$\nabla_\theta J(\theta) = \mathbb{E}_\pi\{\nabla_\theta \log \pi_\theta(a|s) (r + \gamma V(s'))\}. \tag{6-38}$$

This function is the second version of indirect on-policy gradient. The benefit of (6-38) is that its computational burden is often lower than that of (6-36) since the state-value function has only one random variable, i.e., state, while the action-value function has two random variables, i.e., both state and action. The dimensional reduction is also helpful to reduce the variance of policy gradient estimation because a more accurate value approximation is obtained with the same scale of samples.

6.4.2   Indirect Off-Policy Gradient

As discussed in off-policy value approximation, there may be no sufficient amount of data under the target policy. The off-policy strategy allows us to use the data from behavior policies to calculate indirect policy gradients. The starting point of deriving an off-policy gradient is very similar to that of its on-policy counterpart. Both of their gradient derivations start from the primitive formula (6-35), which has a fixed weighting function $d(s)$. In the off-policy version, one does not have access to the target samples and thus needs to choose the SSD of the behavior policy as the weighting function, i.e., $d(s) = d_b(s)$. Therefore, the off-policy version of indirect policy gradient becomes

$$\nabla_\theta J(\theta) = \sum_s d_b(s) \sum_a \pi_\theta(a|s) \nabla_\theta \log \pi_\theta(a|s) Q(s, a). \tag{6-39}$$

Interestingly, one can compute this off-policy gradient (6-39) without introducing any IS ratio. This is because the estimation of action-value function does not need the information of state transition, and the action can be easily sampled from the currently known policy. For example, assume that we have a sample form the behavior policy, i.e., (s_b, a_b, s_b'), where the subscript represents its sampling policy. The next state s_b' is not necessary due to the use of action-value function, and the action a_b is replaced with $a_\pi \sim \pi_\theta(a|s_b)$ to generate action distribution. It is easy to see there is no IS ratio in this off-policy gradient. Different from this action-value version, (6-39) cannot be directly computed in its state-value version, i.e., replacing $Q(s, a)$ with $\mathbb{E}_{s' \sim \mathcal{P}}\{r + \gamma V(s')\}$, as demonstrated in (6-37) and (6-38). This is because the information of state transition must be used in estimating the state-value function. However, $d_b(s)$ and $\pi_\theta(a|s)$ do not match each other: one is from the behavior policy, and the other is the target policy. One more conversion is needed to build a computable off-policy gradient. Applying the IS transformation to (6-39), we have

$$\nabla_\theta J(\theta) = \sum_s d_b(s) \sum_a b(a|s) \frac{\pi_\theta(a|s)}{b(a|s)} \nabla_\theta \log \pi_\theta(a|s) \sum_{s'} p(s'|s, a) (r + \gamma V(s')),$$

and therefore

$$\nabla_\theta J(\theta) = \mathbb{E}_{s \sim d_b, a \sim b} \left\{ \frac{\pi_\theta(a|s)}{b(a|s)} \nabla_\theta \log \pi_\theta(a|s) (r + \gamma V(s')) \right\}. \tag{6-40}$$

This new formula is the indirect off-policy gradient with state-value function. Not surprisingly, this new off-policy gradient suffers from higher variance than that of action-value function. In the formula (6-40), the high-variance issue is attributed to several reasons, including approximation error accumulation in the value function and the large discrepancy in the IS ratio. The errors in function approximation can be accumulated quickly in the bootstrapping process. The states rarely visited by the behavior policy but that are important to the target policy introduce large uncertainties to the IS ratio. The two terms together bring high variance to the estimation of policy gradients. In contrast, (6-39) does not contain any IS ratio, and the mismatching uncertainty between behavior policy and target policy does not exist, which is helpful to reduce the variance of off-policy gradient estimation.

6.4.3　Revisit Better Policy for More Options

One might be interested in asking whether such a greedy search as (6-32) is the exclusive option in the design of policy approximation. The answer is definitely no. As discussed in the previous chapter, the ameliorated definition of better policy shows remarkable advantages over the traditional element-by-element definition. The previous PIM has tabular representation, which is limited to low-dimensional MDPs with finite state and action spaces. Using function approximation, it is possible to search for a parameterized policy in the continuous state and action space. This section will introduce two new designs of PIM optimization problems. One design is based on entropy regularization, in which the traditional PIM loss function is penalized with a special policy entropy. Entropy regularization is used to equip an RL agent with a better exploration ability. The other PIM loss function is based on the expectation-based definition of better policy, which can guarantee a monotonic policy improvement and prevent quick policy changes.

6.4.3.1　Penalize greedy search with policy entropy

One might be interested in the design of the PIM loss function, whose maximization has intrinsic equivalence with the ϵ-greedy search. The importance of selecting an ϵ-greedy search in PIM is its ability to guarantee convergence. On the basis of the standard loss function for tabular PIM (6-31), an additive policy entropy is used to mimic the behavior of randomly selecting actions with a small probability. When extending it to the continuous state space, a new PIM loss function with entropy regularization is designed:

$$J(\theta) = \mathbb{E}_{s \sim d_\pi}\{\pi_\theta(a^*|s) + \lambda \cdot \mathcal{H}(\pi_\theta(\cdot|s))\}$$

with

$$a^* = \arg \max_a Q(s, a), \tag{6-41}$$

$$\mathcal{H}(\pi_\theta(\cdot|s)) = -\int_a \pi_\theta(a|s) \log \pi_\theta(a|s) \, da,$$

where λ is a hyperparameter that balances optimality and exploration. In (6-41), there are two individual terms in the PIM loss function. The first term is used to maximize the probability of choosing the greedy action, while the second term is used to ensure that the learned policy has a certain degree of randomness. To derive the indirect policy gradient, we choose to train a parameterized stochastic policy with a Gaussian distribution, i.e., $\pi_\theta \sim \mathcal{N}(\mu(\theta), \mathcal{K}(\theta))$. Using the Gaussian policy, the first term depends on both the parameterized mean $\mu(\theta)$ and the parameterized covariance $\mathcal{K}(\theta)$, while the second term is only related to the parameterized covariance $\mathcal{K}(\theta)$. Therefore, the entropy-regularized policy gradient is

$$\nabla_\theta J(\theta) = \mathbb{E}_{s \sim d_\pi}\{\nabla_\theta \pi_\theta(a^*|s) + \lambda \cdot \nabla_\theta \log \det(\mathcal{K}(\theta))\}. \tag{6-42}$$

In essence, the formula above directly extends the ϵ-greedy policy from the discrete action space to the continuous action space. Unfortunately, the optimal action a^* must be known in advance before performing any policy gradient calculation. Therefore, $\max Q(s, a)$ can be taken as a standard optimization problem, which relies on a first-order optimization algorithm to find its maximum. Figure 6.8 conceptually illustrates the Gaussian policy after the PIM update, in which a greedy action is searched from the action-value function. The variance term of the Gaussian policy provides a balance between deterministic action and exploration ability.

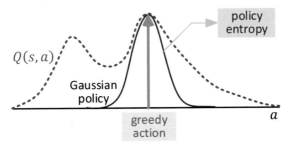

Figure 6.8 Gaussian policy after PIM update

Entropy regularization can be used to encourage the agent to explore more unknown regions and determine more information about the environment. Therefore, an RL agent with this PIM criterion usually has better exploration ability and can be used to achieve higher sample efficiency. However, its policy optimality must be somewhat sacrificed because entropy regularization will lead to a stochastic policy rather than a deterministic policy. The reason is very similar to the ϵ-greedy search since they are the same in nature. A subtle solution is to gradually reduce the hyperparameter λ as it approaches the end of the learning process.

6.4.3.2   Expectation-based definition for better policy

In this section, how to parameterize the expectation-based definition of better policy is discussed. One might see that adding proper regularization into the PIM loss function can introduce additional advantages. The question is what kind of modification is acceptable in the PIM step of indirect RL? The answer is that this modified PIM optimization must still guarantee convergence. The key is whether one can find a new definition about what a better policy is. Whenever talking about "better", we must define a performance measure to compare two policies. Here, let us use the Kullback–Leibler (KL) divergence to measure the distance of two policies:

$$D_{\mathrm{KL}}(\pi_\theta, \pi_{\mathrm{old}}) \stackrel{\text{def}}{=} \int \pi_\theta(a|s) \log \frac{\pi_\theta(a|s)}{\pi_{\mathrm{old}}(a|s)} \, da, \tag{6-43}$$

where θ is the policy parameter to be optimized and π_{old} is the policy before each PIM update. Then, the expectation-based condition is formulated as a constrained optimization problem:

$$\theta = \arg\max_\theta \{\delta - \rho \cdot \mathbb{E}_{s \sim d(s)} D_{\mathrm{KL}}(\pi_\theta, \pi_{\mathrm{old}})\}$$

subject to $\tag{6-44}$

$$\mathbb{E}_{s \sim d(s), a \sim \pi_\theta}\{Q^{\pi_{\mathrm{old}}}(s, a)\} = \delta + \mathbb{E}_{s \sim d(s), a \sim \pi_{\mathrm{old}}}\{Q^{\pi_{\mathrm{old}}}(s, a)\},$$

where δ is the criterion increment and ρ is the penalty constant. The criterion increment is actually a reflection of the distance between the current policy π and the old policy

π_{old}. It is desirable to maximize this criterion increment to maintain the policy improvement ability. The notation $Q^{\pi_{\text{old}}}$ is defined with the superscript "old" for narrative convenience. Actually, $Q^{\pi_{\text{old}}}$ is equal to $Q(s,a)$ in (6-34), which is a known and fixed action-value estimate from the previous PEV. Let us convert (5-60) into an unconstrained optimization problem by replacing the criterion increment with the equality constraint. After conversion, the indirect policy gradient at the current PIM step becomes

$$\nabla_\theta J = \mathbb{E}_{s \sim d(s)} \left\{ \sum_a \nabla_\theta \pi_\theta(a|s) Q^{\pi_{\text{old}}}(s,a) \right\} - \rho \mathbb{E}_{s \sim d(s)} \{ \nabla_\theta D_{\text{KL}}(\pi_\theta, \pi_{\text{old}}) \}.$$

Thus far, the collected samples at hand are from the old policy $d_{\pi_{\text{old}}}$, which is also the target policy. Replacing the weighting function $d(s)$ with the SSD of the old policy $d_{\pi_{\text{old}}}$, we have the following derivation:

$$\approx \mathbb{E}_{s \sim d_{\pi_{\text{old}}}} \left\{ \sum_a \pi_{\text{old}}(a|s) \frac{\nabla_\theta \pi_\theta(a|s)}{\pi_{\text{old}}(a|s)} Q^{\pi_{\text{old}}}(s,a) \right\}$$

$$- \rho \mathbb{E}_{s \sim d_{\pi_{\text{old}}}} \{ \nabla_\theta D_{\text{KL}}(\pi_\theta, \pi_{\text{old}}) \} \tag{6-45}$$

$$= \mathbb{E}_{s \sim d_{\pi_{\text{old}}}, a \sim \pi_{\text{old}}} \left\{ \frac{\nabla_\theta \pi_\theta(a|s)}{\pi_{\text{old}}(a|s)} Q^{\pi_{\text{old}}}(s,a) \right\}$$

$$- \rho \mathbb{E}_{s \sim d_{\pi_{\text{old}}}} \{ \nabla_\theta D_{\text{KL}}(\pi_\theta, \pi_{\text{old}}) \}.$$

To simplify the gradient derivation, let us assume that the parameterized policy obeys a Gaussian distribution. The analytical form of $\nabla_\theta J$ becomes a more concise structure:

$$\nabla_\theta J = \mathbb{E}_{\pi_{\text{old}}} \left\{ \frac{\nabla_\theta \pi_\theta(a|s)}{\pi_{\text{old}}(a|s)} Q^{\pi_{\text{old}}}(s,a) \right.$$

$$+ \rho \nabla_\theta \left[\log \det(\mathcal{K}(\theta)) - \text{tr}\left(\mathcal{K}_{\text{old}}^{-1} \mathcal{K}(\theta) \right) \right. \tag{6-46}$$

$$\left. \left. - (\mu(\theta) - \mu_{\text{old}})^T \mathcal{K}_{\text{old}}^{-1}(\mu(\theta) - \mu_{\text{old}}) \right] \right\},$$

where μ_{old} and $\mathcal{K}_{\text{old}}$ are the mean and covariance of the Gaussian policy, respectively. Note that choosing $d(s) = d_{\pi_{\text{old}}}(s)$ takes advantage of the target data distribution, and the first term on the right-hand side of (6-45) becomes an exact on-policy gradient.

The KL divergence is utilized to avoid substantial parameter changes between two adjacent policies. Slow parameter updating is helpful to stabilize the training process. Readers may find that this updating formula is very similar to the trust region policy optimization (TRPO). Their core mechanisms, however, are completely different. The result in this section comes from the reformulation of the PIM loss function from the new definition of better policy. In this method, the KL divergence of two policies is treated as a penalty term, and the updated formula comes from solving an unconstrained optimization problem. In contrast, TRPO comes from the principle of minorize-maximization optimization. The constrained surrogate function is constructed from the overall RL objective function and the Lagrangian duality optimization is used to solve it.

Nevertheless, their final outcomes are almost identical to each other. The policy gradient (6-45) is exactly the same as the penalty variant of proximal policy optimization (PPO), which is an extended version of TRPO. One can further modify (6-45) with a clipped function, which will become equivalent to the clipped variant of PPO.

6.5   Actor-Critic Architecture from Indirect RL

Standard indirect RL works by alternating two cyclic elements: a complete policy evaluation (PEV) and a complete policy improvement (PIM). The term "complete" means that PEV will iterate infinite times and PIM will find greedy actions. After being generalized with parameterized functions, indirect RL actually evolves into the well-known actor-critic (AC) architecture. The AC architecture is commonly featured with function approximation and gradient-based update. As indicated by its name, an AC algorithm under this architecture has two cyclic components: one is called "actor", and the other is called "critic". The actor controls how the agent behaves with respect to a learned policy, while the critic evaluates the agent's behavior by estimating its value function [6-6][6-7]. Generally, PEV corresponds to the critic update, and PIM corresponds to the actor update. Figure 6.9 illustrates this AC architecture from the viewpoint of indirect RL, in which the same weighting function $d(s)$ is used to build both critic and actor loss functions.

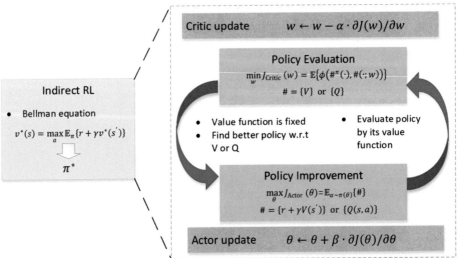

Figure 6.9 Actor-critic architecture from an indirect RL viewpoint

When using function approximation, a complete critic and a complete actor require an infinite number of gradient-based updates at each RL cycle. Complete updates are often impractical due to the high computational burden. To achieve better computational efficiency, most AC algorithms change to use incomplete critic and actor updates. The term "incomplete" means that both PEV and PIM only perform a limited number of gradient-based updates at each RL cycle. Unlike complete critic updates, incomplete critic updates will output imperfect policy evaluation. A similar imperfection in policy improvement is also observed in incomplete actor updates, i.e., actor updates are not able to reach the point of perfect greedy action. Besides, AC with incomplete updates may easily oscillate or even diverge due to bad policy evaluation and inexact greedy search. This issue can become worsen when the environmental exploration is not sufficient. Therefore, the updating numbers of critic and actor at each cycle need to be properly selected to achieve a balance between training stability and computational efficiency.

6.5.1 Generic Actor-Critic Algorithms

Figure 6.10 shows the flow chart of typical model-free AC algorithms. During each cycle, the on-policy strategy is adopted to sample data from the environment. The critic first updates the value parameter, either state-based or action-based, to obtain a good evaluation of each policy. The actor then updates the policy parameter, either stochastic or deterministic, to move closer to greedy actions.

Depending on the types of value function and policy, one can classify the state-of-the-art AC algorithms into four categories. One dimension of this classification depends on the type of value function, either state-value function or action-value function, and the other relies on the policy type, either stochastic policy or deterministic policy. The combination of these two dimensions results in four classes of AC algorithms, which are listed in Table 6.2. Unfortunately, one of these four combinations does not exist in model-free setting, i.e., deterministic policy with state-value function (abbreviated as Det-V). This combination does not exist and its actor gradient has no analytical formula when the environment model is unknown. Therefore, there are only three classes of model-free AC algorithms, i.e., stochastic policy with state-value function (abbreviated as Sto-V), deterministic policy with action-value function (abbreviated as Det-Q), and stochastic policy with action-value function (abbreviated as Sto-Q).

Table 6.2 Classification of AC algorithms from indirect RL

	Action-value	State-value
Deterministic policy	$J_{\text{Critic}} = \mathbb{E}_{s,a}\left\{(q - Q(w))^2\right\}$	$J_{\text{Critic}} = \mathbb{E}_s\left\{(v - V(w))^2\right\}$
	$J_{\text{Actor}} = \mathbb{E}_s\{Q(s, \pi_\theta(s); w)\}$	$J_{\text{Actor}} = \mathbb{E}_{s,s'}\{r + \gamma V(s'; w)\}$
Stochastic policy	$J_{\text{Critic}} = \mathbb{E}_{s,a}\left\{(q - Q(w))^2\right\}$	$J_{\text{Critic}} = \mathbb{E}_s\left\{(v - V(w))^2\right\}$
	$J_{\text{Actor}} = \mathbb{E}_{s,a}\{Q(s, a; w)\}$	$J_{\text{Actor}} = \mathbb{E}_{s,a,s'}\{r + \gamma V(s'; w)\}$

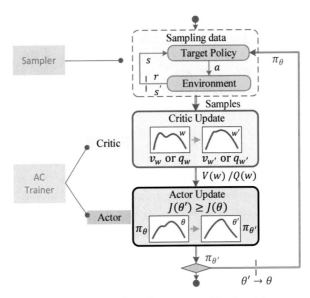

Figure 6.10 Flow chart for actor-critic algorithms

Let us explain why such a model-free deterministic policy gradient does not exist when it combines with state-value function. David Silver (2014) has claimed that deterministic policy gradient has an inherent difference from its stochastic version, even though they look similar at first glance. Fortunately, for a wide class of stochastic policies, the extreme version of stochastic policy gradients is able to become deterministic policy gradients (Theorem 2 in [6-10]). One can assume that a stochastic policy has the form of $\pi_\theta(a|s;\sigma)$, where σ is a term related to the action variance. The required condition of converting the stochastic policy into a deterministic policy is to enforce the action variance σ to become zero. When $\sigma \to 0$, the stochastic policy converges to a delta distribution:

$$\lim_{\sigma \to 0} \sum_a \pi_\theta(a|s;\sigma)\, a = \pi_\theta(s)$$

Therefore, $\pi_\theta(a|s;\sigma)$ can be rewritten in a parameterized form:

$$\pi_\theta(a|s;\sigma) \cong \pi_\theta(a|\sigma,\eta),$$

where $\eta = \pi_\theta(s)$ is the action from a deterministic policy. It can be proven that as $\sigma \to 0$, the stochastic policy gradient converges to a deterministic policy gradient in the action-value case. A short proof is shown as follows:

$$\lim_{\sigma \to 0} \nabla_\theta J\big(\pi_\theta(a|s;\sigma)\big)$$

$$= \lim_{\sigma \to 0} \sum_s d^\gamma_{\pi_\theta(a|s;\sigma)}(s) \sum_a Q^{\pi_\theta(a|s;\sigma)}(s,a) \nabla_\theta \pi_\theta(a|s;\sigma)$$

$$= \lim_{\sigma \to 0} \sum_s d^\gamma_{\pi_\theta(a|s;\sigma)}(s) \sum_a Q^{\pi_\theta(a|\sigma,\eta)}(s,a) \nabla_\theta \pi_\theta(s) \nabla_\eta \pi_\theta(a|\sigma,\eta)$$

$$= \lim_{\sigma \to 0} \sum_s d^\gamma_{\pi_\theta(a|s;\sigma)}(s)\, \nabla_\theta \pi_\theta(s) \sum_a Q^{\pi_\theta(a|\sigma,\eta)}(s,a) \nabla_\eta \pi_\theta(a|\sigma,\eta)$$

By using integration by parts and omitting the boundary term, we have:

$$= \sum_s \lim_{\sigma \to 0} d^\gamma_{\pi_\theta(a|s;\sigma)}(s) \nabla_\theta \pi_\theta(s) \sum_a \nabla_a Q^{\pi_\theta(a|\sigma,\eta)}(s,a)\pi_\theta(a|\sigma,\pi_\theta(s))$$

$$= \sum_s d^\gamma_{\pi_\theta(a|s;0)}(s) \nabla_\theta \pi_\theta(s) \lim_{\sigma \to 0} \sum_a \nabla_a Q^{\pi_\theta(a|\sigma,\eta)}(s,a)\pi_\theta(a|\sigma,\pi_\theta(s))$$

As $\sigma \to 0$, π_θ reduces to a delta distribution:

$$= \sum_s d^\gamma_{\pi_\theta(a|s;0)}(s) \nabla_\theta \pi_\theta(s) \nabla_a Q^{\pi_\theta(s)}(s, \pi_\theta(s))$$

$$= \nabla_\theta J(\pi_\theta(s)).$$

When the randomness in the stochastic policy goes to be infinitely small, the stochastic policy gradient reduces to a deterministic policy gradient in the action-value case. Unfortunately, in the state-value case, the aforementioned successful derivation does not hold again when there is no environment model. This is because $v(s')$ rather than $v(s)$ must be utilized to perform the gradient derivation, and its partial derivative with respect to the current action must contain the transition dynamics from the current state s to the next state s'. According to the chain rule, this partial derivative is

$$\nabla_a v(s') = \frac{\partial s'^{\mathrm{T}}}{\partial a}\frac{\partial v(s')}{\partial s'} = \frac{\partial f^{\mathrm{T}}(s,a)}{\partial a}\frac{\partial v(s')}{\partial s'}. \tag{6-47}$$

It is easy to see that an accurate model $f(s,a)$ is required to calculate the partial derivative of the state-value function. If $\partial f(s,a)/\partial a$ is not accessible, it is impossible to build an analytical version of deterministic policy gradient. Here, the term "analytical" is intentionally used because a model-free Det-V algorithm may still be implemented with sample-based estimation. For example, one can roughly estimate $\partial f(s,a)/\partial a$ with the second-order difference technique and then use this estimation to compute policy gradients. In nature, this kind of numerical approximation does not belong to standard first-order optimization. Moreover, it policy gradient variance must be very high, and the training quality is often not satisfactory.

6.5.2   On-policy AC with Action-Value Function

On-policy AC with action-value is a major branch of model-free AC algorithms. Compared with the state-value function, the action-value function contains more information, including both action evaluation and environment dynamics. The utilization of the action-value function is compatible with both the deterministic policy and the stochastic policy. When equipped with a deterministic policy, the actor gradient becomes

$$\nabla_\theta J_{\text{Actor}} = \mathbb{E}_{s \sim d_\pi}\{\nabla_\theta \pi_\theta(s)\nabla_a Q(s,a;w)\}, \tag{6-48}$$

This formula is referred to as deterministic policy gradient (DPG). This actor gradient is a vector with $l_\theta \times 1$ dimension, in which l_θ is the dimension of policy parameter. Here, $\nabla_\theta \pi_\theta(s) \overset{\text{def}}{=} \partial \pi^{\mathrm{T}}/\partial \theta \in \mathbb{R}^{l_\theta \times m}$ is a matrix, and $\nabla_a Q(s,a;w) \in \mathbb{R}^{m \times 1}$ is a column vector. Note that m is the dimension of the action. Usually, such an actor policy gradient has a much lower variance than its stochastic counterpart because its expectation operator is only taken over the state distribution, i.e., no-action-distribution. All actions can be replaced by the deterministic policy. As a result, its gradient estimation only focuses on the state distribution. One shortcoming is that DPG has no state-value version, and accordingly adding baselines to action-value requires an additional state-value function, which will somewhat increase the computational burden of critic updates.

When equipped with a stochastic policy, the log-derivative trick is a common statistical technique to derive stochastic policy gradients. Unfortunately, this kind of stochastic policy gradient suffers from high variance in the sample-based estimation. A better approach is to utilize the reparameterization trick to output a low-variance stochastic gradient. The reparameterized stochastic policy gradient is

$$\nabla_\theta J_{\text{Actor}} = \nabla_\theta \mathbb{E}_{s \sim d_\pi, a \sim \pi_\theta} \{Q(s, a; w)\}.$$

Reparameterization: $a = g_\theta(s, \epsilon)$

$$(6\text{-}49)$$

$$
\begin{aligned}
\nabla_\theta \mathbb{E}_{s, a \sim \pi_\theta} \{Q(s, a; w)\} &= \nabla_\theta \mathbb{E}_{s, \epsilon \sim p(\epsilon)} \{Q(s, g_\theta(s, \epsilon); w)\} \\
&= \mathbb{E}_{s, \epsilon \sim p(\epsilon)} \{\nabla_\theta Q(s, g_\theta(s, \epsilon); w)\} \\
&= \mathbb{E}_{s, \epsilon \sim p(\epsilon)} \{\nabla_\theta g_\theta(s, \epsilon) \nabla_a Q(s, a; w)\},
\end{aligned}
$$

where $g_\theta(s, \epsilon)$ is a reparameterized policy and ϵ is a random variable with a known distribution $p(\epsilon)$. In essence, the reparameterization of a stochastic policy decomposes the policy into a deterministic function and a random noise term. When choosing a Gaussian policy, i.e., $\pi_\theta(a|s) \sim \mathcal{N}(\mu_\theta(s), \sigma_\theta^2(s))$, its reparameterization becomes the sum of two separable items:

$$g_\theta(s, \epsilon) = \mu_\theta(s) + \epsilon \cdot \sigma_\theta(s),$$

where $\epsilon \sim \mathcal{N}(0,1)$ is the Gaussian random noise. When a stochastic policy is chosen to obey the uniform distribution, i.e., $\pi_\theta(a|s) \sim \mathcal{U}(\theta_{\min}, \theta_{\max})$, it can be reparameterized as

$$g_\theta(s, \epsilon) = \frac{\theta_{\max} + \theta_{\min}}{2} + \epsilon \cdot \frac{\theta_{\max} - \theta_{\min}}{2},$$

where $\epsilon \sim \mathcal{U}(-1,1)$ is a standard uniform distribution. After reparameterization, the stochastic policy gradient becomes very similar to the deterministic policy gradient. This new policy gradient holds the benefit of having no-action distribution. Its variance is much reduced due to the elimination of action randomness in the expectation estimation. In other words, this trick actually sends the information of predefined policy structure into the stochastic policy gradient and builds the new policy gradient with more determinism. As a result, the partial derivative of action-value function is explicitly taken into consideration, which decreases the dependency on noisy measurement data.

The ability to reduce variance depends on how to choose the distribution of random noise. If the chosen distribution matches the reality, excellent variance reduction will be achieved; otherwise, it may block the reparameterized policy to reach its optimum. The pseudocodes of two on-policy AC algorithms are shown in Algorithm 6-1 and Algorithm 6-2, in which a stochastic policy (i.e., Sto-Q) and a deterministic policy (i.e., Det-Q) are trained, respectively. Note that the update frequency defines how many iterations are maximally allowed in either critic updates or actor updates.

Algorithm 6-1: On-policy AC with Q-value and stochastic policy (i.e., Sto-Q)

Hyperparameters: critic learning rate α, actor learning rate β, discount factor γ, critic update frequency n_c, actor update frequency n_a, number of environment resets M, length of each episode B

Initialization: action-value function $Q(s, a; w)$, policy function $\pi(a|s; \theta)$

Repeat (indexed with k)

    (1) <u>Data collection</u>

    Initialize memory buffer $\mathcal{D} \leftarrow \emptyset$

    Repeat M environment resets

        $s_0 \sim d_{\text{init}}(s)$

        For i in $0,1,2,\dots,B-1$ or until episode termination

            $\epsilon \sim \mathcal{N}(0,1)$

            $a_i = g_\theta(s_i, \epsilon) = \mu_\theta(s_i) + \epsilon \cdot \sigma_\theta(s_i)$

            Apply a_i in environment, observe s_{i+1} and r_i

            $\mathcal{D} \leftarrow \mathcal{D} \cup \{(s_i, a_i, r_i, s_{i+1})\}$

        End

    End

    (2) <u>Critic update</u>

    Repeat n_c times

$$\nabla_w J_{\text{Critic}} \leftarrow \frac{1}{|\mathcal{D}|} \sum_{\mathcal{D}} \left(r + \gamma Q(s', a'; w) - Q(s, a; w)\right) \frac{\partial Q(s, a; w)}{\partial w}$$

$$w \leftarrow w - \alpha \cdot \nabla_w J_{\text{Critic}}$$

    End

    (3) <u>Actor update</u>

    Repeat n_a times

$$\nabla_\theta J_{\text{Actor}} \leftarrow \frac{1}{|\mathcal{D}|} \sum_{\mathcal{D}} \nabla_\theta g_\theta(s, \epsilon) \nabla_a Q(s, a; w)$$

$$\theta \leftarrow \theta + \beta \cdot \nabla_\theta J_{\text{Actor}}$$

    End

End

■

Algorithm 6-2: On-policy AC with Q-value and deterministic policy (i.e., Det-Q)

Hyperparameters: critic learning rate α, actor learning rate β, discount factor γ, critic update frequency n_c, actor update frequency n_a, number of environment resets M, length of each episode B, variance of exploration noise σ_ϵ^2

Initialization: action-value function $Q(s, a; w)$, policy function $\pi(s; \theta)$

Repeat (indexed with k)

   (1) Data collection

  Initialize memory buffer $\mathcal{D} \leftarrow \emptyset$

  Repeat M environment resets

     $s_0 \sim d_{\text{init}}(s)$

     For i in $0, 1, 2, \ldots, B - 1$ or until episode termination

       $a_i \leftarrow \pi(s_i; \theta) + \epsilon,\, \epsilon \sim \mathcal{N}(0, \sigma_\epsilon^2)$

       Apply a_i in environment, observe s_{i+1} and r_i

       $\mathcal{D} \leftarrow \mathcal{D} \cup \{(s_i, a_i, r_i, s_{i+1})\}$

     End

  End

   (2) Critic update

  Repeat n_c times

$$\nabla_w J_{\text{Critic}} \leftarrow \frac{1}{|\mathcal{D}|} \sum_{\mathcal{D}} (r + \gamma Q(s', a'; w) - Q(s, a; w)) \frac{\partial Q(s, a; w)}{\partial w}$$

$$w \leftarrow w - \alpha \cdot \nabla_w J_{\text{Critic}}$$

  End

   (3) Actor update

  Repeat n_a times

$$\nabla_\theta J_{\text{Actor}} \leftarrow \frac{1}{|\mathcal{D}|} \sum_{\mathcal{D}} \nabla_\theta \pi(s; \theta) \nabla_a Q(s, a; w)$$

$$\theta \leftarrow \theta + \beta \cdot \nabla_\theta J_{\text{Actor}}$$

  End

End

∎

In the AC implementation, choosing proper hyperparameters is critical to obtaining satisfactory performance. The learning rates need to be manually tuned for each task. A small learning rate leads to a slow training speed. A large learning rate, however, can easily cause instability. When the discount factor is close to 1, an RL agent places more importance on future rewards and usually has a better total average return. Being too long-sighted also poses more challenges in the convergence guarantee. The critic update frequency determines how many updates are performed during each RL iteration. In theory, n_c should be sufficiently large to converge to the true value. The same conclusion suits the selection of the actor update frequency. In practice, this option will result in very low computational efficiency. A high critic update frequency, i.e., n_c=5~20, n_a=1, is often chosen so that an accurate enough value function can be learned, which provides satisfactory guidance for policy search. Sometimes, the same updating frequency, i.e., n_c=n_a=1, is selected for critic and actor, and they both only update once during each RL iteration. In on-policy AC, the batch size is controlled by two hyperparameters: the number of environment resets and the length of each episode. The batch size determines how many samples are used to estimate actor and critic gradients at each RL iteration. Choosing a large batch size effectively reduces the gradient estimation variance but definitely it is not computationally efficient. In the deterministic policy setting, its policy lacks sufficient exploration ability, and additional exploration noise is added to interact with the environment. Proper exploration noise is helpful to reach each corner of the environment and find an optimal policy in the whole state space.

6.5.3   On-policy AC with State-Value Function

Compared with the action-value case, on-policy AC with state-value is less flexible in the policy selection. The stochastic policy and its log-derivative policy gradient must be used in its actor update. The low-variance benefit in deterministic policy gradients or reparameterized policy gradients cannot be taken advantage of. Fortunately, in the state-value case, the best baseline can be added easily to reduce actor gradient variance without introducing extra computational burden. Its single-step computational efficiency is often better because a state-value function has a lower dimension than an action-value function. In the previous section, it has been demonstrated that a stochastic policy gradient can come from greedy search. Here, we follow a new path to derive this gradient. This path results in a weaker version of the Bellman equation, whose parametric derivative is

$$
\begin{aligned}
\nabla_\theta J_{\text{Actor}}(\theta) \ &= \ \nabla_\theta \mathbb{E}_{s,a,s'}\{r + \gamma V(s'; w)\} \\
&= \ \mathbb{E}_{s \sim d_{\pi_\theta}}\left\{\sum_{a,s'} \nabla_\theta \pi_\theta(a|s)(r + \gamma V(s'; w))\right\} \\
&= \ \mathbb{E}_{s \sim d_{\pi_\theta}}\left\{\sum_{a,s'} \pi_\theta(a|s)\nabla_\theta \log \pi_\theta(a|s)\left(r + \gamma V(s'; w)\right)\right\} \\
&= \ \mathbb{E}_{\pi_\theta}\{\nabla_\theta \log \pi_\theta(a|s)\left(r + \gamma V(s'; w)\right)\}.
\end{aligned}
\tag{6-50}
$$

This new path yields the same actor policy gradient as (6-38), but it comes from the right side of a weak Bellman equation. The term "weak" means that its follows the structure of Bellman equation, but the state-value function is not optimal. That is to say, the optimal function is replaced with an estimate of the current state-value function. The major advantage of on-policy AC with state-value function and stochastic policy (i.e., Sto-V) is that it requires fewer samples to learn an accurate state-value function. The pseudocode is illustrated in Algorithm 6-3, in which the stochastic policy must be processed with the log-derivative trick.

Algorithm 6-3: On-policy AC with V-value and stochastic policy

Hyperparameters: critic learning rate α, actor learning rate β, discount factor γ, critic update frequency n_c, actor update frequency n_a, number of environment resets M, length of each episode B
Initialization: state-value function $V(s; w)$, policy function $\pi(a|s; \theta)$
Repeat <u>(indexed with k)</u>

---

(1) Data collection

Initialize memory buffer $\mathcal{D} \leftarrow \emptyset$

Repeat M environment resets

$\qquad s_0 \sim d_{\text{init}}(s)$

$\qquad$ **For** i in $0,1,2,\dots,B-1$ or until episode termination

$\qquad\qquad a_i \sim \pi(a|s_i;\theta)$

$\qquad\qquad$ Apply a_i in environment, observe s_{i+1} and r_i

$\qquad\qquad \mathcal{D} \leftarrow \mathcal{D} \cup \{(s_i, a_i, r_i, s_{i+1})\}$

$\qquad$ **End**

End

(2) Critic update

Repeat n_c times

$$\nabla_w J_{\text{Critic}} \leftarrow \frac{1}{|\mathcal{D}|} \sum_{\mathcal{D}} (r + \gamma V(s';w) - V(s;w)) \frac{\partial V(s;w)}{\partial w}$$

$$w \leftarrow w - \alpha \cdot \nabla_w J_{\text{Critic}}$$

End

(3) Actor update

Repeat n_a times

$$\nabla_\theta J_{\text{Actor}} \leftarrow \frac{1}{|\mathcal{D}|} \sum_{\mathcal{D}} \nabla_\theta \log \pi(a|s;\theta)\,(r + \gamma V(s';w))$$

$$\theta \leftarrow \theta + \beta \cdot \nabla_\theta J_{\text{Actor}}$$

End

End

∎

It is easy to see that both Algorithm 6-1 and Algorithm 6-3 are related to how to learn stochastic policies. In Algorithm 6-1, the utilization of action-value function allows the reparameterization trick to be deployed, which results in lower variance and better learning stability. In contrast, in Algorithm 6-3, the log-derivative trick must be adopted due to the limit of the state-value function. Thus, its policy gradient variance is often larger than that of the action-value function. Even though the state-value estimation has lower complexity than action-value estimation, its high-variance actor gradient somewhat deteriorates the training efficiency.

6.6   Example: Autonomous Car on Circular Road

An autonomous vehicle is capable of guiding itself to operate on structured roads without intervention from a human driver. The car is equipped with various onboard sensors, including radar, LIDAR, GPS, odometer, and inertial measurement unit (IMU), to perceive both itself and its surrounding environment. The advanced self-driving module then interprets the sensory information to search appropriate local path, select proper maneuver, and regulate the vehicle's motion. An autonomous driving system has motion control function in both longitudinal and lateral direction. A typical longitudinal automation function is cruise control, which adjusts the engine torque to enforce the vehicle's speed to track the driver's setup. A typical lateral automation function is lane-keeping control, which keeps the car inside its lane and relieves the driver of steering operations. The real driving environment is full of external disturbances, including strong environmental winds and varying road slopes. Therefore, it is challenging to develop a motion control system with excellent robustness in the presence of large uncertainties. In this example, RL can be used to learn a self-adaptive optimal policy by interacting with a perturbed traffic environment. The ability to resist environmental uncertainties is naturally equipped in the process of training the optimal control policy.

6.6.1   Autonomous Vehicle Control Problem

The autonomous car runs on a dry concrete road. Its inputs are commanded by an electrohydraulic steering actuator and an electronic throttle actuator. Dry roads can offer enough tire adhesion. Extreme conditions, such as drifting on low friction roads, are not considered. The circular road has a single lane, of which the centerline radius is 100 m, and its width is 3.75 m. No surrounding vehicles are in the traffic environment. The goal of vehicle motion control is to minimize the weighted sum of accumulative errors of the lateral position, yaw direction and longitudinal speed as much as possible, as well as control efforts of steering and accelerating operations.

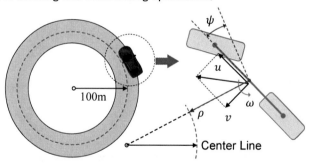

Figure 6.11 Autonomous passenger car on a circular road

To simplify the description of vehicle dynamics, the vehicle state is defined in the polar coordination system. The state space has 5 dimensions: (1) lateral position error ρ, i.e., the distance between the vehicle center of gravity (CG) and the road centerline; (2) heading angle error ψ, i.e., the error between the vehicle heading angle and the road tangent line; (3) longitudinal speed u; (4) lateral speed v; and (5) yaw rate of vehicle body ω. The action space has 2 dimensions: (1) the steering angle of the front wheel δ and (2) the longitudinal acceleration a_x. The state and action are summarized as

$$s = [\rho, \psi, u, v, \omega]^T \in \mathbb{R}^5$$
$$a = [\delta, a_x]^T \in \mathbb{R}^2. \qquad (6\text{-}51)$$

After limiting longitudinal acceleration to a low interval and neglecting rolling/pitching motions, the passenger car is mathematically modeled as a two-wheeled bicycle model, in which its tire dynamics is described by a nonlinear Fiala tire model. In this example, the passenger car has two external disturbances: (1) the road ramp force and (2) the acceleration actuator noise.

Coordinate transformation:

$$\dot{\rho} = -u \sin\psi - v \cos\psi$$
$$\dot{\psi} = \omega - \frac{u \cos\psi - v \sin\psi}{\rho}$$

Bicycle dynamic model:

$$a_x + a_{\text{noise}} = \dot{u} - v\omega$$
$$F_{Y1} \cos\delta + F_{Y2} + F_{\text{ramp}} = m(\dot{v} + u\omega)$$
$$a F_{Y1} \cos\delta - b F_{Y2} = I_{zz}\dot{\omega}$$

Fiala tire model:

$$F_{Y\#} \tag{6-52}$$
$$= \begin{cases} -C_\# \tan\alpha_\# \left(\dfrac{C_\#^2 (\tan\alpha_\#)^2}{27(\mu_\# F_{Z\#})^2} - \dfrac{C_\#|\tan\alpha_\#|}{3\mu_\# F_{Z\#}} + 1 \right), & |\alpha_\#| \leq |\alpha_{\text{max},\#}| \\ \mu_\# F_{Z\#} & , \quad |\alpha_\#| > |\alpha_{\text{max},\#}| \end{cases}$$

$$\alpha_{\text{max},\#} = \frac{3\mu_\# F_{Z\#}}{C_\#}, \quad \mu_\# = \frac{\sqrt{(\mu F_{Z\#})^2 - (F_{X\#})^2}}{F_{Z\#}}$$
$$\# = 1,2$$

External disturbances:

$$F_{\text{ramp}} \sim \mathcal{N}\left(\mu_{\text{ramp}}, \sigma_{\text{ramp}}^2\right)$$
$$a_{\text{noise}} \sim \mathcal{N}\left(\mu_{\text{acce}}, \sigma_{\text{acce}}^2\right),$$

where F_{Y1}, F_{Y2} denotes the lateral tire force, F_{X1}, F_{X2} denotes the longitudinal tire force, F_{Z1}, F_{Z2} denotes the tire load, α_1, α_2 denotes the slip angle, a is the distance between CG and front axle, b is the distance between CG and rear axle, C_1, C_2 denotes the wheel cornering stiffness, μ_1, μ_2 denotes the lateral friction coefficient of wheels, and $\alpha_{\text{max},1}, \alpha_{\text{max},2}$ denotes the tire slip angle when the tires' full sliding behavior occurs. The environment has two kinds of stochastic uncertainties: (1) acceleration actuator uncertainty a_{noise} and (2) road ramp uncertainty F_{ramp}. The longitudinal acceleration randomly fluctuates around its actuator commands. The lateral ramp force is directly added to the lateral force on the center of gravity. Both of these uncertainties are assumed to be Gaussian random noises with known means and variances. The key parameters are shown in Table 6.3.

Table 6.3 Key environment parameters

Parameters	Notation	Values
Desired vehicle speed	u_{exp}	15 m/s
Distance between CG and front axle	a	1.14 m
Distance between CG and rear axle	b	1.40 m

Mass	m	1500 kg
Moment of inertia	I_{zz}	2420 kg · m^2
Front wheel cornering stiffness	C_1	88000 N/rad
Rear wheel cornering stiffness	C_2	94000 N/rad
Longitudinal acceleration uncertainty	μ_{acce}	0 m · s^{-2}
	σ_{acce}	0.05 m · s^{-2}
Road ramp force uncertainty	μ_{ramp}	0 N
	σ_{ramp}	200 N

6.6.2   Design of On-policy Actor-Critic Algorithms

A well-defined reward signal is the prerequisite to designing effective on-policy AC algorithms. With respect to the reward, the lateral position error ρ and heading angle error ψ are penalized with 1-norm or 2-norm to maintain the desirable lane-keeping functionality. The error of longitudinal speed $u - u_{exp}$ is also penalized with 1-norm for the purpose of maintaining the desired vehicle speed. The steering and accelerating "energy" are penalized to reduce the magnitude of actuator operation. Meanwhile, I_{fail} offers an additional punishment if the car collides with the lane boundaries, which is helpful to discourage potential collision behavior. However, if the reward signal is negative most of the time, suicidal collision may be encouraged because an episode can be terminated earlier for a higher return. Hence, a constant positive bonus $c_0 > 0$ is added to roughly maintain zero-centered reward signals. The purpose of this design is to prevent suicidal behaviors caused by collision-related penalties. In summary, the reward signal is defined as

$$r(s) = c_0 - c_\rho|\rho| - c_\psi\psi^2 - c_u|u - u_{exp}| - c_\delta\delta^2 - c_a a_x^2 - I_{fail}$$

$$\text{where } I_{fail} = \begin{cases} 100, & \text{if out of lane} \\ 0, & \text{otherwise} \end{cases}. \tag{6-53}$$

Here, the coefficients $[c_0, c_\rho, c_\psi, c_u, c_\delta, c_a]$ are set to $[10, 1.3, 2, 0.6, 8.2, 0.5]$, which have been manually adjusted to obtain a satisfying trade-off among various self-driving performances. The vehicle control task is formulated as a discrete-time continuing RL problem with discount factor $\gamma = 0.97$. In the following simulations, the initial state is sampled from a predefined initial state distribution:

$$\rho_0 \sim \mathcal{N}(0, 0.7), \psi_0 \sim \mathcal{N}\left(0, \frac{\pi}{18}\right),$$

$$u_0 \sim \mathcal{N}(u_{exp}, 1), v_0 \sim \mathcal{N}(0, 0.1), \tag{6-54}$$

$$\omega_0 \sim \mathcal{N}(0, 0.1),$$

where the subscript "0" represents the initial time. The trained policy, which can be either stochastic or deterministic, controls how the agent behaves. In the stochastic case, a type-I Gaussian policy is used, and its mean and variance are parameterized:

$$a \sim \pi(a|s) = \mathcal{N}(\mu, \sigma^2),$$
$$\mu = \text{NN}(s; \theta_{\text{NN}}) \in \mathbb{R}^2, \tag{6-55}$$
$$\sigma = \exp(\theta_{\text{exp}}) \in \mathbb{R}^2,$$

where $\text{NN}(\cdot, \cdot)$ represents the artificial neural network, $\exp(\cdot)$ is the exponential function, and $\theta = [\theta_{\text{NN}}, \theta_{\text{exp}}]$ is the policy parameter to be learned. Meanwhile, the standard deviation is enforced to be an exponential function to ensure that its value is always positive. During training, actions for environment exploration are sampled from the Gaussian policy. In the deterministic case, a neural network-based policy is adopted to directly output actions:

$$a = \pi(s) = \text{NN}(s; \theta) \in \mathbb{R}^2. \tag{6-56}$$

One common approach to evaluate the performance of a stochastic policy is to convert it into a deterministic policy. This conversion occurs by selecting the peak probability point of the Gaussian distribution. Both state-value function and action-value function are approximated by artificial neural networks:

$$V(s) = \text{NN}(s; \theta)$$
$$Q(s, a) = \text{NN}(s, a; \theta). \tag{6-57}$$

The artificial neural networks are multilayer perceptrons (MLPs) with two hidden layers. Each hidden layer has 128 neurons. Scaled exponential linear units (SELUs) are used as the activation function. In this example, three kinds of on-policy AC algorithms are trained and tested, including stochastic AC with the state-value function (Algorithm 6-3 with $n_c=10$, and $n_a=1$, abbreviated as Sto-V), deterministic AC with the action-value function (Algorithm 6-2 with $n_c=10$, $n_a=1$, and Det-Q), and stochastic AC with the action-value function (Algorithm 6-1 with $n_c=10$, $n_a=1$, and Sto-Q). A high critic update frequency can output a more accurate value function, and it is helpful to stabilize the training process. The total average return (TAR) is utilized to measure policy performance. For fair comparison, 10 initial states are randomly sampled from the initial state distribution (6-54). At each initial state, 5 different episodes are generated independently. The number of steps at each episode is limited to 2000, which is long enough to reach the steady state distribution.

6.6.3   Training Results and Performance Comparison

Here, two kinds of tests are presented to compare the three AC algorithms: Test I is used to compare the convergence speed, and Test II is used to compare the self-driving performance. For Test I, we select three different learning rates for fair comparison. Figure 6.12 shows that a small learning rate leads to a slow training process for all three algorithms. When the learning rate is too large, the training process becomes more oscillatory. Seen from Figure 6.12 (a), bad state-value function cannot provide accurate guidance to policy updates. Its self-driving policy makes little progress at the beginning. The actor begins to work only after the entire value function is well learned. In contrast, action-value function implicitly incorporates environment information, which has a few benefits in fast policy learning at the start. Moreover, by comparing Figure 6.12 (b) and (c), the deterministic policy has lower variance than the stochastic policy. This is because stochastic policy gradient must include the integration over random actions, which easily leads to an inaccurate gradient estimation due to sampling uncertainties.

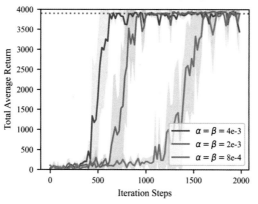

(a) Stochastic + state-value (i.e., Sto-V)

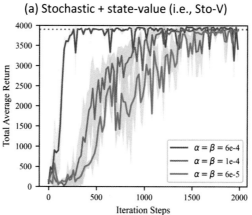

(b) Stochastic + action-value (i.e., Sto-Q)

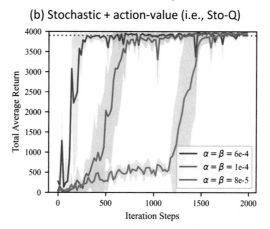

(c) Deterministic + action-value (i.e., Det-Q)

Figure 6.12 Performance comparison at different learning rates

In Test II, the best policy from each AC algorithm in Test I is selected to compare their self-driving performances. The vehicle states and actions are shown in Figure 6.13 and Figure 6.14, respectively. The chosen algorithms and hyperparameters lead to three self-driving policies with relatively small tracking errors in the heading angle, lateral position and longitudinal speed. In all three algorithms, steady-state errors exist both in lateral position error ρ and longitudinal speed u, which comes from the natural results of trade-off among different performance requirements. This is also an evidence that neural network-based policy is a stationary function, which has no integral term to converge zero-order state errors into zeros. In theory, a perfectly learned policy should be optimal, and all three performances should be identical to each other. However, the three trained policies are not perfect, and they show obviously different behaviors. This is because the three algorithms are still very primitive in the RL community, and their algorithmic imperfectness leads to inaccurate learning results. More accurate "optimal" policy need a lot of empirical tricks, like experience replay, parallel exploration, separated target network, etc., to increase training efficiency and stabilize the training process.

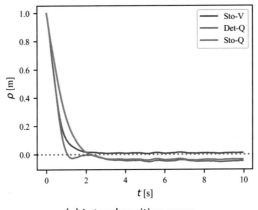

(a) Lateral position error

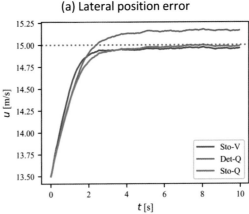

(b) Vehicle longitudinal speed

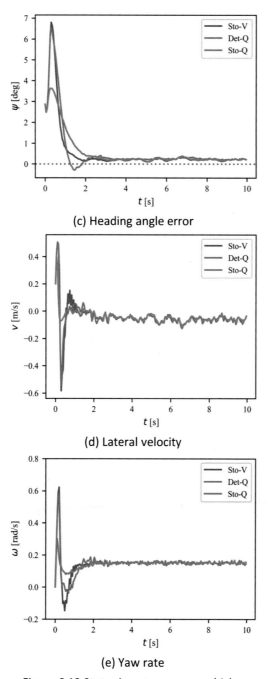

(c) Heading angle error

(d) Lateral velocity

(e) Yaw rate

Figure 6.13 States in autonomous vehicles

The two actions, i.e., steering angle and longitudinal acceleration, eventually converge to their steady-state equilibrium. As seen from Figure 6.14, neither of these actions can settle down to zero due to the noisy uncertainties in the acceleration actuator and road ramp. Interestingly, the stochastic policy with action-value function (i.e., Sto-Q) has the smallest action vibration, while Det-Q and Sto-V have relatively high action fluctuations. We must say this result is not trivial, and which one is better depends on how to choose key hyperparameters and when to stop training. Even though their TARs are very close to each other, their states and actions still have noticeable differences, especially in their transient behaviors. This result means that searching for a perfectly optimal policy is not an easy job in primitive model-free RLs.

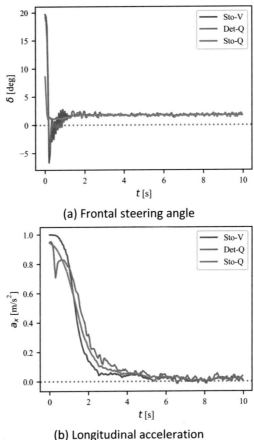

(a) Frontal steering angle

(b) Longitudinal acceleration

Figure 6.14 Actions in autonomous vehicles

6.7   References

[6-1]   Baird L (1995) Residual algorithms: Reinforcement learning with function approximation. ICML, California, USA

[6-2]   Bertsekas D, Tsitsiklis J (1996) Neuro-Dynamic Programming, Athena Scientific, Belmont

[6-3]   Boyan J, Moore AW (1995) Generalization in reinforcement learning: Safely approximating the value function. NeurIPS, Colorado, USA

[6-4]   Buşoniu L, Lazaric A, Ghavamzadeh M, et al (2011) Least-squares methods for policy iteration. In: Adaptation, Learning, and Optimization, vol 12. Springer, Berlin, Heidelberg

[6-5]   Espeholt L, Soyer H, Munos R (2018) Impala: Scalable distributed deep-rl with importance weighted actor-learner architectures. ICML, Stockholm, Sweden

[6-6]   Guan Y, Li SE, Duan J, et al (2021) Direct and indirect reinforcement learning. Intl J of Intelli Syst (36): 4439-4467

[6-7]   Konda VR, Tsitsiklis JN (2000) Actor-critic algorithms. NeurIPS, Colorado, USA

[6-8]   Munos R, Stepleton T, Harutyunyan A, Bellemare MG (2016) Safe and efficient off-policy reinforcement learning. NeurIPS, Barcelona, Spain

[6-9]   Seijen H, Sutton R (2014) True online TD (lambda). ICML, Beijing, China

[6-10]   Silver D, Lever G, Heess N, Degris T, Wierstra D, Riedmiller M (2014) Deterministic policy gradient algorithms. ICML, Beijing, China

[6-11]   Sutton RS, Barto AG (2018) Reinforcement learning: An introduction. MIT press, Cambridge

[6-12]   Szepesvári C (2010) Algorithms for reinforcement learning. Synthesis lectures on AI and mach learn 4 (1):1-103

[6-13]   Tsitsiklis JN, Van Roy B (1997) An analysis of temporal-difference learning with function approximation. IEEE Trans Automatic Control 42 (5): 674-690

[6-14]   Van Hasselt H, Doron Y, Strub F, et al (2018) Deep reinforcement learning and the deadly triad. https://arxiv.org/abs/1812.02648v1

Chapter 7. Direct RL with Policy Gradient

> You cannot teach a man anything;
>
> you can only help him discover it in himself.
>
> -- Galileo Galilei (1564-1642)

Indirect RL, such as Monte Carlo and temporal difference algorithms, depends on Bellman's principle of optimality to compute the optimal policy. In contrast, when using direct RL, a parameterized policy is searched for by optimizing a predefined objective function. This approach does not rely on any optimality condition, but directly takes advantage of some numerical optimization methods. That is why "direct" is named after. One large class of direct RL algorithms belongs to the first-order optimization, and how to calculate their policy gradients plays a central role in this algorithm family. Popular policy gradients include likelihood ratio gradient, natural policy gradient, and deterministic policy gradient. In this family, policy search is performed in the parameter space and ascends along the gradient direction until the maximum is reached. Compared to indirect RL methods, direct RL like gradient-based policy search offers numerous advantages in real-world applications, including convergence guarantee at least to a local optimum, compatibility with various numerical optimization techniques, and ability to handle imperfect Markovian states.

The cornerstone of direct RL methods is to derive an analytical yet structurally concise policy gradient formula. In 1990, Glynn derived a likelihood ratio gradient estimator for discrete-time Markov chains using the score function technique [7-6]. The REINFORCE algorithm was first introduced by Williams in 1992, which is one of the earliest attempts to build useful policy gradients [7-24]. It adjusts the policy parameter along the steepest direction towards a higher average return. In essence, Monte Carlo policy gradient is used in REINFORCE, and the true action-value function is replaced with average return. From the study of Sutton, McAllester, Singh and Mansour in 1999, an approximate formula of likelihood ratio gradient was formally presented. This is the well-known vanilla policy gradient, which suits both on-policy and off-policy RL algorithms and can be combined with both value function and advantage function [7-22]. With reliance on vanilla policy gradient, an on-policy actor-critic algorithm was proposed by Konda and Tsitsiklis in 2000. This algorithm is used to learn a parameterized value function as an auxiliary element to reduce the variance of policy gradient estimation [7-11]. The work of Degris, White and Sutton extended this on-policy algorithm to an off-policy version, in which a behavior policy is used to provide better environment exploration [7-3].

The abovementioned policy gradients must be stochastic and always output a probability distribution over actions. The randomness of actions is one major source of high variance in stochastic policy gradients. In 2014, Silver, Lever and Heess first proposed a deterministic version of policy gradients, which allows deterministic decisions to be made in stochastic environments [7-20]. Deterministic policy gradient (DPG) often has low variance as it removes integration over random actions. On this basis, deep deterministic policy gradient (DDPG) algorithm was proposed in 2015 to extend deep Q-network (DQN) to the continuous action space with a deterministic policy [7-13]. In 2018,

S. E. Li, *Reinforcement Learning for Sequential Decision and Optimal Control*, https://doi.org/10.1007/978-981-19-7784-8_7

Fujimoto et al. further proposed the twin delayed deep deterministic (TD3) algorithm, which is used to calculate a single target with the minimum value of two Q-networks to reduce the overestimation issue [7-5].

One main challenge of direct RL, especially with off-policy gradients, is the easiness of instability in the training process. The key idea to addressing this issue is to avoid adjusting the policy too fast at each step. In 2015, trust region policy optimization (TRPO) was proposed to carry out the minorize-maximization optimization by enforcing a KL divergence constraint on the policy update [7-18]. It has been proven that TRPO can guarantee monotonic policy improvement in a locally convex region. One shortcoming of TRPO is the high computational burden due to computation of the Hessian matrix's inverse. Later, proximal policy optimization (PPO) was presented to use a clipped surrogate function to relieve the high computational complexity [7-19]. The deployment of small-step updates can effectively reduce the issue of potential training instability.

7.1   Overall Objective Function for Direct RL

To discuss how to build direct RL algorithms, we mainly consider continuing tasks with stochastic policies. Luckily, it is possible to unify episodic and continuing tasks through simple conversion. For example, one can assume that all the future rewards are zero after one episode is terminated. This assumption actually converts an episodic task into a continuing task. Even though we mainly talk about stochastic policy, deterministic policy is definitely acceptable in direct RL. A stochastic policy can naturally handle the exploration-exploitation dilemma, while a deterministic policy requires an additional perturbation to strengthen its exploration ability. The two policies look different with respect to their action behaviors, but in fact a deterministic policy can be regarded as the limiting case of a stochastic policy. Regardless of what kind of policy is adopted, we always want to maximize the long-term accumulative reward under a certain initial state distribution. Without loss of generality, we assume the problem starts from time t (note that τ denotes the time from time t to the end):

$$
\begin{aligned}
\max_{\theta} J(\theta) &= \mathbb{E}_{s_t \sim d(s_t)}\{v^{\pi_\theta}(s_t)\} \\
&= \int d(s_t) v^{\pi_\theta}(s_t) \mathrm{d}s_t \\
&= \mathbb{E}_{s_t, a_t, s_{t+1}, \dots \sim \rho_{\pi_\theta}}\left\{\sum_{\tau=t}^{\infty} \gamma^{\tau-t} r_\tau\right\},
\end{aligned}
\tag{7-1}
$$

where $0 < \gamma < 1$ is the discount factor that satisfies the condition of compatibility, $d(s_t)$ is the initial distribution of state s_t, $v^{\pi}(s_t)$ is the state-value function, and ρ_{π_θ} is the joint probability of states and actions in the trajectory. Here, the notation $J(\theta)$ is specially called "overall" objective function to distinguish it from those criteria in PEV and PIM, i.e., critic loss function and actor loss function. The parameterization of stochastic policy is denoted by

$$
\pi_\theta(a|s) \overset{\text{def}}{=} \pi(a|s; \theta).
\tag{7-2}
$$

The trajectory-based definition for state-value and action-value functions is expressed as

$$v^\pi(s) = \mathbb{E}_{a_t, s_{t+1}, a_{t+1}, s_{t+2}, a_{t+2}, \cdots \sim \rho_{\pi_\theta}} \left\{ \sum_{\tau=t}^{\infty} \gamma^{\tau-t} r_\tau \mid s_t = s \right\},$$

$$q^\pi(s,a) = \mathbb{E}_{s_{t+1}, a_{t+1}, s_{t+2}, a_{t+2}, \cdots \sim \rho_{\pi_\theta}} \left\{ \sum_{\tau=t}^{\infty} \gamma^{\tau-t} r_\tau \mid s_t = s, a_t = a \right\},$$

(7-3)

where (s,a) is a given state-action pair at time t. The overall objective function defined above actually reflects the long-term performance of a policy since it measures the weighted return of all possibly generated trajectories. Note that the initial state distribution $d(s_t)$ must be independent of the policy parameter θ. Its usage is equivalent to introducing a weighting function to the state-value function, where more emphasis is placed on the high-occurrence state and less emphasis is placed on the low-occurrence state. We have proven in previous chapters that the initial state distribution does not change the control optimality if the policy to be learned has no structural limitation. In theory, one can maximize $\sum_s v^{\pi_\theta}(s)$ with average initial state distribution or even the state-value function $v^{\pi_\theta}(s)$ at a given state point. In practice, these designs may result in low policy accuracy and unpredictable training instability because of the large mismatch in the occurrence frequencies between weighting states and real states. The stationary state distribution is a more popular choice as the weighting function due to its compatibility with the collected samples in the long run.

7.1.1 Stationary Properties of MDPs

The stationary state distribution is one important property of Markov decision processes. While this kind of state distribution can be seen in many Markov processes, early studies mainly focus on simple cases with finite and countable state spaces, i.e., Markov chain. The Markov chain can be described by a directed graph, whose nodes are discrete states and whose edges are the probabilities of transiting from one state to another. Let us demonstrate how to define the stationary state distribution (SSD). The time series of random state variables are $\{s_t\}, t = 1,2,\cdots,\infty$. Define $\zeta_{i,j}$ as the one-step transition probability from $s_{(i)}$ to $s_{(j)}$ (Figure 7.1) [7-17]:

$$\zeta_{i,j} = \sum_{a \in \mathcal{A}} \pi(a \mid s = s_{(i)}) p(s' = s_{(j)} \mid s = s_{(i)}, a),$$

where $\mathcal{S} = \{s_{(1)}, s_{(2)}, \cdots, s_{(n)}\}$ is the state space, which contains a set of finite elements. The transition probability is governed by both $\pi(a \mid s)$ and $p(s' \mid s, a)$.

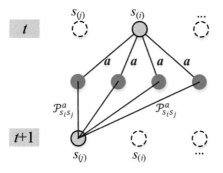

Figure 7.1 One-step transition probability

The Markov chain is a stochastic process, and its state has no explicit analogy to that of a deterministic state space model. The state in a physical system is often deterministic, for example, that in a spring-mass-damping system, but in the Markov chain, the state is essentially a random variable. If one must want to find a deterministic analogy for the Markov chain, the new "state" is the marginal distribution. The marginal distribution of a random variable is

$$d_t(s) = \begin{bmatrix} d_t(s_{(1)}) & d_t(s_{(2)}) & \cdots & d_t(s_{(n)}) \end{bmatrix}^{\mathrm{T}} \in \mathbb{R}^{n\times 1}.$$

Obviously, the marginal distribution is a deterministic value at each time instant, which satisfies

$$d_{t+1} = H_{n\times n} d_t,$$

and $H = \begin{bmatrix} \zeta_{i,j} \end{bmatrix}_{n\times n}$ is the matrix form of the one-step transition probability.

$$H_{n\times n} = \begin{bmatrix} \zeta_{1,1} & \cdots & \zeta_{n,1} \\ \vdots & \ddots & \vdots \\ \zeta_{1,n} & \cdots & \zeta_{n,n} \end{bmatrix}.$$

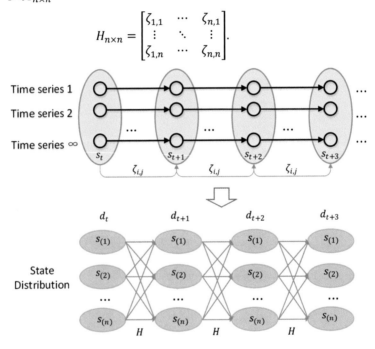

Figure 7.2 Time series of a random variable and its state distribution

Each element of marginal state distribution is equal to $d_t(s_{(i)}) = \Pr\{s_t = s_{(i)}\}$, where $\sum_{i=1}^{n} d_t(s_{(i)}) = 1$. Figure 7.2 shows the connection between the time series of a random variable and its marginal state distribution. The time series comprises the samples that are generated in the Markov chain, while the state distribution represents the occurrence probability of each sample at current time. Most stochastic RLs are assumed to be time-homogeneous, i.e., independent of time t. An important property of the time-homogenous Markov chain is that as $t \to \infty$, the chain will reach an equilibrium, i.e., the stationary state distribution. Here, "stationary" means "remains unchanged with respect to time". A Markov chain is stationary if and only if the marginal distribution of s_t does not change with time t.

- Definition 7.1 [7-2]: The SSD of a time-homogeneous finite MDP is

$$d_\pi(s) = \left[d_\pi\big(s_{(1)}\big) \quad d_\pi\big(s_{(2)}\big) \quad \cdots \quad d_\pi\big(s_{(n)}\big)\right]^{\mathrm{T}},$$

where $\sum_{i=1}^{n} d_\pi\big(s_{(i)}\big) = 1$ and $d_\pi\big(s_{(i)}\big) \geq 0$, satisfy

$$d_\pi(s) = H d_\pi(s).$$

That is,

$$d_\pi\big(s_{(j)}\big) = \sum_{s_{(i)} \in \mathcal{S}} d_\pi\big(s_{(i)}\big) \zeta_{i,j}.$$

When an MDP reaches its stationary status, the state distribution will remain the same and be independent of time t.

- Theorem 7.2 [7-2]: Any finite, irreducible, and ergodic MDP has the following intrinsic properties:

(1) The chain has a unique SSD;

(2) For any $i, j \in \mathcal{S}$, the limit $\lim_{t \to \infty} \Pr\{s_t = s_{(j)} | s_0 = s_{(i)}\}$ exists and is independent of the initial state s_0; and

(3) The SSD is equal to

$$d_\pi\big(s_{(j)}\big) = \lim_{N \to \infty} \frac{\sum_{k=1}^{N} \Pr\{s_k = s_{(j)} | s_0 = s_{(i)}\}}{N} = \lim_{t \to \infty} \Pr\{s_t = s_{(j)} | s_0 = s_{(i)}\},$$

i.e., the convergence to the SSD occurs as $t \to \infty$, regardless of the initial state distribution.

∎

The SSD is interpreted as the limiting case of instantaneous state distribution. In general, a Markov chain may have more than one equilibrium. For an irreducible Markov chain, a SSD exists if and only if all states are positive recurrent. Moreover, the SSD is unique in this condition. Having a unique SSD is critical to a continuing stochastic RL task. This type of SSD under the policy $\pi(a|s)$ satisfies

$$d_\pi\big(s_{(j)}\big) = \sum_{s \in \mathcal{S}} d_\pi(s) \sum_{a \in \mathcal{A}} \pi(a|s) p\big(s' = s_{(j)} | s, a\big),$$

$$j = 1, 2, \cdots, n.$$

(7-4)

The SSD is an equilibrium distribution regardless of the initial distribution. As illustrated in Figure 7.3, the equilibrium of a stochastic RL task is the statistical summary of all the collected samples without any ordering information. Under stationary conditions, the sampling order becomes meaningless since each sample may occur randomly at any temporal position.

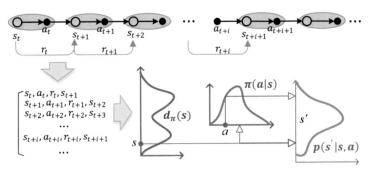

Figure 7.3 Order-free data distribution under stationary conditions

7.1.2   Discounted Objective Function

In a continuing task, the overall objective function must be defined carefully to ensure that its value function is always bounded, even though the number of time steps goes to infinity. The discounted cost provides a simple but intuitive way to achieve this requirement by rapidly decreasing rewards in future stages. This kind of discounted costs arise in many engineering tasks. For example, in welfare management, discounting means that consumption at the current time is preferred (provides greater welfare) than consumption in the future. Therefore, the discount factor represents a trade-off between current and future consumption. In some special tracking tasks, the current error is more important than future error. In this case, the discount factor represents the risk level of state errors at different moments. The fulfillment of reward decreasing ability in the discounted cost depends on the discount factor. The discount factor $0 < \gamma < 1$ holds the following property:

$$\sum_{i=0}^{\infty} \gamma^i = \frac{1}{1-\gamma}.$$

(7-5)

Obviously, the stage reward at the current time carries a greater weighting coefficient than those at future times. As a result, states and actions far in the future have a limited impact on the current return. Using the discount factor, the overall objective function can be written in a summation form:

$$J(\theta)$$

$$= \mathbb{E}_{s_t, a_t, \ldots \sim \rho_{\pi_\theta}} \left\{ \sum_{\tau=t}^{\infty} \gamma^{\tau-t} r_\tau \right\}$$

$$= \sum_{s_t} d(s_t) \sum_{a_t} \pi_\theta(a_t|s_t) \sum_{s_{t+1}} p(s_{t+1}|s_t, a_t)$$

$$\sum_{a_{t+1}} \pi_\theta(a_{t+1}|s_{t+1}) \sum_{s_{t+2}} p(s_{t+2}|s_{t+1}, a_{t+1}) \cdots \sum_{\tau=t}^{\infty} \gamma^{\tau-t} r_\tau$$

$$= \sum_{s} p(s_t = s|\pi_\theta) \sum_{a_t} \pi_\theta(a_t|s) \sum_{s_{t+1}} p(s_{t+1}|s, a_t) r_t$$

$$+ \sum_{s} p(s_{t+1} = s|\pi_\theta) \sum_{a_{t+1}} \pi_\theta(a_{t+1}|s) \sum_{s_{t+2}} p(s_{t+2}|s, a_{t+1}) \cdots \sum_{\tau=t+1}^{\infty} \gamma^{\tau-t} r_\tau.$$

Then, roll forward until infinity, and we obtain the collective form of the overall objective function:

$$= \sum_{\tau=t}^{\infty} \gamma^{\tau-t} \sum_{s} p(s_\tau = s|\pi_\theta) \sum_{a} \pi_\theta(a|s) \sum_{s'} p(s'|s, a) r(s, a, s')$$

$$= \frac{1}{1-\gamma} \sum_{s} d_{\pi_\theta}^\gamma(s) \sum_{a} \pi_\theta(a|s) \sum_{s'} p(s'|s, a) r(s, a, s'), \tag{7-6}$$

where $d_{\pi_\theta}^\gamma(s)$ is called the discounted state distribution, defined as

$$d_{\pi_\theta}^\gamma(s) = (1-\gamma) \sum_{\tau=t}^{\infty} \gamma^{\tau-t} p(s_\tau = s|\pi_\theta). \tag{7-7}$$

The definition of discounted state distribution contains a special coefficient $1-\gamma$ because any distribution should satisfy the normalization condition, i.e., $\sum d(s) = 1$. It is easily observed that the collective form of discounted objective function contains three layers of summations (or integrals in the continuous space). The inner layer is the summation over the next state distribution, which is governed by an environment dynamics. The middle layer is the summation over the action space, which is dominated by a stochastic policy. The outer layer is the summation over the state space, which is described by the discounted state distribution. The three layers form a structure of hierarchical summation, which is helpful to derive likelihood ratio gradients.

7.1.3   Average Objective Function

A common replacement for the discounted cost is the average cost, which is used to keep the cumulative return inside a bounded range by dividing it into the total number of time stages. There is a noticeable difference between an average criterion and a discounted criterion. In the average objective function, near-term rewards are treated equally with long-term rewards. A typical example is fuel-saving vehicle automation, in which engine fuel consumption at different time instants should be equally treated. The fuel consumption at this moment is equally expensive as that in the following instants. The average objective function is defined as

$$J(\theta) \;=\; \lim_{N \to \infty} \frac{1}{N} \mathbb{E}_{\pi_\theta} \left\{ \sum_{\tau=t}^{t+N-1} r_\tau \right\}$$

$$= \sum_s d_{\pi_\theta}(s) \sum_a \pi_\theta(a|s) \sum_{s'} p(s'|s,a)\, r(s,a,s').$$

(7-8)

The second equation follows the third condition in Theorem 7.2, where $d_{\pi_\theta}(s)$ becomes the SSD, i.e.,

$$d_{\pi_\theta}(s) = \lim_{\tau \to \infty} p(s_\tau = s | \pi_\theta).$$

The goal of direct RL is to find the best policy parameter by directly maximizing either the discounted objective function (7-6) or the average objective function (7-8). Usually, such an optimization problem contains different types of stochasticity, either in the environment dynamics or in the stochastic policy. Therefore, direct RL with a parameterized policy naturally becomes a stochastic optimization problem. The policy parameter of this problem can be iteratively searched for using mature numerical optimization algorithms such as derivative-free optimization (e.g., finite-difference) and first-order optimization (e.g., stochastic gradient descent).

7.2   Likelihood Ratio Gradient in General

The analytical policy gradient is badly needed in direct RLs with continuous state and action spaces. When facing a stochastic policy with parametric probability distribution, the parameter search with its policy gradient is able to push up the probabilities of good actions that lead to a higher return and push down the probabilities of bad actions that lead to a lower return. In fact, both stochastic policy gradient and deterministic policy gradient can be derived, but they have obvious differences in terms of estimation accuracy and computational efficiency. In the stochastic policy gradient, both state and action spaces are integrated. This stochastic gradient is equipped with good exploration ability but often has high variance. In contrast, only the state space is integrated in the deterministic policy gradient. This deterministic gradient has high accuracy and low computational burden but it naturally loses good exploration ability. In this chapter, how to derive stochastic policy gradient will be discussed first, followed by that of deterministic policy gradient. Most of our discussion will focus on discounted optimal control problems.

7.2.1   Gradient Descent Algorithm

The gradient descent method is a first-order optimization algorithm that is used to minimize some criteria by iteratively moving along the steepest descent direction. As a tradition, we still use the term "gradient descent" when referring to this algorithm even though the maximization operation, instead of minimizer, is preferred in the RL community. The policy search for a local maximum is performed by ascending along the policy gradient:

$$\theta \leftarrow \theta + \beta \cdot \nabla_\theta J(\theta), \tag{7-9}$$

where $\nabla_\theta J(\theta)$ is the policy gradient and β is the learning rate. This type of learner is referred to as "direct" because no optimality condition is involved. If the gradient estimation is unbiased, its convergence to a local minimum can be guaranteed with a few mild conditions on the learning rate. Since the early 1990s, the RL community has witnessed how to use different kinds of policy gradients to train an RL agent. In early studies of this area, the actor-only algorithms were numerically brittle and often resulted in unacceptably high variance. One significant improvement is the introduction of the vanilla policy gradient (R Sutton et al., 1999), which has become a cornerstone in today's actor-critic algorithms [7-22]. The vanilla policy gradient is the simplified version of likelihood ratio gradient that comes from the perspective of stochastic optimization. Its derivation involves two main tricks: (1) the score function estimator and (2) the factor cancellation from temporal causality. In the literature, there are two basic ways to derive vanilla policy gradients. One way is to use the trajectory concept, in which the expectation of discounted return is directly optimized. The other way is to use the cascading concept, in which the weighted expectation of state-value function is optimized. Even though these two methods are different with respect to their derivations, their final results are exactly the same, that is the famous likelihood ratio gradient.

7.2.2   Method I: Gradient Derivation from Trajectory Concept

In this section, the vanilla policy gradient from the trajectory concept will be derived. For the sake of conciseness, we consider MDPs with finite state and action spaces. As a result, we can use "$\sum \#$" to replace "$\int \#$" whenever it is necessary to simplify the usage of mathematical notations. Let us define $\mathcal{T}$ to represent a trajectory of random samples, i.e., $\mathcal{T} \overset{\text{def}}{=} \{s_t, a_t, \dots, s_{t+n}, a_{t+n}, \dots\}$. The discounted return can be written as a function of the trajectory $G(\mathcal{T}) = \sum_{\tau=t}^{\infty} \gamma^{\tau-t} r(s_\tau, a_\tau)$. Hence, the discounted objective function in the trajectory form is

$$J(\theta) = \mathbb{E}_{\mathcal{T} \sim \rho_{\pi_\theta}}\{G(\mathcal{T})\} = \sum_{\mathcal{T}} \rho_{\pi_\theta}(\mathcal{T}) G(\mathcal{T}), \tag{7-10}$$

where $\rho_{\pi_\theta}(\mathcal{T})$ is the joint probability of states and actions in the trajectory form:

$$\rho_{\pi_\theta}(\mathcal{T}) = d(s_t) \prod_{\tau=t}^{\infty} \pi_\theta(a_\tau|s_\tau) p(s_{\tau+1}|s_\tau, a_\tau).$$

The primitive first-order derivative of discounted objective function with respect to the policy parameter θ is

$$\begin{aligned}
\nabla_\theta J(\theta) &= \sum_{\mathcal{T}} \nabla_\theta \rho_{\pi_\theta}(\mathcal{T}) G(\mathcal{T}) \\
&= \sum_{\mathcal{T}} \rho_{\pi_\theta}(\mathcal{T}) \frac{\nabla_\theta \rho_{\pi_\theta}(\mathcal{T})}{\rho_{\pi_\theta}(\mathcal{T})} G(\mathcal{T}) \\
&= \mathbb{E}_{\mathcal{T} \sim \rho_{\pi_\theta}}\{\nabla_\theta \log \rho_{\pi_\theta}(\mathcal{T}) \cdot G(\mathcal{T})\},
\end{aligned}$$

where $\nabla_\theta \log \rho(\mathcal{T}) = \nabla_\theta \rho(\mathcal{T})/\rho(\mathcal{T})$ is called score function. The score function is defined as the partial derivative of the log-likelihood function. One of its important properties is that its expected value is always equal to zero, i.e., $\sum_{\mathcal{T}} \rho(\mathcal{T})\nabla_\theta \log \rho(\mathcal{T}) = 0$. Hence, the score function allows us to use the famous log-derivative trick to rewrite the primitive first-order derivative. Some terms in the joint probability are independent of the policy parameter, and hence, we can simplify the score function estimator:

$$\nabla_\theta \log \rho_{\pi_\theta}(\mathcal{T}) \;\; = \nabla_\theta \log d(s_t) + \nabla_\theta \sum_{\tau=t}^{\infty} (\log \pi_\theta(a_\tau|s_\tau) + \log p(s_{\tau+1}|s_\tau, a_\tau))$$

$$= \sum_{\tau=t}^{\infty} \nabla_\theta \log \pi_\theta(a_\tau|s_\tau).$$

Now, the analytical formula of likelihood ratio gradient becomes

$$\nabla_\theta J(\theta) = \mathbb{E}_{\mathcal{T} \sim \rho_{\pi_\theta}} \left\{ \sum_{\tau=t}^{\infty} \nabla_\theta \log \pi_\theta(a_\tau|s_\tau) \cdot \sum_{\tau=t}^{\infty} \gamma^{\tau-t} r_\tau \right\}. \tag{7-11}$$

The log-derivative trick is actually a straightforward application of the chain rule. This trick allows us to switch between two representations of a given distribution, i.e., normalized probability and logarithmic probability. The switching ability forms the basis of score function estimation and helps us to derive an analytical formula of stochastic policy gradient. This is also the reason why the "likelihood ratio gradient" is named after. Looking carefully at the score function, its denominator is $\rho_{\pi_\theta}(\mathcal{T})$, which represents the joint probability of states and actions in the trajectory. The joint probability must not be zero; otherwise, the score function will become infinitely large. Therefore, the derivation of likelihood ratio gradient only suits the stochastic policy but not the deterministic policy.

7.2.2.1 Simplification by temporal causality

The aforementioned likelihood ratio gradient has very complicated operations, including the summation and multiplication of a series of factors. Luckily, some factors are allowed to be cancelled because they have no causal connection. For example, the current-step policy has no causal connection with the rewards received in previous time steps. In other words, future action selection has no influence on what has already happened. This temporal noncausality can help us greatly simplify the final formula of likelihood ratio gradient (7-11). The temporal noncausality also introduces a few benefits such as a low computational burden and high accuracy in sample-based estimation.

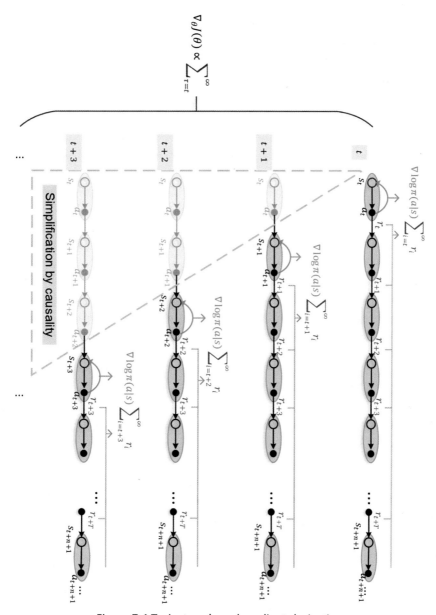

Figure 7.4 Trajectory-based gradient derivation

Figure 7.4 shows the causal disconnection in the trajectory-based derivation. The horizontal axis represents the trajectory in which all states and actions are random variables. The vertical axis represents the time steps in the summation operation. At each time step τ, a gradient factor consists of two subparts: (a) $\nabla_\theta \log \pi_\theta (a_\tau | s_\tau)$ and (b) the accumulative reward starting from time τ. The gradient factors of all time steps are summed, leading to the first-order derivative of the overall objective function:

$$\nabla_\theta J(\theta)$$

$$= \mathbb{E}_{\mathcal{T} \sim \rho_{\pi_\theta}} \left\{ \sum_{\tau=t}^{\infty} \nabla_\theta \log \pi_\theta(a_\tau|s_\tau) \cdot \left(\sum_{i=t}^{\tau-1} \gamma^{i-t} r_i + \sum_{i=\tau}^{\infty} \gamma^{i-t} r_i \right) \right\} \tag{7-12}$$

$$= \mathbb{E}_{\mathcal{T} \sim \rho_{\pi_\theta}} \left\{ \sum_{\tau=t}^{\infty} \nabla_\theta \log \pi_\theta(a_\tau|s_\tau) \gamma^{\tau-t} \sum_{i=\tau}^{\infty} \gamma^{i-\tau} r_i \right\}.$$

7.2.2.2   Vanilla policy gradient with state-action format

Thus far, the likelihood ratio gradient has been written into a simplified trajectory format, i.e., the expectation over a trajectory distribution. However, it can be further converted to a state-action format, where the expectation is defined over the joint distribution of the initial state-action pair. The conversion is similar to (7-6). The second subpart of each gradient factor in (7-12) is replaced with the action-value function, yielding

$$\nabla_\theta J(\theta) = \sum_s \sum_{\tau=t}^{\infty} \gamma^{\tau-t} p(s_\tau = s|\pi_\theta) \sum_a \nabla_\theta \pi_\theta(a|s) q^{\pi_\theta}(s, a).$$

By extracting the discounted state distribution $d_{\pi_\theta}^{\gamma}(s)$ from the equation above, we have

$$\nabla_\theta J(\theta) = \frac{1}{1-\gamma} \sum_s d_{\pi_\theta}^{\gamma}(s) \sum_a \nabla_\theta \pi_\theta(a|s) q^{\pi_\theta}(s, a). \tag{7-13}$$

We call (7-13) "vanilla policy gradient" because it is very basic and has no assumptive simplification. This formula is actually a "true" policy gradient from the viewpoint of direct RL. The existence of a discounted state distribution means that the beginning steps matter more and the later steps matter less.

7.2.3   Method II: Gradient Derivation from Cascading Concept

Different from the previous trajectory concept, this section will derive the vanilla policy gradient from the cascading concept. This derivation starts from the second form of the overall objective function in (7-1). This equivalent form allows us to use the technique of cascading derivation. In the trajectory-based derivation, gradient factors are fully expanded at first and then reasonably concealed by temporal causality. In contrast, the cascading derivation expands the gradient factors step by step until infinity. The temporal causality is naturally embedded inside each cascading expansion. Hence, the derivation of its likelihood ratio gradient becomes

$$\nabla_\theta J(\theta)$$

$$= \nabla_\theta \sum_{s_t} d(s_t) v^{\pi_\theta}(s_t)$$

$$= \nabla_\theta \sum_{s_t} d(s_t) \sum_{a_t} \pi_\theta(a_t|s_t) q^{\pi_\theta}(s_t, a_t)$$

$$= \sum_{s_t} d(s_t) \sum_{a_t} [\nabla_\theta \pi_\theta(a_t|s_t) q^{\pi_\theta}(s_t, a_t) + \pi_\theta(a_t|s_t) \nabla_\theta q^{\pi_\theta}(s_t, a_t)]$$

$$= \sum_{s_t} d(s_t) \sum_{a_t} \left[\nabla_\theta \pi_\theta(a_t|s_t) q^{\pi_\theta}(s_t, a_t) \right.$$

$$\left. + \pi_\theta(a_t|s_t) \gamma \nabla_\theta \sum_{s_{t+1}} p(s_{t+1}|s_t, a_t) v^{\pi_\theta}(s_{t+1}) \right]$$

$$= \sum_{s_t} d(s_t) \sum_{a_t} \nabla_\theta \pi_\theta(a_t|s_t) q^{\pi_\theta}(s_t, a_t) + \gamma \sum_{s_{t+1}} d(s_{t+1}|\pi_\theta) \nabla_\theta v^{\pi_\theta}(s_{t+1}).$$

Then, roll forward until infinity, and we have

$$\nabla_\theta J(\theta)$$

$$= \sum_{s} \sum_{\tau=t}^{\infty} \gamma^{\tau-t} p(s_\tau = s|\pi_\theta) \sum_{a} \nabla_\theta \pi_\theta(a|s) q^{\pi_\theta}(s, a)$$

$$= \sum_{\tau=t} \sum_{s_\tau} d(s_\tau) \sum_{a_\tau} \pi_\theta(a_\tau|s_\tau) \nabla_\theta \log \pi_\theta(a_\tau|s_\tau) \gamma^{\tau-t} q^{\pi_\theta}(s_\tau, a_\tau) \qquad (7\text{-}14)$$

$$= \frac{1}{1-\gamma} \sum_{s} d_{\pi_\theta}^{\gamma}(s) \sum_{a} \nabla_\theta \pi_\theta(a|s) q^{\pi_\theta}(s, a).$$

This formula is exactly equal to "vanilla policy gradient" in (7-13). This mathematical derivation, which originated from Richard Sutton and his colleagues in 1999, provided the first computable policy gradient for the discounted objective function [7-22]. Meanwhile, Sutton's version of vanilla policy gradient laid a solid theoretical foundation for various advanced policy gradient algorithms, including trust region policy optimization (TRPO), proximal policy optimization (PPO) and soft actor-critic (SAC).

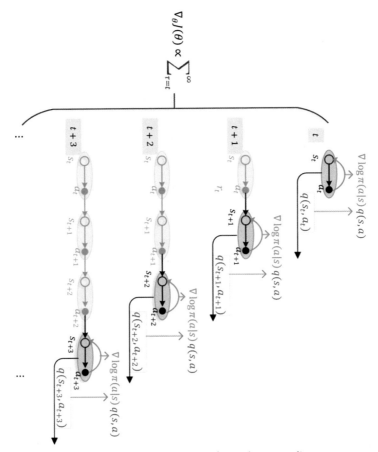

Figure 7.5 Policy gradient derivation from the cascading concept

Figure 7.5 illustrates how to derive the vanilla policy gradient in a cascading way. The vertical axis represents the time steps in the summation operation. At each time step, a gradient factor is calculated to maximize the future return. Instead of using the accumulative reward, the action-value function is repeatedly utilized in the cascading derivation, in which the information of temporal causality is subtly integrated into each expansion step. Eventually, the overall policy gradient is produced by summing all the gradient factors over time.

7.3   On-Policy Gradient

In the setting of on-policy RL, attempt is made to improve the same policy that is used to explore the environment. In this situation, we try to maximize the discounted objective function (7-6) using the gradient descent technique. To date, vanilla policy gradient is still not useful because there is no access to the discounted state distribution. Moreover, it is often too expensive to compute the integrals over the whole state and action spaces. Instead, we rely on the sample-based estimation to deal with the computation of vanilla policy gradient. A computable on-policy gradient should be in the expectation form of SSD under the target policy.

7.3.1   On-Policy Gradient Theorem

We only focus on continuing tasks with discount factor. In the vanilla policy gradient, its discounted state distribution means that the beginning steps matter more and the later steps matter less. In most continuing tasks, equally treating samples at different times is preferred so that the RL agent does not only focus on the beginning steps and neglect the future steps. The underlying idea can effectively simplify the sample-based estimation. The solution is easy, that is to replace the discounted state distribution with the stationary state distribution (SSD). In theory, exact equivalence between the two distributions happens only in two special cases. The first case is $d(s_t) = d_{\pi_\theta}(s)$, i.e., initial state distribution is exactly equal to the SSD. The second case is that $d(s_t)$ is not stationary, but the discount factor $\gamma \to 1$. In the first case, because $d(s_t) = d_{\pi_\theta}(s)$, we have the following equality:

$$d_{\pi_\theta}^\gamma(s) = (1-\gamma)\sum_{\tau=t}^{\infty} \gamma^{\tau-t} p(s_\tau = s|\pi_\theta) = (1-\gamma)\sum_{\tau=t}^{\infty} \gamma^{\tau-t} d_{\pi_\theta}(s) = d_{\pi_\theta}(s),$$

where $d_{\pi_\theta}(s)$ is the SSD under the current policy. When selecting $d_{\pi_\theta}(s)$ as the weighting function, one can remarkably reduce the computational complexity of vanilla policy gradient. In the second case, the discount factor is forced to become one, and we have the following relation in the limiting condition:

$$
\begin{aligned}
\lim_{\gamma \to 1} d_{\pi_\theta}^\gamma(s) &= \lim_{\gamma \to 1} \frac{d_{\pi_\theta}^\gamma(s)}{\sum_s d_{\pi_\theta}^\gamma(s)} \\
&= \lim_{\gamma \to 1} \lim_{N \to \infty} \frac{\sum_{\tau=t}^{t+N} \gamma^{\tau-t} p(s_\tau = s|\pi_\theta)}{\sum_s \sum_{\tau=t}^{t+N} \gamma^{\tau-t} p(s_\tau = s|\pi_\theta)} \\
&= \lim_{N \to \infty} \frac{\sum_{\tau=t}^{t+N} p(s_\tau = s|\pi_\theta)}{N+1} \\
&= d_{\pi_\theta}(s).
\end{aligned}
\tag{7-15}
$$

Even though exact equivalence holds only in the limiting condition of discount factor, this kind of distribution replacement often works very well over a wide range, for example, $0.95 < \gamma < 1.0$. Especially when a sufficiently large amount of samples are collected, the discounted state distribution will be very close to the stationary one. After performing the distribution replacement, we can build an on-policy gradient in the following form:

$$
\begin{aligned}
\nabla_\theta J(\theta) &\approx \frac{1}{1-\gamma} \sum_s d_{\pi_\theta}(s) \sum_a \pi_\theta(a|s) \nabla_\theta \log \pi_\theta(a|s) q^{\pi_\theta}(s,a) \\
&\propto \mathbb{E}_{s \sim d_{\pi_\theta}} \left\{ \mathbb{E}_{a \sim \pi_\theta} \{ q^{\pi_\theta}(s,a) \nabla_\theta \log \pi_\theta(a|s) \} \right\} \\
&= \mathbb{E}_{\pi_\theta} \{ q^{\pi_\theta}(s,a) \nabla_\theta \log \pi_\theta(a|s) \}.
\end{aligned}
\tag{7-16}
$$

The resulting on-policy gradient is quite appealing because no information on the ergodic state distribution or environment model must be known. Simply collecting data under the current policy is able to provide an estimation of on-policy gradient. Moreover, this formula is an unbiased estimation. Similar to the quadrature of an integral, sample-based estimation can be used to attain high accuracy with a set of discrete thick points. As a result, (7-16) is the basis for a variety of model-free RL algorithms. Figure 7.6 briefly shows how to estimate on-policy gradient, in which $\mathcal{D}_\pi$ is the dataset under the policy to be optimized. The notation $|\mathcal{D}_\pi|$ represents the sample size and $Q(s, a)$ is the estimate of action-value function.

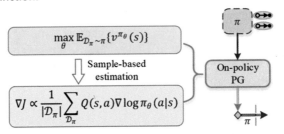

Figure 7.6 On-policy gradient estimation

Even though on-policy gradient is unbiased, it often has very high variance if the sample collection is not sufficient. One remedy is the baseline technique, which is to reduce the gradient variance while still maintaining the benefit of unbiased estimation. The baseline technique is an effective way to improve the quality of sample-based estimation. Basically, it introduces a subtraction operation to reduce the potential uncertainty in the multiplication of action-value function and logarithmic score function.

7.3.2   Extension of On-Policy Gradient

There are three main variants of on-policy gradients. In the first variant, action-value function is replaced with the Monte-Carlo average estimation. In the second variant, a baseline is added to reduce the estimation variance. The baseline is a zero-mean term that has the ability to distinguish between good actions and bad actions. In the third variant, action-value function is replaced with state-value function. This replacement is justified by the self-consistency condition. The on-policy gradient with state-value function is very popular because it can naturally combine the minimum-variance baseline without introducing any additional computational burden.

7.3.2.1   Monte Carlo policy gradient

In theory, accurate policy gradient needs to compute a true value function. However, the true value function is not always accessible in practical RL problems. The simplest solution is to replace the action-value function with its Monte Carlo estimate. The new formula is called "Monte Carlo policy gradient", which uses the average of long-term returns to replace the true action value in (7-16). Using this replacement, we have a primitive policy gradient algorithm:

$$\theta \leftarrow \theta + \beta \cdot \text{Avg}\{G_t | s_t, a_t\} \nabla \log \pi_\theta(a_t | s_t). \tag{7-17}$$

This algorithm is also called REINFORCE (Williams, 1992) [7-24]. It is easy to see that (7-17) does not include any parameterized value function, and accordingly, REINFORCE is a special actor-only algorithm. Usually, Monte Carlo policy gradient has very high variance and it is very ineffective in handling large-scale tasks. The pseudocode of REINFORCE is shown in Algorithm 7-1, in which an episodic task is explored to train a stochastic policy and each episode must have a termination point.

Algorithm 7-1: REINFORCE

Hyperparameters: learning rate β, discount factor γ, number of environment resets M, number of episodes B

Initialization: policy function $\pi(a|s; \theta)$

Repeat (indexed with k)

    (4) Data collection

    Initialize memory buffer $\mathcal{D} \leftarrow \emptyset$

    Repeat M environment resets

        $s_0 \sim d_{\text{init}}(s)$

        $a_0 \sim \pi(a|s_0; \theta)$

        For i in $1, 2, \ldots, B$

            Sample an episode τ_i until termination

            $\tau_i \leftarrow \{s_0, a_0, r_0, \cdots, s_{T-1}, a_{T-1}, r_{T-1}, s_T\}$

            $G_i(s_0, a_0) \leftarrow \sum_{t=0}^{T-1} \gamma^t r_t$

        End

        $\bar{G}(s_0, a_0) \leftarrow \text{Avg}\{G_i(s_0, a_0), i = 1, 2, \cdots, B\}$

        $\mathcal{D} \leftarrow \mathcal{D} \cup \{(\bar{G}, s_0, a_0)\}$

    End

    (5) Update with MC policy gradient

    $$\nabla_\theta J \leftarrow \frac{1}{|\mathcal{D}|} \sum_{\mathcal{D}} \bar{G}(s, a) \nabla \log \pi(a|s; \theta)$$

    $\theta \leftarrow \theta + \beta \cdot \nabla_\theta J$

End

∎

There are two major steps in the REINFORCE algorithm. In the first step, environment exploration is started from random initial state-action pairs. From the same initialization point, multiple trajectories are generated from environment interaction, and their returns are averaged as the estimate of the true action-value function. In the second step, the MC policy gradient is calculated by taking sample-based estimation over all available gradient points from diverse state-action pairs. The overall gradient estimate is utilized to update the policy parameter with a fixed learning rate.

7.3.2.2   Add baseline to reduce variance

The baseline technique has been widely used in a variety of model-free RL algorithms, including REINFORCE (1992)[7-24], A3C (2016) [7-15], and TRPO (2015) [7-18]. A proper baseline can help to choose a better updating direction and strengthen the convergence speed of policy improvement. The action with a Q-value that is higher than the baseline is regarded as a good action, and its probability will be increased in successive updates. In contrast, the probability of a bad action, which has a lower Q-value, will be decreased in successive updates. Therefore, more accurate gradient estimation will be obtained by choosing good actions and removing bad actions. The on-policy gradient with a general baseline is

$$\nabla_\theta J(\theta) \propto \mathbb{E}_{\pi_\theta}\{(q^{\pi_\theta}(s,a) - \zeta(s))\nabla \log \pi_\theta(a|s)\}, \tag{7-18}$$

where $\zeta(s)$ is a baseline function with proper input dimension. This formula is equivalent to the previous version (i.e., unbiased estimate) because the expectation of the subtracted quantity is equal to zero:

$$
\begin{aligned}
\mathbb{E}_{\pi_\theta}\{\zeta(s)\nabla \log \pi_\theta(a|s)\} &= \sum_s d(s) \sum_a \pi_\theta(a|s) \cdot \zeta(s) \frac{\nabla \pi_\theta(a|s)}{\pi_\theta(a|s)} \\
&= \sum_s d(s)\zeta(s)\nabla \sum_a \pi_\theta(a|s) \\
&= \sum_s d(s)\zeta(s)\nabla 1 \\
&= 0.
\end{aligned}
\tag{7-19}
$$

It can be inferred from this proof that $\zeta(s)$ can be any function, but it must be independent of action. Even a random variable can be selected as $\zeta(s)$ so long as it does not vary with action. Otherwise, the baseline cannot be separated from the expectation operator, and the subtracted quantity cannot equal zero. One may be interested in what kind of baseline is the best in terms of variance reduction. There is a famous relationship between the variance and the mean of a random variable, i.e., $\mathbb{D}\{X\} = \mathbb{E}\{X^2\} - \mathbb{E}\{X\}^2$. After adding a baseline, the variance reduction level is equal to

$$
\begin{aligned}
\Delta\mathbb{D} &= \mathbb{D}\{\nabla J_{\text{BL}}\} - \mathbb{D}\{\nabla J\} \\
&= \mathbb{D}_{\pi_\theta}\{(q^{\pi_\theta}(s,a) - \zeta(s))\nabla \log \pi_\theta\} - \mathbb{D}_{\pi_\theta}\{q^{\pi_\theta}(s,a)\nabla \log \pi_\theta\} \\
&= -\mathbb{E}_{\pi_\theta}\{(\nabla \log \pi_\theta)^2\}\mathbb{E}_{\pi_\theta}\{(2v^{\pi_\theta}(s) - \zeta(s))\zeta(s)\}.
\end{aligned}
$$

The variance reduction level is a quadratic function of the baseline variable. When $\zeta(s)=v^{\pi_\theta}(s)$, the best level of variance reduction is reached:

$$\Delta\mathbb{D}_{\min} = -\mathbb{E}_{\pi_\theta}\{(\nabla \log \pi_\theta)^2\}\mathbb{E}_{\pi_\theta}\left\{\left(v^{\pi_\theta}(s)\right)^2\right\}. \tag{7-20}$$

Therefore, the best choice of a baseline is the state-value function. The use of state-value function as a baseline actually leads to the well-known advantageous learning architecture. In this architecture, $A(s,a) \overset{\text{def}}{=} q(s,a) - v(s)$ is called the advantage function. The emergence of advantage function provides a noticeable benefit to variance reduction. This is because the error between $q(s,a)$ and $v(s)$ is generally smaller than $q(s,a)$ itself, and its amplification on the uncertainty of logarithmic policy gradient is much reduced. However, such a primitive design needs to take care of both action-value function and state-value function. As a result, the computational burden is actually doubled. A practical way to solve this problem is to replace the action-value function with the state-value function to avoid additional computational burden.

7.3.2.3   Replace action-value with state-value

When taking a baseline to reduce variance, a vanilla policy gradient with only state-value function is more popular in real-world applications. It has the potential to provide RL with the best variance reduction ability and an unbiased gradient estimation. Compared with an action-value function, a state-value function has a lower input dimension. Therefore, to reach the same level of accuracy, the estimation of state-value function needs fewer samples than that of action-value function. The self-consistency relationship between the two value functions is

$$q^{\pi_\theta}(s, a) = \mathbb{E}_{s'}\{r + \gamma v^{\pi_\theta}(s')\}. \tag{7-21}$$

A new on-policy gradient is obtained by substituting (7-21) into (7-16):

$$\nabla_\theta J(\theta) \propto \mathbb{E}_{\pi_\theta}\{(r + \gamma v^{\pi_\theta}(s'))\nabla \log \pi_\theta(a|s)\}. \tag{7-22}$$

This is the version of on-policy gradient with state-value function. Similar to that with action-value function, the best baseline function can be added to (7-22). When $\zeta(s) = v^{\pi_\theta}(s)$, the best variance reduction ability is obtained:

$$\nabla_\theta J(\theta) \propto \mathbb{E}_{\pi_\theta}\{(r + \gamma v^{\pi_\theta}(s') - v^{\pi_\theta}(s))\nabla \log \pi_\theta(a|s)\}. \tag{7-23}$$

In essence, a state-value function describes the average level of all actions. An advantage function represents the extra return that a certain action may lead to. Therefore, such a policy gradient with advantage function will increase the probability of actions that are beyond the average level and decrease the probability of below-average actions. Interestingly, the advantage function exactly becomes one-step TD error in (7-23), i.e.,

$$A(s, a) = r + \gamma v(s') - v(s).$$

Thus far, we have two ways to look at the origin of advantage function. From the perspective of one-step TD error, it originates from the bootstrapping mechanism. From the perspective of baseline function, it comes from the best variance reduction ability. The same advantage function can be generated from either of the two approaches. This kind of coincidence bridges the gap between variance reduction and bootstrapping technique. One natural extension of advantage function is to replace the one-step TD error with multistep TD error or even TD-lambda error, which usually offers better training stability and faster convergence speed. This kind of extension is justified because the true action-value function can be always replaced by any length of multistep TDs since their expectations are the same. More samples in multistep TDs mean that more information from environment interaction is used, and the approximation errors from history value estimates can be better corrected.

7.3.3   How Baseline Really Works

A useful baseline function can be any form that does not directly depend on actions. With the best baseline choice, i.e., state-value function, this type of advantage function can be treated as a one-step TD error. Figure 7.7 shows the benefits of applying the best baseline. In this example, the autonomous car has learned to run on a circular road at a certain longitudinal speed. Here, two kinds of on-policy AC algorithms with state-value function and stochastic policy are utilized to compare their policy performance. The same hyperparameter settings are used, except that one does not have any baseline (i.e., on-policy AC) and the other has the best baseline (i.e., on-policy A2C). In both of them, policy performance is measured by the total average return (TAR). It is easy to see that on-policy A2C is more stable than on-policy AC. This is because the low-variance policy gradient in on-policy A2C needs much fewer samples to achieve the same level of accuracy. Moreover, under the same learning rates ($\alpha = \beta = 2 \times 10^{-3}$), on-policy A2C has a much faster convergence speed than on-policy AC. In addition, satisfactory convergence rate is obtained in on-policy A2C even though its learning rate is very small. In contrast, there exists a long wait time before the total average return begins to rise in on-policy AC.

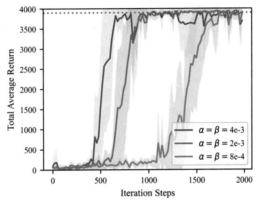

(a) Without a baseline (i.e., on-policy AC)

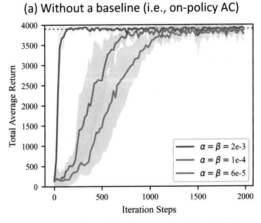

(b) With a baseline (i.e., on-policy A2C)

Figure 7.7 Comparison of total average return

Figure 7.8 and Figure 7.9 compare actions and states of the two algorithms, respectively. The two actions include the commands for longitudinal acceleration and steering angle, and the five states are yaw rate, lateral velocity, heading angle error, lateral position error, and longitudinal speed. Generally, their stochastic policies cannot exactly reach equilibrium due to their intrinsic randomness. This kind of randomness, however, helps to better explore the environment and learn a close-to-optimal policy.

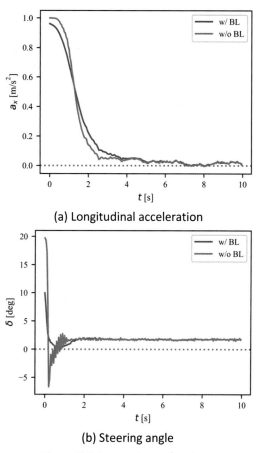

(a) Longitudinal acceleration

(b) Steering angle

Figure 7.8 Comparison of actions

The curves of actions and states are drawn with one final policy to study their transient and steady-state behaviors. It is observed from Figure 7.8 that the two algorithms both output stable actions, but their optimal policies are not identical to each other. Similar results can be found in the state curves in Figure 7.9. In addition, both yaw rate and lateral velocity are very noisy because they are directly influenced by environmental noise and external disturbance. It is difficult for a stationary policy to smooth these noisy fluctuations because the policy does not hold any filtering ability. The curves of heading angle error, lateral position error, and longitudinal speed are much smoother than the first two states because their influence from external randomness is smoothed by the integrator in the environment dynamics.

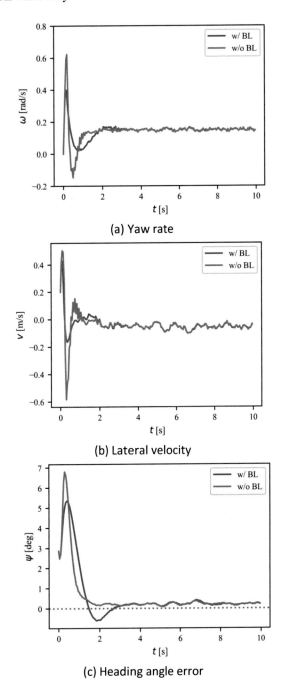

(a) Yaw rate

(b) Lateral velocity

(c) Heading angle error

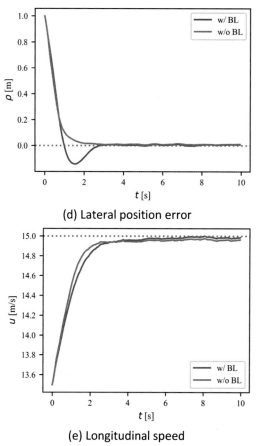

(d) Lateral position error

(e) Longitudinal speed

Figure 7.9 Comparison of states

7.4   Off-Policy Gradient

Generating data with only the target policy is not very efficient for policy updates. Due to the low interaction efficiency in a sampler, on-policy RL may lack useful data at each cycle. Even worse, if there is not enough supportive data to output a better policy, the algorithm may easily lose convergence. Parallel exploration somewhat alleviates the need for new data but comes at the cost of occupying heavy computational resource consumption. Therefore, it is important for an RL agent to have the ability to utilize historical data from previous policies. In recent years, off-policy gradient technique has drawn increasing attention due to its potential to largely increase sample efficiency. In 2012, Degris, White and Sutton completed an early study in which parallel off-policy gradients were proposed in the form of tabular representation [7-3]. Their study has motivated a few advanced off-policy algorithm designs, including emphatic actor-critic [7-14], off-policy deterministic policy gradient (off-policy DPG) [7-20], actor-critic with experience replay (ACER) [7-23], and interpolated policy gradient (IPG) [7-8].

7.4.1   Off-Policy True Gradient

The objective of off-policy RL is to train the target policy using not only the newly collected data, but also those historical data from behavior policies. Strictly speaking, the derivation of a theoretically solid off-policy gradient should start from the primal objective function (7-1), and its vanilla formula should remain the same form as that in the on-policy setting (7-13). Since (7-13) is sampled with the target policy, the expectation of policy gradients must be rewritten with the SSD of a behavior policy. Therefore, the importance sampling (IS) technique should be applied to harmonize the distributional discrepancy between the target policy and the behavior policy. After performing such an IS transformation, an analytical off-policy gradient is readily derived:

$$\nabla_\theta J(\theta) = \mathbb{E}_{\mathcal{T} \sim b} \left\{ \sum_{\tau=t}^{\infty} \nabla_\theta \log \pi_\theta(a_\tau|s_\tau) \left(\prod_{k=t}^{\tau} \frac{\pi_\theta(a_k|s_k)}{b(a_k|s_k)} \right) \gamma^{\tau-t} q^{\pi_\theta}(s_\tau, a_\tau) \right\}. \qquad (7\text{-}24)$$

This formula is called "off-policy true gradient". It comes from the primal criterion and has no sacrifice that comes from the state distribution replacement. However, this true gradient is not useful in practice due to its extremely high variance. The IS transformation introduces the cascading multiplication of a series of IS ratios. Each quotient factor may have a large uncertainty when the behavior policy is not well recorded or some actions have close-to-zero probabilities. The multiplication of highly uncertain IS ratios can greatly amplify the estimation error in the action-value function and make the overall gradient estimation very inaccurate.

7.4.2   Off-Policy Quasi-Gradient

Computing an off-policy gradient is a subtle task because it involves both action evaluation and distribution selection. With sparse reward signals, accurately estimating return is not always easy, and certain value estimation error exists inevitably. In addition, the used historical samples might belong to several different behavior policies, rather than only one behavior policy. Currently, the most widely used off-policy gradient is the one introduced by Degris, White and Sutton (2012) [7-3]. To avoid high uncertainties in the multiplication of cascading IS ratios, the overall RL objective function is modified to utilize the behavior state distribution as the weighting function:

$$J(\theta) = \mathbb{E}_{s \sim d_b}\{v^{\pi_\theta}(s)\}. \qquad (7\text{-}25)$$

The resulting gradient of (7-25) will be called "off-policy quasi-gradient". Compared with the true gradient, the quasi-gradient is biased but has much lower variance. After taking the derivative of (7-25), we have the first-step expansion:

$$\nabla_\theta J(\theta) = \mathbb{E}_{s \sim d_b} \left\{ \sum_{a_t} [\nabla_\theta \pi_\theta(a_t|s_t) q^{\pi_\theta}(s_t, a_t) + \pi_\theta(a_t|s_t) \nabla_\theta q^{\pi_\theta}(s_t, a_t)] \right\}.$$

Then, roll forward exactly like the on-policy gradient derivation, and we have

$$\nabla_\theta J(\theta) \propto \mathbb{E}_{s \sim d_{\pi_\theta}^\gamma(s|d_b), a \sim \pi_\theta}\{q^{\pi_\theta}(s,a)\nabla_\theta \log \pi_\theta(a|s)\}, \qquad (7\text{-}26)$$

The key of successful derivation is to replace the discounted state distribution with a stationary state distribution (SSD) to match sample-based estimation. Note that

$d_{\pi_\theta}^\gamma(s|d_b)$ is the discounted state distribution conditioned on d_b. The conditional probability means that its associated data samples come from the behavior policy rather than the target policy. The definition for $d_{\pi_\theta}^\gamma(s|d_b)$ and its connection with $d_{\pi_\theta}^\gamma(s|s_t)$ are expressed as

$$
\begin{aligned}
d_{\pi_\theta}^\gamma(s|d_b) &= (1-\gamma) \sum_{\tau=t}^\infty \gamma^{\tau-t} p(s_\tau = s|\pi_\theta, d_b) \\
&= (1-\gamma) \sum_{s_t} d_b(s_t) \sum_{\tau=t}^\infty \gamma^{\tau-t} p(s_\tau = s|\pi_\theta, s_t) \\
&= \sum_{s_t} d_b(s_t) d_{\pi_\theta}^\gamma(s|s_t).
\end{aligned}
$$

One challenge of this definition is that its each term $p(s_\tau = s|\pi_\theta, s_t)$ is computationally intractable as there is no easy access to sufficient data samples under the target policy in the off-policy setting. A subtle solution is to assume that the conditioned discounted state distribution $d_{\pi_\theta}^\gamma(s|d_b)$ is almost equal to the SSD under the behavior policy, i.e., $d_b(s)$. This assumption can simplify (7-26) to be the expectation form of the behavior policy:

$$
\begin{aligned}
\nabla J(\theta) &\approx \sum_s d_b(s) \sum_a \nabla_\theta \pi_\theta(a|s) q^{\pi_\theta}(s, a) \\
&= \mathbb{E}_{s \sim d_b} \left\{ \sum_a b(a|s) \frac{\pi_\theta(a|s)}{b(a|s)} q^{\pi_\theta}(s, a) \frac{\nabla \pi_\theta(a|s)}{\pi_\theta(a|s)} \right\} \\
&= \mathbb{E}_{s \sim d_b} \left\{ \mathbb{E}_{a \sim b} \left\{ \frac{\pi_\theta(a|s)}{b(a|s)} q^{\pi_\theta}(s, a) \nabla \log \pi_\theta(a|s) \right\} \right\} \\
&= \mathbb{E}_b \left\{ \frac{\pi_\theta(a|s)}{b(a|s)} q^{\pi_\theta}(s, a) \nabla \log \pi_\theta(a|s) \right\}.
\end{aligned}
\tag{7-27}
$$

This formula is named as "off-policy quasi-gradient". Here, $\pi_\theta(a|s)/b(a|s)$ is equivalent to the one-step IS ratio, but its origin differs from the importance sampling technique. Without loss of generality, we keep the primitive form of this quasi-gradient and do not write $\pi_\theta(a|s)/b(a|s)$ into traditional notation of one-step IS ratio like ρ_t. Clearly, this quasi-gradient can eliminate the cascading multiplication of multiple IS ratios, which is helpful in avoiding the unpredictable uncertainties in the sample-based estimation. Compared to off-policy true gradient, the off-policy quasi-gradient is computationally efficient, and its three elements, including an IS ratio, an action-value function and a current-step policy, are easy to calculate. For better efficiency, the one-step IS ratio can be simultaneously stored in a memory buffer at the moment when storing other variables, such as state, action and reward.

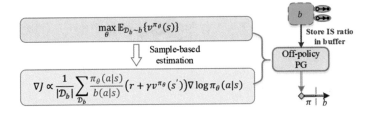

Figure 7.10 Off-policy gradient estimation

By replacing action-value function with state-value function, we have the following relationship:

$$\nabla J(\theta) \quad = \mathbb{E}_{s\sim d_b, a\sim b}\left\{\frac{\pi_\theta(a|s)}{b(a|s)}\mathbb{E}_{s'\sim\mathcal{P}}\{r+\gamma v^{\pi_\theta}(s')\}\nabla\log\pi_\theta(a|s)\right\}$$

$$= \mathbb{E}_{s\sim d_b, a\sim b, s'\sim\mathcal{P}}\left\{\frac{\pi_\theta(a|s)p(s'|s,a)}{b(a|s)p(s'|s,a)}(r+\gamma v^{\pi_\theta}(s'))\nabla\log\pi_\theta(a|s)\right\}$$

$$= \mathbb{E}_b\left\{\frac{\pi_\theta(a|s)}{b(a|s)}(r+\gamma v^{\pi_\theta}(s'))\nabla\log\pi_\theta(a|s)\right\}$$

How to estimate the off-policy quasi-gradient is briefly illustrated in Figure 7.10, in which $\mathcal{D}_b$ is the dataset under the behavior policy. The estimation variance can be reduced by adding a baseline function. The variance reduction mechanism is the same as that of the on-policy version [7-22]. The best baseline technique naturally suits this off-policy formula, which will not introduce any additional computational burden. Obviously, if multistep TD is used to replace the action-value function, the IS ratios from the future states, including $s_{t+2}, s_{t+3}, \cdots, s_{t+n}$, will appear. A series of new IS ratios are needed to build a compatible off-policy gradient. There are great benefits to using off-policy gradients for offline training, especially when an RL agent only has a single stream of collected samples. In today's deep RL, learning from historical data is known as experience replay, which is a popular trick to stabilize the approximation of deep neural network. The use of off-policy gradients also enables learning from other forms of data sources, including expert demonstrations, non-optimal controllers, and even random control behaviors, so long as their behavior policies are known.

7.5   Actor-Critic Architecture from Direct RL

The actor-critic (AC) architecture has a separate memory structure to represent a policy and a value function, respectively. The update of the policy parameter is known as "actor" because it seeks to select better actions, and the update of the value parameter is known as "critic" because it criticizes action selection [7-12]. This kind of AC architecture can be derived from both direct RL and indirect RL. From the perspective of indirect RL, the actor corresponds to parameterized PIM, while the critic corresponds to parameterized PEV. From the perspective of direct RL, the actor corresponds to policy gradient-based update, and the critic is its embedded value function in the policy gradient estimation [7-9]. A typical AC architecture with vanilla policy gradient is shown in Figure 7.11. In this figure, policy gradient is the dominating part in the AC architecture, and "critic" is an auxiliary element to support accurate gradient estimation. Whether the resulting AC algorithm converges or not depends on the property of first-order optimization rather than the updating mechanism of a strictly better policy.

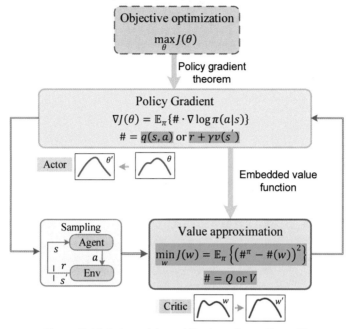

Figure 7.11 Actor-critic architecture from direct RL

7.5.1   Off-policy AC with Advantage Function

The advantage actor-critic (A2C) algorithm emerges if one chooses the best baseline, i.e., state-value function, in traditional AC methods. An off-policy A2C algorithm can utilize historical data, and all the experience is replayed to estimate actor and critic gradients. This algorithm consists of an off-policy actor and an off-policy critic. The actor had better use accurate information of critic evaluation to pursue good actions and avoid bad actions. The more accurate the critic is, the better stability the AC algorithm has. The pseudocode of off-policy A2C is demonstrated in Algorithm 7-2, in which a replay buffer stores the probability of selecting a certain action and the IS ratio is calculated to address the discrepancy between target and behavior policies.

Algorithm 7-2: Off-policy A2C

Hyperparameters: critic learning rate α, actor learning rate β, discount factor γ, number of environment resets M, episode length N, mini-batch size $|\mathcal{B}|$

Initialization: state-value function $V(s; w)$, policy function $\pi(a|s; \theta)$, memory buffer $\mathcal{D} \leftarrow \emptyset$

Repeat (indexed with k)

    (1) Data collection

    Repeat M environment resets

Generate an initial state $s_0 \sim d_{\text{init}}(s)$

For i in $0, 1, 2, \ldots, N$ or until episode termination

$\quad a_i \sim b(a|s_i; \theta)$

$\quad p_i \leftarrow b(a_i|s_i; \theta)$

Apply a_i in environment and observe s_{i+1} and r_i

$\quad \mathcal{D} \leftarrow \mathcal{D} \cup \{(s_i, a_i, r_i, s_{i+1}, p_i)\}$

End

End

(2) Re-sampling from memory buffer

Randomly select a mini-batch $\mathcal{B} \subset \mathcal{D}$

Sweep mini-batch $\mathcal{B}$

$\quad p_i \leftarrow \pi(a_i|s_i; \theta)/p_i, i \in \mathcal{B}$

End

(3) Critic update

$$\nabla_w J_{\text{Critic}} \leftarrow \frac{1}{|\mathcal{B}|} \sum_{\mathcal{B}} \rho \cdot \left(r + \gamma V(s'; w) - V(s; w) \right) \frac{\partial V(s; w)}{\partial w}$$

$w \leftarrow w - \alpha \cdot \nabla_w J_{\text{Critic}}$

(4) Actor update

$$\nabla_\theta J_{\text{Actor}} \leftarrow \frac{1}{|\mathcal{B}|} \sum_{\mathcal{B}} \rho \cdot \nabla_\theta \log \pi(a|s; \theta) \left(r + \gamma V(s'; w) - \zeta(s) \right)$$

$\theta \leftarrow \theta + \beta \cdot \nabla_\theta J_{\text{Actor}}$

End

∎

Even though equipped with experience replay, training an off-policy agent still requires massive throughput and a large amount of data. Luckily, the AC architecture allows the use of parallel computation to accelerate the overall sampling speed, as shown in Figure 7.12. A well-known example is the asynchronous advantage actor-critic (A3C) algorithm [7-15], in which each thread has an individual sampler and an individual learner. All the learners collectively train one global value function and one global policy. The sampler collects data samples from an individually running environment. Each learner has a local critic and a local actor to compute the value and policy gradients in parallel. The value function and policy are updated in an asynchronous manner; that is, no synchronization is needed when parallel learners are utilized.

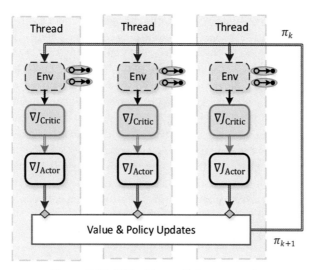

Figure 7.12 A3C with parallel computation

In the beginning, standard A3C was only combined with an on-policy strategy. As a result, historical experience cannot be used in standard A3C. Combined with the importance sampling technique, Google DeepMind has developed a scalable distributed off-policy A2C algorithm called IMPALA, which can be scaled to thousands of parallel machines without sacrificing data efficiency or resource utilization [7-4]. The core idea of IMPALA is to decouple samplers from learners. Since the samplers no longer need to wait for the learners to stop, a very high throughput can be achieved in IMPALA.

7.5.2   Off-policy Deterministic Actor-Critic

Before the emergence of deterministic policy gradients, most model-free RL algorithms utilized stochastic policies and the policy parameter was adjusted along the expected direction of stochastic policy gradient. There have been a few successful applications using the stochastic policy gradient, but its large variance is still an intrinsic bottleneck to achieve an accurate optimal solution. In 2014, David Silver and his colleagues demonstrated that a deterministic policy gradient is a limiting case of its stochastic counterpart, as the policy randomness is enforced to become zero [7-20]. The deterministic policy gradient (DPG) has a particularly appealing benefit. It can be estimated more efficiently than stochastic policy gradient. Let us define the objective function with a deterministic policy:

$$J(\theta) \stackrel{\text{def}}{=} \mathbb{E}_{s \sim d(s)}\{q^{\pi_\theta}(s, \pi_\theta(s))\}. \tag{7-28}$$

The derivation of DPG needs to satisfy several regularity conditions, which must assume that all involved functions and their derivatives are continuous and bounded. The final result of this derivation is formally expressed as

$$\nabla J(\theta) \approx \mathbb{E}_{s \sim d_b/s \sim d_\pi}\{\nabla_\theta \pi_\theta(s) \nabla_a q^{\pi_\theta}(s, a)|_{a=\pi_\theta(s)}\}. \tag{7-29}$$

Here, $\nabla_\theta \pi_\theta(s) \stackrel{\text{def}}{=} \partial \pi^{\mathrm{T}}/\partial \theta \in \mathbb{R}^{l_\theta \times m}$ is a matrix, and $\nabla_a q(s, a) \in \mathbb{R}^{m \times 1}$ is a column vector. In the deterministic policy, each action is solely determined by state, and action selection is no longer a random behavior. Hence, off-policy DPG does not explicitly contain any IS ratio. The off-policy DPG shares the same formula as on-policy DPG. The notation "≈" can become "=" if one of the following conditions holds: (1) $d(s)$ is exactly equal to d_b or d_π, or (2) $\gamma \rightarrow 1$. The DPG does not look like the stochastic version at first glance. For a wide class of stochastic policies, DPG is indeed a special limiting case of a stochastic policy gradient. This proof can be finished by a chain rule with a stochastic policy that contains a deterministic part and a parameter-independent variance term. The off-policy AC with DPG (i.e., off-policy DPG) is demonstrated in Algorithm 7-3, in which its critic also eliminates the utilization of the IS ratios. The reason is that each sample only contains the current state s, the current action a and the next state s', and the next action a' is sampled from the current behavior policy. The triplet (s, a, s') does not need an explicit IS ratio in the action-value estimation. The critic update is built with stochastic optimization, whose goal is to minimize the error between Q-target and Q-value under the behavior policy.

Algorithm 7-3: Off-policy DPG

Hyperparameters: critic learning rate α, actor learning rate β, discount factor γ, number of environment resets M, episode length N, mini-batch size $|\mathcal{B}|$

Initialization: action-value function $Q(s, a; w)$, policy function $\pi(s; \theta)$, and memory buffer $\mathcal{D} \leftarrow \emptyset$

Repeat (indexed with k)

    (1) Data collection

    Repeat M environment resets

        Generate an initial state $s_0 \sim d_{\text{init}}(s)$

        For i in $0, 1, 2, \ldots, N$ or until episode termination

            $a_i \sim b(a|s_i; \theta)$

            Apply a_i in environment and observe next state s_{i+1} and reward r_i

            $\mathcal{D} \leftarrow \mathcal{D} \cup \{(s_i, a_i, r_i, s_{i+1})\}$

        End

    End

    (2) Re-sampling from memory buffer

    Randomly select a mini-batch $\mathcal{B} \subset \mathcal{D}$

    (3) Critic update

    $a' = \pi(s'; \theta)$

$$\nabla_w J_{\text{Critic}} \leftarrow \frac{1}{|\mathcal{B}|} \sum_{\mathcal{B}} (r + \gamma Q(s', a'; w) - Q(s, a; w)) \frac{\partial Q(s, a; w)}{\partial w}$$

    $w \leftarrow w - \alpha \cdot \nabla_w J_{\text{Critic}}$

    (4) Actor update

$$\nabla_\theta J_{\text{Actor}} \leftarrow \frac{1}{|\mathcal{B}|} \sum_{\mathcal{B}} \nabla_\theta \pi(s; \theta) \nabla_a Q(s, a; w)$$

$$\theta \leftarrow \theta + \beta \cdot \nabla_\theta J_{\text{Actor}}$$

End

∎

From a computational viewpoint, there is a crucial difference between stochastic policy gradients and deterministic policy gradients. In the stochastic case, the policy gradient integrates over both state and action spaces, whereas in the deterministic case, the policy gradient only integrates over the state space. As a result, computing DPG requires much fewer samples, especially if the action space has a high dimension. It is worth noting that (7-29) has two partial derivatives that come from action-value function and policy function. This kind of partial derivative is often computed by automatic differentiation tools in TensorFlow or PyTorch. In general, adequate exploration cannot be achieved when only behaving with a deterministic policy, and only suboptimal solutions can be found. On-policy DPG is only suitable for environments that already have good self-exploration ability. In general, the off-policy version is preferred because it can add an exploration term to build various behavior policies.

7.5.3   State-Of-The-Art AC Algorithms

The study by Barto, Sutton and Anderson in 1983 might be the starting point of modern AC algorithms. In their study, actors and critics were called "associative search elements" and "adaptive critic elements", respectively [7-1]. The functionality of these elements has almost no difference from today's understanding about actor updates and critic updates. After this study, one can see a clear path of how advanced AC algorithms have evolved from their early versions. The two extremes of AC architecture are critic-only method and actor-only method. The critic-only method (i.e., Q-learning and its variants) relies only on the fixed-point iteration mechanism, and it has the ability to utilize off-policy experience replay. For large-scale tasks, one needs to resort to a certain optimization technique to obtain the solution of a parameterized value function. The actor-only method (i.e., REINFORCE) must work with gradient-based updates, in which the policy gradient is computed without reliance on any value function. Both discrete and continuous actions can be addressed, but naive policy gradients often suffer from high variance, especially when samples are not sufficient.

The actor-critic (AC) architecture is one of the most popular algorithm families in today's RL community. Typical algorithms that use this architecture include deep deterministic policy gradient (DDPG), advantage actor-critic (A2C), asynchronous advantage actor-critic (A3C), soft actor-critic (SAC), distributional soft actor-critic (DSAC), twin delayed DDPG (TD3), trust region policy optimization (TRPO), and proximal policy optimization (PPO). In Table 7.1, state-of-the-art AC algorithms are summarized and classified according to five criteria, including policy type (e.g., stochastic and deterministic), value function type (e.g., state-value and action-value), critic update mechanism, policy update mechanism, and on-policy/off-policy.

Table 7.1 Summary of Actor-Critic Algorithms

Algorithm	Policy	Value	Critic Update	Actor Update	On/Off

DDPG	D	Q	TD-based	Vanilla PG	Off
TRPO	S	V	TD-based	Natural PG	On
PPO	S	V	TD-based	Clipped PG	On
TD3	D	Q	Clipped Double Q	Vanilla PG	Off
D4PG	D	Q	Discrete Distributional Q-TD	Vanilla PG	Off
ACKTR	S	V	TD-based	Natural PG	On
A2C/A3C	S	V	TD-based	Vanilla PG	On
Off-PAC	S	V	TD-based	Vanilla PG	Off
ACER	S	Q	TD-based	Vanilla PG	Off
IMPALA	S	V	TD based	Vanilla PG	Off
Soft Q-learning	S	Q	Soft Q-iteration	Soft PG	Off
SAC	S	Q	Clipped Double-Q	Soft PG	Off
DSAC	S	Q	Continuous Distributional Q-TD	Soft PG	Off

* S-Stochastic policy; D-Deterministic policy; Q-Action-value function; V-State-value function; PG-Policy gradient

In the critic, an advantage function evaluates how good an action is by comparing it with average action behavior. From this comparison, high variance in the gradient estimate can be reduced. An interesting observation is that the critic target is a one-step TD, which makes an advantage function actually a one-step TD error. The target can naturally be extended to a multistep TD target. Even though it appears to be a trivial replacement, a multistep TD target provides better stability and convergence speed than one-step TD. The advantage function with a multistep TD target is defined as

$$\delta_V^{\text{TD}(n)}(s) \stackrel{\text{def}}{=} \underbrace{G_{t:t+n-1} + \gamma^n V^\pi(s_{t+n})}_{n-\text{step TD target}} - V^\pi(s_t).$$

Here, $s_t = s$ is taken as the initial state. The formula above is also called the n-step TD error. Table 7.2 lists a few TD(n)-based critic criteria, including both on-policy and off-policy criteria. Let us point out that a semi-gradient is preferred in a TD-based critic since its learning process is more stable than its true gradient. The reason is easy to understand. The true gradient that optimizes the TD error often has multiple local minima and has difficulty staying at one point because of stochastic optimization behaviors. In contrast, a semi-gradient can solve the issue of unstable behavior by fixing the target value for some amount of time.

Table 7.2 n-step TD-based critic update

	On-policy	Off-policy
V	$J_{\text{Critic}} = \mathbb{E}_s\left\{\left(R^{(n)} - V(s_t; w)\right)^2\right\}$ $R^{(n)} = G_{t:t+n-1} + \gamma^n V(s_{t+n}; w)$	$J_{\text{Critic}} = \mathbb{E}_s\left\{\left(\rho_{t:t+n-1}R^{(n)} - V(s_t; w)\right)^2\right\}$ $R^{(n)} = G_{t:t+n-1} + \gamma^n V(s_{t+n}; w)$
Q	$J_{\text{Critic}} = \mathbb{E}_{s,a}\left\{\left(R^{(n)} - Q(s_t, a_t; w)\right)^2\right\}$ $R^{(n)} = G_{t:t+n-1} + \gamma^n Q(s_{t+n}, a_{t+n}; w)$	$J_{\text{Critic}} = \mathbb{E}_{s,a}\left\{\left(\rho_{t+1:t+n-1}R^{(n)} - Q(s_t, a_t; w)\right)^2\right\}$ $R^{(n)} = G_{t:t+n-1} + \gamma^n Q(s_{t+n}, a_{t+n}; w)$

The policy gradients can be divided into several categories according to the policy type (either stochastic or deterministic policies), type of value function (either state-value or action-value functions), and on-policy/off-policy. As shown in Table 7.3, a deterministic policy gradient with a state-value function does not truly exist in the model-free setting. Accordingly, there are only six kinds of analytical policy gradients, as shown in Table 7.3 and Table 7.4. Note that in state-value-based AC, its actor loss function can also be replaced with n-step TD error. A balance can be achieved between MC with high variance and one-step TD with high bias when this replacement is made. In contrast, in action-value-based AC, $Q(s, a)$ is almost the only choice to introduce the critic information into the actor loss function. Even though it is rarely used in practice, $A(s, a) = Q(s, a) - \mathbb{E}_a\{Q(s, a)\}$ is correct in theory to build an action-value-based advantage function.

Table 7.3 On-policy gradient for actor update

	Stochastic	Deterministic
V	$\nabla J_{\text{Actor}} = \mathbb{E}_\pi\left\{\delta_V^{\text{TD}(n)}(s)\nabla_\theta \log \pi_\theta(a\|s)\right\}$	
Q	$\nabla J_{\text{Actor}} = \mathbb{E}_{s\sim d_\pi, a\sim\pi}\{Q(s, a)\nabla_\theta \log \pi_\theta(a\|s)\}$	$\nabla J_{\text{Actor}} = \mathbb{E}_{s\sim d_\pi}\{\nabla_\theta \pi_\theta(s)\nabla_a Q(s, a)\}$

Table 7.4 Off-policy gradient for actor update

	Stochastic	Deterministic
V	$\nabla J_{\text{Actor}} = \mathbb{E}_b\left\{\rho_{t:t+n-1}\delta_V^{\text{TD}(n)}\nabla_\theta \log \pi_\theta(a\|s)\right\}$	
Q	$\nabla J_{\text{Actor}} = \mathbb{E}_{s\sim d_b, a\sim b}\left\{\dfrac{\pi_\theta(a\|s)}{b(a\|s)}Q(s, a)\nabla_\theta \log \pi_\theta(a\|s)\right\}$	∇J_{Actor} $= \mathbb{E}_{s\sim d_b}\{\nabla_\theta \pi_\theta(s)\nabla_a Q(s, a)\}$

In addition to TD-based critic and vanilla PG, other critic and policy gradients are briefly discussed as follows. The clipped double-Q critic is used to address the overestimation issue. The minimum between the two alternative Q-value estimates is taken to perform the target calculation. In discrete distributional Q-TD, the Bellman equation is applied to learn an approximate discrete value distribution, highlighting the advantage that value distribution is able to stabilize the training process. Continuous distributional Q-TD extends the former approach to learn a continuous value distribution by truncating the difference between the target distribution and current distribution to prevent gradient explosion, indicating that a continuous value distribution can be used to improve the value estimation accuracy. To improve the compositionality that transfers skills among tasks, soft Q-iteration has been proposed to obtain a solution to the Bellman equation by iteratively updating the estimates of value functions in the maximum entropy RL framework.

7.6   Miscellaneous Topics in Direct RL

The main task of direct RL is to search for a parameterized policy that maximizes a scalar performance index. It is essentially a stochastic optimization problem that is governed by the environment dynamics. Similar to deterministic optimization, stochastic optimization can be classified into three types: derivative-free optimization, first-order optimization and second-order optimization. Derivative-free optimization algorithms, e.g., finite difference and evolutionary algorithm, do not use any analytical gradient information and thus have relatively slow convergence. First-order optimization algorithms are fast and efficient due to their cooperation with searching along the steepest descending direction. Gradient-based policy search in RL is able to quickly increase the chance of taking good actions, and decrease the chance of taking bad actions. The likelihood ratio gradient and natural policy gradient are two well-known examples. Second-order optimization algorithms explicitly utilize the second derivative (i.e., Hessian matrix) to search the parameter space. In theory, this approach has the fastest convergence speed but often suffers from high computational burden and low robustness to external disturbances. Table 7.5 and Table 7.6 classify a few model-free RL algorithms from the viewpoints of optimization order, type of objective function (i.e., primal objective function, minorize-maximization optimization, and entropy regularization), and type of policy parameterization.

Table 7.5 Direct RL with a stochastic policy

		Derivative-free optimization	First-order optimization	Second-order optimization
Objective function	Primal	--	Vanilla PG[#]	--
	MM optimization	--	Natural PG[#] (TRPO)	--
	Entropy regularization	--	Soft PG[&]	--

\#- Log-derivative trick; &- reparameterization trick

Table 7.6 Direct RL with deterministic policy

		Derivative-free optimization	First-order optimization	Second-order optimization
Objective function	Primal	FD; EA	Deterministic PG	--
	MM optimization	--	--	--
	Entropy regularization	--	--	--

* FD-Finite difference; EA-evolutionary algorithm

7.6.1   Tricks in Stochastic Derivative

The field of reinforcement learning is filled with random variables, especially in those procedures related to stochastic optimization. One classic task is to compute the expected gradient of a loss function. Accurate gradient computation is also very useful in many other fields like statistical learning and signal processing. For example, in variational inference, the expected derivative of evidence-lower-bound (ELBO) loss needs to be carefully calculated. Consider a random variable $z \sim p_\theta$, where p_θ is a parametric distribution. Given a function $h(z)$, for which we wish to compute its expected gradient, the quantity of interest is

$$\nabla_\theta \mathbb{E}_{z \sim p_\theta}\{h(z)\} = \sum_z \nabla_\theta p_\theta(z) h(z). \tag{7-30}$$

Obviously, this type of derivative formula is not in the expectation form and thus cannot be used to directly deploy sample-based estimation. In model-free RL, the set of collected samples is an exclusive choice in representing the information of environment dynamics. Considering the randomness of sampling behavior, sample-based estimation must be performed with the expectation form of policy gradients. The log-derivative trick and reparameterization trick are two popular approaches that allow us to derive expected policy gradients.

7.6.1.1   Log-derivative trick

The derivation of expected gradients involves how to manipulate probabilities, which are often represented as either a normalized probability or as a logarithmic probability. The log-derivative trick provides a useful ability to switch between these two probabilistic representations. In principle, this trick is simply an application of differentiation rule:

$$\nabla_\theta \mathbb{E}_{z \sim p_\theta}\{h(z)\} = \sum_z p_\theta(z) \frac{\nabla_\theta p_\theta(z)}{p_\theta(z)} h(z) = \mathbb{E}_{z \sim p_\theta}\{\nabla_\theta \log p_\theta(z)\, h(z)\}. \tag{7-31}$$

This trick makes it possible to estimate a stochastic gradient through sample average. Moreover, it places no restriction on the property of $h(z)$ and does not require $h(z)$ to be differentiable. One shortcoming of this trick is that the resulting estimation often suffers from high variance, which is caused by the uncertain denominator of logarithmic operation. The stochastic policy may output actions with very small probability. When the probability of an action is very close to zero, its logarithmic operation goes to infinity, and any error associated with this action is undoubtedly amplified.

7.6.1.2   Reparameterization trick

The reparameterization trick, which provides an alternative way to rewrite stochastic policy gradients into an expectation form, utilizes a technique called random variate substitution. In other words, a complex random variable is replaced with the deterministic transformation of a simple random variable. Let us assume that the random variable z is reparameterized with a structured function:

$$z \sim p_\theta \to z = g_\theta(\epsilon), \epsilon \sim p(\epsilon), \tag{7-32}$$

where g_θ is a given deterministic function and ϵ is a new random variable. The reparameterized gradient in the expectation form becomes

$$\nabla_\theta \mathbb{E}_{z\sim p_\theta}\{h(z)\} = \nabla_\theta \mathbb{E}_{\epsilon\sim p(\epsilon)}\{h(g_\theta(\epsilon))\} = \mathbb{E}_{\epsilon\sim p(\epsilon)}\{\nabla_\theta h(g_\theta(\epsilon))\}. \qquad (7\text{-}33)$$

In a broad sense, this reparameterization trick subtly separates determinacy and stochasticity in the stochastic policy gradient. The reparameterized gradient has been observed to have lower variance than that with the log-derivative trick. The low variance is probably achieved because some determinism is inserted into the stochastic policy. If the deterministic structure is fortunately selected to match the real distribution, the reparameterized policy can significantly improve the quality of gradient estimation.

7.6.2 Types of Numerical Optimization

The AI community has witnessed a fast-growing trend in which well-posed RL tasks are formulated into various optimization problems. This perspective provides a promising future that equips RL with mature optimization tools, for example, proximal optimization, minorize-maximization optimization, and entropy regularization. Moreover, these formulations allow RL to be easily extensible to different types of tasks, such as input and state constraints, robust control, and multi-agent control. Even though modern numerical optimization is powerful, sequential decisions still raise severe challenges regarding how to tailor RL for more effective optimization. This section briefly presents optimization-based RL as a window for readers to dig deeper in this field. Figure 7.13 illustrates three types of stochastic optimization, i.e., derivative-free optimization, first-order optimization, and second-order optimization. As a popular technique in RL, the trust region constraint is also graphically demonstrated, including its basic mechanism of confining the update length and shifting the searching direction.

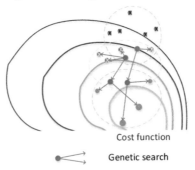

Cost function

Genetic search

(a) Derivative-free optimization

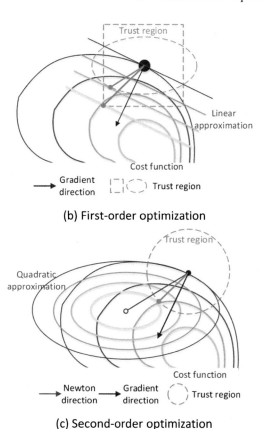

(b) First-order optimization

(c) Second-order optimization

Figure 7.13 Types of stochastic optimization

7.6.2.1   Derivative-free optimization

Derivative-free optimization (DFO), which is also called zeroth-order or black-box optimization, refers to numerical optimization algorithms that do not rely on analytical gradient information. One famous example of DFO is the genetic algorithm, whose mutation operators are inspired by natural evolution, e.g., mutation, crossover and selection. The zeroth-order optimization enables the ability to handle various problems, such as non-convexity, non-differentiability, and ill-conditioned optimization. A straightforward way to apply DFO to RL is to ignore the underlying MDP structure and to directly search in a parameter space. In the RL community, a resampling and updating procedure is adopted that only uses function values to implement this technique. In each iteration, several random perturbations on the current parameter are generated with a predesigned rule (e.g., sampled from a certain distribution). According to the evaluation results of the perturbed policy, its offsprings are produced in the next iteration. These steps are repeated until the stopping condition is satisfied. The simplest method of DFO is to perturb the policy parameter along each unit direction. Based on this idea, the simplest version of finite difference (FD) estimator is built:

$$\widehat{\nabla J}(\theta) = \frac{1}{n}\sum_{i=1}^{n}\frac{J(\theta + \sigma\epsilon_i) - J(\theta)}{\sigma}\epsilon_i, \tag{7-34}$$

where $\epsilon_i \in \mathbb{R}^n$ is the i-th unit direction and $\sigma \in \mathbb{R}$ is a small scalar perturbation. This method perturbs each parameter dimension individually and therefore scales poorly as the dimension of searching space increases. Another example of DFO is called the evolutionary strategy (ES), in which the perturbation direction from an isotropic Gaussian distribution is sampled. Consider a Gaussian smoothed objective function:

$$J_G(\theta(\mu)) = \mathbb{E}_{\theta \sim \mathcal{N}(\mu,\sigma^2)}\{J(\theta)\},$$

where μ is the mean and σ is the standard deviation. The mean μ will also become the current parameter θ after the derivation of zeroth-order gradient is finished. The zeroth-order gradient is equal to

$$\nabla_\mu J_G = \nabla_\mu \mathbb{E}_{\theta \sim \mathcal{N}(\mu,\sigma^2)}\{J(\theta)\}$$

$$= \mathbb{E}_{\theta \sim \mathcal{N}(\mu,\sigma^2)}\{\nabla_\mu \log p_\mu(\theta) J(\theta)\},$$

where $p_\mu(\theta)$ is the probability density function (PDF) of the perturbed distribution:

$$p_\mu(\theta) = \frac{1}{(2\pi\sigma^2)^{n/2}}\exp\left(-\frac{(\theta - \mu)^{\mathrm{T}}(\theta - \mu)}{2\sigma^2}\right).$$

Now we can write the log-derivative term as $\nabla_\mu \log p_\mu(\theta) = (\theta - \mu)/\sigma^2$. The perturbed policy is reparametrized as a linear function $\theta = \mu + \sigma\epsilon$, in which ϵ is a random variable and obeys the Gaussian distribution with zero mean, i.e., $\epsilon \sim \mathcal{N}(0, I_{n \times n})$. Here, the previous mean variable μ actually has an identical physical meaning as the policy parameter θ. Replacing the mean variable μ with the policy parameter θ, the zeroth-order gradient finally becomes

$$\nabla J_G(\theta) = \frac{1}{\sigma}\mathbb{E}_{\epsilon \sim \mathcal{N}(0,I)}\{J(\theta + \sigma\epsilon)\epsilon\}.$$

Clearly, such a zeroth-order gradient can be viewed as the weighted sum of several random descending directions. As a fake "gradient" estimator, the zeroth-order gradient only uses the information of criterion values but not any first-derivative information. Thus, it still belongs to derivative-free optimization. In addition, this estimator can subtract a baseline function $J(\theta)$ to reduce variance. A practical zeroth-order gradient with the baseline is expressed as

$$\widehat{\nabla J}(\theta) \approx \frac{1}{m\sigma}\sum_{i=1}^{m}(J(\theta + \sigma\epsilon_i) - J(\theta))\epsilon_i, \tag{7-35}$$

where $\{\epsilon_i\}_{i=1,2,\cdots,m}$ is sampled from an isotropic Gaussian distribution $\mathcal{N}(0, I_{n \times n})$ and the standard deviation σ is a fixed hyperparameter. For better exploration, there are a few parameter adaptation methods that automatically adjust both the mean and covariance of the Gaussian distribution, for example, cross entropy method (CEM) and covariance matrix adaptation evolutionary strategies (CMA-ES). Usually, these adaptive methods have better generalization ability but with a few sacrifices in computational efficiency.

7.6.2.2   First-order optimization

First-order optimization is the most popular algorithm for large-scale RLs. Stochastic gradient descent (SGD) is one of its basic versions. Since only a mini-batch is allowed to be used at each cycle, SGD has a relatively small computational cost per iteration. Generally, SGD requires careful fine-tuning of a few hyperparameters to achieve good performance, including learning rate and mini-batch size. One major drawback is that naive SGD may struggle with local optimum or saddle points that occur in nonconvex problems. Today, many advanced SGD algorithms have been proposed, including momentum-based SGD, RMSProp, and Adam [7-10][7-21]. Instead of using only the current-step gradient, the gradient information of the past steps is accumulated to determine a new direction in momentum-based SGD. RMSProp is used to automatically adjust the learning rate and choose a different learning rate for each parameter, which partially removes the need for manual adjustment. Basically, Adam, which is the combination of momentum-based SGD and RMSProp, inherits the advantages of these two methods and has become one of the most prevalent algorithms in deep learning. One may ask why other first-order derivatives, such as conjugate gradients, are not quite as popular in stochastic optimization. Different from deterministic optimization, the conjugate gradient almost receives no benefit from the conjugate property of search directions. In model-free RL, any policy gradient encounters noisy uncertainty and random behaviors due to stochastic environments. Consequently, the precise conjugate property can hardly hold again, which considerably deteriorates the superiority of conjugate search.

7.6.2.3   Second-order optimization

A well-known second-order optimization is the Newton-Raphson method, which utilizes the second-order derivative to achieve a quadratic convergence rate. The Newton-Raphson method can be understood as how to approximate the original objective function by a truncated second-order Taylor series and repeatedly solve a sequence of cascading quadratic programming (QP) subproblems. The second-order Taylor expansion of $J(\theta)$ can first be derived and then the original optimization can be replaced with truncated Taylor series. From this replacement, a standard QP is formed at each iteration:

$$\max_{\Delta\theta} g^{\mathrm{T}}\Delta\theta + \frac{1}{2}\Delta\theta^{\mathrm{T}}F\Delta\theta, \tag{7-36}$$

where $\Delta\theta$ is the parameter increment, $g = \nabla_\theta J(\theta)$ denotes the first-order derivative, and $F = \nabla_\theta^2 J(\theta)$ denotes the second-order derivative (i.e., Hessian matrix). The QP subproblem has an analytical solution:

$$\Delta\theta^* = F^{-1}g = [\nabla_\theta^2 J(\theta)]^{-1}\nabla_\theta J(\theta). \tag{7-37}$$

The Newton-Raphson method adopts this type of optimal increment as the updating direction: $\theta \leftarrow \theta + \Delta\theta^*$. The first-order derivative in (7-37) can be expressed as a likelihood ratio gradient. The major challenge is how to more accurately and efficiently compute the Hessian matrix and its inversion. Using a similar idea as likelihood ratio gradients, one can obtain the analytical formula of Hessian matrix from the trajectory concept. Usually, this inverse Hessian matrix is hard to compute mathematically. Here, we assume that $q^{\pi_\theta}(s, a)$ and $d^\gamma_{\pi_\theta}(s)$ are fixed, and they do not change with the policy parameter θ. Accordingly, the simplified Hessian matrix can be derived:

$$\nabla^2_\theta J(\theta) \approx \sum_s d_\pi(s) \sum_a \nabla^2_\theta \pi_\theta(a|s) q^{\pi_\theta}(s, a). \tag{7-38}$$

Even so, this simplified Hessian matrix still has a high computational burden. Its matrix inversion may become computationally inefficient in high-dimensional tasks. In some pioneering studies, there have been several attempts to apply the Broyden-Fletcher-Goldfarb-Shanno (BFGS) algorithm to the inverse Hessian computation, but corresponding experiments have shown that its performance improvement is very limited in deep RL [7-16]. Although there is a potential to achieve fast convergence, second-order optimization is actually not well recognized in the today's RL community.

7.6.3   Some Variants of Objective Function

The objective function is the criterion on which an RL agent learns how to behave optimally. It is a user-provided reinforcement signal that accumulates from immediate rewards. In some special tasks like Chinese Go, reward signals can be very sparse, and useful feedback is received only after the end of each episode. In this situation, the RL agent does not obtain enough reinforcement. Reward shaping is an empirical trick to enhance sparse rewards. Existing approaches such as potential-based reshaping technique need to make use of human knowledge. One shortcoming is that the transformation of human knowledge into numeric rewards cannot be perfect due to human cognitive bias. Different from reward shaping, this section focuses on a high-level topic on RL criterion design, which is how to formulate an intrinsically motivated objective function. Some variants of RL overall objective function, such as surrogate function optimization and entropy regularization, are able to output higher learning stability and better exploration ability.

7.6.3.1   Surrogate function optimization

Severe training instability occurs in large-scale problems with high-capacity policy parameterization. Minorize-maximization (MM) optimization is used to optimize policies with a guaranteed monotonic improvement property. The basis of this optimization is to recursively optimize the surrogate function, i.e., a lower bound of the primal objective function. It has been proven that optimizing the surrogate function will either improve the value of original objective function or at least leave it unchanged. This is an excellent property to maintain the property of monotonic policy improvement in reinforcement learning. The key to applying this technique is to construct a proper surrogate function.

By making several assumptions, Schulman et al. (2016) proposed the trust region policy optimization (TRPO) algorithm with the support of MM optimization [7-18]. TRPO still selects (7-1) as the criterion to be maximized. At each step, the policy to be updated is denoted as π, and its last policy is denoted as π_{old}. The following surrogate inequality is used to find the lower bound of original objective function, of which the right-hand side is an available surrogate function:

$$J(\pi) \geq L_{\pi_{\text{old}}}(\pi) - C \cdot D_{\text{KL}}^{\max}(\pi_{\text{old}}, \pi),$$
$$C = 2\epsilon\gamma/(1-\gamma)^2,$$

where $J(\pi)$ is the primal objective function, $D_{\text{KL}}^{\max}(\cdot,\cdot)$ is the maximum KL divergence, C is a penalty coefficient, and $L_{\pi_{\text{old}}}(\pi)$ is a local approximate function to $J(\pi)$:

$$L_{\pi_{\text{old}}}(\pi) = J(\pi_{\text{old}}) + \sum_s d_{\pi_{\text{old}}}^{\gamma}(s) \sum_a \pi(a|s) A^{\pi_{\text{old}}}(s,a).$$

where $d_{\pi_{\text{old}}}^{\gamma}(s)$ is the discounted state distribution under the last policy π_{old}, and $A^{\pi_{\text{old}}}(s,a)$ is its corresponding advantage function. By the sandwich inequality, the monotonic improvement of the learned policy can be guaranteed by alternately optimizing the surrogate function, i.e., the right-hand side of the surrogate inequality:

$$\max_{\pi}\{L_{\pi_{\text{old}}}(\pi) - C \cdot D_{\text{KL}}^{\max}(\pi_{\text{old}}, \pi)\}.$$

The second term in the surrogate function can be viewed as a penalty to the local approximate function. From this viewpoint, its first-order derivative with respect to the policy parameter results in the well-known "natural policy gradient". Compared to naive policy gradient, natural policy gradient limits the step size of each policy update and reduces the high sensitivity to large-scale policy parameterization. With a trust region constraint from the KL divergence, monotonic policy updates with nontrivial step sizes can be achieved [7-7].

7.6.3.2   Entropy regularization

As a conservative learner, an RL agent is more likely to inherit positive behaviors and is not likely to try new actions. In fact, other actions may contain better rewards, but they will never be explored if there is no sufficient intrinsic excitation. Accordingly, RL may easily stick to a local optimum due to the lack of sufficient exploration. In most RL algorithms, one can use entropy regularization to encourage exploration and try to find the global optimum. Policy entropy is defined as the predictability of actions in an agent. Its value is closely related to the uncertainty about what action will yield the highest cumulative reward in the long run. The more random a policy is, the higher its entropy becomes, and vice versa. The RL objective function with entropy regularization has two basic forms:

$$J_{\mathcal{H}}(\pi) = \mathbb{E}_{s_t \sim d(s_t)}\left\{\sum_{i=t}^{\infty} \gamma^{i-t}\left(r_i + \alpha\mathcal{H}(\pi(\cdot|s_i))\right)\right\},$$

and

$$J_{\mathcal{H}}(\pi) = \mathbb{E}_{s_t \sim d(s_t)}\left\{\mathbb{E}_{\pi}\left\{\sum_{i=t}^{\infty} \gamma^{i-t}r_i\right\} + \alpha\mathcal{H}(\pi(\cdot|s_t))\right\},$$

where α is called the temperature parameter. The first form adds the policy entropy into each reward signal, while the second form penalizes the overall value function. The action distribution with the highest entropy is the one that best explores the knowledge of an unknown environment. Hence, the utilization of entropy regularization encourages exploring every corner of the environment dynamics. In addition, some knowledge from previously learned policy can be reused in transfer learning, as opposed to starting with no prior knowledge. Owing to better exploration ability, the RL agent will be more robust to abnormal behaviors or rare events.

Previous chapters have discussed how to equip RL with an entropy function, but it has noticeable difference from entropy regularization in this section. Previously, entropy is only deployed into the actor loss function. The actor-level regularization is essentially an extension of ϵ-greedy search, whose rationality depends on whether such a design can still guarantee convergence. In this section, entropy is an element of the overall objective function. Its main purpose is to increase the exploration ability of the trained policies. Moreover, this design does not change the convergence property of the RL algorithms.

7.7   References

[7-1]   Barto A, Sutton R, Anderson C (1983) Neuronlike adaptive elements that can solve difficult learning control problems. IEEE Trans Syst, Man, and Cyber, 13(5): 834-846

[7-2]   Brémaud P (2013) Markov chains: Gibbs fields, Monte Carlo simulation, and queues. Springer, New York

[7-3]   Degris T, White M, Sutton R (2012) Off-policy actor-critic. ICML, Scotland, UK

[7-4]   Espeholt L, Soyer H, Munos R et al (2018) Impala: Scalable distributed deep-RL with importance weighted actor-learner architectures. ICML, Stockholm, Sweden

[7-5]   Fujimoto S, Hoof H, Meger D (2018) Addressing function approximation error in actor-critic methods. ICML, Stockholm, Sweden

[7-6]   Glynn P (1990) Likelihood ratio gradient estimation for stochastic systems. Communications of ACM, 33(10): 75–84

[7-7]   Grondman I, Busoniu L, Lopes G et al (2012) A survey of actor-critic reinforcement learning: Standard and natural policy gradients. IEEE Trans Syst, Man, and Cyber, 42(6): 1291-1307

[7-8]   Gu S, Lillicrap T, Ghahramani Z (2017) Interpolated policy gradient: Merging on-policy and off-policy gradient estimation for deep reinforcement learning. NeurIPS, California, USA

[7-9]   Guan Y, Li SE, Duan J, et al (2021) Direct and indirect reinforcement learning. Intl J of Intelli Syst (36): 4439-4467

[7-10]   Kingma D, Ba J (2015) Adam: A method for stochastic optimization. ICLR, California, USA

[7-11]   Konda V, Tsitsiklis J (2002) Actor-critic algorithms. NeurIPS, Vancouver, Canada

[7-12]   Konda V, Tsitsiklis J (2003) On actor-critic algorithms. SIAM J on Control and Optimization, 42 (4): 1143-1166

[7-13]   Lillicrap T, Hunt J, Pritzel A, et al. (2015) Continuous control with deep reinforcement learning. ICLR, Puerto Rico

[7-14]   Maei H (2018) Convergent actor-critic algorithms under off-policy training and function approximation. https://arxiv.org/abs/1802.07842

[7-15]   Mnih V, Badia A, Mirza M et al (2016) Asynchronous methods for deep reinforcement learning. ICML, New York City, USA

[7-16]   Rafati J, Marica R (2020) Quasi-Newton optimization methods for deep learning applications. In: Deep Learning Applications, 1098. Springer, Singapore

[7-17]   Ross S, Kelly J, Sullivan R et al (1996) Stochastic processes. John Wiley & Sons, Inc., New York

[7-18]   Schulman J, Levine S, Abbeel P et al (2015) Trust region policy optimization. ICML, Lille, France

[7-19]   Schulman J, Wolski F, Dhariwal P, et al (2017) Proximal policy optimization algorithms. Available via arXiv. https://arxiv.org/abs/1707.06347

[7-20]   Silver D, Lever G, Heess N et al (2014) Deterministic policy gradient algorithms. ICML, Beijing, China

[7-21]   Sutskever I, Martens J, Dahl G et al (2013) On the importance of initialization and momentum in deep learning. ICML, Georgia, USA

[7-22]   Sutton R, McAllester D, Singh S, et al. (1999) Policy gradient methods for reinforcement learning with function approximation. NeurIPS, Colorado, USA

[7-23]   Wang Z, Bapst V, Heess N et al (2017) Sample efficient actor-critic with experience replay. ICLR, Toulon, France

[7-24]   Williams R (1992) Simple statistical gradient-following algorithms for connectionist reinforcement learning. Mach learn, 8(3): 229-256

Chapter 8.   Approximate Dynamic Programming

> The saddest aspect of life right now is that
> science gathers knowledge faster than
> society gathers wisdom.
>
> -- Isaac Asimov (1920-1992)

Approximate dynamic programming (ADP), which can be used to control complex dynamical systems, is drawing increasing attention in engineering practice. In this field, both infinite-horizon and finite-horizon control tasks are discussed under the assumption that perfect deterministic models are known. These control tasks are generally formulated as optimal control problems (OCPs), and their controller implementation is computationally inefficient when using the way of online receding horizon optimization. As a replacement of online receding optimizer, ADP refers to a class of offline learning methods that are utilized to compute the approximate solutions in the whole state space. Basically, there are two kinds of ADP algorithms, i.e., value iteration ADP and policy iteration ADP. Both of them belong to indirect RL family since the computation of their solutions relies on the optimality condition from Bellman's principle.

The concept of ADP was first introduced in the pioneering study of Bellman and Dreyfus (1959) with the purpose of overcoming the curse of dimensionality in exact DP. In their study, a set of Legendre polynomials was used to approximate the cost functional [8-2]. From the late 1970s, Paul Werbos suggested parameterizing the continuous state space and reconstructing an approximate DP problem in the parameter space [1-29][1-28]. Compared to exact DP, ADP sacrifices the solution accuracy so as to reduce the computational burden. In the early 1980s, neural networks (NNs) began to emerge in the parameterization of ADP actor and critic. To learn a nonlinear optimal policy, Barto, Anderson, and Sutton (1982) used a two-layer NN, in which a suitable representation is learned in the first layer, and optimal actions are the output of the second layer [8-1]. Hampson (1983) extended this two-layer method to multilayer neural networks, and both policy and value function were learned in the actor-critic architecture [8-5]. Thus far, there was no clear understanding on what was the key connection between learning in unknown environments and controlling with known models.

In the early 1990s, ADP studies accelerated remarkably since inherent connections between model-based ADP and model-free RL became apparent. Bertsekas and Tsitsiklis (1996) provided an excellent summary of up-to-date studies and clarified a few terminological misunderstandings in previous studies. In their book, ADP was referred to as neuro-dynamic programming, from which a few prominent mathematical theories of discrete-time and infinite-horizon ADP algorithms were established [8-4]. Since then, ADP has become an effective bridge to connect optimal control and reinforcement learning. The former has a tradition to implement optimal action of current state point with receding horizon optimization, while the latter seeks to learn an optimal policy in the whole state space through interactions with the environment.

S. E. Li, *Reinforcement Learning for Sequential Decision and Optimal Control*, https://doi.org/10.1007/978-981-19-7784-8_8

After a few successful studies and applications in the last decade, the term "approximate dynamic programming (ADP)" has become the specific name of model-based RL that works in a deterministic environment. For example, Murray (2002) proposed a value iteration ADP algorithm for nonlinear continuous-time systems. His study has proven that an admissible initial policy is not required for value iteration ADP, and its closed-loop stability is guaranteed by policy optimality and value finiteness [8-8]. In the study by Derong Liu and his colleagues in 2009, a nonquadratic cost function was used for policy iteration ADP to solve discrete-time problems with input constraints. It has been proven that constrained value iteration can converge to its optimum, which is the infimum of all possible value functions obtained through admissible control sequences [8-12]. Recently, an online version of policy iteration ADP was constructed by Modares and Lewis (2013) for continuous-time systems with unknown input constraints. Two neural networks were simultaneously updated under the actor-critic structure, in which one neural network represents the bounded policy and the other represents the state-value function. An additional neural network was trained online in conjunction with actor and critic to identify unknown environment dynamics [8-7]. Today, ADP has become an effective tool to solve complex optimal control problems in nonlinear and constrained environments. The learned control policy serves as an efficient online controller, and its implementation significantly increases the real-time performance of optimal control.

8.1   Discrete-time ADP with Infinite Horizon

The focus of modern feedback control is mainly on digital controllers. Although almost all physical systems work in the continuous-time domain, their digital controllers must run in a discrete-time manner. This discrepancy usually requires engineers to perform time discretization on continuous-time systems and formulate discrete-time OCPs. The theory of consistent approximation provides a necessary condition under which the solution of increasingly accurate discretized OCPs converges to that of continuous-time problems.

Time-invariant nonlinear systems cover a large number of control scenarios in the engineering world, and the construction of accurate dynamic models is the basic requirement for digital computation. A tractable OCP must have the following two properties: (1) its environment model has the Markovian property and (2) it has a cost function that is structurally separable over time steps. After time discretization, a deterministic dynamic model naturally satisfies the Markov property. The cost function is composed of additive utility functions, which are also separable because each utility function occurs at the time instant. The discrete-time OCP with infinite horizon is defined as

$$\min_{\{u_t, u_{t+1}, \cdots, u_\infty\}} V(x) = \sum_{i=0}^{\infty} l(x_{t+i}, u_{t+i}),$$

(8-1)

subject to

$$x_{t+1} = f(x_t, u_t),$$

where t is the time step, $u \in \mathcal{U} \subset \mathbb{R}^m$ is the action, $x \in \mathcal{X} \subset \mathbb{R}^n$ is the state, $V(x)$ is the cost function with a given initial state x, $l(x, u) \geq 0$ is called the utility function, and

$f(\cdot,\cdot)$ is the environment model. Here, X, U are the sets of admissible states and actions, respectively. In this problem definition, $V(x)$ is also called the state-value function, and $l(x, u)$ serves as the reward signal that occurs in model-free RLs. In addition, $f(x, u)$ must be deterministic to be compatible with the undiscounted sum of utility functions. Following assumptions are considered:

- Assumption 8.1: The plant dynamic model has an exclusive equilibrium $x = 0, u = 0$, i.e., $f(0,0) = 0$. The model can be an analytical function, a neural network, or even a MATLAB/Simulink model as long as its partial derivative $\partial f(x, u)/\partial u$ is available.
- Assumption 8.2: The utility function satisfies $l(x, u) > 0$ except at the equilibrium $(0,0)$, where $l(x, u) = 0$. The utility function can be in any form, either quadratic or highly nonlinear, but it must assure that $\partial l(x, u)/\partial u$ and $\partial l(x, u)/\partial x$ are available.

There may be multiple equilibriums in the physical environment. However, we can still use Assumption 8.1 to facilitate the theoretical analysis because each equilibrium is exclusive in its local state region. One benefit of discrete-time setup is that difference equations can be recursively processed, which facilitates the digital computation of self-consistency condition and Bellman equation. In fact, this kind of problem formulation covers a variety of real-world control problems, such as product manufacturing, inventory control, transportation management, multi-arm robot control, and vehicle automation.

8.1.1   Bellman Equation

The self-consistency condition, which states that a separable cost function is equivalent to the sum of the utility function in the current step and the cost function in the next step (Figure 8.1). This condition is an inherent property that stays behind the Markovian and separable structure of discrete-time OCPs. For brevity, (x, u) is used to represent (x_t, u_t), and (x', u') is used to represent (x_{t+1}, u_{t+1}) in the following context. The self-consistency condition is stated as

$$V(x) = l(x, u) + V(x'), \forall x \in X. \tag{8-2}$$

This recursive condition describes the inherent connection of three key elements in the whole state space, including state-value function, utility function and environment model.

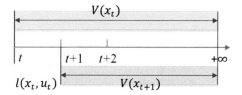

Figure 8.1 Self-consistency condition

The optimal cost function is defined as

$$V^*(x) \stackrel{\text{def}}{=} V^{\pi^*}(x) = \min_{\{u_t, u_{t+1}, \cdots, u_\infty\}} V(x), \tag{8-3}$$

where $\{u_t, u_{t+1}, \cdots, u_\infty\}$ is the action sequence starting from time t. Recall the fact that in discrete-time OCPs, the discrete-time model has the Markov property, and its cost function is structurally separable over time steps. These two properties can be utilized to separate the current-time utility function from its successors, which builds the structural basis for applying Bellman's principle of optimality. The importance of these two properties is shown from the first step to the second step in the derivation of the following Bellman equation:

$$V^*(x) = \min_{\{u_t, u_{t+1}, \cdots, u_\infty\}} \{l(x_t, u_t) + V(f(x_t, u_t))\}$$

$$= \min_{u_t} \{l(x_t, u_t) + V^*(f(x_t, u_t))\} \tag{8-4}$$

$$= \min_{u} \{l(x, u) + V^*(f(x, u))\}.$$

In general, the Bellman equation is the sufficient and necessary condition of optimality for discrete-time OCPs. Its advantage lies in the fact that one only needs to search for a single optimal action at the current time if the next-step optimal value is known. In theory, discrete-time Bellman equation is a backward-in-time problem. The optimal actions can be calculated using backward recursion (if the final cost is already known), and then they are implemented with forward rollouts. This kind of control process is very similar to how a human being chooses his behaviors. As joked by Dimitri Bertsekas in his book, life can only be understood by looking backward, but it must be lived going forward. To plan the current optimal action, we need to look ahead at the eventual cost that we want to incur. However, looking forward to the far future is too computationally expensive. The benefit of Bellman equation is that it only requires to look one step ahead, which avoids the high computational complexity of searching all future actions.

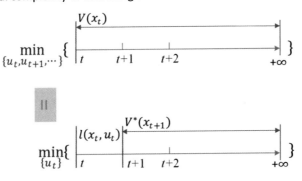

Figure 8.2 Bellman equation

Violations of the Markov property may be common in the real world, such as moving-average filter, time-delayed control, and correlated disturbance control. In many cases, this violation can be addressed by reformulating a high-dimensional augmented problem. The general guideline is to add all the information needed into an enlarged state space and use this new state to select optimal actions. Definitely, state augmentation is not a free lunch, and it always comes at a price: the augmented problem has more complex state and action spaces, and needs higher computational power.

8.1.1.1   Existence of Bellman equation and optimal policy

Accurately calculating optimal policy is the central task of ADP algorithms. One might ask why the solution from the Bellman equation is the optimal policy of original OCP. This equivalence depends on the two critical properties, i.e., Markov property of environment dynamics and separable property of cost function. These two properties allow an optimization problem to be broken into a series of multistage subproblems, in which the application of Bellman's principle is theoretically complete. A more advanced analysis begins from two kinds of optimizers:

$$u^* = \arg\min_u\{\cdot\},$$

$$\pi^* = \arg\min_\pi\{\cdot\}.$$

These two optimizers look very similar except for their subscripts. They actually come from two different OCPs, i.e., open-loop OCP and closed-loop OCP. In the former OCP, an open-loop optimization is performed. A sequence of open-loop actions $\{u_t, u_{t+1}, \cdots, u_\infty\}$ are optimized without considering their connection with the current state or historical states. For the latter OCP, a closed-loop optimization must be performed. The closed-loop policy must have a confined structure; for example, a stationary policy is the structural mapping from the current state to the current action, i.e., $u = \pi(x)$. Hence, any optimal policy must be searched within the structural limitation (as demonstrated in Figure 8.3). Once such an optimal policy is found, it can be implemented as an online feedback controller.

Let us revisit the relationship of optimal state-action pairs and Bellman equation. Similar to static optimization, Bellman equation (8-4) is essentially an open-loop optimality condition of dynamic optimization. When a structured control policy is utilized, this open-loop optimization becomes a closed-loop optimization:

$$\pi^*(x) = u^* = \arg\min_u\{l(x,u) + V^*(x')\}, \forall x \in \mathcal{X}, \tag{8-5}$$

where $\pi^*(\cdot)$ is an optimal policy that satisfies the structured mapping from the state space $\mathcal{X}$ to the action space $\mathcal{U}$:

$$u = \pi(x), \forall x \in \mathcal{X}. \tag{8-6}$$

The stationary policy (8-6) is a structurally constrained function, in which its action must be a static mapping of a certain state. The existence of this optimal policy means that it can perfectly represent the solution of open-loop Bellman equation.

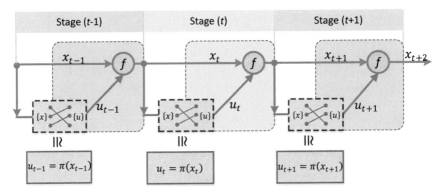

Figure 8.3 Existence analysis for optimal policy

Now let us explain the equivalence of closed-loop and open-loop Bellman equations. The open-loop Bellman equation has only open-loop action sequences, and there is no limitation on its action freedom. It is easy to see that if the open-loop Bellman equation is computable, such an optimal open-loop trajectory exists:

$$(x_t, u_t^*), (x_{t+1}, u_{t+1}^*), (x_{t+2}, u_{t+2}^*), \cdots, (x_\infty, u_\infty^*).$$

From the supervised learning perspective, one can inspect the following two cases: (1) there is a unique optimal action for each fixed state. Hence, (x, u^*) is a strict one-to-one mapping, and the stationary function $u^* = \pi^*(x)$ can be fitted perfectly; (2) there are multiple optimal actions at a fixed state. One can arbitrarily select one of them as the optimal action and still construct a well-defined mapping from state to action. Clearly, the arbitrarily selected action does not sacrifice any control optimality. In summary, the stationary policy can perfectly reproduce the solution of open-loop Bellman equation. The closed-loop Bellman equation is formally expressed as

$$V^*(x) = \min_{u=\pi(x)} \{l(x, u) + V^*(f(x, u))\}. \tag{8-7}$$

Obviously, if the structural constraint imposed on $\pi(x)$ is stricter than "stationary", for example, a linear function, its closed-loop Bellman equation will sacrifice a certain amount of control optimality. To avoid unnecessary performance loss, $\pi(x)$ is always assumed to have a good approximation ability throughout this book. The deep neural network has a powerful fitting ability to any continuous function, which is usually taken as the feedback control policy in most complex problems. Thus, one can disregard the discrepancy between open-loop optimization and closed-loop optimization.

8.1.1.2   Why initial state is NOT considered

One might notice that there are a few differences between the problem definition of ADP and that of RL. One difference is that in the definition of ADP, no initial state distribution is included as the weighting coefficient. Intuitively, it seems as though the optimal policy should be associated with the weighting coefficient. In fact, the optimal policy at a special initial state distribution is also optimal for other initial state distributions. In the previous chapter on model-free RL, the inherent reason has been explained, i.e., if $\pi(s)$ is completely free, $\pi^*(s)$ becomes independent of the initial state distribution. Hence, the result of minimizing (8-1) at each initial state is equivalent to that of minimizing the weighted integral of the cost function:

$$\min_{u} V(x), x \in \mathcal{X} \Leftrightarrow \min_{u} \int d(x)V(x)\mathrm{d}x, \tag{8-8}$$

In practice, a parameterized function such as artificial neural network, if properly designed, has almost infinite approximation ability. Thus, the weighting influence on the control optimality can be neglected in the problem definition. When implementing a model predictive controller (MPC), we actually do not consider any initial state distribution. This is because MPC utilizes an open-loop optimizer due to its receding horizon control mechanism. The optimal action is online calculated at each state point in a receding horizon manner.

8.1.2   Discrete-time ADP Framework

The aim of ADP is to find the solution of closed-loop Bellman equation. Similar to stochastic DP, most ADP algorithms can be classified into two types: policy iteration ADP and value iteration ADP. Both of them can be unified into the classic actor-critic (AC) framework with two alternating elements, i.e., policy evaluation (PEV) and policy improvement (PIM). In this framework, PEV seeks to calculate a new value function, while PIM aims to find a new policy. More specifically, in policy iteration ADP, PEV iterates an infinite number of times. In value iteration ADP, PEV stops after only one iteration, and an explicit intermediate policy is not necessary. Their comparison is listed in Table 8.1. Usually, policy iteration ADP has higher iteration efficiency, and it is often more stable than value iteration ADP. Here, iteration efficiency is a new performance measure that suits for model-based RL methods. It is defined as how many cycles are required when a certain level of policy performance is reached. Policy iteration ADP can be applied both online and offline, but an admissible initial policy must be specified. The online implementation is available because its outcomes are a series of policies with recursive stability. In contrast, in value iteration ADP, an admissible initial policy is not necessary. This approach can be more suitable for cases where the initial guess is poor.

Table 8.1 Policy iteration ADP vs. value iteration ADP

	Policy iteration ADP	Value iteration ADP
PEV iteration	Infinite steps	One step
Initial policy	Must be admissible	No admissibility requirement
Is each intermediate policy admissible?	Yes	No guarantee
Iteration efficiency	High, but every iteration is more computationally demanding	Low
Online/Offline	Online/Offline	Mostly offline

The flow chart of policy iteration ADP is illustrated in Figure 8.4. Its structure is similar to that of model-free AC algorithms but requires a known environment model. In the ADP field, the environment model is often deterministic rather than stochastic. The deterministic characteristic of system dynamics provides a few simplifications in the problem formulation. For example, the reward signal (or utility function) does not need to contain the measurement of the next state because the next state is solely determined by the current state and action. The discount factor can be removed from the cost function because both state and action can reach zero equilibrium and the bounded state-value function is accessible. However, such simplifications, especially the removal of discount factor, may introduce a few difficulties in the convergence guarantee of ADP algorithms, which require more careful hyperparameter tuning and some special treatments, including zero boundary condition and high-order value approximation.

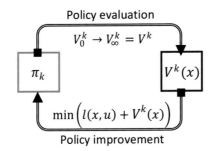

(a) Cycle of PEV and PIM

$$\pi_0 \xrightarrow{\text{PEV}} V^0 \xrightarrow{\text{PIM}} \pi_1 \xrightarrow{\text{PEV}} V^1 \xrightarrow{\text{PIM}} \quad \cdots \quad \xrightarrow{\text{PIM}} \pi^* \xrightarrow{\text{PEV}} V^*$$

(b) Convergence to the optimum

Figure 8.4 Iterative procedure of PEV and PIM

For a deterministic model, PEV seeks to numerically solve the self-consistency condition. Careful readers might see that without the discount factor, the self-consistency condition does not have a unique solution. One can add an arbitrary constant to the value functions on both sides, but this does not change the equivalence of the self-consistency condition. Usually, we enforce $V^k(x)$ to be zero when $x = x_{\text{equ}}$ for the purpose of ensuring a unique solution. This assumption builds a supplementary condition, i.e., zero boundary condition, to the traditional self-consistency condition. Since the self-consistency condition is a special algebraic equation, one can use a fixed point iteration scheme to design its iterative PEV algorithm. The simplest way to achieve this goal is to apply the Picard iteration:

Repeat until $j \to \infty$

$$V_{j+1}^k(x) \leftarrow l(x, \pi_k) + V_j^k(x') - V_j^k(x_{\text{equ}}), \forall x \in X \tag{8-9}$$

End

where k denotes the main iteration, π_k is the current action from the intermediate policy $\pi_k(x)$, and $V^k(x)$ is the value estimate at the k-th main iteration. Note that $V^k(x)$ is actually an abbreviation of the following notation:

$$V^k(x) \stackrel{\text{def}}{=} V^{\pi_k}(x).$$

It superscript k means that $V^k(x)$ is the estimate of state-value function under the policy π_k, rather than an estimate of k-th value iteration. In most regulating tasks, x_{equ} is zero, i.e., zero equilibrium. Without subtracting the constant value at the equilibrium state, (8-9) might become very unstable during iteration. Similar to stochastic DP, PIM aims to find a better policy under the "weak" Bellman equation:

$$\pi_{k+1}(x) = \arg \min_u \{l(x,u) + V^k(x')\}, \forall x \in \mathcal{X}. \tag{8-10}$$

Here, the term "weak" means that (8-10) is not from the Bellman's principle but a nonoptimal version of the Bellman equation. In the weak Bellman equation, the optimal value function is substituted with the value estimate at each main iteration step. The minimization of the weak Bellman equation is essentially a variant of greedy search under the condition that the system model is known. Even though it is structurally similar to stochastic DP with a discount factor, model-based ADP has a few new practical challenges, such as easily losing convergence and low policy accuracy. One major reason is that standard ADP does not contain any discount factor, and its convergence ability and training stability are largely sacrificed due to the lack of the γ-contraction property.

8.1.3 Convergence and Stability

Almost all ADP designs involve two entangled properties, i.e., convergence and stability. More specifically, convergence is the abbreviation of algorithm convergence, and stability is that of closed-loop stability. Their conceptual connection is illustrated in Figure 8.5.

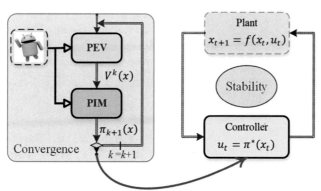

Figure 8.5 Comparison of convergence and stability

Convergence is critical to a useful ADP algorithm. It describes whether the algorithm can gradually shift from its intermediate policy π_k to the optimal solution π^*. The proof is to examine whether $V^k(x)$ is monotonically improved with respect to the main iteration step k for all $x \in \mathcal{X}$, i.e.,

$$V^k(x) \geq V^{k+1}(x), \forall x \in \mathcal{X}.$$

Different from convergence, stability is of importance to closed-loop control systems. In a narrow sense, stability only refers to whether the optimal policy $\pi^*(x)$ can stabilize the plant. The proof is to select the optimal state-value function $V^*(x)$ as the Lyapunov candidate and examine whether $V^*(x)$ is monotonically decreasing with respect to time t, i.e.,

$$V^*(x_t) \geq V^*(x_{t+1}), \forall x_t \in X.$$

Let us clarify again that convergence is related to the main iteration step while stability is related to time. Obviously, stability is said with a feedback control policy, and convergence is related to its training algorithm. In fact, these two properties cannot be clearly separated, and they have an inherently entangled connection. The potential of achieving stability is the prerequisite of convergence guarantee. An ADP algorithm has no chance to converge to an unstable policy. As long as an intermediate policy becomes unstable, there is no way to perform successful environment interactions. Even so, stability does not mean the policy is optimal. An imperfect algorithm may stop at a stable but bad policy, and its convergence is meaningless due to the lack of optimality. The relationship between algorithm convergence and closed-loop stability becomes more complex in the condition of noisy measurements and uncertain models.

8.1.3.1   Closed-loop Stability

For unconstrained OCP, closed-loop stability is equivalent to the policy admissibility. A control policy is said to be admissible if and only if its closed-loop system is stable. Clearly, an inadmissible policy must lead to unstable states. When some states diverge, there is no way to build a finite state-value function. Hence, an admissible policy also implies that its corresponding state-value function is bounded. When there is no bounded value function, neither self-consistency condition nor Bellman equation has a feasible solution. Therefore, to continue successful environment iterations, recursive stability must be a necessary property for policy iteration ADP algorithms. The recursive stability describes a kind of self-sustaining property that the next policy π_{k+1} is admissible when the current policy π_k is admissible. In the following context, we will prove that the property of recursive stability exists in standard policy iteration ADP algorithms.

- Theorem 8.1: In policy iteration ADP, each intermediate policy π^k is admissible if the initial policy is admissible.

Proof:

We first assume that π_k is admissible, i.e., its corresponding state-value function is finite. To examine whether π_{k+1} is admissible, we select $V^k(x)$ as the Lyapunov function, which is positive definite when $l(x, u) \geq 0$. Because π_{k+1} is optimal for the weak Bellman equation at the k-th step, we have

$$l(x, \pi_{k+1}) + V^k(f(x, \pi_{k+1})) = \min_u \{l(x, u) + V^k(f(x, u))\}.$$

Since π_k is not necessarily optimal for the same weak Bellman question, we have

$$V^k(x) = l(x, \pi_k) + V^k(f(x, \pi_k)) \geq \min_u \{l(x, u) + V^k(f(x, u))\}.$$

Note that the utility function is positive definite, i.e., $l(x, u) \geq 0$. Combining the two formulas above, we have

$$V^k(x) \geq l(x, \pi_{k+1}) + V^k(f(x, \pi_{k+1})) \geq V^k(f(x, \pi_{k+1})) = V^k(x').$$

Therefore, $V^k(x)$ is monotonically decreasing with respect to time t:

$$V^k(x) \geq V^k(x'), \forall x \in \mathcal{X}. \tag{8-11}$$

Obviously, equality holds only at equilibrium. Note that x' in (8-11) is governed by π_{k+1}, and hence, π_{k+1} is admissible around the equilibrium.

∎

- Theorem 8.2: The closed-loop system under π^* is asymptotically stable in policy iteration ADP.

Proof:

Let us select the optimal value $V^*(x)$ as a Lyapunov function, which is positive definite. Since the optimal policy π^* also satisfies the self-consistency condition,

$$V^*(x) = l(x, \pi^*) + V^*(f(x, \pi^*)) = l(x, \pi^*) + V^*(x').$$

Considering that $l(x, u) \geq 0$, we have

$$V^*(x) \geq V^*(x'), \forall x \in \mathcal{X}.$$

The equality holds only at equilibrium. Therefore, the closed-loop system under π^* is asymptotically stable.

∎

It is easy to see that the positive definiteness of utility function $l(x, u)$ is critical to the maintenance of recursive stability. The property of recursive stability allows deploying a series of intermediate policies into the closed-loop system without any instability concern. Obviously, the prerequisite is that the initial policy must be admissible as policy iteration ADP needs a stable starting point. Moreover, since $l(0,0) = 0$ only occurs at the equilibrium, the equality in (8-11) holds at the equilibrium, i.e., $x = 0, u = 0$. According to Lyapunov's second method, this kind of closed-loop system is asymptotically stable. In contrast, value iteration ADP does not output any intermediate policy, and accordingly the property of recursive stability is not quite necessary. Even so, its final policy should be admissible for the sake of convergence guarantee to the optimal result.

8.1.3.2   Convergence of ADP

Stability does not imply convergence. That is, the admissibility of a policy does not mean that its successive policy is better. A special but reasonable example is that the next policy remains the same as the current one. Every policies are admissible, but there is no improvement in two successive policies. Before applying ADP, one needs to carefully examine whether its convergence can be guaranteed during iteration. The proof is divided into two key steps. The first step is to prove the value decreasing property, i.e.,

$$V^k(x) \geq V^{k+1}(x), \forall x \in \mathcal{X},$$

and the second step is to prove that it will reach the optimum, i.e.,

$$V^\infty(x) = V^*(x), \forall x \in \mathcal{X}.$$

We must clarify that $V^\infty(x)$ and $V^*(x)$ have different meanings and they originate from different definitions. The former notation is from the ADP iteration, and the latter notation is the solution of the Bellman equation. Because of their different origin, there is a potential risk that $V^\infty(x)$ exists but it is not equal to $V^*(x)$. That is, convergence does not mean optimality. The following proof will exclude this kind of undesirable possibility. Here, we assume that $V^*(x)$ is the unique solution of the Bellman equation.

- Theorem 8.3: For $x \in X$, $V^k(x)$ is monotonically decreasing with respect to k in policy iteration ADP, i.e.,

$$V^k(x) \geq V^{k+1}(x), \forall x \in X.$$

Proof:

For an arbitrary initial state $x \in X$, generate a sequence of states with π_{k+1}

$$x, x', x'', x''', \cdots, x^{(\infty)}$$

Since π_{k+1} is admissible, the state sequence tends toward zero, i.e., $x^{(\infty)} = 0$. This result is ensured by the fact that $l(0,0) = 0$ only occurs at equilibrium. Considering the weak Bellman equation in PIM, we have

$$V^k(x) = l(x, \pi_k) + V^k(f(x, \pi_k)) \geq l(x, \pi_{k+1}) + V^k(f(x, \pi_{k+1}))$$

By combining PEV at the $(k+1)$-th main iteration, we have

$$V^k(x) \geq V^{k+1}(x) - V^{k+1}(f(x, \pi_{k+1})) + V^k(f(x, \pi_{k+1}))$$
$$= V^{k+1}(x) - V^{k+1}(x') + V^k(x')$$

Therefore, the relationship between V^k and V^{k+1} becomes

$$V^{k+1}(x) - V^k(x) \leq V^{k+1}(x') - V^k(x') \leq \cdots \leq V^{k+1}(x^{(\infty)}) - V^k(x^{(\infty)})$$

Considering the same zero boundary conditions $V^k(0) = 0$ and $V^{k+1}(0) = 0$ in each self-consistency condition, we have

$$V^k(x^{(\infty)}) = V^{k+1}(x^{(\infty)}) = 0$$

Finally, we conclude the value decreasing property.

∎

It is easy to see that there is a very different convergence mechanism between ADP with average cost and stochastic DP with discounted cost. With a discount factor, the Bellman operator has the property of γ-contraction. Unfortunately, ADP has no discount factor, and the property of γ-contraction does not exist. Its convergence must rely on the positive definiteness of utility function and the determinism of environment dynamics. Meanwhile, the zero boundary condition, i.e., $V^k(0) = 0$, plays a central role in the proof of convergence guarantee. In addition, when PIM stops, $V^k(x)$ converges to $V^*(x)$ in policy iteration ADP, i.e.,

$$\lim_{k \to \infty} V^k(x) = V^*(x).$$

When PIM stops, any two successive policies are the same. Considering how weak Bellman optimization is performed in each PIM, the value functions of these two successive policies obey the same Bellman equation, and such an improvement-stopping policy naturally becomes optimal according to the optimality condition.

8.1.4   Convergence in Inexact Policy Iteration

In policy iteration ADP, fixed-point iterator is often used in the PEV step to solve the self-consistency condition, leading to an inner iteration inside each PEV. In the standard version, the inner iteration must be repeated infinite times, which is very computationally inefficient. Similar to stochastic DP with discount factor, value iteration ADP and policy iteration ADP can be unified into a generalized policy iteration (GPI) framework with incomplete policy evaluations. In the incomplete PEV iteration, the fixed-point solver only repeats a finite number of steps, and the inner convergence to the true value function is not required. The incomplete PEV with N iterations is

Repeat j until $N - 1$

$$V_{j+1}^k(x) \leftarrow l(x, \pi_k) + V_j^k(x'), \forall x \in X \tag{8-12}$$

End

The policy evaluation outputs the state-value function after N-step iteration, that is, $V^k(x) = V_N^k(x)$. The final value $V^k(x)$ will initialize the value function of next PEV, i.e., $V_0^{k+1}(x) = V^k(x)$. The weak Bellman equation is still minimized by PIM. By combining (8-12) and (8-10), a new ADP algorithm called inexact policy iteration (abbreviated as inexact PI) is built. In the literature, this algorithm is referred to as other names, including modified PI and optimistic PI. When $N = 1$, inexact PI is equivalent to the standard value iteration ADP, and when $N \rightarrow \infty$, inexact PI becomes a standard policy iteration ADP. The convergence of inexact PI is proven as follows.

- Assumption 8.3: The following initialization is assumed to hold in inexact PI:
$$V^0(x) \geq 0 , V^0(x) \geq \min_u\{l(x, u) + V^0(x')\}, \forall x \in X.$$

- Theorem 8.4: If Assumption 8.3 holds, $V^k(x)$ is monotonically decreasing in inexact PI.

Proof:

According to Assumption 8.3, we have the following inequality:
$$V^0(x) \geq \min_u\{l(x, u) + V^0(x')\} = l(x, \pi_1) + V^0(f(x, \pi_1)),$$

Rolling out one-step forward yields

$$l(x, \pi_1) + V^0(x') \geq l(x, \pi_1) + l(x', \pi_1(x')) + V^0(x'')$$

$$= \sum_{i=0}^{1} l\left(x^{(i)}, \pi_1(x^{(i)})\right) + V^0(x^{(2)})$$

$$\geq \sum_{i=0}^{1} l\left(x^{(i)}, \pi_1(x^{(i)})\right) + l\left(x^{(2)}, \pi_1(x^{(2)})\right) + V^0(x^{(3)})$$

$$\cdots$$

$$\geq \sum_{i=0}^{N-1} l\left(x^{(i)}, \pi_1(x^{(i)})\right) + V^0(x^{(N)})$$

$$= V^1(x).$$

By repeating incomplete PEV for N steps, we have

$$V^1(x) = \sum_{i=0}^{N-1} l\left(x^{(i)}, \pi_1(x^{(i)})\right) + V^0(x^{(N)})$$

$$\geq l(x, \pi_1) + \sum_{i=1}^{N} l\left(x^{(i)}, \pi_1(x^{(i)})\right) + V^0(x^{(N+1)})$$

$$= l(x, \pi_1) + V^1(x'),$$

Note that $x^{(1)}, x^{(2)}, \cdots, x^{(N)}$ are all generated by the first policy π_1. It is easy to obtain that

$$l(x, \pi_1) + V^1(x') \geq \min_u \{l(x, u) + V^1(x')\}.$$

Combining the results shown above, we have the following inequalities:

$$V^0(x) \geq V^1(x),$$
$$V^1(x) \geq \min_u \{l(x, u) + V^1(f(x, u))\}.$$

It is easy to see that $V^1(x) \geq 0$, since $l(x, u) \geq 0$. By rolling forward to the k-th step, we have

$$V^k(x) \geq \min_u \{l(x, u) + V^k(x')\} \geq V^{k+1}(x), \forall x \in \mathcal{X}.$$

We can conclude that $\{V^k(x), k = 0,1,2, \cdots, \infty\}$ is monotonically decreasing.

∎

The most proper coefficient N needs to be experimentally chosen, and a moderate iteration number, i.e., $N = 5 \sim 10$, usually works very well in most continuing tasks. Interested readers can refer to [8-3] for detailed discussion of inexact PI. The result of incomplete PEV iteration bypasses the strict demand for recursive stability, which was previously taken as the basis of maintaining the finiteness of value function. The extension from complete PEV to incomplete PEV actually builds a new perspective to analyze ADPs with function approximation. Perfect approximation of value function seems to be not necessary since convergence guarantee does not need complete policy evaluations. Certain approximation error in the critic update is also tolerable so long as it does not affect the convergence of overall ADP algorithm.

8.1.5   Discrete-time ADP with Parameterized Function

ADP is a powerful tool for solving complex OCPs. Tabular ADP algorithms suffer from large memory demands and time-consuming computations. The urgent need for solving large-scale problems has drawn increasing attention to the generalization technique. After parameterizing both policy and value function, some representational accuracy must be sacrificed. This sacrifice is generally acceptable since in real-world, we usually seek to find "good enough" policies instead of exact solutions. Similar to the actor-critic architecture in model-free RL, the update of value parameter in ADP is still called as "critic", and the update of policy parameter is called as "actor". Deep neural network is the most popular representation for better approximation accuracy. That is why ADP is also called "neuro-dynamic programming" in some literature. The flow chart of parameterized ADP is shown in Figure 8.6. It contains three layers of iteration cycles. The outer cycle is where the main iteration between PEV and PIM occurs. The middle cycle is where policy updates or value updates occur. The inner cycle is where backpropagation through the depth of deep neural networks occurs.

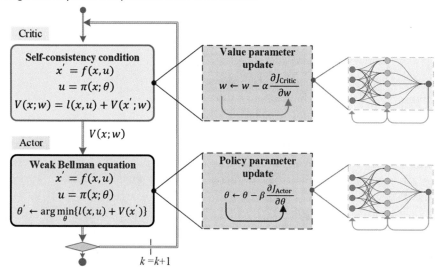

Figure 8.6 Policy iteration ADP with parameterized functions

For the sake of computational convenience, the value and policy approximations are often formulated into numerical optimization problems. At the k-th main iteration, the deterministic policy $u = \pi(x; \theta)$, in which θ is the policy parameter, is known from the previous cycle. In the critic, the parameter w of the value function $V(x; w)$ is updated until convergence. The essence of critic updates is to search for the solution of a parametrized self-consistency condition. Instead of directly solving the algebraic equation, one needs to convert it into an optimization problem:

$$J_{\text{Critic}}(w) = \frac{1}{2}\big(l(x, u) + V(x'; w) - V(x; w)\big)^2.$$

where   (8-13)

$$x' = f(x, u),$$
$$u = \pi(x; \theta).$$

The minimization of this problem results in an extremum, i.e., $J_{\text{Critic}} = 0$, where the self-consistency condition holds. Obviously, such an extremum is not unique, and $V(x; w) +$ Const is always an available value function, where Const $\in \mathbb{R}$ is an arbitrary constant. The nonuniqueness of extrema might cause unpredictable and unstable updates. One popular trick is to use a semi-gradient to update the value parameter instead of attempting to utilize true gradient updates. By applying first-order optimization to (8-13), its semi-gradient becomes

$$\nabla J_{\text{Critic}} = \frac{\partial J_{\text{Critic}}}{\partial w} = -\big(l(x, u) + V(x'; w) - V(x; w)\big)\frac{\partial V(x; w)}{\partial w}. \qquad (8\text{-}14)$$

Here, $\nabla J_{\text{Critic}} \in \mathbb{R}^{l_w}$ is called the critic gradient, and l_w is the dimension of the critic parameters. It is semi-gradient because $V(x'; w)$ is assumed to be fixed in the gradient derivation. This kind of critical updates can enable a stable learning process due to its structural similarity to that of the one-step bootstrapping mechanism. Another trick to address the nonuniqueness of extrema is to embed the zero boundary condition into the critic loss function:

$$J_{\text{Critic}}(w) = \frac{1}{2}\left\{\left(l(x,u) + V(x'; w) - V(x; w)\right)^2 + \rho V(0; w)^2\right\}$$

where $\rho \in \mathbb{R}^+$ is a coefficient to balance the strength from the zero boundary condition. With this kind of critic loss function, one can still select a semi-gradient to perform critic updates. Obviously, the critic gradient in model-based ADP is very similar to that in model-free RL except for a slight difference in how the environment is sampled. The analytical model is utilized in ADP instead of a real environment. Partial derivative information such as $\partial f^{\text{T}}(x,u)/\partial u$ is actually not used in the critic gradient. Therefore, there are no obvious benefit in the critic updates of ADP even though an accurate system model is known.

The purpose of actor updates is to find a better policy. The weak Bellman equation is actually a replacement of greedy search in the condition that an analytical model is perfectly known. The actor loss function is selected to be the right side of weak Bellman equation, which naturally builds an optimization problem:

$$J_{\text{Actor}}(\theta) = l(x,u) + V(x'; w),$$

where

$$x' = f(x,u),$$
$$u = \pi(x; \theta),$$

(8-15)

where $V(\cdot; w)$ and $l(\cdot, \cdot)$ are known at the k-th main iteration step, and $\pi(x; \theta)$ has unknown parameters. Applying the first-order optimization to (8-15), the actor gradient of J_{Actor} is equal to

$$\begin{aligned}
\nabla J_{\text{Actor}} &= \frac{\partial J_{\text{Actor}}}{\partial \theta} = \frac{\partial u^{\text{T}}}{\partial \theta}\frac{\partial l}{\partial u} + \frac{\partial u^{\text{T}}}{\partial \theta}\frac{\partial x'^{\text{T}}}{\partial u}\frac{\partial V}{\partial x'} \\
&= \frac{\partial \pi^{\text{T}}(x; \theta)}{\partial \theta}\left(\frac{\partial l(x,u)}{\partial u} + \frac{\partial f^{\text{T}}(x,u)}{\partial u}\frac{\partial V(x'; w)}{\partial x'}\right).
\end{aligned}$$

(8-16)

Here, $\nabla J_{\text{Actor}} \in \mathbb{R}^{l_\theta}$ is called the actor gradient, and l_θ is the dimension of actor parameters. Compared to the critic gradient, the actor gradient in ADP can greatly benefit from the model information. This is mainly due to the usage of two partial derivatives, $\partial f^{\text{T}}(x,u)/\partial u$ and $\partial l(x,u)/\partial u$. Their usage can effectively reduce the computational complexity to estimate actor gradients. With the same computational resources, the quality of actor gradients is significantly increased using model-based ADP compared with its model-free counterpart. The existence of environment model provides huge benefits in terms of enhancing computational efficiency compared to interacting with the real environment.

Algorithm 8-1: Infinite-horizon ADP

Hyperparameters: critic learning rate α, actor learning rate β, critic update frequency n_c, actor update frequency n_a, number of environment resets M

Initialization: state-value function $V(x; w)$, policy function $\pi(x; \theta)$.

Repeat (indexed with k)

(1) Use environment model

Initialize memory buffer $\mathcal{D} \leftarrow \emptyset$

Repeat M environment resets

$x \sim d_{\text{init}}(x)$

$u \leftarrow \pi(x; \theta)$

$x' \leftarrow f(x, u)$

Calculate $l, V, \partial V/\partial w, \partial V/\partial x, \partial l/\partial u, \partial f^{\mathrm{T}}/\partial u$, and $\partial \pi^{\mathrm{T}}/\partial \theta$

$$\mathcal{D} \leftarrow \mathcal{D} \cup \left\{ \left(l, V, \frac{\partial V}{\partial w}, \frac{\partial V}{\partial x}, \frac{\partial l}{\partial u}, \frac{\partial f^{\mathrm{T}}}{\partial u}, \frac{\partial \pi^{\mathrm{T}}}{\partial \theta} \right) \right\}$$

End

(2) Critic update

Repeat n_c times

$$\nabla_w J_{\text{Critic}} \leftarrow -\frac{1}{M} \sum_{\mathcal{D}} \left(l(x, u) + V(x'; w) - V(x; w) \right) \frac{\partial V(x; w)}{\partial w}$$

$$w \leftarrow w - \alpha \cdot \nabla_w J_{\text{Critic}}$$

End

(3) Actor update

Repeat n_a times

$$\nabla_\theta J_{\text{Actor}} \leftarrow \frac{1}{M} \sum_{\mathcal{D}} \frac{\partial \pi^{\mathrm{T}}(x; \theta)}{\partial \theta} \left(\frac{\partial l(x, u)}{\partial u} + \frac{\partial f^{\mathrm{T}}(x, u)}{\partial u} \frac{\partial V(x'; w)}{\partial x'} \right)$$

$$\theta \leftarrow \theta - \beta \cdot \nabla_\theta J_{\text{Actor}}$$

End

End

∎

The pseudocode of an infinite-horizon ADP algorithm is shown in Algorithm 8-1, in which the environment model must be used to calculate a few necessary components for critic and actor gradients. A practical infinite-horizon ADP algorithm has at least three layers of iteration cycles (including the backpropagation through the depth of neural network). In the second layer, there are two separated iterative updates: one is the update of value parameter to minimize the critic loss function and the other is the update of policy parameter to minimize the actor loss function. When ADP is generalized with deep neural networks, the third layer is the backpropagation computation in the depth direction. In many tasks, this three-layer ADP algorithm is not very computationally efficient. Fortunately, one can introduce incomplete critic and actor updates into parameterized

ADP. For example, by enforcing $n_c = n_a = 1$ in Algorithm 8-1, the second iteration layer can be removed, which often results in much better training efficiency. It has been proven that such a layer reduction can still guarantee the convergence property of overall ADP algorithm.

8.2   Continuous-time ADP with Infinite Horizon

Feedback controllers must be built to deal with real-world systems. For engineers, most physical systems work in the continuous-time domain. The problem formulation of continuous-time OCPs is similar to that of discrete-time OCPs; however, sums will be replaced by integrals, finite differences by derivatives, and difference equations by differential equations. Although a continuous-time system can always be discretized into a discrete-time system, significant theoretical merits in the mathematical proof and controller synthesis are still preserved. In a continuous-time system, the cost functional to be minimized is

$$\min_{u(\tau),\tau\in[t,\infty)} V\big(x(t)\big) = \int_t^\infty l\big(x(\tau),u(\tau)\big)\,\mathrm{d}\tau,$$

(8-17)

subject to

$$\dot{x}(t) = f\big(x(t),u(t)\big),$$

where t is the current time, $V(\cdot)$ is the cost functional, which is set to zero at equilibrium, τ is the virtual time, x, u are the state and action, and $l(x,u) \geq 0$ is the utility function. The dynamic model $f(x,u)$ is controllable, and $f(0,0) = 0$ is the equilibrium. The utility function is positive definite $l(x,u) > 0$ except at equilibrium, where $x = 0, u = 0$. The optimal value function of continuous-time OCP satisfies a first-order partial differential equation called the Hamilton-Jacobi-Bellman (HJB) equation. In optimal control theory, the HJB equation provides a necessary and sufficient condition for the optimality of an action with respect to a cost functional. Once the HJB equation is solved, an optimal control law can be built by minimizing its Hamiltonian.

8.2.1   Hamilton-Jacobi-Bellman (HJB) Equation

The HJB equation is the continuous-time analog to the discrete-time Bellman equation. It is also an extension of the classic Hamilton-Jacobi equation from Hamiltonian mechanics to general optimal control problems. The continuous-time HJB equation is suitable for many variational problems, including the brachistochrone problem, and can be seamlessly applied to a broad spectrum of optimal control problems. Here, this equation will be derived from the first-order approximation of the Taylor series expansion. Taking the derivative of both sides of (8-17) with respect to time t, we have the following self-consistency condition in continuous-time domain:

$$\frac{\partial V(x)}{\partial x^{\mathrm{T}}} f(x,u) = -l(x,u).$$

(8-18)

Here, $\partial V(x)/\partial x^{\mathrm{T}} \in \mathbb{R}^{1\times n}$ is a row vector with n dimensions. The derivation of the optimality condition for continuous-time systems requires a few tactical assumptions. For example, the state-value function is continuously differentiable with respect to both state and time. Let us assume that t has a small perturbation to $t + \mathrm{d}t$, and the perturbed state at $t + dt$ is $x(t) + dx(t)$. Therefore, the continuous-time value function can be divided into two separate stages:

$$V\big(x(t)\big) = \int_t^{t+dt} l\big(x(\tau), u(\tau)\big)d\tau + V\big(x(t) + dx(t)\big).$$

Applying Bellman's principle of optimality to this two-stage cost functional, we have a quasi-Bellman equation:

$$V^*\big(x(t)\big) = \min_{\substack{u(\tau) \\ t \le \tau \le t+dt}} \left\{ \int_t^{t+dt} l\big(x(\tau), u(\tau)\big)d\tau + V^*\big(x(t) + dx(t)\big) \right\}.$$

Use the Taylor expansion for the perturbed optimal value function, such as $V^*\big(x(t) + dx(t)\big)$, and recall the fact that $V^*\big(x(t)\big)$ is only a functional of state x for the infinite horizon. Then,

$$V^*\big(x(t) + dx(t)\big) = V^*\big(x(t)\big) + \frac{\partial V^*\big(x(t)\big)}{\partial x^{\mathrm{T}}(t)} dx(t) + \mathcal{O}\big(dx(t)\big).$$

Considering that $dx(t) = \dot{x}(t)dt$, $dt \to 0$ and $\tau \to t$, we have

$$V^*\big(x(t)\big) = \min_{u(t)} \left\{ l\big(x(t), u(t)\big)dt + \frac{\partial V^*\big(x(t)\big)}{\partial x^{\mathrm{T}}} \dot{x}(t)dt + V^*\big(x(t)\big) \right\}.$$

Since $V^*\big(x(t)\big)$ is independent of action u and time τ in the infinite-horizon problem, the optimality condition becomes

$$\min_{u(t)} \left\{ l\big(x(t), u(t)\big) + \frac{\partial V^*\big(x(t)\big)}{\partial x^{\mathrm{T}}} f\big(x(t), u(t)\big) \right\} = 0. \tag{8-19}$$

This formula is the well-known Hamilton-Jacobi-Bellman (HJB) equation. In optimal control theory, the HJB equation gives a necessary and sufficient condition for the optimality of an action with respect to a loss function. This equation is a natural result of the theory of dynamic programming, which was invented by Richard Bellman. Its connection to the Hamilton-Jacobi equation from classical physics was first drawn by Rudolf Kalman. In discrete-time problems, the corresponding difference equation is usually referred to as the Bellman equation, which can also be generalized to stochastic systems, in which case the HJB equation is a second-order elliptic partial differential equation. Define the Hamilton function H as

$$H\left(x, u, \frac{\partial V^*(x)}{\partial x}\right) \stackrel{\text{def}}{=} l(x, u) + \frac{\partial V^*(x)}{\partial x^{\mathrm{T}}} f(x, u). \tag{8-20}$$

Therefore, the optimal action is equal to

$$\pi^*(x) = \arg\min_u H\left(x, u, \frac{\partial V^*(x)}{\partial x}\right), \tag{8-21}$$

where $\pi^*(\cdot)$ is an optimal policy mapping from the state space $\mathcal{X}$ to the action space $\mathcal{U}$. Clearly, a simpler optimality condition is allowed for discrete-time OCPs compared with continuous-time OCPs. In general, the HJB equation is a nonlinear partial differential equation (PDE) in the value function, which means that its solution is the value function itself. Once this solution is known, it can be used to obtain the optimal control by taking the minimizer of the associated Hamiltonian involved in the HJB equation. Due to the real-world complexity, this information often provides limited utility for analytically building an optimal policy. However, the correctness of policy guesses can be examined using this relatively inexpensive method. One major drawback is that the HJB equation only accepts classical solutions for a sufficiently smooth value function, which is not guaranteed in some special situations. Instead, a viscosity solution is usually needed, in which conventional derivatives are replaced by set-value subderivatives.

8.2.2   Continuous-time ADP Framework

In nature, the HJB equation is a highly nonlinear partial differential equation (PDE). Traditionally, PDEs can be solved by three kinds of numerical methods: (1) directly approximating derivatives by difference quotients; (2) approximating infinite-dimensional spaces by finite-dimensional spaces; and (3) approximating PDEs at sampling points and finite basis expansion. Unfortunately, none of these methods works very well for highly nonlinear PDEs such as the HJB equation. Using any of these three methods for a challenging application can easily lead to very poor results.

Continuous-time ADP refers to a class of iterative algorithms that are used to effectively solve the HJB equation. An early version of this type of algorithm was introduced by Bertsekas and Tsitsiklis with the proper implementation of artificial neural networks. In this type of algorithm, there is a mitigation strategy to reduce the impact of large dimensionality; the memorization of the complete function in the whole space is replaced with that of sole neural network parameters. The continuous-time version of policy iteration ADP is demonstrated in Figure 8.7. It contains two cyclic steps similar to its discrete-time counterpart. The aim of the PEV step is to let the weak Hamiltonian equal zero, while that of the PIM step is to minimize the weak HJB equation:

$$l(x, \pi_k) + \frac{\partial V^k(x)}{\partial x^{\mathrm{T}}} f(x, \pi_k) = 0,$$

$$\pi_{k+1} \leftarrow \arg\min_u \left\{ l(x, u) + \frac{\partial V^k(x)}{\partial x^{\mathrm{T}}} f(x, u) \right\}.$$

(8-22)

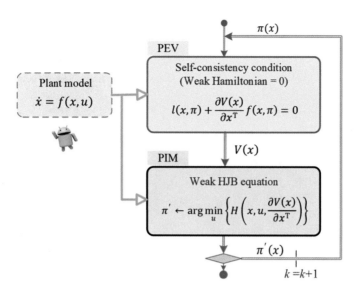

Figure 8.7 Continuous-time policy iteration ADP

Note that $V(0) = 0$ is a necessary condition in continuous-time ADP. One reason is that similar to that in the discrete-time domain, $H = l + \nabla^{\mathrm{T}} V \cdot f = 0$ is essentially a first-order derivative condition, and it does not contain any information about the initial condition or terminal condition. This condition fixes the value function at equilibrium to a special setting point. The high efficiency of policy iteration ADP is attributed to its intrinsic Newton-Raphson mechanism, i.e., repeatedly solving a simple first-order ordinary differential equation (ODE) and a low-dimensional optimization problem. Even though continuous-time ADP is promising as a numerical solver, previous studies have suggested that accurately finding an analytical optimal solution is still very challenging in the continuous-time domain, especially for large-scale nonlinear problems.

8.2.3 Convergence and Stability

Similar to the discrete-time version, convergence and stability in the continuous-time domain are two different but inherently entangled properties. Stability refers to whether a control policy such as $u = \pi(x)$ can stabilize the plant. The key of the proof is to select a positive definite function $V(x)$ for $\forall x \in \mathcal{X}$ as the Lyapunov candidate and examine whether its time derivative is negative semidefinite, i.e.,

$$\frac{\mathrm{d}V(x)}{\mathrm{d}t} \leq 0, \forall x \in \mathcal{X}.$$

Convergence describes whether ADP can gradually shift its intermediate policy π_k to the optimum π^*. The key of the proof is to examine whether $V^k(x)$ is monotonically decreasing with the main iteration step k for all $x \in \mathcal{X}$. The proofs of these two properties in continuous-time are very similar to the proofs for these properties in the discrete-time version. Continuous-time behavior poses a few new characteristics in terms of positive-definiteness examination and state-value function comparison.

8.2.3.1   Closed-loop stability

An initial policy π^0 must be admissible to guarantee that its corresponding initial value is finite, i.e., $V^0(x) < \infty$. An inadmissible policy will inevitably make the state diverge, and thus $V^0(x)$ easily goes to infinity. We will prove that $\pi_k, k = 1,2,\cdots,\infty$, are recursively admissible. The recursive stability property plays a central role in the effective deployment of policy iteration algorithms. This property allows the safe exploration of the environment (either actual or virtual) without being concerned that some intermediate policies are not admissible. This result can also be found in the discrete-time version of policy iteration ADP.

- Theorem 8.5: Policy π_{k+1} is admissible if π_k is admissible in the continuous-time version of policy iteration ADP.

Proof:

First, assume that π_k is admissible. For its successive policy π_{k+1}, we choose $V^k(x)$ as the Lyapunov function, which is naturally positive definite except at equilibrium. Since π_{k+1} is optimal for the weak HJB equation at the k-th step, we have the following condition:

$$H\left(x, \pi_{k+1}, \frac{\partial V^k(x)}{\partial x}\right) = \min_u \left\{ l(x,u) + \frac{\partial V^k(x)}{\partial x^{\mathrm{T}}} f(x,u) \right\}.$$

Moreover, π_k is not necessarily optimal for the $(k+1)$-th step value function, yielding

$$\min_u \left\{ l(x,u) + \frac{\partial V^k(x)}{\partial x^{\mathrm{T}}} f(x,u) \right\} \le l(x, \pi_k) + \frac{\partial V^k(x)}{\partial x^{\mathrm{T}}} f(x, \pi_k) = 0.$$

The equality on the right holds due to the realization of the k-th PEV step. By combining the two formulas above, we obtain

$$H\left(x, \pi_{k+1}, \frac{\partial V^k(x)}{\partial x}\right) \le 0.$$

Furthermore,

$$\frac{dV^k(x)}{dt} = H\left(x, \pi_{k+1}, \frac{\partial V^k(x)}{\partial x^{\mathrm{T}}}\right) - l(x, \pi_{k+1}) \le 0 - l(x, \pi_{k+1}) \le 0.$$

By Lyapunov's second method, π_{k+1} is admissible in the sense of Lyapunov stability. In addition, "equality" only holds when PEV stops and $l(0,0) = 0$ at equilibrium. Thus, the closed-loop system under π_{k+1} is asymptotically stable around the equilibrium.

∎

It is easy to see that the stability analysis on these intermediate policies also applies to the optimal policy, whose closed-loop system is asymptotically stable in the sense of Lyapunov stability. Lyapunov's second method provides an effective tool to study the stability property of equilibrium points in a dynamical system. The idea behind this approach is to examine how the scalar Lyapunov function changes as the system state evolves over time. The term "second" is in contrast with Lyapunov's first method, in which only the stability of an equilibrium point is considered by studying the behavior of the linearized system at that point. The positive definiteness of utility function $l(x, u)$ is also the basis of Lyapunov stability in infinite-horizon OCP. Intuitively, the scalar utility function is viewed as a generalized energy. A closed-loop system is said to be stable if the generalized energy monotonically decreases over time. If this energy function only has one global minimum, it strictly decreases along with all nonequilibrium solutions.

8.2.3.2   Convergence of ADP

The admissibility of an optimal policy is obviously independent of its searching algorithm. The optimal policy is admissible if and only if the HJB equation has a solution or its original OCP is tractable. In contrast, the intermediate policies depend on what kind of ADP algorithm is chosen. In general, both value function and policy will converge to their optima in policy iteration ADP. Its value decreasing property relies on the fact that the time derivative of the value function is smaller than zero, and moreover, two successive derivatives are monotonically decreasing. After applying the Newton-Leibniz formula, their integral will monotonically decrease, too.

- Theorem 8.6: For any $x \in \mathcal{X}$, $V^k(x)$ is monotonically decreasing with respect to the main iteration step k, i.e.,

$$V^{k+1}(x) \leq V^k(x), \forall x \in \mathcal{X}$$

Proof:

By using PEV (8-22), it can be found that $V^{k+1}(x)$ satisfies

$$\frac{\partial V^{k+1}(x)}{\partial x^{\mathrm{T}}} f(x, \pi_{k+1}) = -l(x, \pi_{k+1}),$$

$$V^{k+1}(0) = 0.$$

(8-23)

The zero boundary condition in (8-23) is supplementary information for the continuous-time self-consistency condition, whose purpose is to ensure the uniqueness of the value function in PEV. Taking the derivatives of $V^k(x)$ and $V^{k+1}(x)$ with respect to time t and leveraging their connection, we obtain the following inequalities:

$$\frac{\mathrm{d}V^k(x)}{\mathrm{d}t} = H\left(x, \pi_{k+1}, \frac{\partial V^k(x)}{\partial x}\right) - l(x, \pi_{k+1}) \leq -l(x, \pi_{k+1}),$$

$$\frac{\mathrm{d}V^{k+1}(x)}{\mathrm{d}t} = \frac{\partial V^{k+1}(x)}{\partial x^{\mathrm{T}}} f(x, \pi_{k+1}) = -l(x, \pi_{k+1}).$$

Since $l(x, u) \geq 0$, it is obvious that

$$\frac{\mathrm{d}V^k(x)}{\mathrm{d}t} \leq \frac{\mathrm{d}V^{k+1}(x)}{\mathrm{d}t} \leq 0.$$

From the Newton-Leibniz formula, we have the following equality when π_k is admissible:

$$V(x(t)) = V(x(\infty)) - \int_t^\infty \frac{\mathrm{d}V(x(\tau))}{\mathrm{d}\tau} \mathrm{d}\tau.$$

Considering the property of recursive stability, the policy π_{k+1} is also admissible. Hence, $x(\infty) \to 0$ as time goes to infinity. The zero boundary condition of each value function outputs

$$V^k\big(x(\infty)\big) = V^{k+1}\big(x(\infty)\big) = 0$$

Therefore, we have the following inequality:

$$V^{k+1}(x) \le V^k(x), \forall x \in \mathcal{X}.$$

∎

8.2.4   Continuous-time ADP with Parameterized Function

For continuous-time systems, both value function and policy must be parameterized to facilitate numerical computation. The gradient descent technique is utilized to update both the actor parameter and critic parameter to minimize their respective loss functions. The aim of critic updates is to solve the continuous-time self-consistency condition. This condition only provides the first-order derivative information and should include a supplementary zero boundary condition. At each main iteration, the parameterized state-value function needs to be enforced to become zero at the equilibrium, i.e., $V(0; w) = 0$. Otherwise, its solution may gradually deviate from the true value and eventually result in a very large approximation error. The continuous-time critic minimizes the squared sum of the Hamiltonian and zero-value function:

$$J_{\text{Critic}}(w) = \frac{1}{2}\left(H^2\left(x, u, \frac{\partial V(x; w)}{\partial x}\right) + \rho V^2(0; w)\right).$$

The true gradient of the critic is

$$\nabla J_{\text{Critic}} = H\left(x, u, \frac{\partial V}{\partial x}\right)\frac{\partial^2 V(x; w)}{\partial x^{\mathrm{T}}\partial w} f(x, u) + \rho\frac{\partial V(0; w)}{\partial w},$$

$$\frac{\partial^2 V}{\partial x^{\mathrm{T}}\partial w} \overset{\text{def}}{=} \frac{\partial}{\partial w}\left(\frac{\partial V}{\partial x^{\mathrm{T}}}\right) \in \mathbb{R}^{l_w \times n}. \tag{8-24}$$

Here, $\nabla J_{\text{Critic}} \in \mathbb{R}^{l_w}$ is called the critic gradient, and l_w is the dimension of the critic parameters. Obviously, the critic gradient is a full gradient, but a second-order derivative must be calculated. This high-order derivative is more computationally inefficient than its discrete-time counterpart. This is an evidence that higher computational complexity exists in continuous-time critic updates. The actor is to minimize the "weak" HJB equation:

$$J_{\text{Actor}}(\theta) = l\big(x, \pi(x; \theta)\big) + \frac{\partial V(x; w)}{\partial x^{\mathrm{T}}} f\big(x, \pi(x; \theta)\big),$$

where $V(\cdot; w)$ becomes a known variable after the critic update, and $\pi(x; \theta)$ has unknown policy parameter θ. The actor gradient of J_{Actor} is

$$\nabla J_{\text{Actor}} = \frac{\partial \pi^{\mathrm{T}}(x; \theta)}{\partial \theta}\left(\frac{\partial l(x, u)}{\partial u} + \frac{\partial f^{\mathrm{T}}(x, u)}{\partial u}\frac{\partial V(x; w)}{\partial x}\right), \tag{8-25}$$

Here, $\nabla J_{\text{Actor}} \in \mathbb{R}^{l_\theta}$ is called the actor gradient, and l_θ is the dimension of the actor parameters. The continuous-time actor gradient is not identical to its discrete-time version. We have found that the second-order derivative in (8-24) is very computationally inefficient. An interesting observation is that in continuous-time ADP, one can parameterize the partial derivative of state-value function rather than the state-value function itself. This approach is called dual heuristic programming (DHP) in the literature

[8-9], and it is featured with the generalization of value gradient function. The rationality of this choice is justified because both critic and actor gradients only need the first-order derivative information of the value function. The DHP algorithm is not only correct in theory but is also very effective in terms of convergence guarantee and stability enhancement.

8.2.5   Continuous-time Linear Quadratic Control

Optimal control of a linear system is particularly important to understand the theoretical completeness of ADP. One of the most influential methods is the linear quadratic (LQ) regulator, whose performance index is quadratic and whose dynamic model is linear. LQ may be the most well-developed control theory in the optimal control field. Its resulting optimal policy has many attractive properties, including elegant linear structure, ease of in-site adjustment, and solid convergence guarantee. Although many environments may be nonlinear, their dynamics can be regulated to the neighborhood of an equilibrium where the linearization becomes increasingly valid. Consider a continuous-time quadratic cost functional:

$$V\big(x(t)\big) = \int_t^\infty \big(x^T(\tau)Qx(\tau) + u^T(\tau)Ru(\tau)\big)\, d\tau,$$

(8-26)

subject to

$$\dot{x}(t) = Ax(t) + Bu(t),$$

where $u \in \mathbb{R}^m$ is the action, $x \in \mathbb{R}^n$ is the state, (A, B) is stabilizable, and $Q \geq 0$ and $R > 0$ have proper dimensions. Its minimization has been found to have a strong connection with the algebraic Riccati equation. The algebraic Riccati equation is an important tool that is used to analyze the optimality condition and deduce the optimal feedback law. We already know that its optimal policy has a linear and stationary structure:

$$u = -\frac{1}{2}R^{-1}B^T\frac{\partial V^*(x)}{\partial x} = -R^{-1}B^TP^*x,$$

With this linear policy, the optimal state-value function has a quadratic structure:

$$V^*(x) = x^TP^*x$$

where P^* is called the optimal Riccati matrix. Its solution comes from a continuous-time algebraic Riccati equation:

$$Q + P^*A + A^TP^* - P^*BR^{-1}B^TP^* = 0.$$

It can be shown that the Riccati equation has a unique positive definite solution if (A, B) is stabilizable. The history of Riccati equation can be traced back to Count Riccati's original letter to his friend in 1720. His major concern was how to study two time-dependent scalar differential equations. The importance of Riccati equation has led to the development of considerable research activity in linear Hamiltonian systems, optimal control problems, optimal filtering and estimation, etc. In infinite-horizon LQ control, one cares about all the utility functions through the distant future. However, an instantaneous optimal action must be chosen knowing that one will also behave optimally in the future. The instantaneous optimal action can be calculated using the solution of the algebraic Riccati equation and the current state observation. Here, we will demonstrate how ADP works for infinite-horizon LQ control. First, let us parameterize $V^k(x)$ at the k-step main iteration to be a quadratic function:

$$V^k(x) = x^\mathrm{T} P^k x, \tag{8-27}$$

where P^k is the value parameter, i.e., intermediate Riccati matrix. The critic update then becomes

$$
\begin{aligned}
P_{j+1}^k &= P_j^k - \alpha H\left(x, \pi_k, \frac{\partial V\left(x; P_j^k\right)}{\partial x}\right) x f^\mathrm{T} \\
&= P_j^k - \alpha\left(x^\mathrm{T} Q x + u^\mathrm{T} R u + 2 f^\mathrm{T} P_j^k x\right) x f^\mathrm{T}
\end{aligned}
$$

where $H \in \mathbb{R}$ is the Hamiltonian, f^T is short for $f^\mathrm{T}(x, u)$, and α is the learning rate. Note that $x f^\mathrm{T} \in \mathbb{R}^{n \times n}$ is a second-order partial derivative matrix with proper dimensions. The update of Riccati matrix is based on the information of the second-order partial derivative of the quadratic state-value function. When $j \to \infty$, we have $P^k = P_\infty^k$, and PEV stops. For the actor loss function, its minimization can be simplified as an analytical solution:

$$\pi_{k+1}(x) = -\frac{1}{2} R^{-1} B^\mathrm{T} \frac{\partial V^k(x)}{\partial x} = -R^{-1} B^\mathrm{T} P^k x. \tag{8-28}$$

When $k \to \infty$, $P^\infty = P^*$ becomes the optimal Riccati matrix, that is, the solution of the continuous-time algebraic Riccati equation. In addition, such an optimal controller relies on complete accessibility to the full state observation. In many examples, accurate state measurements are either technologically inaccessible or too expensive. It is necessary to estimate the state from noisy sensor measurements. Under some well-defined conditions, a proper state estimate can be used in conjunction with an optimal linear feedback law without any negative impacts on stability and convergence.

The aforementioned ADP algorithm builds an in-site automatic approach to finding the optimal gain of LQ controller. The use of this algorithm reduces the amount of work done by an engineer to optimize the feedback controller. Even so, engineers still need to specify most cost function hyperparameters, e.g., Q and R, and compare their control results with the specified design goals. Often, this means that the controller construction will be an iterative process in which the "optimal" control gain is in-site adjusted by this automatic algorithm and the hyperparameters are empirically judged to produce a more desirable behavior.

8.3   How to Run RL and ADP Algorithms

The environment model plays a central role in training and implementing a policy. In model-free RL, data can be collected from either a real environment or an environment model, while in model-based RL, one must rely on the modeling information. One might want to know which is more computationally efficient: model-free RL (abbreviated as "RL") or model-based ADP (abbreviated as "ADP"). The differentiability of the environment model allows the model information to be more efficiently utilized in ADP, while in RL, one must dig out the gradient information from the roll-out data. Therefore,

ADP often has a more accurate gradient estimation, and its cycle number is much smaller than that of RL. This is why people generally believe ADP is more efficient than RL. This conclusion is right in most cases, but its explanation is not easy. The true reason is slightly complicated, especially when no perfect model is available.

8.3.1   Two Kinds of Environment Models

Models are the knowledge of the environment dynamics. Building a model must focus on the vital characteristics of environment dynamics that are intellectually tractable. Modeling techniques can have various levels of abstraction, expressed in a diverse list of modeling languages and toolsets. In general, there are two kinds of abstractive models: (1) high-fidelity simulation models and (2) simplified analytical models. Figure 8.8 illustrates these two models of a passenger car. Both of them belong to the mathematical abstraction of the real car. The former is more accurate but has no analytical formula. The latter is less accurate but featured with differentiability. Table 8.2 shows their compatibility in RLs and ADPs. In the following context, whenever we discuss model-based methods, we refer to a "model" that must be differentiable.

(a) Real-world environment

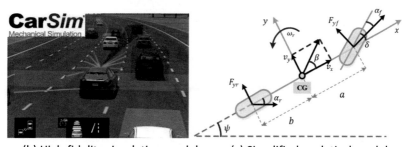

(b) High-fidelity simulation model   (c) Simplified analytical model

Figure 8.8 An autonomous vehicle represented by two kinds of models

A high-fidelity simulation model is based on code-level simulation software and has no explicit mathematical formula. A real environment is simulated by using computer software, and its simulation results are very close to those of the actual system. Collecting data in the real world is expensive and time-consuming. Moreover, it is difficult to guarantee the safety of environment interactions. Because the model builds a virtual environment, its data generation is less expensive than that of the real environment. Therefore, a high-fidelity model is a risk-free environment for trial-and-error learners that can more efficiently generate a large number of data samples.

A simplified analytical model is featured with explicit mathematical formulas, for example, ordinary differential equations (ODEs) that capture the major characteristics of environment dynamics. The differentiability of this model is one critical difference as compared with a high-fidelity simulation model. This property allows ADP to use the model derivatives to more efficiently compute the policy gradients. The shortcoming of this model is that the gradient accuracy depends on how accurate the model is. In reality, model mismatch inevitably exists, and the quality of gradient estimation may be largely sacrificed in an imperfect model.

Table 8.2 Suitability of the real environment and its models

Environment	Model-free RL	Model-based ADP
(a) Real-world environment	●	○
(b) High-fidelity simulation model	●	○
(c) Simplified analytical model	●	●

8.3.2   Implementation of RL/ADP Algorithms

As discussed before, a model can be a good replacement of the real environment. The simplified analytical model can be used in both model-free RL and model-based ADP. In the former, the model serves as an environment to explore new data. In the latter, the model provides its partial derivative information to facilitate accurate gradient calculation. When such an analytical model is available, one might be interested in knowing which path is more computationally efficient. Let us compare the two policy gradient algorithms, i.e., RL with a stochastic policy and ADP with a deterministic policy, to demonstrate their differences with respect to training efficiency. In RL with state-value, the stochastic policy must be learned because its deterministic gradient does not truly exist. Without loss of generality, it is assumed that RL has the following actor and critic gradients:

$$\nabla J_{\text{Critic}}^{\text{RL}} = -\mathbb{E}_s \left\{ \left(V^{\text{target}}(s) - V(s; w) \right) \frac{\partial V(s; w)}{\partial w} \right\},$$

$$\nabla J_{\text{Actor}}^{\text{RL}} = \mathbb{E}_{s,a,s'} \left\{ \nabla_\theta \log \pi_\theta(a|s) \left(r + \gamma V(s'; w) \right) \right\},$$

where V^{target} is the target value. The actor gradient of RL must be estimated on the basis of three random variables, including current state, current action and next state. In contrast, the actor gradient of ADP can decompose the state-value to the current-step reward and the next-step state value. With the support of a differentiable model, both the reward function and its subsequent state are analytical functions of actions. Hence, ADP has the following critic and actor gradients:

$$\nabla J_{\text{Critic}}^{\text{ADP}} = -\mathbb{E}_x \left\{ \left(V^{\text{target}}(x) - V(x; w) \right) \frac{\partial V(x; w)}{\partial w} \right\},$$

$$\nabla J_{\text{Actor}}^{\text{ADP}} = \mathbb{E}_x \left\{ \frac{\partial \pi^{\text{T}}(x; \theta)}{\partial \theta} \left(\frac{\partial l(x, u)}{\partial u} + \frac{\partial f^{\text{T}}(x, u)}{\partial u} \frac{\partial V(x'; w)}{\partial x'} \right) \right\}.$$

Whether to utilize the information of model differentiability (including that of the reward) is a key feature to distinguish model-based ADP from model-free RL. It is easy to see that

RL and ADP share the same critic gradient, which means their update efficiencies are identical. But their actor gradients are very different: ADP benefits greatly from the analytical model and reward information, i.e., their partial derivatives with respect to action, which can significantly reduce the gradient estimation error caused by action randomness and imperfect value estimation. However, high gradient accuracy in ADP comes at the expense of high computational complexity. The three derivatives, i.e., $\partial l/\partial u$, $\partial f/\partial u$ and $\partial V/\partial x'$, introduce some additional burdens to the actor-gradient computation.

8.3.2.1   State initialization sampling (SIS)

We must clarify that ADP uses the partial derivative information to compute critic and actor gradients, but it does not mean that there is no sampling process. Both model-free RL and model-based ADP require samples from environment interaction. In general, there are two basic sampling techniques to collect samples, i.e., state initial sampling (SIS) and state rollout forward (SRF). In SIS, a specific initial distribution at the current time in the state space is manually selected, for example, uniform state distribution or Gaussian state distribution. For RL, one needs to roll out at least one step forward to collect the information of the next state, which starts from the chosen initial state distribution. For ADP, one does not need to roll out the model because the partial derivative already provides the necessary information to compute gradients.

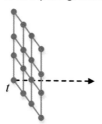

Figure 8.9 State initialization sampling (SIS)

8.3.2.2   State rollout forward (SRF)

Another sampling technique involves rolling out the model with the current policy from a fixed initial state. The state trajectory, including both states and actions, is collected to support the computation of critic and actor gradients. In RL, one needs to use each state-action pair in the trajectory, while in ADP, one only needs the state distribution behind the trajectory. Compared to SIS, SRF has the advantage that its state distribution is much closer to the stationary state distribution.

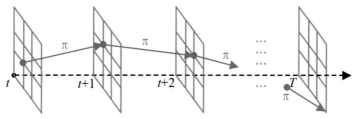

Figure 8.10 State rollout forward (SRF)

A more popular sampling technique is to combine SIS and SRF, i.e., selecting a batch of random initial states and then rolling them out simultaneously. The generated batch trajectories are very convenient for matrix operations. The exploration capability can also be enhanced with random state initialization, especially when both the environment and policy are deterministic.

8.3.3   Training Efficiency Comparison: RL vs. ADP

Now, we can formally discuss the training efficiency of model-free RL and model-based ADP. The overall training efficiency depends on two key factors: (1) the computational complexity of each main iteration and (2) the quality of gradient estimation. The former mainly determines the running time of each RL/ADP iteration, denoted as $T_\text{iteration}$. Its running time is determined by a sampler, an actor learner and a critic learner. The latter mainly determines how many iterations will be needed before reaching a certain performance level. With the same size of batch data, RL has a much lower computational complexity at each main iteration, while ADP often has higher gradient estimation quality.

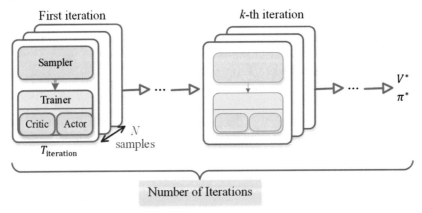

Figure 8.11 Total running time in the training process

Let us first discuss the computational complexity of each main iteration. In today's AI community, neural networks have been widely used to approximate value functions and policies. For RL, each critic update is controlled by three components: V^target, $V(x; w)$, and $\partial V/\partial w$. Two components, V^target and $V(x; w)$, require very similar running time, which is equal to the forward propagation time of value network T_FD^V. The running time of $\partial V/\partial w$ is equal to the backward propagation time of value network T_BP^V. The actor update is determined by two components: $\nabla_\theta \log \pi_\theta$ and $V(s'; w)$. The computation of $V(s'; w)$ is completed in the previous critic update, and hence, its running time does not need to be counted again. The running time of $\nabla_\theta \log \pi_\theta$ is equal to $T_\text{FD}^\pi + T_\text{BP}^\pi$, where T_FD^π is the forward propagation time of the policy network, and T_BP^π is the backward propagation time.

For ADP, the complexity of each critic update is exactly equivalent to that of RL. Their overall training efficiency mainly differs in how actor gradients are computed. Different from RL, each actor update in ADP is dominated by four components: $\partial l/\partial u$, $\partial V/\partial x'$, $\partial f/\partial u$, and $\partial \pi/\partial \theta$. The computation of $\partial V/\partial x'$ includes a forward propagation and a backward propagation of the value network. The former is completed in the previous critic update, so only the latter needs to be counted. The running times of $\partial l/\partial u$ and $\partial f/\partial u$ are denoted as T^l and T^f, respectively. Similar to the actor updates in RL, the computation of $\partial \pi/\partial \theta$ also includes a forward propagation and a backward propagation of the policy network. Usually, computing first-order derivatives of the reward and model are very fast, and we can neglect T^l and T^f in the complexity analysis. An easy observation is that each actor update in ADP needs more running time than that in RL. Table 8.3 shows the computational efficiency of each main iteration in RL and ADP, where N is the number of samples, n_c is the frequency of critic updates, and n_a is the frequency of actor updates.

Table 8.3 Computational efficiency of each main iteration

	RL	ADP
Interaction	$T_{\mathrm{env}} \cdot N$	$T_{\mathrm{model}} \cdot N$
Critic	$(2T_{\mathrm{FD}}^V + T_{\mathrm{BP}}^V) \cdot n_c N$	$(2T_{\mathrm{FD}}^V + T_{\mathrm{BP}}^V) \cdot n_c N$
Actor	$(T_{\mathrm{FD}}^\pi + T_{\mathrm{BP}}^\pi) \cdot n_a N$	$(T_{\mathrm{FD}}^\pi + T_{\mathrm{BP}}^\pi + T^f + T^l + T_{\mathrm{BP}}^V) \cdot n_a N$

Although more running time is required in ADP to compute each actor gradient than in RL, its gradient quality is much better than that in RL. The reason is that in ADP, the partial derivative information is directly embedded into actor-gradient estimation, which reduces the stochasticity and sparsity of randomly collected action samples. Accordingly, it reduces the occurrence of high-variance gradients that come from imperfect measurements and environmental randomness. Obviously, a high-quality actor gradient is helpful to reduce the overall cycle number. The reduction in the interaction number can compensate for the high running time at each main iteration. In summary, model-based ADP is usually faster than model-free RL.

8.4 ADP for Tracking Problem and its Policy Structure

Both ADP and MPC can be regarded as numerical solvers of general OCPs. One of their apparent differences is how to find and implement optimal actions. Most ADP algorithms work offline since the environment model is already known. The optimal policy is first trained in the whole state space, and this policy is then implemented as an online controller. The high online efficiency is one of its key benefits since tedious optimization moves from the online stage to the offline stage. Different from ADP, MPC calculates a series of optimal actions by receding horizon optimization. The time-consuming online optimizer must be adopted in MPC, and its closed-loop action is almost equal to the true solution of open-loop OCP (except for some optimization errors). In general, some performance loss is observed in ADP because its control policy is usually approximated by a parameterized function. The approximate policy always has an intrinsically parametric structure. For example, in a linear static controller, its action is proportional to the state. Together with the imperfection of offline ADP algorithm, an improper policy structure may largely sacrifice the optimality of the trained policy. Therefore, it is important to properly select the policy structure so as to train a more accurate policy.

8.4.1   Two Types of Time Domain in RHC

The optimal tracker, which covers a broad class of control tasks, requires carefully matching the problem definition and policy selection. Without loss of generality, standard infinite-horizon tracking problems are defined to compare various types of policy structures. The receding horizon control (RHC) perspective is used to explain how to choose proper policies for optimal trackers. Even though it has not been explicitly stated, there are actually two kinds of time domains under the RHC perspective: (a) the real-time domain in the physical world and (b) the virtual-time domain in the predictive horizon (shown in Figure 8.12). Take traditional MPC as an example: its problem optimization and action implementation work in the two temporal domains, respectively. The optimal action sequence is optimized in the virtual-time domain (also called the predictive horizon in most MPCs), and the first optimal action is implemented in the real-time domain.

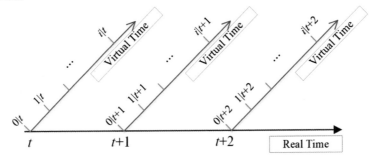

Figure 8.12 Real-time domain vs. virtual-time domain

The separation of real-time domain and virtual-time domain provides us a useful perspective to understand how ADP works. In the literature, almost no ADP algorithm clearly distinguishes these two domains. We could see that both problem optimization and policy implementation were defined in the real-time domain. In fact, one can always separate these two tasks into two time domains. The optimal policy is trained in the virtual-time domain, and it is implemented in the real-time domain (Figure 8.13). This perspective allows rewriting ADP into an MPC-like problem, which helps us to design more efficient sampling behavior and select better policy structure.

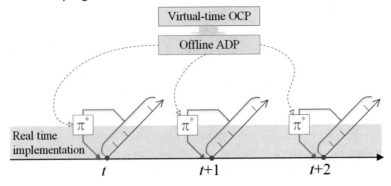

Figure 8.13 Receding horizon control for ADP

8.4.2 Definition of Tracking ADP in Real-time Domain

The performance in tracking ADP is determined by its criterion design. Let us consider tracking OCPs with infinite-horizon cost function, which is defined from $t = 0$ to $t = \infty$. Traditionally, these OCPs are only defined in the real-time domain, and there is no virtual-time domain. For example, the cost function that tracks a time-varying reference is

$$V(x_t, X^R) = \sum_{i=0}^{\infty} l(x_{t+i}^R - x_{t+i}, u_{t+i}), \tag{8-29}$$

where t is the real-time step, and $X^R = \{x_t^R, x_{t+1}^R, \cdots, x_\infty^R\}$ is the reference from the real world. As a standard tracking control problem, the reference trajectory is not completely fixed but dynamically changes with respect to time. The open-loop sequence of its optimal actions is $\{u_t^*, u_{t+1}^*, \cdots, u_\infty^*\}$, which creates the best open-loop performance in the real-time domain. The optimal value function with open-loop actions is denoted as $V_{\text{Open}}^*(x)$. The open-loop Bellman equation becomes

$$V_{\text{Open}}^*(x_t, X^R) = \min_u \{l(x_t^R - x_t, u) + V_{\text{Open}}^*(x_{t+1}, X^R)\}. \tag{8-30}$$

The optimal value $V^*(x_t, X^R)$ is a function of both the current state x_t and the infinitely long reference X^R. At time t, the optimal action u_t^* is determined not only by the current state, but also by all the future reference signals. Here, two popular policies are trained to examine their closed-loop performances:

$$u_t = \pi(x_t, x_t^R), \tag{8-31}$$

$$u_t = \varphi(x_t, x_t^R, x_{t+1}^R, x_{t+2}^R, \cdots, x_\infty^R). \tag{8-32}$$

Both of them are stationary functions. The former policy has the current state and reference signal as the input. In the latter policy, one needs to input all the future reference signals. By minimizing (8-29) with these two policies, the "optimal" results, which are denoted as π^* and φ^*, can be found. The "optimal" state-value functions, which are denoted as $V_\pi^*(x)$ and $V_\varphi^*(x)$, represent the best closed-loop performance. One may recognize that $\pi(\cdot)$ in (8-31) is widely used in tracking controllers. Its popularity may let people mistakenly believe that this policy perfectly minimizes the tracking criterion in (8-29). In fact, this idea is not correct. A short explanation is that $\pi(\cdot)$ is an overly structured policy, and there is not enough freedom to allow it to carry various possibilities of time-varying references. In contrast, $\varphi(\cdot)$ is the best choice in optimal tracking control and it has the potential to mimic optimal actions that correspond to all possible references. This kind of freedom, however, comes at the cost of high-dimensional input and high training complexity.

8.4.3 Reformulate Tracking ADP in Virtual-time Domain

To better understand the behaviors of aforementioned two policies, let us convert the real-time OCP to an MPC-like problem that works in the virtual-time domain. The infinite-horizon MPC is not very popular in engineering, but its problem definition is more consistent with that of previous ADP. With the support of the RHC perspective, tracking OCP (8-29) is rewritten into an infinite-horizon MPC-like problem. Depending on the selection of virtual-time reference, two kinds of virtual-time MPCs are defined: one is called the full-horizon tracker, and the other is called the first-point tracker. At time t, the full-horizon tracker in the virtual-time domain is

$$J^{\varphi}_{\text{MPC}} = \sum_{i=0}^{\infty} l\left(x^R_{i|t} - x_{i|t}, u_{i|t}\right), \tag{8-33}$$

where t in $(i|t)$ denotes the real time and i denotes the virtual time at time t. Note that $x^R_{0|t}, x^R_{1|t}, \cdots, x^R_{\infty|t}$ are the varying references in the predictive horizon. One important feature of such an MPC-like problem is that its reference $x^R_{i|t}$ varies with virtual time rather than real time. This domain separation property means that the virtual-time reference can differ from the real-time reference, which provides us with huge flexibility to build various ADP trackers. The common choice is to select virtual-time reference signals that exactly equal the corresponding real-time reference signals:

$$x^R_{i|t} = x^R_{t+i}, i \in \mathbb{N}.$$

After minimizing (8-33), the first optimal action is taken as the control input in the real-time domain. This type of design naturally results in a full-horizon policy, i.e., φ-policy:

$$u(t) = \varphi\left(x_{0|t}, x^R_{0|t}, x^R_{1|t}, \cdots, x^R_{\infty|t}\right),$$

In addition to the current-time state, i.e., $x_{0|t} = x_t$, the input of $\varphi(\cdot)$ includes all the virtual-time reference signals, i.e., $x^R_{0|t}, x^R_{1|t}, \cdots, x^R_{\infty|t}$. This tracking ADP is equipped with φ-policy and it actually builds a full-horizon tracker. When a perfect model is available, its closed-loop optimality is equal to the open-loop optimality. Similarly, let us define the first-point tracker in the virtual-time domain. At time t, one of its MPC-like cost functions becomes

$$J^{\pi}_{\text{MPC}} = \sum_{i=0}^{\infty} l\left(x^R_{0|t} - x_{i|t}, u_{i|t}\right), \tag{8-34}$$

where $x^R_{0|t}$ is the first-step virtual-time reference. Its implementation is equivalent to equipping ADP with the π-policy (i.e., the first-point tracker). This kind of MPC only takes the first reference point x^R_t as the policy input. In other words, tracking ADP with π-policy has a hidden limitation, i.e., the virtual-time reference trajectory is assumed to remain unchanged:

$$x^R_{i|t} = x^R_t, i \in \mathbb{N}. \tag{8-35}$$

Note that (8-35) is not the only choice of ADP with π-policy, and there may be other virtual reference designs according to the user's preference. For example, if a fixed prediction model $x^R_{i|t} = g(x^R_t)$ is chosen, $x^R_{i|t}$ is only determined by x^R_t. This selection still outputs an optimal first-point policy but has different tracking performance from (8-35).

The comparison of first-point tracker and full-horizon tracker is shown in Figure 8.14. In ADP with π-policy, the first reference point is always used in the virtual-time domain. This kind of tracker actually becomes an optimal regulator with varying "equilibrium". The resultant optimal regulator takes the tracking error as the new state. Since the first reference point still changes as the real time evolves, this kind of regulator keeps a certain tracking ability. However, its optimality should be much weaker than that of a well-trained φ-policy. Therefore, one can easily conclude that

$$V_\pi^*(x) \geq V_\varphi^*(x) = V_{\text{Open}}^*(x), \forall x \in \mathcal{X}.$$

Briefly speaking, the policy structure can place a large amount of limitation on the freedom to reach the open-loop optimality. The first-point policy in (8-31) has some unnecessary constraints, while the full-horizon policy in (8-32) defines a completely free policy, posing almost no constraint on the optimal action.

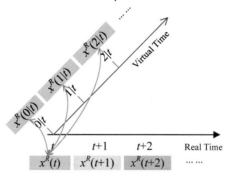

(a) Policy π (i.e., first-point tracker)

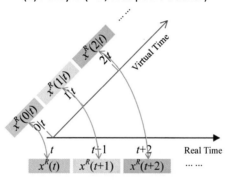

(b) Policy φ (i.e., full-horizon tracker)

Figure 8.14 Virtual-time references with different policies

A more illustrative example is shown in Figure 8.15, in which an autonomous car tries to track a desired trajectory at an intersection. This figure shows the reference trajectory in the virtual-time domain at a specified real-time moment. At the virtual-time domain, only the first reference point is tracked when using π-policy. In contrast, all the reference signals are tracked when using φ-policy. Some performance loss is observed in the first-point tracker since its policy structure is not completely free. The level of performance sacrifice depends on how dynamic the reference varies in the real world. Even so, one can still use the π-policy as an acceptable tracker if its performance loss is not significant. Even though the φ-policy is a perfect choice in performance guarantee, its high structural complexity requires more computational resources. A natural choice is to select a middle candidate between π-policy and φ-policy, which can strike a balance between tracking optimality and computational efficiency. Usually, a structured policy that takes finite-horizon reference points as the input is good enough to build an approximate "optimal" tracker.

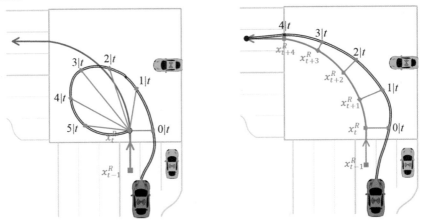

(a) Policy π (i.e., first-point tracker)   (b) Policy φ (i.e., full-horizon tracker)

Figure 8.15 Virtual-time optimizer in ADP with different policies

8.4.4   Quantitative Analysis with Linear Quadratic Control

As a special case of OCP, linear quadratic (LQ) control has some interesting properties. Traditionally, both LQ trackers and LQ regulators are solved with Riccati equations, and they use static linear functions as optimal control laws. For the LQ regulator, the state is required to be kept around the equilibrium, which usually refers to zero. For the LQ tracker, the state is required to follow a reference trajectory. Some reference trajectories can be highly dynamic, and others may change slowly or even remain constant. One might have observed that today's control community often easily misuses the two optimal controllers, and many engineers do not distinguish what is an optimal regulator and what is an optimal tracker. One common mistake is to take the tracking error as a new state and build an LQ regulator as the optimal tracker. This design still has some kind of tracking ability, but such a "fake" tracker is not optimal with respect to the cost function to be minimized. Surprisingly, this kind of "fake" optimal tracker still has good tracking performance in many tasks. This might be the reason why similar designs, which are wrong in theory but useful in practice, are regularly seen in engineering. To clarify their differences, let us consider the following finite-horizon LQ tracking problem:

$$V(x_t, X^R) = \sum_{i=t}^{T} \left((x_i^R - x_i)^\mathsf{T} Q (x_i^R - x_i) + u_i^\mathsf{T} R u_i \right),$$

subject to

$$x_{t+1} = A x_t + B u_t,$$

where $u \in \mathbb{R}^m$ is the action, $x \in \mathbb{R}^n$ is the state, $x^R \in \mathbb{R}^n$ is the reference signal, and $(T - t)$ is the length of the predictive horizon. The real-time reference is denoted as $X^R = \{x_t^R, x_{t+1}^R, \cdots, x_T^R\}$. Assume that the pair (A, B) is controllable, and $Q \geq 0, R > 0$ are the two weighting matrices with proper dimensions. The Bellman equation is

$$V^*(x_t, X^R) = \min_u \{(x_t^R - x_t)^\mathrm{T} Q(x_t^R - x_t) + u_t^\mathrm{T} R u_t + V^*(x_{t+1}, X^R)\},$$

The optimal value function $V^*(x_t, X^R)$ is selected as a quadratic function with three unknown coefficients:

$$V^*(x_t, X^R) = x_t^\mathrm{T} P_t x_t + 2x_t^\mathrm{T} \beta_t + \alpha_t, \tag{8-36}$$

The information from the time-varying reference is embedded into the three time-varying coefficients, i.e., P_t, β_t and α_t. The optimal policy finally becomes

$$u_t^* = -R^{-1}B^\mathrm{T}(P_{t+1}x_{t+1} + \beta_{t+1}). \tag{8-37}$$

This feedback controller is governed by three time-varying coefficients. Compared to the regulator version, a feedforward term is added to reflect the varying reference. Substituting optimal action u_t^* into $V^*(x_t, X^R)$, we have the following three recursive formulas:

$$P_t = Q + A^\mathrm{T} P_{t+1} A - A^\mathrm{T} P_{t+1} B(R + B^\mathrm{T} P_{t+1} B)^{-1} B^\mathrm{T} P_{t+1} A,$$

$$\beta_t = (A^\mathrm{T} - A^\mathrm{T} P_{t+1} B(R + B^\mathrm{T} P_{t+1} B)^{-1} B^\mathrm{T}) \beta_{t+1} - Q x_t^R, \tag{8-38}$$

$$\alpha_t = \alpha_{t+1} + x_t^{R^\mathrm{T}} Q x_t^R - \beta_{t+1}^\mathrm{T} B(R + B^\mathrm{T} P_{t+1} B)^{-1} B^\mathrm{T} \beta_{t+1},$$

This equation is known as the differential Riccati equation for finite-horizon LQ tracking control problems. It is easy to see that P_t is only determined by the first Riccati equation, which forms the static feedback gain for the optimal regulator. The coefficient β_t is controlled by the time-varying reference, which forms the feedforward gain to fulfill the tracking ability. Note that the feedforward gain β_t in (8-37) is reduced to zero in an LQ regulator because the desired trajectory becomes zero in (8-38). How to solve (8-38) is critical to quantitatively analyzing the influence of policy structures in various LQ tracking problems, including both finite-horizon and infinite-horizon.

8.4.4.1   Case I: Recursive solutions for finite-horizon problems

Let us consider finite-horizon LQ problems with terminal conditions. Here, the terminal time for the finite-horizon LQ tracker is time T. Assume that at time T, the optimal value function is set as zero, i.e., $V^*(x_T, x_T^R) = 0$. Considering the arbitrariness of the final state, we have the following terminal boundary condition:

$$P_T = 0, \beta_T = 0, \alpha_T = 0.$$

The three coefficients (P_t, β_t, α_t) can be recursively calculated with (8-38) in a backward manner. Optimal action u_t^* contains the feedforward term β_{t+1}, which implies that the optimal policy is affected by the reference after time t. Therefore, the optimal control law for the finite-horizon tracker must be in the structure of $u_t = \varphi(x_t, x_t^R, x_{t+1}^R, \cdots, x_T^R)$, where the reference before time t has no effect on the successive optimal actions. Therefore, such a static policy $u_t = \pi(x_t, x_t^R)$ is not completely free, and its real-time performance must be sacrificed to some extent depending on how dynamic the varying reference is.

8.4.4.2   Case II: Steady-state solutions for infinite-horizon problems

An infinite-horizon LQ tracker is very different from its finite-horizon counterpart. For any infinite-horizon problems, we can regard the differential Riccati equation (8-38) as a discrete-time dynamic system. Its three state variables are the three coefficients (P_t, β_t, α_t). Such a Riccati dynamic system must be stable in order to ensure the existence of a steady-state solution. Of course, a stable Riccati equation means its three state variables will converge to their equilibrium, i.e., the steady-state solution. When reaching the steady-state status, we have

$$P = P_t = P_{t+1} = \cdots = P_\infty,$$
$$\beta = \beta_t = \beta_{t+1} = \cdots = \beta_\infty, \qquad (8\text{-}39)$$
$$\alpha = \alpha_t = \alpha_{t+1} = \cdots = \alpha_\infty,$$

where P, β and α are steady-state coefficients. Substituting (8-39) into (8-38), we have the following equality:

$$\beta^T B(R + B^T PB)^{-1}B^T\beta = \beta^T D^T Q^{-1}D\beta,$$
$$D \overset{\text{def}}{=} A^T - A^T PB(R + B^T PB)^{-1}B^T - I.$$

This equality holds if either $\beta = 0$ or the middle weighting matrices of two sides equal each other, i.e., $B(R + B^T PB)^{-1}B^T = D^T Q^{-1}D$. The equivalency of the two middle matrices is further reduced to an equality condition $Q = P$. Substituting $Q = P$ into the first equation in (8-38), i.e., the discrete-time algebraic Riccati equation, we have the following equality:

$$B(R + B^T QB)^{-1}B^T = Q^{-1}. \qquad (8\text{-}40)$$

Recall the fact that $Q > 0$. If $\dim(u) < \dim(x)$, we have the rank equality: $\text{rank}(Q^{-1}) = \text{rank}(Q) = \dim(x)$. By calculating the matrix ranks on both sides of (8-40), we have the following inequality:

$$\text{rank}(B(R + B^T QB)^{-1}B^T)$$
$$\leq \min\{\text{rank}(B), \text{rank}((R + B^T QB)^{-1})\} \qquad (8\text{-}41)$$
$$\leq \dim(x) = \text{rank}(Q^{-1}).$$

Recall the fact that $R > 0$. If $\dim(x) = \dim(u)$, we have $|R| > 0$. By calculating the determinants on both sides of (8-40), we have the following inequality:

$$|B(R + B^T QB)^{-1}B^T| \;=\; \frac{|B|^2}{|R + B^T QB|}$$
$$\leq \frac{|B|^2}{|R| + |B^T QB|} = \frac{|B|^2}{|R| + |Q||B|^2} \qquad (8\text{-}42)$$
$$< \frac{|B|^2}{|Q||B|^2} = |Q^{-1}|.$$

Here, $|B| \neq 0$ because (A, B) must be controllable. If $|B| = 0$, we also have $|B(R + B^T QB)^{-1}B^T| = 0 < |Q^{-1}|$. The inequalities in (8-41) and (8-42) contradicts (8-40), and therefore, (8-40) cannot be true. Therefore, only the first choice remains, i.e., $\beta = 0$. This contradiction eventually leads to the strong condition that

$$x_t^R = x_{t+1}^R = \cdots = x_\infty^R = 0.$$

The analysis above indicates that zero is the exclusive equilibrium of a linear control system. One can easily conclude that the optimal solution exists if and only if the reference is fixed to zero. In this case, zero equilibrium is said to be a self-harmonized reference. In theory, the LQ tracker with infinitely long horizon does not truly have an optimal solution if its reference is not self-harmonized. The reason is easy to understand: without a self-harmonized reference, the steady-state error is always nonzero, and there is no way to output a bounded value function. Therefore, one can only have an infinite-horizon LQ regulator, and it is unrealistic to design an infinite-horizon LQ tracker.

In engineering practice, one may easily see that many tracking controllers work very well even though they do not truly consider the harmonization of reference trajectories. For example, people may build an "optimal" tracker by viewing its tracking error as the new state to be regulated. Certainly, this kind of design is a mistake in theory and the tracker is not truly optimal. This mistake comes from the negligence of an implicit, but mandatory, condition that this new state must have zero equilibrium. The fix-to-equilibrium condition actually enforces the optimal tracker to degenerate into a standard regulation problem. Therefore, any design about infinite-horizon LQ tracker is just an illusion. In most cases, its tracking ability comes from the behavior of a zero-equilibrium regulator, and moreover this tracking behavior is limited by the hidden harmonization condition.

The hidden condition of zero equilibrium also explains why some finite-horizon trackers are better than their infinite-horizon versions. People might think that for optimal control, the finite-horizon version is always worse than the infinite-horizon version. This belief is based on the understanding that the former is a local optimizer (i.e., short-horizon optimization in MPC), while the latter is a global optimizer (i.e., full-horizon optimization in LQ control). In fact, the mandatory condition on self-harmonized references largely affects the ability of an infinite-horizon tracker to handle a fast varying reference. An infinite-horizon tracker that is theoretically incomplete often has worse optimality than its finite-horizon version.

8.4.5 Sampling Mechanism Considering Reference Dynamics

One key issue of an infinite-horizon tracker is that its optimal policy does not exist if the reference signal is not self-harmonized. Even though this analysis comes from the linear quadratic version, its result applies to general tracking control problems. The phenomenon of nonharmonized references can be commonly seen in engineering but has rarely been discussed in details. Take vehicle longitudinal automation as an example. A car has at least two states in the longitudinal direction: one is its longitudinal position, and the other is its longitudinal speed. The following reference is said to be not self-harmonized if the car is required to stay at a fixed position but still maintain a nonzero speed. Obviously, the two goals conflict with each other. One cannot simultaneously maintain both zero position error and nonzero vehicle speed. In this situation, the utility function will never settle to zero, and the cost function will diverge to infinity.

To address the nonharmonized reference issue, one has to train a quasi-optimal policy with finite-horizon reference inputs, neither only one reference point nor infinitely long reference. The basic idea is to build a splitting cost function as the replacement of the original infinite-horizon criterion. This splitting technique provides high flexibility to balance tracking performance and computational efficiency. To understand its working mechanism, the splitting cost function should be defined and analyzed in the virtual-time domain. The virtual-time horizon at time t is divided into two sections: one is from $0|t$ to $N-1|t$ (with N steps), and the other is from $N|t$ to $\infty|t$. As illustrated in Figure 8.16, the first section keeps the real-time reference to ensure desirable tracking ability, which also serves as the finite-dimensional input to the quasi-optimal policy. How to choose the second section is the key to building a self-harmonized reference. There are two potential options, either keeping the last reference point of the first section or switching to the known equilibrium. Keeping an arbitrary constant reference is obviously not self-harmonized because nonzero stationary error may still exist, and its cost function is unbounded. In contrast, the known equilibrium is a more reasonable choice because its stationary error will go to zero and the finite-horizon tracking policy becomes trainable with previous infinite-horizon ADP.

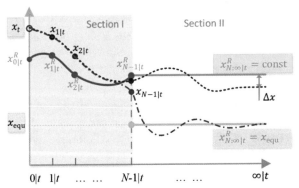

Figure 8.16 Splitting cost function in the virtual-time domain

Thus far, a new splitting reference is built in the virtual-time domain, but one needs to collect samples in the real-time domain. The relationship of reference dynamics and sampling mechanism must be thoroughly disclosed in order to build a theoretically complete ADP algorithm. Let us revisit this new OCP from the perspective of augmented dynamics. In fact, any reference modification (incl. splitting it into two sections) can be viewed as introducing special reference dynamics into the original environment. The reference dynamics is described by a reference prediction model:

$$x_{i+N|t}^{R} = g\left(x_{i|t}^{R}, x_{i+1|t}^{R}, \cdots, x_{i+N-1|t}^{R}\right)$$

where $g(\cdot)$ represents the one-step transition of the reference dynamics. In this model, the future reference signal is exactly determined by N-step historical reference signals. Therefore, a new tracking control problem is built in the form of the criterion-splitting OCP, which is subject to both the original environment model and the reference prediction model. How to predict the future reference can be a quite diversified task. One can choose his own reference dynamics depends on his own preference or practical requirement. For example, when the self-harmonized equilibrium is chosen as the second section in the splitting cost function (Figure 8.16), its reference prediction model is equivalent to

$$g_{t+N} \overset{\text{def}}{=} g\left(x_{i|t}^R, x_{i+1:i+N-1|t}^R\right) = x_{\text{equ}}.$$

where g_{t+N} is the predicted reference point. One can also select proper reference dynamics to build the aforementioned first-point tracker. When $g(\cdot) = x_{i|t}^R$ and $N = 1$ are chosen, the reference prediction model becomes an identity mapping, i.e., $x_{i|t}^R = x_t^R, i \in \mathbb{N}$. This will enforce the criterion-splitting OCP to become a first-point tracker. Under this perspective, one can reconstruct the tracking version of Bellman equation to facilitate the design of criterion-splitting ADP algorithms. Due to the introduction of reference dynamics, the new state-value function reduces to a function of only $(N + 1)$ input elements:

$$V^*\left(x_{0|t}, x_{0:N-1|t}^R\right) \overset{\text{def}}{=} \min_{u_{0:\infty|t}} \sum_{i=0}^{\infty} l\left(x_{i|t}^R - x_{i|t}, u_{i|t}\right)$$

Naturally, the tracking version of Bellman equation becomes

$$V^*\left(x_{0|t}, x_{0:N-1|t}^R\right)$$

$$= \min_{u_{0:\infty|t}} \left\{ l\left(x_{0|t}^R - x_{0|t}, u_{0|t}\right) + \sum_{i=1}^{\infty} l\left(x_{i|t}^R - x_{i|t}, u_{i|t}\right) \right\}$$

$$= \min_{u_{0|t}} \left\{ l\left(x_{0|t}^R - x_{0|t}, u_{0|t}\right) + \min_{u_{1:\infty|t}} \left\{ \sum_{i=1}^{\infty} l\left(x_{i|t}^R - x_{i|t}, u_{i|t}\right) \right\} \right\}$$

$$= \min_{u_{0|t}} \left\{ l\left(x_{0|t}^R - x_{0|t}, u_{0|t}\right) + V^*\left(x_{1|t}, x_{1:N|t}^R\right) \right\}$$

Based on this new Bellman equation, its minimization is performed with respect to the current action but depends on both the current state and N-step reference points. Optimizing this problem naturally yields a finite-horizon policy with the following structure:

$$u_t^* = \pi \underbrace{\left(x_{0|t}, x_{0:N-1|t}^R\right)}_{\text{virtual time}} = \pi \underbrace{\left(x_t, x_{t:t+N-1}^R\right)}_{\text{real time}}. \tag{8-43}$$

This policy, which contains $N + 1$ inputs, is quasi-optimal to the original real-time references. Compared to the full-horizon tracker, it has a finite number of inputs, whose computational complexity is much reduced. Meanwhile, its input has more than one reference point, which should have better tracking ability than the first-point tracker. One might have observed that this finite-horizon policy does not hold the Markov property in the original environment. To maintain the Markov property, we can regard the reference prediction model as an additional part of the augmented environment. Therefore, the augmented state is a combination of states from both original environment dynamics and reference dynamics:

$$\bar{x}_{i|t} = \left[x_{i|t}, \underbrace{x_{i|t}^R, x_{i+1|t}^R, \cdots, x_{i+N-1|t}^R}_{\text{Reference}}\right]^{\text{T}}, i \in \mathbb{N}$$

The augmented environment model is

$$\bar{x}_{i+1|t} = \begin{bmatrix} x_{i+1|t} \\ x_{i+1|t}^R \\ \cdots \\ x_{i+N-1|t}^R \\ x_{i+N|t}^R \end{bmatrix} = \begin{bmatrix} f(x_{i|t}, u_{i|t}) \\ x_{i+1|t}^R \\ \cdots \\ x_{i+N-1|t}^R \\ g(x_{i|t}^R, x_{i+1|t}^R, \cdots, x_{i+N-1|t}^R) \end{bmatrix} = f_{\text{aug}}(\bar{x}_{i|t}, u_{i|t}). \qquad (8\text{-}44)$$

The augmented reward signal remains the same as before:

$$\bar{l}_{i|t} = l(x_{i|t}^R - x_{i|t}, u_{i|t})$$

Fortunately, the new state has the Markov property in the augmented environment. Let us return to the real-time domain to learn how to collect samples to solve this splitting problem. As demonstrated in Figure 8.17, the augmented state is an $(N + 1)$-element vector containing one original state and N reference points. The next augmented state has one special component from the reference prediction model, that is, the last element g_{t+N}. Through the last element in augmented samples, the reference sends its dynamic information to the tracking Bellman equation and eventually becomes an embedded feature of the trained policy. Figure 8.17 shows the connection of samples from the augmented environment and the original environment. The aforementioned sampling mechanism is equivalent to first rolling forward N-step real reference signals in the original environment and then collecting the $(N + 1)$-th step reference point from the reference prediction model rather than from real reference signals.

One sample in augmented env.
$$\bar{x}_t = [x_t, x_t^R, x_{t+1}^R, \cdots, x_{t+N-1}^R]$$
$$\bar{l}_t = l(x_t, x_t^R)$$
$$\bar{x}_{t+1} = [x_{t+1}, x_{t+1}^R, \cdots, x_{t+N-1}^R, g_{t+N}]$$

Look forward N-steps in tracking env.
$$\text{env} = \{x_t, x_{t+1}\}$$
$$\text{ref} = \{x_t^R, x_{t+1}^R, \cdots, x_{t+N-1}^R, g_{t+N}\}$$
$$\text{reward} = l(x_t, x_t^R)$$

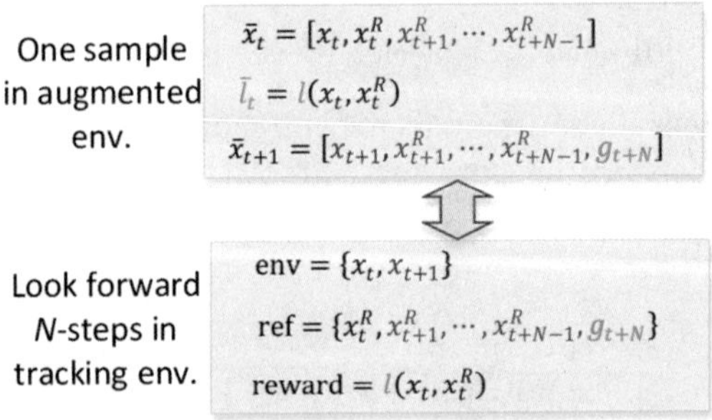

Figure 8.17 Samples in augmented and original environments

Now we know the connection between choosing the virtual-time reference and sampling mechanism in the real environment. The reference prediction model provides us with great flexibility to address the nonharmonized reference issue. Figure 8.18 lists three kinds of real-time references. Reference (a) ends with an arbitrary constant. If this constant is not self-harmonized, its value function has no possibility of being bounded. Both (b) and (c) are periodic signals, and they are not able to output bounded value functions. Their infinite-horizon ADP algorithms do not work very well if one directly samples these real-time references. To achieve self-harmonization, one can convert them into virtual-time references with the aforementioned splitting technique.

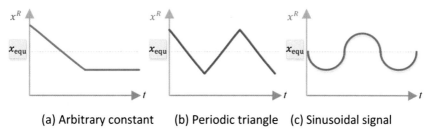

(a) Arbitrary constant   (b) Periodic triangle   (c) Sinusoidal signal

Figure 8.18 Three kinds of real-time references

Figure 8.19 illustrates the sampling mechanism along the virtual-time references. Each virtual-time reference is required to switch to the self-harmonized equilibrium after following exactly N steps of the real-time reference. Hence, a few new, but self-harmonized, reference trajectories are implicitly built due to the introduction of switching behaviors. Together with random initial states and rollouts in the original environment, each virtual-time reference trajectory is able to generate a sequence of augmented samples. Each augmented sample contains $N+1$ reference points and 2 original states. The first sample is demonstrated in Figure 8.19(a), and its successive sample is shown in Figure 8.19(b). Obviously, to achieve sufficient exploration, the initial state needs to start from every possible position. Moreover, the virtual-time references must contain every N-element section of the real-time reference to track each part of the real reference. The information from new reference dynamics will be embedded into the trained policy as sufficient samples are collected.

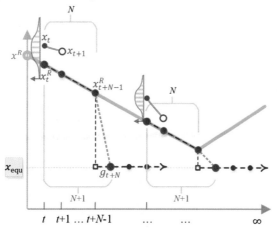

(a) First sample of each virtual-time reference

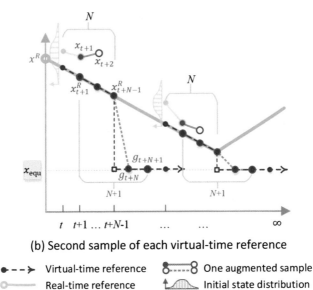

(b) Second sample of each virtual-time reference

●- - ➤ Virtual-time reference   ⎚----⎚ One augmented sample
○—— Real-time reference   Initial state distribution

Figure 8.19 Sampling mechanism for self-harmonized virtual references

What will happen if one insists on using the real-time reference for sample collection? In this situation, we may have either single or multiple reference trajectories. With a single reference trajectory, the solution of tracking OCP can be reduced to a time-dependent policy, whose input does not contain any reference signal. That is because future return is solely determined by the current state and its temporal position. Obviously, its generalizability ability is not satisfactory since only one fixed reference is tracked. With multiple reference trajectories, a certain reference transition distribution is implicitly contained, but this information stays behind the given references. After being sufficiently trained, the hidden information about reference transition distribution will sent into the policy parameters. This policy will have better generalizability since multiple reference trajectories can cover more possible reference distribution. Of course, the prerequisite is that these real-time references are self-harmonized, and their state-value functions can be bounded.

8.5   Methods to Design Finite-Horizon ADPs

When converting an OCP into a dynamic programming problem, we need to be careful about the temporal horizon of state and action variables. In infinite-horizon optimal control, we need to ensure a finite-valued cost function; otherwise, there is no way to build its convergent ADP algorithm. Finite-horizon optimal control is of particular significance in the reduction of algorithm complexity because its value function is easily bounded and its policy gradient derivation obeys a cascading structure. In today's control community, the majority of optimal controllers are implemented in model predictive control (MPC). This approach is particularly popular in process control sectors such as refining and petrochemicals. MPC has several advantages compared with classic linear controllers, including suitability for nonlinear systems without reliance on model linearization, explicit consideration of physical or operational constraints, and fitness for both set-point regulation and reference tracking problems. However, in MPC, an online optimization must be performed at each control step, which leads to a heavy computational burden and a long calculation time. Therefore, the application of conventional MPC is very limited to slow dynamic processes with a large sampling period or low-dimensional systems that are equipped with fast computational devices.

8.5.1   Basis of Finite-Horizon ADP Algorithm

In a continuous-time OCP, one basic approach to obtaining the optimum is to compute its sufficient and necessary optimality condition, i.e., the Hamilton-Jacobi-Bellman (HJB) equation. Unlike the infinite-horizon HJB equation, the finite-horizon version contains a time-dependent value function, and its partial derivative with respect to time is very difficult to address. Hence, this kind of HJB equation often becomes computationally intractable when strong nonlinearities and high-dimensional spaces are encountered. The finite-horizon ADP algorithm defined in the discrete-time domain is more useful in practice. Different from its infinite-horizon counterpart, the utilization of finite-horizon ADP has a few additional properties, including a time-dependent value function, less sensitivity to non-discounted returns, and an intrinsic multistage policy structure.

The problem formulation of finite-horizon OCP still requires two principal features: (1) the system dynamics have the Markov property and (2) the cost function has the separable property. The deterministic state-space model naturally satisfies the Markov property, and the operational cost is separable because its utility function incurs individually at each time step. A general discrete-time OCP with finite horizon is

$$\min_{u_t,\dots,u_{t+N-1}} V(x, X^R) = \sum_{i=0}^{N-1} l(x_{t+i}, x_{t+i}^R, u_{t+i}),$$

(8-45)

subject to

$$x_{t+1} = f(x_t, u_t),$$

where t is the initial time, N is the horizon length, $X^R = x_{t:t+N-1}^R$ is the reference signal, and $V(x, X^R)$ is the cost function under the initial state $x_t = x$. This formulation builds a fixed-horizon OCP, which is in accordance with the common definition of MPC. In this definition, the utility function $l(\cdot, \cdot)$ is not necessarily positive definite because the sum of finite-number reward signals is easily bounded. According to whether the reference exists, finite-horizon OCPs are classified into two common types: (1) finite-horizon optimal regulator and (2) finite-horizon optimal tracker.

(1) In the finite-horizon optimal regulator, the state reference is enforced to be its equilibrium, i.e., $x_{t:t+N-1}^R = x_{equ}$. An optimal regulator is equivalent to the tracker that takes the equilibrium as the reference signal. In many cases, this equilibrium is zero, i.e., $x_{equ} = 0$. Zero equilibrium is common in linear control systems, in which global stability is achievable around the equilibrium.

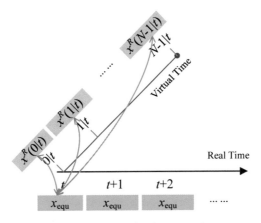

Figure 8.20 Finite-horizon regulator

(2) In the finite-horizon optimal tracker, the reference signal is allowed to vary with time. Here, its cost function must be a function of the initial state and the reference signal. The finite-horizon optimal tracker has two subtypes: first-point tracker and full-horizon tracker. From the RHC perspective, only the first-step reference signal at each virtual-time problem should be tracked when using the first-point tracker, and the varying reference signals in the overall virtual-time horizon need to be tracked when using the full-horizon tracker. The difference between these trackers is illustrated in Figure 8.21.

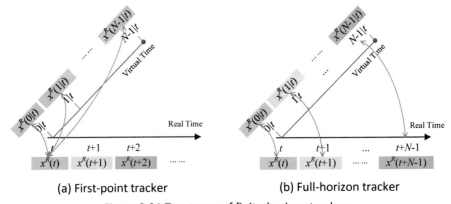

(a) First-point tracker   (b) Full-horizon tracker

Figure 8.21 Two types of finite-horizon trackers

In the finite-horizon problem, the standard Bellman equation no longer exists, and its optimality condition must be replaced with a multistage Bellman equation. Moreover, the number of stages depends on the length of the predictive horizon. Traditionally, the solution of this multistage Bellman equation is computed in a stage-by-stage backward manner, which is also referred to as exact dynamic programming. In finite-horizon ADP, an alternative algorithm family is built to solve this Bellman equation, whose working mechanism comes from first-order optimization and function generalization.

Generalizing the policy with a proper parameterized function is critical to building efficient feedback controllers. The generalization basis can be any linear or nonlinear function, such as a polynomial function, radial basis function, or artificial neural network.

Since optimal action depends on both the current state and future reference signals, a time-dependent structure in tracking OCPs must be selected for the optimal policy.

(1) The policy structure for an optimal regulator is

$$u_{t+i} = \pi(x_t, N - i; \theta), i = 0, 1, \dots, N - 1, \tag{8-46}$$

where $\theta \in \mathbb{R}^{l_\theta}$ is the policy parameter to be optimized. Different from the infinite-horizon version, a finite-horizon regulator must have a time-dependent structure, whose working horizon spans from time t to time $t + N - 1$. In theory, using this time-dependent policy, the cost function can be perfectly minimized in the optimal regulator.

(2) The policy structure for an optimal tracker is

$$u_{t+i} = \pi(x_t, x^R_{t+i:t+N-1}; \theta_i), i = 0, 1, \dots, N - 1, \tag{8-47}$$

where $\theta_0 \in \mathbb{R}^{l_0}, \theta_1 \in \mathbb{R}^{l_1}, \dots, \theta_{N-1} \in \mathbb{R}^{l_{N-1}}$ are the policy parameters to be trained. Due to the existence of varying references, no fixed-parameter policy can carry all potential optimal actions. The policy parameters must change with respect to time to adapt various kinds of reference signals. The optimal parameters can be searched through either value-based ADP or policy-based ADP. In the former, optimal state-value functions are first calculated, and then, the corresponding optimal policy is found. In the latter, no value function must be known; however, the policy parameter is iteratively updated with the gradient descent technique.

8.5.2   Optimal Regulator

The optimal regulation problem, in which the state is forced to remain around a known equilibrium, is quite common in the feedback control community. For example, the cruise control system of a passenger car is a typical regulator problem. The cruise controller continuously adjusts the engine fuel supply according to powertrain states and future road conditions so that the car is able to maintain a user-defined speed. Unlike the infinite-horizon version, the Bellman equation of finite-horizon OCP contains a time-dependent value function. The optimal time-dependent value function is defined as

$$V^*(x, t) = \min_{u_t, u_{t+1}, \dots, u_T} \sum_{i=t}^{T} l(x_i, u_i). \tag{8-48}$$

in which T is the terminal time, and $t \in \{0, 1, 2, \cdots, T\}$ is the starting time. In essence, the definition of (8-48) is different from that in (8-45). In (8-48), a fixed-terminal-time OCP is defined, while in (8-45), a fixed-horizon OCP is defined. In the previous definition, the interval between initial time and terminal time is fixed, i.e., the horizon length is fixed. In this definition, the initial time varies, but the terminal time is fixed. Hence, its horizon length changes with the varying initial time. Although they look structurally different, these two definitions can become equivalent at some special time instants. When $t = 0$, the fixed-terminal-time OCP and fixed-horizon OCP are exactly equivalent to each other. This kind of equivalence in their problem formulations means that their optimal solutions are the same at time $t = 0$, and one can use the optimal action of one problem to replace that of the other problem. In this situation, it is required that

$$T = N - 1.$$

The merit of defining a fixed-terminal-time OCP is that its varying starting time is a time-dependent variable, which allows the multistage Bellman equation to be derived. In contrast, no Bellman equation exists in the fixed-horizon OCP because its horizon length is fixed and there is no variable to carry the recursive structure. Considering the variation of starting time, the Bellman equation of finite-horizon OCP has multiple temporal stages, and the number of temporal stages is equal to the length of the predictive horizon. The multistage Bellman equation is

$$V(x,t) = \min_{u}\{l(x,u) + V(x',t+1)\},$$
$$V(x,t+1) = \min_{u}\{l(x,u) + V(x',t+2)\},$$

$$\dots\ \dots \tag{8-49}$$

$$V(x,T-1) = \min_{u}\{l(x,u) + V(x',T)\},$$
$$V(x,T) = \min_{u}\{l(x,u)\}.$$

One may be interested in knowing how to compute the optimal policy of finite-horizon OCP. One straightforward idea is to parameterize the time-dependent value function as $V(x,t;w_t)$, $V(x,t+1;w_{t+1})$, $V(x,t+2;w_{t+2})$, ..., and $V(x,T;w_T)$ and then apply the value iteration technique to find the optimal value parameters. The optimal policy is obtained by minimizing each optimal state-value function. This kind of value parameterization often leads to super high-dimensional problems and has very low computational efficiency. The finite-number sum of utility functions provides an alternative but more effective approach to update the policy parameter. No state-value function must be explicitly known; however, it relies on the cascading computation of finite-horizon policy gradients, whose derivation utilizes the chain rule of derivative calculation. For the optimal regulator, the updating formula of the policy parameter is

$$\theta \leftarrow \theta - \alpha\frac{\mathrm{d}V(x)}{\mathrm{d}\theta}, \tag{8-50}$$

where α is the updating step and $\mathrm{d}V(x)/\mathrm{d}\theta \in \mathbb{R}^{l_\theta}$ is a deterministic policy gradient. Different from the derivation of multistage Bellman equation, the policy to be optimized must rely on a fixed horizon; otherwise this policy cannot be trained as a receding horizon controller. That is to say the calculation of finite-horizon policy gradient needs to go back to the problem formulation of fixed-horizon OCP. Considering the equivalence of fixed-terminal-time OCP and fixed-horizon OCP at time $t = 0$, one can use the multistage Bellman equation in (8-49) (by assuming its time starts from $t = 0$ and ends at $T = N - 1$) to compute this finite-horizon policy gradient. In this Bellman equation, both initial time and terminal time are fixed, and the horizon length is fixed to be the constant N. For the sake of cascading derivation, we rewrite (8-46) into an equivalent policy structure:

$$u_{t+i} = \pi(x_{t+i}, N - i; \theta), i = 0, 1, \dots, N - 1,$$

where N is the length of the fixed-time horizon. The equivalence of this policy comes from the temporal causality between actions and states. Obviously, a state in the future is only determined by its precedent actions, but a precedent optimal action has no connection with any future states. Since the overall cost function is equal to the summation of separable utility functions, one easily has

$$\frac{dV(x)}{d\theta} = \sum_{i=0}^{N-1} \frac{dl(x_{t+i}, u_{t+i})}{d\theta}.$$ (8-51)

The first-order derivative of each utility function $dl/d\theta$ is calculated with a directed acyclic graph. Taking the derivative of $l(x_{t+i}, u_{t+i})$ as an example, its cascading graph of derivative calculation is shown in Figure 8.22. This graph starts from the node of $l(x_{t+i}, u_{t+i})$ and ends at the node of θ. Each arrow represents a specific partial derivation. The derivative calculation must go through all possible paths in this directed acyclic graph to cover all possible derivative paths. Even though the path graph looks complicated, such a derivation can be simply built with a group of recursive formulas.

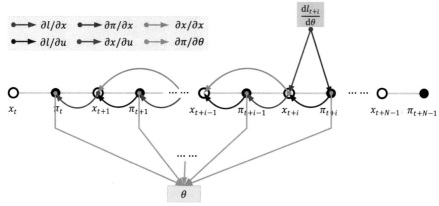

Figure 8.22 Cascading graph of the utility gradient for the optimal regulator

For narrative brevity, we introduce the following simplified notations:

$$l_{t+i} \stackrel{\text{def}}{=} l(x_{t+i}, u_{t+i}),$$
$$f_{t+i} \stackrel{\text{def}}{=} f(x_{t+i}, u_{t+i}),$$
$$\pi_{t+i} \stackrel{\text{def}}{=} \pi(x_{t+i}, N - i; \theta).$$

The structural regularity of directed acyclic graph can help us rewrite (8-51) as a group of recursive formulas to reduce the computational complexity. Hence, the policy gradient of each utility function becomes

$$\frac{dl_{t+i}}{d\theta} = \frac{dx_{t+i}^{T}}{d\theta} \frac{\partial l_{t+i}}{\partial x_{t+i}} + \frac{du_{t+i}^{T}}{d\theta} \frac{\partial l_{t+i}}{\partial u_{t+i}}.$$ (8-52)

Furthermore, we define two immediate variables:

$$\phi_{t+i} \stackrel{\text{def}}{=} \frac{dx_{t+i}^{T}}{d\theta}, \psi_{t+i} \stackrel{\text{def}}{=} \frac{du_{t+i}^{T}}{d\theta} = \frac{d\pi_{t+i}^{T}}{d\theta}.$$

Thus far, we have the following recursive formulas for finite-horizon gradient calculation:

$$\phi_{t+i} = \phi_{t+i-1} \frac{\partial f_{t+i-1}^{T}}{\partial x_{t+i-1}} + \psi_{t+i-1} \frac{\partial f_{t+i-1}^{T}}{\partial u_{t+i-1}},$$

$$\psi_{t+i} = \phi_{t+i} \frac{\partial \pi_{t+i}^{T}}{\partial x_{t+i}} + \frac{\partial \pi_{t+i}^{T}}{\partial \theta},$$ (8-53)

where $\phi_t = 0$ is the initial variable. Combining (8-51), (8-52) and (8-53), the overall policy gradient becomes

$$\frac{dV(x)}{d\theta} = \sum_{i=0}^{N-1} \left(\phi_{t+i} \frac{\partial l_{t+i}}{\partial x_{t+i}} + \psi_{t+i} \frac{\partial l_{t+i}}{\partial u_{t+i}} \right). \tag{8-54}$$

In this method, there is no need to calculate any state-value function, and each separable utility function is used to calculate each components of policy gradients. The advantages of finite-horizon policy gradient come from the finite-number utility function. This is a special advantage of finite-horizon ADP that works in the discrete-time domain. It does not waste any computational resource to estimate a parameterized time-dependent value function. In contrast, continuous-time ADP with finite horizon can be more complicated because its HJB equation is actually a nonlinear PDE due to the existence of time-dependent value function.

Algorithm 8-2: Finite-horizon ADP regulator

Hyperparameters: learning rate α, predictive horizon N, number of environment resets M

Initialization: policy function $\pi(x, \#; \theta)$

Repeat

  (1) Use environment model

  Initialize memory buffer $\mathcal{D} \leftarrow \emptyset$

  Repeat M environment resets

    $x_0 \sim d_{\text{init}}(x)$

    $u_0 = \pi(x_0, N; \theta)$

    $\phi_0 = 0$

    $\psi_0 = \partial \pi^{\mathrm{T}}(x_0, N; \theta)/\partial \theta$

    For i in $1, 2, \ldots, N-1$

      //Rollout with model f and policy π

      $x_i = f(x_{i-1}, u_{i-1})$

      $u_i = \pi(x_i, N-i; \theta)$

      $\text{Calculate} \left(l, V, \dfrac{\partial l}{\partial u}, \dfrac{\partial l}{\partial x}, \dfrac{\partial f^{\mathrm{T}}}{\partial u}, \dfrac{\partial f^{\mathrm{T}}}{\partial x}, \dfrac{\partial \pi^{\mathrm{T}}}{\partial x}, \dfrac{\partial \pi^{\mathrm{T}}}{\partial \theta} \right)_i$

      $\phi_i = \phi_{i-1} \left(\dfrac{\partial f^{\mathrm{T}}}{\partial x} \right)_{i-1} + \psi_{i-1} \left(\dfrac{\partial f^{\mathrm{T}}}{\partial u} \right)_{i-1}$

      $\psi_i = \phi_i \left(\dfrac{\partial \pi^{\mathrm{T}}}{\partial x} \right)_i + \dfrac{\partial \pi^{\mathrm{T}}(x_i, N-i; \theta)}{\partial \theta}$

  End

$$\frac{dV(x_0)}{d\theta} = \sum_{i=0}^{N-1} \left(\phi_i \left(\frac{\partial l}{\partial x}\right)_i + \psi_i \left(\frac{\partial l}{\partial u}\right)_i \right)$$

$$\mathcal{D} \leftarrow \mathcal{D} \cup \left\{ \frac{dV(x_0)}{d\theta} \right\}$$

End

(2) Actor update

$$\nabla_\theta J_{ADP} \leftarrow \frac{1}{M} \sum_{\mathcal{D}} \frac{dV(x_0)}{d\theta}$$

$$\theta \leftarrow \theta - \alpha \cdot \nabla_\theta J_{ADP}$$

End

∎

8.5.3 Optimal Tracker with Multistage Policy

In the optimal tracking problem, the state variables are controlled to follow the desired reference. Instead of maintaining zero equilibrium or a predefined constant, this reference often varies with time. For example, an automated car is required to have excellent path tracking ability. This path is in a local region, starting from the current position and ending at a front position. The appropriate steering operation is applied to guide the vehicle along a given trajectory. Usually, the path tracker is formulated as a finite-horizon OCP, which minimizes the error between the true path and the reference path. Unlike any optimal regulator, a finite-horizon optimal tracker must have a multistage policy. The stage number of this policy is equal to the number of the predictive horizon. Moreover, each stage must have different input dimensions and different parameters. This is because optimal tracking actions are related to both the current state and the remaining reference signals. Therefore, a single fixed policy cannot recommend all possible actions in the predictive horizon. As a structural variant of (8-47), the multistage policy can be written into another form:

$$u_t = \pi(x_t, x_{t:t+N-1}^R; \theta_0),$$
$$u_{t+1} = \pi(x_{t+1}, x_{t+1:t+N-1}^R; \theta_1), \tag{8-55}$$
$$\cdots$$
$$u_{t+N-1} = \pi(x_{t+T-1}, x_{t+N-1}^R; \theta_{N-1}),$$

where $\{\theta_0, \theta_1, \ldots, \theta_{N-1}\}$ are the policy parameters of the N stages. Those policy parameters can either share the same dimension or have their own dimensions. In essence, (8-47) and (8-55) are equivalent to each other. In this new form, the $(t+i)$-th state and remaining reference signals are chosen as the policy inputs, which is beneficial to derive the finite-horizon policy gradient. During training, the $(t+i)$-th state can be calculated by rolling out the environment model from the current state x_t. The structure of the multistage policy is illustrated in Figure 8.23, in which each neural network has different input dimensions and different policy parameters.

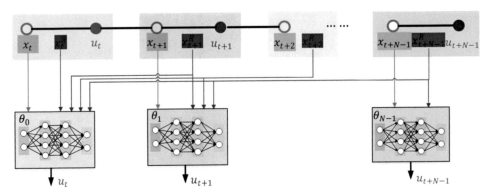

Figure 8.23 Multiple policy structure for the optimal tracker

In the multistage policy, the input dimension of each approximate function, including that of policy parameters, depends on how many remaining reference signals are used. It is easy to see that those stages in the front have more inputs and those stages in the back have fewer inputs. The input dimension gradually decreases as the stage position moves forward. Even though equipped with a multistage policy, the calculation of finite-horizon policy gradient in the optimal tracker is very similar to that of an optimal regulator. The overall policy gradient is equal to the gradient sum of all utility functions, i.e.,

$$\frac{dV(x, X^R)}{d\theta_j} = \sum_{i=0}^{N-1} \frac{dl(x_{t+i}, x_{t+i}^R, u_{t+i})}{d\theta_j}, j \in \{0,1, \cdots, N-1\} . \tag{8-56}$$

Similar to any optimal regulator, the gradient of each utility function is calculated by a directed acyclic graph. Taking the gradient calculation of $l(x_{t+i}, u_{t+i})$ to θ_j as an example, its graph is shown in Figure 8.24. Similar to before, this graph starts from the node of $l(x_{t+i}, u_{t+i})$ and ends at the node of θ_j. Each arrow represents a specific partial derivation. The gradient calculation must go through all possible paths in the directed graph. The number of steps in the multistage policies is equal to the length of the predictive horizon N, and each parameter θ_j $(j = 0,1, ..., N-1)$ represents a single policy for each step.

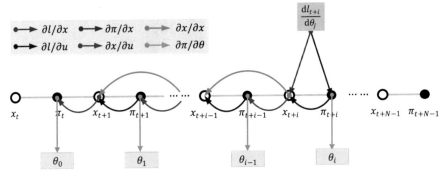

Figure 8.24 Cascading graph of the utility gradient for the optimal tracker

For narrative brevity, the following simplified notations are introduced:

$$l_{t+i}(x_{t+i}^R) \stackrel{\text{def}}{=} l(x_{t+i}, x_{t+i}^R, u_{t+i}),$$
$$f_{t+i} \stackrel{\text{def}}{=} f(x_{t+i}, u_{t+i}),$$
$$\pi_{t+i}(\theta_i) \stackrel{\text{def}}{=} \pi(x_{t+i}, x_{t+i:t+N-1}^R; \theta_i).$$

The symbol $\pi_{t+i}(\theta_i)$ means that this policy is different in both its structure and parameters. The subscript $t + i$ under π means that its policy structure is different with time, while the subscript i under θ means that its parameters are different with time. Hence, the first-order derivative of each utility function becomes

$$\frac{dl(x_{t+i}, x_{t+i}^R, u_{t+i})}{d\theta_j} = \frac{dx_{t+i}^T}{d\theta_j} \frac{\partial l_{t+i}(x_{t+i}^R)}{\partial x_{t+i}} + \frac{du_{t+i}^T}{d\theta_j} \frac{\partial l_{t+i}(x_{t+i}^R)}{\partial u_{t+i}} . \tag{8-57}$$

By defining two intermediate variables:

$$\phi_{t+i}^j = \frac{dx_{t+i}^T}{d\theta_j} = \frac{df_{t+i-1}^T}{d\theta_j}, \psi_{t+i}^j = \frac{du_{t+i}^T}{d\theta_j} = \frac{d\pi_{t+i}^T(\theta_i)}{d\theta_j} .$$

We have the recursive formula to calculate the cascading derivative of each utility function:

$$\phi_{t+i}^j = \phi_{t+i-1}^j \frac{\partial f_{t+i-1}^T}{\partial x_{t+i-1}} + \psi_{t+i-1}^j \frac{\partial f_{t+i-1}^T}{\partial u_{t+i-1}},$$
$$\psi_{t+i}^j = \phi_{t+i}^j \frac{\partial \pi_{t+i}^T(\theta_i)}{\partial x_{t+i}} + \frac{\partial \pi_{t+i}^T(\theta_i)}{\partial \theta_j} , \tag{8-58}$$

in which $\phi_t^j = 0$ is the initial value. In addition, this recursive formula requires the following two conditions to remain consistent with the narrative style in the optimal regulator:

$$\frac{\partial \pi^T(x_{t+i}, x_{t+i:t+N-1}^R; \theta_i)}{\partial \theta_j} = 0, \text{if } i \neq j,$$
$$\phi_{t+i}^j = \frac{dx_{t+i}^T}{d\theta_j} = 0, \text{if } i \leq j.$$

Therefore, we can simplify the overall policy gradient to be a recursive formula:

$$\frac{dV(x, X^R)}{d\theta_j} = \sum_{i=0}^{N-1} \left(\phi_{t+i}^j \frac{\partial l_{t+i}(x_{t+i}^R)}{\partial x_{t+i}} + \psi_{t+i}^j \frac{\partial l_{t+i}(x_{t+i}^R)}{\partial u_{t+i}} \right). \tag{8-59}$$

The updating rule of each parameter in the multistage policy is

$$\theta_0 \leftarrow \theta_0 - \alpha_0 \frac{dV(x, X^R)}{d\theta_0} ,$$
$$\theta_1 \leftarrow \theta_1 - \alpha_1 \frac{dV(x, X^R)}{d\theta_1} , \tag{8-60}$$
$$\cdots$$
$$\theta_{N-1} \leftarrow \theta_{N-1} - \alpha_{N-1} \frac{dV(x, X^R)}{d\theta_{N-1}} .$$

Note that in most tasks, the parameters in the multistage policy do not have the same dimension and change with the number of reference signals. Therefore, each policy parameter should be handled individually. In addition, the computations of finite-horizon policy gradients at different stages are strongly coupled with each other, which means that all the parameters θ_j ($j = 0,1,\dots,N-1$) must update simultaneously without any spatial or temporal order. Therefore, there is no way to only optimize the first parameter without optimizing others, which is in accordance with our understanding of multivariable optimization. In multivariable optimization, only optimizing one or some variables, but leaving other variables free, is not a realistic technique.

Algorithm 8-3: Finite-horizon ADP tracker

Hyperparameters: learning rate α, predictive horizon N, number of environment resets M

Initialization: policy function $\pi(x, x_{1:N}^R; \theta_0), \pi(x, x_{2:N}^R; \theta_1),\cdots, \pi(x, x_N^R; \theta_{N-1})$

Repeat

    (1) Use environment model

    Initialize memory buffer $\mathcal{D} \leftarrow \emptyset$

    Repeat M environment resets

        $x_0 \sim d_{\text{init}}(x)$

        $u_0 = \pi(x_0, x_{1:N}^R; \theta_0)$

        $\phi_0^j = 0, j = 0,1,\dots N-1$

        $\psi_0^j = \partial \pi^{\text{T}}(x_0, x_{1:N}^R; \theta_0)/\partial \theta_j, j = 0,1,\dots N-1$

        For i in $1,2,\dots,N-1$

            //Rollout model and policy

            $x_i = f(x_{i-1}, u_{i-1})$

            $u_i = \pi(x_i, x_{i+1:N}^R; \theta)$

            $\text{Calculate}\left(l, V, \dfrac{\partial l}{\partial u}, \dfrac{\partial l}{\partial x}, \dfrac{\partial f^{\text{T}}}{\partial u}, \dfrac{\partial f^{\text{T}}}{\partial x}, \dfrac{\partial \pi^{\text{T}}}{\partial x}, \dfrac{\partial \pi^{\text{T}}}{\partial \theta}\right)_i$

            For j in $0,1,\dots,N-1$

$$\phi_i^j = \phi_{i-1}^j \left(\frac{\partial f^{\text{T}}}{\partial x}\right)_{i-1} + \psi_{i-1}^j \left(\frac{\partial f^{\text{T}}}{\partial u}\right)_{i-1}$$

$$\psi_i^j = \phi_i^j \left(\frac{\partial \pi^{\text{T}}}{\partial x}\right)_i + \frac{\partial \pi^{\text{T}}(x_i, x_{i+1:N}^R; \theta_i)}{\partial \theta_j}$$

        End

    End

$$\frac{dV(x_0)}{d\theta_j} = \sum_{i=0}^{N-1}\left(\phi_i^j \left(\frac{\partial l}{\partial x}\right)_i + \psi_i^j \left(\frac{\partial l}{\partial u}\right)_i\right), j = 0,1,\dots N-1$$

$$\mathcal{D} \leftarrow \mathcal{D} \cup \left\{ \frac{dV(x_0)}{d\theta_0}, \frac{dV(x_0)}{d\theta_1}, \cdots, \frac{dV(x_0)}{d\theta_{N-1}} \right\}$$

End

(2) Actor update

For j in $0,1,\ldots,N-1$

$$\nabla_{\theta_j} J_{\text{ADP}} \leftarrow \frac{1}{M} \sum_{\mathcal{D}} \frac{dV(x_0)}{d\theta_j}$$

$$\theta_j \leftarrow \theta_j - \alpha \cdot \nabla_{\theta_j} J_{\text{ADP}}$$

End

End

∎

8.5.4   Optimal Tracker with Recurrent Policy

One interesting feature of optimal tracker is that its multistage policy can be simplified to a simple recurrent function. Different from the multistage policy that uses varying-dimensional parameters, a recurrent function can share the same parameter. Recurrent neural network (RNN) is a popular type of recurrent function that has been widely used to fit the sequential data. Due to its hidden memory flow, it behavior is very different from an MLP. In RNN, the feedback information is taken from previous states to affect the current output. Another distinguishing characteristic of RNNs is that they share the same parameters across each layer of the network. The training of RNN requires the backpropagation through time (BPTT) algorithm, which is slightly different from traditional backpropagation for a normal neural network. Liu et al. (2022) proposed a recurrent model predictive control (RMPC) algorithm to build a finite-horizon optimal tracking controller [8-6]. In this study, the recurrent neural network is applied to approximate the optimal tracking policy. Compared with the multistage policy, the recurrent policy can choose proper cycles of recurrent computation to maximally utilize the given computational resources. As illustrated in Figure 8.25, the recurrent structure allows the remaining references to be fed in a step-by-step manner.

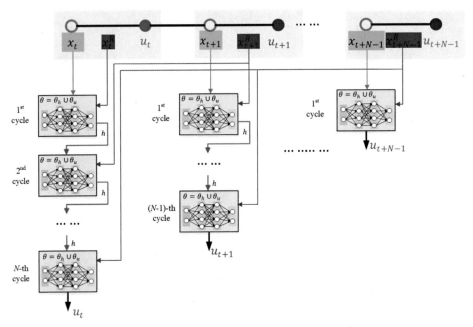

Figure 8.25 Decomposed policy with the recurrent function

With the recurrent function, the policy structure is expressed as

$$u_{t+i} = \pi^{N-i}(x_{t+i}, x_{t+i:t+N-1}^R; \theta),$$
$$i = 0,1, \dots, N-1.$$
(8-61)

More specifically, each recurrent cycle at time step i is mathematically denoted as

$$\pi^c(x_{t+i}, x_{t+i:t+N-1}^R; \theta) = \sigma_u(h_c; \theta_u),$$
$$h_c = \sigma_h(x_{t+i}, x_{t+i+c-1}^R, h_{c-1}; \theta_h),$$
$$c = 0,1, \dots, N-i,$$
(8-62)

where c is the number of recurrent cycles, h_c is the hidden state, and π^c is the recurrent policy at the c-th cycle. All recurrent cycles share the same parameter θ, which is a combination of two unknown parameters $\theta = \theta_h \cup \theta_u$. The notations σ_h and σ_u represent the hidden function and output function, respectively. The policy in (8-55) is equal to the output of the recurrent function at the final cycle. Once its policy gradient is obtained, the parameter can be updated by

$$\begin{bmatrix} \theta_h \\ \theta_u \end{bmatrix} \leftarrow \begin{bmatrix} \theta_h \\ \theta_u \end{bmatrix} - \alpha \frac{dV(x, X^R)}{d\theta},$$
(8-63)

where α is the step size and $dV(x, X^R)/d\theta$ is the policy gradient. It is equal to

$$\frac{dV(x, X^R)}{d\theta} = \sum_{i=0}^{N-1} \frac{dl_{t+i}(x_{t+i}^R)}{d\theta}$$
(8-64)

$$= \sum_{i=0}^{N-1} \left(\frac{dx_{t+i}^{\mathrm{T}}}{d\theta} \frac{\partial l_{t+i}(x_{t+i}^{R})}{\partial x_{t+i}} + \frac{du_{t+i}^{\mathrm{T}}}{d\theta} \frac{\partial l_{t+i}(x_{t+i}^{R})}{\partial u_{t+i}} \right),$$

in which $l_{t+i}(x_{t+i}^{R}) \overset{\text{def}}{=} l(x_{t+i}, x_{t+i}^{R}, u_{t+i})$, $f_{t+i} \overset{\text{def}}{=} f(x_{t+i}, u_{t+i})$, and $\pi^{N-i}(\theta) \overset{\text{def}}{=} \pi^{N-i}(x_{t+i}, x_{t+i:t+N-1}^{R}; \theta)$ are introduced for narrative brevity.

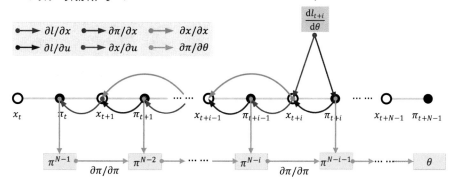

Figure 8.26 Cascading graph for tracker with recurrent function

By defining two intermediate variables:

$$\phi_{t+i} = \frac{dx_{t+i}^{\mathrm{T}}}{\partial \theta} = \frac{df^{\mathrm{T}}(x_{t+i-1}, u_{t+i-1})}{d\theta},$$

$$\psi_{t+i} = \frac{du_{t+i}^{\mathrm{T}}}{d\theta} = \frac{d[\pi^{N-i}(x_{t+i}, x_{t+i:t+N-1}^{R}; \theta)]^{\mathrm{T}}}{d\theta}.$$

Their recursive formula is given as

$$\phi_{t+i} = \phi_{t+i-1} \frac{\partial f_{t+i-1}^{\mathrm{T}}}{\partial x_{t+i-1}} + \psi_{t+i-1} \frac{\partial f_{t+i-1}^{\mathrm{T}}}{\partial u_{t+i-1}},$$

$$\psi_{t+i} = \phi_{t+i} \frac{\partial [\pi^{N-i}(\theta)]^{\mathrm{T}}}{\partial x_{t+i}} + \frac{\partial [\pi^{N-i}(\theta)]^{\mathrm{T}}}{\partial \theta},$$

$$\frac{\partial [\pi^{N-i}(\theta)]^{\mathrm{T}}}{\partial \theta} = \frac{\partial [\pi^{1}(\theta)]^{\mathrm{T}}}{\partial \theta} \prod_{k=N-i}^{2} \frac{\partial [\pi^{k}(\theta)]^{\mathrm{T}}}{\partial \pi^{k-1}(\theta)}, \qquad (8\text{-}65)$$

where $\phi_{t} = 0$ represents the initialization. Therefore, (8-64) can be simplified as:

$$\frac{\partial V(x, x^{R})}{\partial \theta} = \sum_{i=0}^{N-1} \left(\phi_{t+i} \frac{\partial l_{t+i}(x_{t+i}^{R})}{\partial x_{t+i}} + \psi_{t+i} \frac{\partial l_{t+i}(x_{t+i}^{R})}{\partial u_{t+i}} \right). \qquad (8\text{-}66)$$

Compared with that of the previous multistage policy, the recursive formula of this new policy gradient has an added path that originates from the recurrent function. An interesting property of perfectly trained recurrent policy is that its recurrent cycle can stop at any time step but some level of criterion optimality can be still guaranteed. For example, if the recurrent policy stops after only one cycle, it can output an action that is optimal for one-step predictive horizon. If it stops after two cycles, its outputs are optimal for two-step predictive horizon. Interested readers can refer to [8-6] for mathematical proof. Therefore, the use of recurrent policy provides the ability to automatically adjust the predictive horizon to perform real-time optimal control. The length of predictive horizon depends on how many computational resources are available at current control step. Obviously, the cycle number cannot be extended to infinity because its policy accuracy will drop quickly due to the accumulation of numerical errors.

8.6   Example: Lane Keeping Control on Curved Road

Fatigue and distraction are the most common causes of unintentional drifting from driving lanes. As one of technical solutions, the lane-keeping system (LKS) actively helps human drivers keep their vehicles staying inside the desired lane. By definition, LKS is not a fully autonomous system but a driver assistance system. It uses a camera behind the windshield to monitor the markings of the driving lane and continuously detect any unintentional drifting. If an impending unintentional drift is detected, LKS will activate the steering actuator to aid the driver in keeping the car inside the lane.

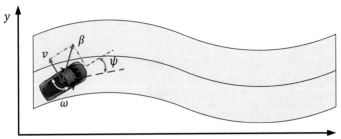

Figure 8.27 Lane-keeping control on a curved road

Lateral motion control is one indispensable functionality in LKS and it is often formulated as an optimal control problem. Basically, it aims to determine a sequence of valid steering commands that make the vehicle track the lane centerline. Its control challenges come from several reasons, such as the nonlinearity of tire dynamics, the complexity of external disturbances and high demand on real-time control. The infinite-horizon ADP algorithm can be applied to first find an optimal policy and then implement it in the feedback loop of lane-keeping control scenario.

8.6.1   Lane Keeping Problem Description

This section mainly discusses how to control the car to track the lane centerline. The car is automatically steered to keep inside a curved road, as shown in Figure 8.27. The vehicle lateral dynamics is described by a bicycle model with 4 states: y − vehicle lateral position at CG; v − vehicle lateral speed at CG; ψ − vehicle heading angle; and ω − yaw rate:

$$x = [y, v, \psi, \omega]^{\mathrm{T}} \in \mathbb{R}^4.$$

The control input is the steering angle of the front wheels:

$$u = \delta \in \mathbb{R}.$$

The vehicle lateral dynamics is described as

$$\dot{x} = f(x, u), \tag{8-67}$$

$$f(x, u) = \begin{bmatrix} u_{long} \sin \psi + v \cos \psi \\ m^{-1}\left(F_{y1} \cos \delta + F_{y2}\right) - u_{long}\omega \\ \omega \\ I_{zz}^{-1}\left(aF_{y1} \cos \delta - bF_{y2}\right) \end{bmatrix}.$$

The longitudinal speed is assumed to be constant, i.e., $u_{long} = 5$ m/s. The tire model and its key parameters are listed in the previous chapter. The lane centerline to be followed is in a sinusoidal shape with spatial frequency $\omega_{Road} = 1/10$ m^{-1} and maximal amplitude $A_{Road} = \pm 1$m. The tracking targets are the desired lateral position y_t^R and desired yaw angle ψ_t^R, which are compressed into the reference x_t^R; i.e.,

$$x_t^R = \begin{bmatrix} y_t^R \\ 0 \\ \psi_t^R \\ 0 \end{bmatrix} = \begin{bmatrix} |A_{Road}| \sin\left(\omega_{Road} x_{long}\right) \\ 0 \\ \arctan\left(|A_{Road}|\omega_{Road} \cos\left(\omega_{Road} x_{long}\right)\right) \\ 0 \end{bmatrix}, \tag{8-68}$$

where x_{long} is the travel distance in the longitudinal direction. The vehicle dynamic model is discretized by backward Euler difference with sampling time $\Delta t = 0.02$ sec. Generally, the shorter the sampling time is, the more accurate the discretized model becomes. However, a very short sampling time can result in a high computational burden. A subtle balance should be achieved between model accuracy and computational complexity. In engineering practice, an empirical rule is to select the sampling time $\Delta t = (1/10 \sim 1/5) \cdot T_{fast}$, where T_{fast} is the rising time of the fastest dynamics in the controlled plant.

8.6.2   Design of ADP Algorithm with Neural Network

Dry asphalt pavement is a very common working condition, and an LKS is usually designed to run on high adhesion coefficient roads. Here, a quadratic performance index is chosen:

$$\min_{u_t, u_{t+1}, \cdots, u_\infty} J(x_t, x_t^R) = \sum_{i=0}^{\infty} \left\{ (x_t^R - x_{t+i})^T Q (x_t^R - x_{t+i}) + u_{t+i}^T R u_{t+i} \right\}. \tag{8-69}$$

Each utility function is quadratic, in which $Q = \text{diag}[0.1, 0, 1, 0]$ and $R = 1$. Clearly, this utility function defines a first-point tracker, in which the virtual-time reference in each predictive horizon remains "constant". Note that the constant value dynamically changes with real time, and accordingly the tracking ability is still kept. From the RHC perspective, all future states in the predictive horizon are enforced to track a fixed point. This kind of tracker is essentially a regulator because one can view the first reference point as a special equilibrium. It is equivalent to regarding the tracking error as a new state in a regulation problem. The corresponding policy is

$$u_{t+i} = \pi(e_{t+i}),$$
$$e_{t+i} \overset{\text{def}}{=} [y_t^R - y_{t+i}, v_{t+i}, \cos(\psi_t^R - \psi_{t+i}), \sin(\psi_t^R - \psi_{t+i}), \omega_{t+i}]^T,$$

where e_t is the tracking error, i.e., the new state in the regulation problem. The policy and value function are both approximated by fully connected neural networks. In the policy network, the hyperbolic tangent function is applied to bound actions in a reasonable range. In the value network, a rectified linear unit is used to prevent the values from decreasing below zero. Both of them have 3 hidden layers, and each hidden layer has 256 neurons. Due to unstable initialization, divergence may easily occur in ADP with one-step look-ahead strategy. Here, a multistep look-ahead strategy is utilized to stabilize the updates of actor and critic.

8.6.3   Training Results of ADP and MPC

For a fair comparison, three finite-horizon MPCs are utilized to compute optimal actions. The length of their predictive horizons is denoted as N_P, in which $N_p = \{20,60,150\}$ are selected as the three controllers. These MPCs are all formulated to track the first reference point at each control moment. The comparison results are shown in Figure 8.28, in which MPC with $N_p = \{20,60,150\}$ is shortened to MPC-20, MPC-60, and MPC-150.

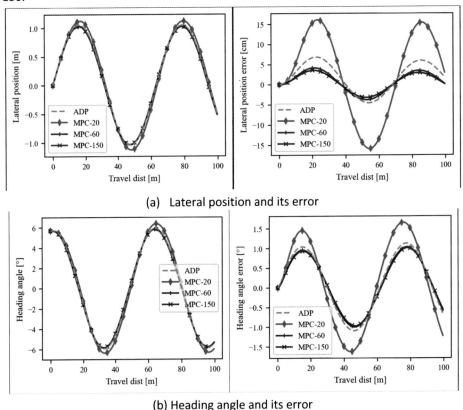

(a)   Lateral position and its error

(b) Heading angle and its error

Figure 8.28 Comparison of states in ADP and MPC

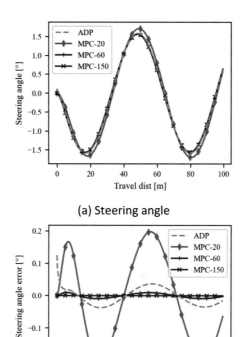

(a) Steering angle

(b) Steering angle error

Figure 8.29 Comparison of control input in ADP and MPC

In this example, MPC-150 is taken as the baseline controller (as an approximation of MPC with $N_p = \infty$) to compare the performance of ADP, MPC-20, and MPC-60. Figure 8.28 shows their real-time trajectories of action and state. The marker points are drawn every 50 points (i.e., every 1 sec) to distinguish different controllers. It can be seen that MPC-60 is much closer to MPC-150 than MPC-20. This is because the shorter the predictive horizon becomes, the more performance loss it will have. In theory, perfect ADP should have the same result as an infinite-horizon first-point tracker (i.e., MPC with $N_p = \infty$). The optimal policy could serve as a replacement of online MPC. In practice, a certain discrepancy must exist in the result of real ADP because of its errors from function approximation and algorithm imperfectness. Table 8-4 shows the relative errors in the real-time domain between ADP and MPCs. In this example, ADP demonstrates good similarity to MPC-150 with a maximum relative error less than 4.0% and a mean relative error less than 1.0% in both states and actions.

Table 8.4 Relative errors in the real-time domain between ADP and MPC-150

	Steering Angle	Lateral Position	Heading Angle
Mean Error	0.74%	0.88%	0.69%
Max Error	3.94%	1.66%	1.18%

The relative errors between ADP and MPC are defined as

$$e_{max} = \max\left\{\frac{|X_{ADP}(t) - X_{MPC}(t)|}{X_{MPC}^{max} - X_{MPC}^{min}}\right\} \times 100\%,$$

$$e_{mean} = \text{mean}\left\{\frac{|X_{ADP}(t) - X_{MPC}(t)|}{X_{MPC}^{max} - X_{MPC}^{min}}\right\} \times 100\%,$$

where X_{ADP} is a specific variable in ADP, X_{MPC} denotes the same variable in MPC, and X_{MPC}^{max}, X_{MPC}^{min} denote its maximum and minimum values in MPC. The two performance indices actually measure the relative errors between ADP and MPC. One may ask why we do not select absolute errors as the performance measure. The absolute errors are more straightforward but may cause some concerns in comparison fairness. Taking the lateral position error as one example, the same 10-centimeter absolute error when tracking a sinusoidal centerline with 1-meter amplitude is quite different from that when tracking a sinusoidal centerline with 10-meter amplitude.

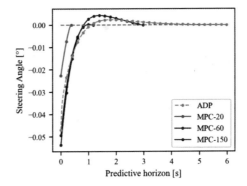

(a) Steering angle in the predictive horizon

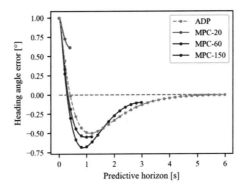

(b) Heading angle error in the predictive horizon

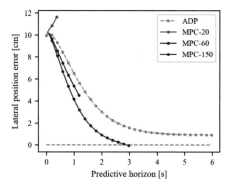

(c) Lateral position error in the predictive horizon

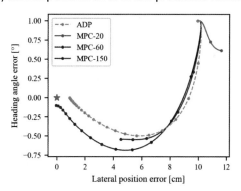

(d) Phase plot in the predictive horizon

Figure 8.30 Comparison of ADP and MPC in the predictive horizon

Table 8.5 Relative errors in the predictive horizon between ADP and MPC-150

	Steering Angle	Lateral Position	Heading Angle
Mean Error	3.91%	1.32%	6.70%
Max Error	20.98%	1.77%	17.50%

Even though high accuracy is obtained in the real-time domain, an in-depth inspection is still necessary in the predictive horizon to examine how far ADP is from MPC. In Figure 8.30, the behaviors of ADP and MPC are compared in the predictive horizon. The dot points are drawn every 10 points (i.e., 0.2 sec) to indicate how far each time step can move. It is observed that high accuracy does not appear again in the predictive horizon. Compared with those in the real-time domain, the state and action of ADP in the predictive horizon have much larger discrepancies from MPC-150. Table 8.5 shows the relative errors of ADP and MPC-150 in the predictive horizon. The steering angle has a maximum relative error about 21% and a mean relative error about 4.0%. Previous studies have pointed out that more accurate ADP algorithms need a few advanced empirical tricks, including deep neural networks, delayed update targets, and varying learning rates.

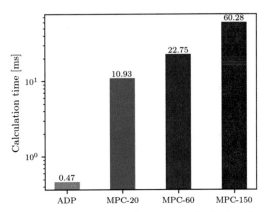

Figure 8.31 Comparison of average calculation time

To compare the online calculation efficiency, we implement ADP and MPCs on the same control unit (Intel Core i7 with 2.80 GHz). More specifically, the implementation of ADP is to deploy its neural network-based policy into the control unit. The policy network has 3 hidden layers with 256 neurons, whose input dimension is 5 (i.e., lateral position error, lateral velocity, cosine of heading angle error, sine of heading angle error, yaw rate), and the output dimension is 1 (i.e., steering angle). The online calculation time of ADP is equal to the forward inference time of the neural network. The Interior Point OPTimizer (IPOPT) is chosen as the numerical solver of three MPCs, and there is only one thread for each online optimizer. IPOPT is an efficient software package for large-scale nonlinear optimization [8-13]. The online calculation time of MPC is determined by the calculation speed of the IPOPT optimizer. Figure 8-30 compares the single-step computational efficiency. It is observed that the calculation time of MPC increases with the length of the predictive horizon, whose average single-step calculation time ranges from 10 ms to 60 ms. The single-step calculation time of ADP is only about 0.47 ms, which is almost one hundred times faster than MPC-150. This result shows that great computational benefit can be achieved when deploying a trained optimal policy as a real-time controller.

8.7   References

[8-1]   Barto A, Anderson C, Sutton R (1982) Synthesis of nonlinear control surfaces by a layered associative search network. Biological Cybernetics 43(3): 175-185

[8-2]   Bellman R, Dreyfus S (1959) Functional approximations and dynamic programming. Mathematics of Computation 13: 247–251

[8-3]   Bertsekas D (2017) Value and policy iterations in optimal control and adaptive dynamic programming. IEEE Trans Neural Network & Learning Syst 28(3): 500-509

[8-4]   Bertsekas D, Tsitsiklis JN (1996) Neuro-dynamic programming. Athena Scientific, Belmont

[8-5]   Hampson S (1983) A neural model of adaptive behavior. Dissertation, University of California

[8-6]   Liu Z, Duan J, Wang W, et al (2022) Recurrent model predictive control: Learning an explicit recurrent controller for nonlinear systems. IEEE Trans Industrial Electronics 69(10): 10437-10446

[8-7]   Modares H, Lewis F, Naghibi-Sistani M (2013) Adaptive optimal control of unknown constrained-Input systems using policy iteration and neural networks. IEEE Trans Neural Networks 24(10): 1513-1525

[8-8]   Murray J, Cox C, Lendaris G, et al (2002) Adaptive dynamic programming. IEEE Trans Syst, Man, and Cybernetics, 32(2): 140-153

[8-9]   Vamvoudakis K, Lewis F, Hudas G (2012) Multi-agent differential graphical games: Online adaptive learning solution for synchronization with optimality. Pergamon, Oxford

[8-10]   Werbos P (1992) Approximate dynamic programming for real-time control and neural modeling. Van Nostrand Reinhold, New York

[8-11]   Werbos, P (1990) Consistency of HDP applied to a simple reinforcement learning problem. Neural Network 3(2): 179-189

[8-12]   Zhang H, Luo Y, Liu D (2009) Neural-network-based near-optimal control for a class of discrete-time affine nonlinear systems with control constraints. IEEE Trans Neural Networks 20(9): 1490-1503

[8-13]   Wächter A, Biegler L (2006) On the implementation of an interior-point filter line-search algorithm for large-scale nonlinear programming. Math Program 106: 25–57

Chapter 9.   State Constraints and Safety Consideration

Research is what I'm doing

when I don't know what I'm doing!

-- Wernher von Braun (1912-1977)

To date, great progress has been made in reinforcement learning (RL) and approximate dynamic programming (ADP), in which a parameterized policy is trained in the whole state space and deployed into closed loop for sequential decision or feedback control. Today, controlling a real-world system with input and state constraints has drawn increasing attention due to practical needs, such as actuator saturation, operating limits, and safety guarantee. In an optimal control system with actuator saturation, the optimal policy without constraint considerations inevitably sacrifices the control performance or even becomes unstable in some conditions. Therefore, equipping RL/ADP with the ability to handle constrained behaviors is of practical significance in both training process and controller implementation.

Compared to the treatment of input constraints, how to deal with state constraints is much more challenging, especially for those problems with hard state constraints. There are several physical reasons why hard state constraints occur, and the constraint violation may cause severe economic loss or fatalities. For instance, the vehicle-to-vehicle distance must be larger than zero in an autonomous driving system, and any violation of this distance bound will result in unwanted collision accidents [9-18]. After formulating an OCP with state constraints, there are three basic constrained RL/ADP methods that can be used to find its optimal policy, including penalty function method, Lagrange multiplier method, and feasible descent direction method. In the penalty function method, a penalty term is added to the cost function to penalize the states that cause constraint violations. In the Lagrange multiplier method, duality theory is used to determine the lower bound of primal problems. In the feasible descent direction method, an updating direction that are both feasible and descent must be found, and the constrained OCP is separated into a series of locally convex optimization subproblems.

It is easy to see that how to handle constrained OCP is analogous to that of static optimization. The constraint management of the former, however, is a more complex and systematic task. The major complexity results from the necessity to maintain the recursive feasibility of the learned policy. The phenomenon of infeasibility occurs when the constrained OCP has no solution due to an overly tight state confinement. In this situation, there is no available policy that is capable of satisfying the strict constraint. The infeasibility issue becomes even more complicated when a useful policy must work in an endlessly feasible region. Therefore, in constrained RL/ADP, an optimal policy and its feasible working region must be simultaneously learned and identified, respectively. In reality, these two subtasks are strongly coupled with one another, which forces us to design a new three-element learning architecture called actor-critic-scenery (ACS). In the ACS architecture, the actor update and critic update still maintain their functionalities, which are policy improvement and policy evaluation, respectively. Meanwhile, the goal

S. E. Li, *Reinforcement Learning for Sequential Decision and Optimal Control*, https://doi.org/10.1007/978-981-19-7784-8_9

of the scenery element is to update a newly added feasibility function to determine an available endlessly feasible region.

To guarantee safety is the ultimate goal of constrained RL/ADP. In this field, ensuring safety is equivalent to enforcing the agent's states to strictly obey the hard constraints. Obviously, safety is more necessary in the real world than a virtual environment, especially when environment exploration in unknown regions is performed. There are two basic modes to train a safe policy that satisfies the hard state constraints. In the first mode, a safe policy is trained offline in a virtual environment and then implemented as a real-world controller. In the second mode, the safe policy is simultaneously trained and deployed in the real-world environment. When a perfect environment model is known, either model-free or model-based ACS algorithm can be designed to train the safe policy and identify its feasible working region. When no perfect model is available, one should rely on real experience to compensate for the model's deficiency. Due to strict safety considerations, one must carefully explore the real-world environment by gradually extending the endlessly feasible region from a known region to an unknown region. Equipped with a special hybrid gradient from both environment model and collected experience, a mixed ACS algorithm has the potential to ensure absolute safety during environment interactions.

9.1   Methods to Handle Input Constraint

The optimal control of nonlinear systems is popular in engineering practice. An input constraint such as actuator saturation naturally creates a nonlinear control problem. There are two basic methods to deal with input constraints, i.e., saturated policy function and penalized utility function. In the former, a saturation function is used to confine the actual output of an unconstrained policy. This method is easy to use and only requires updating along the direction of the saturated policy gradient [9-24]. In the latter, each utility function is augmented with a penalty term to restrict actions outside of the upper and lower boundaries. Both exterior and interior penalty functions can be used to ensure the satisfaction of the input constraints. In particular, the integral penalty function can be specially chosen to obtain a bounded optimal solution, which bridges the gap between saturation policy function method and penalized utility function method [9-1][9-15].

9.1.1   Saturated Policy Function

An unconstrained policy can be equipped with the bounding ability by simply adding a saturation function to its output. The term saturation is often used to describe the flattened behavior at two ends of a mathematical function. A function is referred to as saturated if it approaches constant values as the input variable goes to negative or positive infinity. On the basis of an unconstrained policy, the saturated policy is defined as

$$u = \phi(\pi(x; \theta)), \tag{9-1}$$

where $u \in \mathbb{R}^{m \times 1}$ is the bounded action, $\pi(x; \theta) \in \mathbb{R}^{m \times 1}$ is an unconstrained policy, and $\phi(\cdot)$ is a saturation function with the element-by-element mapping ability. Each element in the action vector is restricted into the unit interval $[-1,1]$. Note that $\phi(\cdot)$ is a one-to-one mapping from the unconstrained action space to the constrained action space. Some popular saturation functions include sigmoid function, hyperbolic tangent function (tanh), inverse tangent function (arctan), and softsign function, which are listed as follows.

(1) Sigmoid function:

$$\phi(z) = \frac{1}{1 + \exp(-z)}.$$

(2) Hyperbolic tangent function (tanh):

$$\phi(z) = \frac{\exp(z) - \exp(-z)}{\exp(z) + \exp(-z)}.$$

(3) Inverse tangent function (arctan):

$$\phi(z) = \tan^{-1}(z).$$

(4) Softsign function:

$$\phi(z) = \frac{z}{1 + |z|}.$$

The saturation function is an element-by-element mapping. Here, a one-dimensional action (i.e., $m = 1$) is taken as an example to demonstrate how to derive the saturated policy gradient. With the chain rule, the saturated policy gradient becomes

$$\frac{\partial u}{\partial \theta} = \frac{\partial \phi(z)}{\partial z} \cdot \frac{\partial \pi}{\partial \theta}.$$

Here, $u \in \mathbb{R}$ is a scalar, $\partial\phi/\partial z$ is a scalar, $\partial u/\partial \theta \in \mathbb{R}^{l_\theta \times 1}$ is a column vector and $\partial \pi/\partial \theta \in \mathbb{R}^{l_\theta \times 1}$ is a column vector, in which l_θ is the dimension of the policy parameter. A high-dimensional action can be treated in a similar element-by-element manner. For a policy that is approximated by an artificial neural network, its output layer can simply take one saturation function as the activation function. This type of saturated output layer is used to compress the neural response into a closed set. One shortcoming is that these activation functions in the outer layer display strong limiting behaviors. Their gradients become very small at the two boundaries, which requires a large amount of iterations to train the saturated policy. The zero-centered activation function such as tanh is often recommended for a neural network-based policy because it is more effective in eliminating the gradient vanishing phenomenon.

9.1.2   Penalized Utility Function

Penalty regularization is a class of constrained optimization methods that are used to convert a constrained optimization into an unconstrained optimization. There are two basic penalty methods: exterior penalty method and interior penalty method. In the first method, the added penalty term becomes effective only after the input constraint is unfortunately violated. Its optimal solution usually has a small constraint violation. Thus, this method does not guarantee a perfect constrained solution and is more suitable for those cases in which a slight violation is acceptable. In the second method, the added penalty term will go to infinity as the action approaches the boundary of the input constraint. The optimal solution is always bounded but is slightly worse than the true constrained optimum. In both methods, an appropriate penalty coefficient must be manually selected to balance the requirements of optimality guarantee and constraint satisfaction. In general, the exterior penalty method can start from any initial point, while the interior penalty method requires an admissible initial guess [9-6][9-9].

Here, a special interior penalty function is introduced to modify each utility function in the overall objective function. The modified utility function has an additive penalty term, which is the product of a penalty function and its penalty coefficient. The new utility function $l_{\mod}(x, u)$ is expressed as

$$l_{\mod}(x, u) = l(x, u) + \rho \cdot \varphi(u), \qquad (9\text{-}2)$$

where $\rho > 0$ is the penalty coefficient and $\varphi(u) \in \mathbb{R}$ is the interior penalty function. This type of new utility function can be used in the search for an admissible controller that keeps its command within the actuator limits. The Bellman equation with the modified utility function becomes

$$V^*(x) = \min_u \{l_{\mod}(x, u) + V^*(x')\}.$$

An interior penalty maintains a hard input constraint, while an exterior penalty has a soft input constraint. Interestingly, the aforementioned saturated policy has a hard input constraint. A subtle choice for the interior penalty function can build the connection between saturated policy function and penalized utility function. As a special case, the following integral function can be selected as an interior penalty:

$$\varphi(u) = \int_0^u \left(\phi^{-1}(v) - \rho^{-1} \frac{\partial l(x, v)}{\partial v} \right) dv, \qquad (9\text{-}3)$$

where $\phi(\cdot)$ is a monotonically increasing saturation function and its inverse function is denoted as $\phi^{-1}(\cdot)$. Here, ϕ^{-1} is still an element-by-element function that maps each action from $[-1, 1]$ to $(-\infty, +\infty)$. The second term in the integrand is designed to compensate for the partial derivative of the utility function. With this integral penalty function, actions will be naturally bounded in the unit interval $[-1, 1]$. Either of the two options, $u \to -1$ or $u \to +1$, will lead to an infinitely large penalty, namely, $\varphi \to +\infty$, which is definitely unacceptable in a minimization problem. The optimal control law is then derived according to the stationary condition of optimality:

$$\frac{\partial V^*(x)}{\partial u} = 0, \text{ when } u = u^*.$$

Therefore, when $u = u^*$, we have

$$\frac{\partial (l_{\mod} + V^*(x'))}{\partial u} = \rho \phi^{-1}(u) + \frac{\partial f^{\mathrm{T}}}{\partial u} \frac{\partial V^*(x')}{\partial x'} = 0.$$

Finally, the optimal action must be constrained by a predefined saturation function:

$$u^* = \phi \left(-\frac{1}{\rho} \frac{\partial f^{\mathrm{T}}}{\partial u} \frac{\partial V^*(x')}{\partial x'} \right).$$

Therefore, by designing an interior penalty function in the form of (9-3), a previously free policy will be constrained by a saturation function. In other words, the saturated policy is equivalent to adding an integral interior penalty to the utility function. The added term describes the errors between the immediate reward signal and the integral of the inverse saturation function. This property is a mathematical bridge that connects the saturated policy function and penalized utility function.

9.2   Relation of State Constraint and Feasibility

Satisfying state constraints is a more complex task in sequential decision and feedback control than handling input constraints. Two critical questions need to be answered before any admissible policy is found: (1) how to formulate an OCP with state constraints and (2) how to design a constrained reinforcement learner. The first question appears to be simple. However, it is intrinsically complex due to the conversion from real-world constraints to virtual-time constraints. The same real-world constraint can be reshaped into various virtual-time constraints, as well as their corresponding constrained OCPs, and their optimal policies have very different feasibility properties. If the virtual-time constraint is not properly reshaped, there may be no satisfactory policy, regardless of the chosen optimization technique. Early attempts at reshaping constraint can be traced back to the studies of model predictive control (MPC). Even though it is not explicitly stated, the problem formulation of MPC is filled with various kinds of constraint reshaping tricks. A well-known example is to only pose constraints on some steps in the prediction horizon, while the remaining steps are left to be free [9-29][9-31]. Another reshaping example is the control barrier function (CBF), which is developed from set invariance theory. Usually, CBF needs to confine only a few predictive steps, thus it reduces the high computational burden in long-horizon constraints [9-3][9-4]. Although its constraint length is short, CBF often has a strong ability in avoiding potential constraint violation [9-32]. As for the second question, constrained optimization must be adjusted to match reinforcement learners, which may be either model-free or model-based. Constrained RL/ADP algorithms are roughly classified into three categories: (1) penalty function method; (2) Lagrange multiplier method; and (3) feasible descent direction (FDD) method. The early study of constrained RL dates back to the study by Gaskett (2003), who used a penalty function to solve the cliff-walking problem and stop the walker from falling off the cliff [9-13]. Recently, the gradient projection technique has been widely used to handle safe constraints, in which an unsafe action will be projected back into the safe set that is defined in the local constraint [9-41]. In addition to gradient projection, there has been increasing attention on the Lagrange multiplier method, in which a minimax optimization problem is solved using a dual ascent updating mechanism [9-12].

Compared with input constraints, the management of state constraints is a more systematic task because of its intrinsic entanglement with the infeasibility phenomenon. In engineering, state constraints may arise from several reasons, such as safety concerns, physical boundaries, and performance limitations. Most real-world decision or control tasks are continuing, and its state constraints must be satisfied from the current time to infinity. Without loss of generality, the state constraint in the real-time domain is typically expressed as

$$h(x_{t+i}) \leq 0, i = 1, 2, \dots, \infty, \tag{9-4}$$

where $h(\cdot) \in \mathbb{R}$ is the constraint function. In continuing tasks, the real-world constraint has infinite number of scalar inequalities. In engineering practice, many state constraints are in the form of a vector function rather than a scalar function. In mathematics, a vector constraint can be converted to a scalar constraint by introducing either summation or maximization to integrate all the elements into one scalar value. The vector constraint function can be directly managed, but one must be careful in dealing with its matrix-level differentiation. As an analogy to the relationship between reward signal and utility function, the output of a scalar constraint function is called the risk signal:

$$c_t = h(x_t).$$

The risk signal $c_t \in \mathbb{R}$ indicates how far the state deviates from its constraint boundaries. The constraint function and its risk signal build a risk-awareness system, which is very similar to a reward system. The risk signal is analogous to the reward signal, while the constraint function is analogous to the utility function. Many constrained RL/ADP actually regards the risk signal as a negative part of the reward system. This idea can greatly simplify the algorithm design, but it lacks theoretical guarantee of constraint satisfaction. In this kind of algorithm design, a certain level of performance loss must be sacrificed because one has to compromise between optimality pursuit and constraint satisfaction. In addition, the constraint function may either be known or unknown in an actual environment. If $h(\cdot)$ is unknown, the risk signal must be observable and needs to be collected from repeated environment interactions. If $h(\cdot)$ has a known explicit form, observing the risk signal is not necessary, and the risk level can be calculated from the observed state and action. In this situation, the differentiability of constraint function provides a few additional benefits in model-based ADP, such as better training stability and higher computational efficiency.

9.2.1   Understanding State Constraint in Two Domains

The complexity of controlling a physical system arises from the conflict between the pursuit of optimal criteria and the guarantee of constraint satisfaction. General analytical tools, such as the calculus of variations and Pontryagin minimum principle, are not suitable for optimal control tasks with state constraints. Some controllers such as MPC allow us to handle state constraints in the way of online optimization. However, if there is no subtle method for accelerating, their implementation is very limited due to the well-known high computational burden issue. Compared to MPC, a parameterized policy trained from model-free RL or model-based ADP can serve as a more efficient controller. Usually, its application is featured with a two-step procedure: first, the parameterized policy is trained offline, and then, the policy is implemented as an online controller.

To explain how to reshape real-world constraints, it is necessary to recall the working mechanism of receding horizon control (RHC). In essence, the RHC scheme has two temporal domains, i.e., virtual-time domain and real-time domain. The trained policy is implemented in the real-time domain, while the offline trainer is deployed in the virtual-time domain. In fact, constrained OCPs for RL/ADP are always defined in the virtual-time domain, and their state constraints come from the real-world requirements. The key of reshaping constraints in virtual-time OCPs is that its associated optimal policy must satisfy the real-time constraint. A traditional but very popular strategy is known as the pointwise state constraint, in which (9-4) is rewritten as a series of constraints at every virtual-time step:

$$h(x_{i|t}) \leq 0, i = 1,2, \dots, \infty. \tag{9-5}$$

Note that although looking similar in structure, (9-4) and (9-5) are not defined in the same time domain. The former is defined in the real-time domain, and the latter is defined in the virtual-time domain. Their temporal domains are distinguished by their subscripts, of which $t + i$ represents the $(t + i)$-th step in the real-time domain, and $i|t$ represents the i-th point in the virtual-time domain starting from time t.

Before going forward, we need to clarify an interesting observation. The continuing control tasks have no termination, and their state constraints as the one in (9-4) should be satisfied from the current time to infinity. However, a virtual-time constraint such as the one in (9-5) does not need to be posed at every step in the virtual-time domain. That is to say, except at some special time instances, the real-time domain and virtual-time domain can have completely different state constraints. For example, the pointwise constraint (9-5) is not an exclusive choice in the problem formulation of virtual-time OCPs. This domain separating perspective provides us with great flexibility to build various virtual-time state constraints, as well as their constrained OCPs.

In the RHC procedure, it is observed that only the first action $u^*_{0|t}$ is implemented in the real-time domain, and the remainder of optimal actions are useless in real-time control. Given the first optimal action, the next state x_{t+1} is a one-step transition from the current state, and it must satisfy the real-time constraint. Hence, only the next virtual-time state, i.e., $x_{1|t}$, needs to be confined in the OCP formulation. This observation leads to the following virtual-time constraint:

$$h(x_{1|t}) = h(x_{t+1}) \leq 0, \tag{9-6}$$

which is called the simplest constraint. This concept of the simplest constraint is illustrated in Figure 9.1. In this formulation, the remaining virtual-time state constraints $h(x_{i|t}) \leq 0, i = 2,3, \dots, \infty$, do not necessarily hold since avoiding their corresponding constraint violations is not the goal of the feedback controller in the real-time domain.

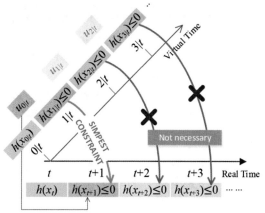

Figure 9.1 State constraints in receding horizon control

Consider the analysis above, we can define the simplest constrained OCP in the virtual-time domain (starting from real time t):

$$\min_{u} V(x) = \sum_{i=0}^{N-1} l\left(x_{i|t}, u_{i|t}\right),$$

subject to

$$x_{i+1|t} = f\left(x_{i|t}, u_{i|t}\right), \qquad\qquad (9\text{-}7)$$

$$x \overset{\text{def}}{=} x_{0|t} = x_t$$

with the simplest constraint:

$$h\left(x_{1|t}\right) \le 0,$$

where $u \in \mathbb{R}^m$ is the action, $x \in \mathbb{R}^n$ is the state, $f(\cdot,\cdot)$ is the environment model in the discrete-time domain, and N represents the horizon length of cost function. When $N < \infty$, (9-7) is a finite-horizon problem in the virtual-time domain, and when $N = \infty$, (9-7) becomes an infinite-horizon problem in the virtual-time domain. This is the most concise reshaping strategy, and it allows to train a useful policy, i.e., the one that holds the ability of minimal constraint satisfaction. The simplest constraint poses the least limitation on criterion optimization; thus, its policy is the closest candidate to the true optimum that originates from the corresponding unconstrained OCP. In addition to its structural conciseness, the simplest constraint has a few additional benefits, such as low computational burden, since most inequalities are removed from the virtual-time OCP.

9.2.2   Definitions of Infeasibility and Feasibility

Since the simplest constraint is both structurally concise and computationally efficient, one may question why such an ideal constraint is rarely seen in engineering practice. To answer this question, we need to formally introduce an important concept: "infeasibility". Infeasibility is an intrinsic property of OCP with state constraints. Formally, "infeasibility" describes the phenomenon that a constrained OCP in the virtual-time domain has no feasible solution, i.e., there is no solution to satisfy the virtual-time state constraint. Usually, the simplest constraint has the worst ability to guarantee feasibility in the long run because it does not provide sufficient confinement to the future state trajectory. The lack of future feasibility will largely limit the usefulness of the learned policy. This is the key reason why the simplest constraint is not popular in practice. To better explain how the infeasibility phenomenon occurs, we define X_{Cstr} as the set of constrained states:

$$X_{\text{Cstr}} \overset{\text{def}}{=} \{x | h(x) \le 0\}.$$

At each time, the simplest constraint can only enforce a single state $x_{1|t}$ to stay inside X_{Cstr}, while a multistep pointwise constraint can confine more points in the virtual-time domain. Figure 9.2 shows an infeasible RHC procedure with a 4-step pointwise constraint. Here, the environment model is assumed to be perfect, i.e., the action and state trajectories are the same in the virtual-time and real-time domains.

At time t and time $t + 1$, there exists a sequence of optimal actions to ensure that the virtual-time state constraint is satisfied, i.e., $x_{1|t}, \dots, x_{4|t} \in X_{\text{Cstr}}$, and $x_{1|t+1}, \dots, x_{4|t+1} \in X_{\text{Cstr}}$. Since the model is perfect, the first optimal action can transfer the current state to a new state, which is still inside X_{Cstr} in the real-time domain. At time $t + 2$, we have $x_{1|t+2}, x_{2|t+2}, x_{3|t+2} \in X_{\text{Cstr}}$, but $x_{4|t+2} \notin X_{\text{Cstr}}$. Therefore, there is no feasible solution because any action will move the last virtual-time state outside of X_{Cstr}. This is how the infeasibility issue occurs. More precisely, the virtual-time states $x_{1:4|t+2}$ do not actually exist since there is no solution at time $t + 2$. This example explains what the infeasibility issue is and how it occurs. We have observed that infeasibility means that the virtual-time OCP has no solution rather than that there is no admissible action in the real-time domain. For instance, at time $t + 2$, an admissible action is still available because the resulting next state is still inside the constrained region. However, the associated virtual-time OCP has no feasible solution because it has a 4-step prediction horizon, rather than only the first step. Briefly, this infeasible behavior attributes to the infeasibility of the used optimizer rather than that of the feedback controller.

Let us qualitatively analyze the relation between state constraint and optimizer feasibility. Generally, for the pointwise constraints, the more state points are constrained in the virtual-time domain, the more feasible the optimal policy will become. For example, there is no limit on the future virtual states in a one-step constraint. The successive states are more likely to deviate from the constraint boundary, and in return long-term infeasibility occurs more easily. In contrast, a larger number of future virtual states are limited in a multistep state constraint, which increases the penalty of the out-of-boundary behaviors in the future. It is helpful to decrease the occurrence probability of infeasible behaviors. However, confining more state points will result in more sacrifice in the policy optimality. Therefore, one must properly reshape the virtual-time state constraints to strike a balance between optimality sacrifice and feasibility guarantee.

Infeasibility has been a long-standing challenge in constrained optimal control, including model predictive control (MPC). In early MPC studies, a few strategies, including horizon enlargement and constraint relaxation, were developed to eliminate the infeasibility issue. The constraint relaxation can be implemented to convert some hard state constraints into soft state constraints. Nevertheless, it is not easy to decide which hard constraint can be dealt with and how much it should be relaxed. Even worse, some hard constraints can never be relaxed if they are determined by the physical system itself. Enlarging the prediction horizon is an effective approach to reduce infeasibility, but it will introduce more unknown variables into the optimization problem. The more unknown variables there are, the higher computational burden is posed into receding horizon optimization.

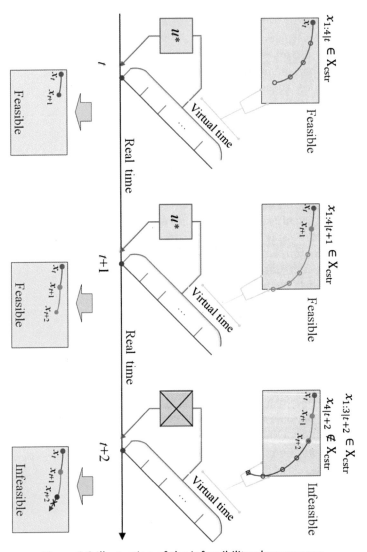

Figure 9.2 Illustration of the infeasibility phenomenon

The above infeasibility analysis helps us to understand some misleading puzzles in finite-horizon MPC. It is well known that short-horizon MPC could easily lose the closed-loop stability. By posing some stability conditions, including terminal state constraint [9-37] and terminal penalty constraint [9-11], the closed-loop stability can be enhanced in short-horizon MPC. In the literature, the disadvantages of stability enhancement are rarely discussed, and most stability proofs are based on the assumption that receding horizon optimization is mysteriously feasible. This assumption, however, is not always realistic. The truth is that the stability condition adds more state constraints, and in most cases it not only sacrifices the control optimality but also reduces the ability to guarantee feasibility. Worse still, complete feasibility may lose if the stability conditions are too tight, and no feasible solution is found. In other words, stability enhancement is not a free lunch and it is at the cost of optimality sacrifice and feasibility decrease.

9.2.3 Constraint Types and Feasibility Analysis

The domain separation perspective from RHC gives us high flexibility to reshape various virtual-time constraints. The most popular choices are pointwise constraints and barrier constraints. Let us elaborate on how these constraints affect the occurrence of infeasible behaviors. A multistep pointwise constraint in the virtual-time domain is

$$h(x_{i|t}) \leq 0, \qquad \forall i = 1, 2, \dots, n,$$

and a multistep barrier constraint in the virtual-time domain is

$$\left(h(x_{i+1|t}) - h(x_{i|t})\right) + \lambda h(x_{i|t}) \leq 0, \qquad \forall i = 1, 2, \dots, n,$$

where n denotes the length of virtual-time constraint, i.e., the number of scalar inequalities. Clearly, the state constraints above are defined in the virtual-time domain. Traditionally, we do not distinguish the differences between virtual-time domain and real-time domain. The virtual-time subscript $x_{i|t}$ is replaced with the real-time subscript x_{t+i} even in most sections of this book. Nevertheless, readers must keep in mind that constrained OCPs to be optimized must be defined in the virtual-time domain rather than the real-time domain.

9.2.3.1 Type I: pointwise constraints

A pointwise constraint describes how the virtual-time state is individually constrained at each time step. Depending on the inequality relationship between n and N, it has two basic variants: (a) short-horizon pointwise constraint $(n < N)$ and (b) full-horizon pointwise constraint $(n = N)$. The simplest constraint can be viewed as a special case of short-horizon constraint in the case of $n = 1$. The full-horizon pointwise constraint further contains a finite-horizon version $(N < \infty)$ and an infinite-horizon version $(N = \infty)$. The finite-horizon version has a finite number of constrained inequalities, while the infinite-horizon version has an infinite number of inequalities. These versions are suitable for finite-horizon OCP and infinite-horizon OCP, respectively:

Finite pointwise constraint (full-horizon):

$$h(x_{i|t}) \leq 0, \qquad \forall i = 1, 2, \dots, N.$$

Infinite pointwise constraint (full-horizon):

$$h(x_{i|t}) \leq 0, \qquad \forall i = 1, 2, \dots, \infty.$$

(9-8)

Both constraints can lead to well-known infeasibility issues. They are called full-horizon constraints because the number of virtual-time inequalities is equal to the length of real-time problem horizon (i.e., that of cost function). An interesting observation is that in a reinforcement learner, the occurrence of infeasible optimization depends on where the state position is located. In the state space, some state points are feasible, and others are infeasible. A state is said to be feasible if there exists at least one solution for the constrained OCP at this state point. Based on this definition, the collection of all feasible states is called the "feasible region". Obviously, the state that is outside of X_{Cstr} cannot be feasible, and a feasible region must be a subset of X_{Cstr}. Simply defining a feasible region is not enough to understand the complexity of infeasible behaviors. We need to classify two more useful concepts: initially feasible region (IFR, denoted as X_{Init}) and endlessly feasible region (EFR, denoted as X_{Edls}).

- Definition 9.1: An initially feasible region (IFR) represents a region in which all the states are feasible. There is no guarantee that their successive states are also feasible.
- Definition 9.2: An endlessly feasible region (EFR) represents a region in which there exists a policy to ensure that both the initial states in this region and their successive states are feasible.

Table 9.1 Three kinds of virtual-time OCPs

| | | Cost function | |
		Infinite horizon	Finite horizon
Constraint	Infinite length	$n = N = \infty$	--
	Finite length	$n < N = \infty$	$n \leq N < \infty$

According to the lengths of cost function and state constraint, one can classify virtual-time OCPs into three categories, as listed in Table 9.1. In the following context, we will mainly discuss two special problem definitions: one is the finite-horizon cost function (i.e., $n = N < \infty$), and the other is the infinite-horizon cost function (i.e., $n = N = \infty$). Both are assumed to have full-horizon pointwise constraints. For the sake of narrative conciseness, the former is called "finite-horizon OCP", and the latter is called "infinite-horizon OCP". If not specially pointed out, we always assume to have a full-horizon pointwise constraint whenever talking about the virtual-time OCP.

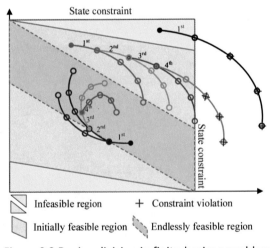

Figure 9.3 Region division in finite-horizon problems

For an arbitrary continuing task, both infinite-horizon OCP and finite-horizon OCP can be formulated in the virtual-time domain to train useful policies. The two kinds of feasible regions, i.e., IFR and EFR, are intrinsic properties of constrained OCPs and their existence is independent of the used RL/ADP algorithm. It is easy to see that for a finite-horizon

OCP, the IFR and EFR are exclusively determined by the environment dynamics and state constraints. Figure 9.3 illustrates the relation between IFR and EFR in finite-horizon OCPs. It is easy to see that EFR must be a subset of IFR; i.e.,

$$X_{\mathrm{Edls}} \subseteq X_{\mathrm{Init}} \subseteq X_{\mathrm{Cstr}}.$$

Ideally, one should train a policy that works in the region of state constraint. Of course, this dream cannot come true due to the infeasibility issue in continuing tasks. If a state stays in the IFR but not the EFR, there is no policy to guarantee that its successive states are feasible. Therefore, a useful policy must work in the endlessly feasible region. The trained policy cannot simply be deployed into real world without specifying which state region it works in.

One interesting observation is that for an infinite-horizon OCP, EFR is strictly equivalent to IFR (Figure 9.4). In this situation, X_{Cstr} contains only two subregions: an infeasible region and an endlessly feasible region. The latter can be simply referred to as feasible region. In this case, EFR becomes an intrinsic property of constrained OCP that only depends on two elements, i.e., environment dynamics and its state constraints. That is to say it has no connection with the optimal criterion. This conclusion can be easily deduced by contradiction. As a result of this special property, the difficulty of analyzing the feasibility behaviors is remarkably reduced in infinite-horizon OCPs.

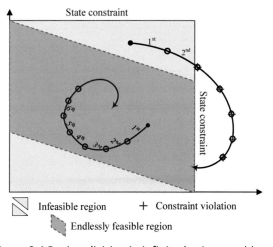

Figure 9.4 Region division in infinite-horizon problems

In today's MPC field, an EFR is always be calculated before the implementation of MPC since the stability and recursive feasibility are not ensured by the mechanism of receding horizon control. Besides, accurately knowing the EFR is of particular significance in constrained RL/ADP algorithms, especially for continuing tasks. The learned policy is physically meaningful only when it works in an EFR. Controlling a state in the infeasible region and IFR (but not in EFR) will inevitably lead to infeasible optimization immediately or in the long run. In addition, to search for a useful policy, the reinforcement learner must sufficiently explore the state space. Any exploration starting from infeasible points will cause severe safety issues. Therefore, a practical constrained RL/ADP algorithm should simultaneously output an EFR and an optimal policy that is defined in this region.

Obviously, our aim is to find the largest EFR. The reason is easy to understand: the larger the EFR is, the wider region an RL agent can safely work in. The first question might be which virtual-time constraint has the largest EFR among all the available constraints. The answer is the full-horizon pointwise constraint because its feasibility examination is based on the first-hand information without any discount on the constraint satisfaction.

- Theorem 9.1: For an infinite-horizon OCP, the full-horizon pointwise constraint has the largest EFR among all the available virtual-time constraints.

Proof:

With a perfect environment model, the optimal control sequences in real-time domain and virtual-time domain are identical in the infinite-horizon OCP. This identical property attributes to the trivial structure of full-horizon pointwise constraint. The EFR of this constraint is denoted as X_{Edls}^{PW}, which holds in the real-time domain. We assume that a new design of virtual-time constraint exists in the virtual-time OCP, which can introduce a new EFR $X_{new} \not\subset X_{Edls}^{PW}$ in the real-time domain.

According to the definition of EFR, for an arbitrary initial state $x \in X_{new}$, there exists a policy π in the real-time domain that ensures all the successive states are feasible. Due to the identical property, the virtual-time OCP at the initial state x is also feasible. There exists at least one solution π to satisfy the full-horizon pointwise constraint in the virtual-time domain. This analysis in return means that $x \in X_{Edls}^{PW}$, i.e., $X_{new} \subseteq X_{Edls}^{PW}$. This brings a contradiction to our assumption.

■

A constrained RL/ADP algorithm should fulfill two central tasks: one is to find an optimal policy, and the other is to identify its corresponding feasible region. In the latter, the largest one is preferred. The largest EFR does not correspond to a unique policy, and a variety of admissible policies can work in the largest EFR. A short explanation is that various actions can stabilize the environment dynamics. One can arbitrarily select any stable actions to build an admissible policy. A variety of different stable policies can be built, but they all work in the same EFR. Among all the stable policies, the one that minimizes the criterion is called the optimal policy. This optimal policy has the property of recursive feasibility only in the endlessly feasible region. Theorem 9.1 also hints us a fact that the recursive feasibility is a kind of ability defined in the real-time domain, rather than in the virtual-time domain. In practice, the largest EFR is not always accessible because one may not choose the full-horizon pointwise constraint in the problem formulation. A subset of the largest EFR, whose virtual-time constraint may introduce a few other benefits like high computational efficiency, is commonly seen in practical algorithm design. To avoid misunderstanding in the following context, we always refer to X_{Edls} as the largest EFR and X as one of its subsets that still holds the recursive feasibility property.

9.2.3.2   Type II: barrier constraints

Usually, one wants to choose a virtual-time constraint that can enlarge the region of recursive feasibility as much as possible. In theory, the full-horizon pointwise constraint has the largest endlessly feasible region (EFR). However, it has too many inequality constraints, which will lead to a very high computational burden. The simplest pointwise constraint has only one inequality, but it considerably shrinks the size of feasible region. As a better choice, the barrier constraint can provide a balance between feasibility enlargement and computational complexity. The multistep barrier constraint is defined as

$$B\left(x_{i|t}, x_{i+1|t}\right) \overset{\text{def}}{=} \left(h\left(x_{i+1|t}\right) - h\left(x_{i|t}\right)\right) + \lambda h(x_{i|t}) \leq 0,$$
$$\forall i = 0, 1, \dots, n - 1,$$

where $B(\cdot, \cdot)$ is called the barrier function, λ is called the barrier parameter, and n is the number of scalar inequalities in the virtual-time barrier constraint. Usually, $n \ll N/\infty$ is selected to reduce the number of inequalities, as well as the computational burden [9-25][9-27]. The barrier function actually defines a series of cascading inequalities. The property of the barrier constraint is analyzed in [9-22], and the cascading condition is given as follows:

$$h\left(x_{i|t}\right) \leq (1 - \lambda)^i h\left(x_{0|t}\right).$$

The selection of barrier parameter λ is critical to its constraint strength, i.e., how strongly it can prevent the controlled state from violating the constraints. Figure 9.5 shows the upper bounds of a few barrier conditions. When $\lambda \in (0,1)$, the barrier constraint has a stronger confining ability than the pointwise constraint. In this situation, the future constraint function is smaller than its predecessor with an exponentially decreasing rate. When $\lambda = 1$, it reduces to a standard pointwise constraint. The barrier constraint becomes too strict when $\lambda \in (-\infty, 0]$, and it becomes a physically meaningless constraint when $\lambda \in (1, +\infty)$.

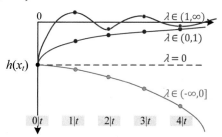

Figure 9.5 Influence of the barrier coefficient

For the same horizon length, a short-horizon barrier constraint is generally more conservative than a short-horizon pointwise constraint. The equivalent confining strength requires fewer inequalities in the design of barrier constraints. As illustrated in Figure 9.6, an n-step barrier constraint has a larger EFR than an n-step pointwise constraint. In this figure, the EFR from the n-step barrier constraint is denoted as $X_{\text{Edls}}^{\text{BR}}(n)$ and that from the n-step pointwise constraint is denoted as $X_{\text{Edls}}^{\text{PW}}(n)$. In the case of finite-horizon OCPs, both of their feasible regions must be smaller than the largest EFR from the full-horizon pointwise constraint, which is denoted as $X_{\text{Edls}}^{\text{PW}}(N)$. Note that n is the length of virtual-time constraint, and N is the length of cost function horizon. Moreover, $n < N$. However, a too strong barrier constraint may have negative impact on the feasibility because an overly tight constraint will largely reduce the feasible region. A reasonable design for short-horizon barrier constraint is to only confine the next state, which is helpful to reduce the computational burden while still maintaining excellent recursive feasibility.

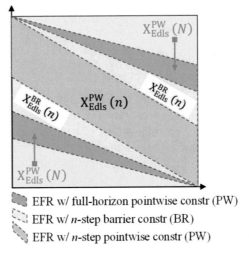

Figure 9.6 Inclusion relations of three kinds of EFRs

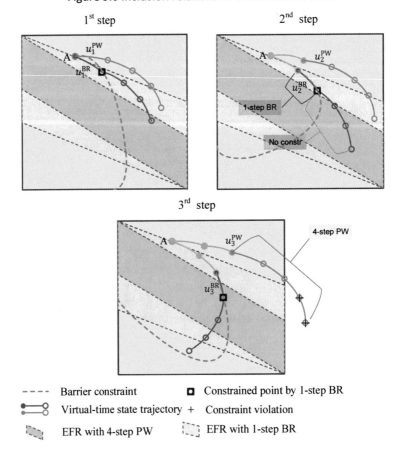

Figure 9.7 Graphical explanation of barrier constraint

Figure 9.7 illustrates how a barrier function enlarges the feasible region. In this example, the virtual-time OCP is defined in the infinite horizon, i.e., $N = \infty$. It has two kinds of virtual-time constraints defined in the short horizon: one is a 4-step pointwise constraint, and the other is a 1-step barrier constraint. State A is a starting point whose recursive feasibility is examined under the two constraints. This point has no recursive feasibility in the setting of 4-step pointwise constraint, from which actions u^{PW} are too weak to drag every successive states inside the feasible region. In contrast, the barrier function has a much stronger ability that could prevent the next state from violating the original constraint. The actions from the 1-step barrier constraint are denoted as u^{BR}. With a series of stronger actions, a previously infeasible point under 4-step pointwise constraint becomes recursively feasible. Accordingly, a properly designed barrier constraint has the potential to enlarge the endlessly feasible region. Of course, its strong confining ability is a double-edged sword, which must come at the cost of a certain optimality sacrifice. Even worse, a too tight barrier constraint might shrink the feasible region instead of enlarging it.

9.2.4   Example of Emergency Braking Control

Infeasibility is a common phenomenon when controlling a system with state constraints. Here, an emergency braking control problem is taken to demonstrate some key concepts of state constraints and feasibility. Emergency braking control is a safety assistance system in autonomous cars. It commonly uses radar or camera to perceive dangerous objects in the front of the car and activates the braking system to avoid vehicle-to-obstacle collisions. The system automatically begins to brake if the driver does not take any action or does not do so quickly enough.

Figure 9.8 Emergency braking control scenario

Consider a simplified vehicle longitudinal model:

$$\begin{bmatrix} d_{t+1} \\ v_{t+1} \end{bmatrix} = \begin{bmatrix} 1 & -\Delta t \\ 0 & 1 \end{bmatrix} \begin{bmatrix} d_t \\ v_t \end{bmatrix} + \begin{bmatrix} 0 \\ \Delta t \end{bmatrix} a_t,$$

where d is the distance to the static obstacle, v is the longitudinal speed, a is the longitudinal deceleration, and Δt is the sampling time. Here, we select $\Delta t = 0.1s$. The state variable is $s = [d, v]^{\mathrm{T}}$, and the action variable is the longitudinal deceleration, which is bounded by

$$a \in [a_{\mathrm{Brk}}, 0],$$

where a_{Brk} is the maximum braking deceleration. Here, $a_{\mathrm{Brk}} = -10 \ \mathrm{m/s^2}$ is taken as the maximum braking ability. In this example, the performance index to be minimized is defined in a finite horizon:

$$\min_{\pi} J = \sum_{i=0}^{N-1} a_{t+i}^2,$$

where J is the performance index. Its goal is to reduce the braking energy. For safety concerns, it is required that the vehicle-to-obstacle distance be larger than a safe threshold, i.e.,

$$d_{t+i} \geq d_{safe}, i = \{1,2,\cdots,\infty\},$$

where $d_{safe} = 0$ is chosen as the minimum distance. In this example, the minimum distance constraint cannot be violated in any circumstances. The full-horizon constraint is used to reshape the virtual-time state constraints. In the following analysis, two types of virtual-time constraints, i.e., full-horizon pointwise constraint (9-9) and full-horizon barrier constraint (9-10) are built:

$$d_{i|t} \geq d_{safe}, i = \{1,2,\cdots,N\}, \tag{9-9}$$

$$\left(d_{i|t} - d_{i+1|t}\right) + \lambda\left(d_{safe} - d_{i|t}\right) \leq 0, i = \{1,2,\cdots,N\}. \tag{9-10}$$

Here, model predictive control (MPC) is used to perform receding horizon optimization and analyze the feasibility in emergency braking scenarios. Its goal is to minimize the finite-horizon performance index while satisfying the minimum distance constraint. The braking controller is essentially a static function of the vehicle-to-obstacle distance and longitudinal speed. The controller is named as MPC-N, where N is the length of the problem horizon (related to both cost function and state constraint). The open-source solver of nonlinear optimization IPOPT is used to calculate their optimal braking commands.

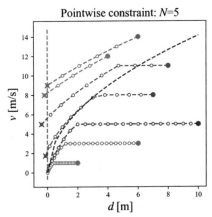

Figure 9.9 State trajectories of MPC-5 with pointwise constraint

Figure 9.9 shows some state trajectories of MPC-5 in the setting of full-horizon pointwise constraint. In each trajectory, the initial state is indicated by a solid point, and the following states are indicated by hollow points. The cross point indicates that the state violates the constraint and a vehicle-to-obstacle collision occurs. The red line is the boundary of the analytical EFR in the case of infinite horizon, which represents the largest EFR in this braking control task. When the vehicle brakes with maximum deceleration, the stopping distance is the smallest:

$$d_{min} = 0.5 \frac{v_0^2}{|a_{Brk}|}.$$

Therefore, $X_{Edls} = \{(d,v)|d \geq 0.5v^2/|a_{Brk}|\}$ is the analytical region in which the vehicle can stop before a collision occurs. In the outside of the analytical region, the vehicle always fails to avoid vehicle-to-obstacle collisions. Through careful consideration, there are two kinds of state trajectories inside the analytical region. Some trajectories fail to stop due to the short-sighted control policy in MPC-5, while other trajectories stop at a distance of zero without causing any collision. Therefore, the largest EFR from the infinite horizon not only contains the EFR of MPC-5 but also contains some part of the initially feasible region of this controller.

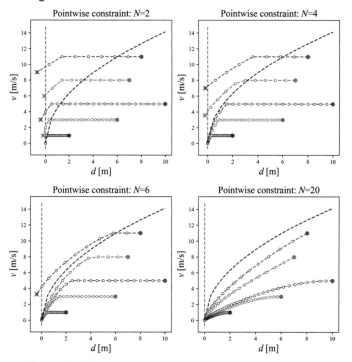

Figure 9.10 State trajectories with pointwise constraints

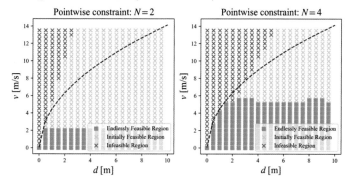

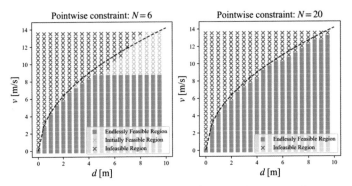

Figure 9.11 Regional division with pointwise constraints

Figure 9.10 shows some state trajectories of MPC with different length of pointwise constraints. As the constraint horizon becomes longer, more state trajectories satisfy the distance constraint. Figure 9.11 shows the feasibility of different state regions. The regional division is conducted by testing the feasibility of each initial state. It is easy to observe that when the constraint horizon becomes longer, both infeasible region and EFR become larger. When the constraint horizon is long enough, the EFR of MPC-20 is almost the same as the analytical EFR of the infinite horizon. The same conclusion is true for the infeasible region. This result also indicates that the initially feasible region exactly becomes an EFR when MPC has an infinitely long horizon.

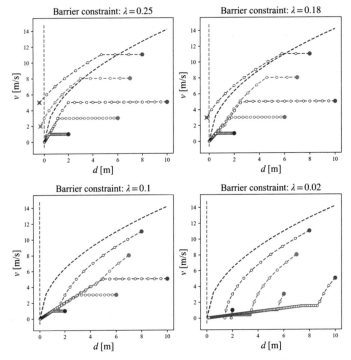

Figure 9.12 State trajectories with barrier constraints

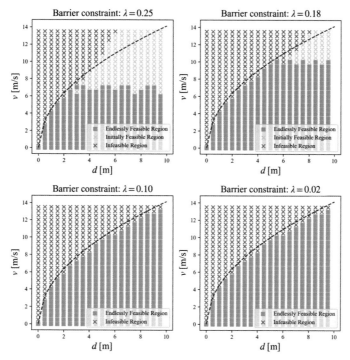

Figure 9.13 Regional division with barrier constraints

A similar simulation is conducted for MPCs with a 2-step barrier constraint. The barrier coefficient λ is the key parameter to adjust the confining strength. As shown in Figure 9.12, when the barrier coefficient decreases, more state trajectories become feasible. This is because the smaller the barrier coefficient is, the more conservative the barrier constraint is. Figure 9.13 shows the regional division by testing the feasibility of different state points. It is easy to observe that when the barrier coefficient decreases, the EFR of barrier-constrained MPCs becomes larger. When $\lambda = 0.1$, the EFR is almost the same as the largest EFR from the infinite horizon OCP. Compared with the previous test, the same feasible region requires a pointwise constraint with 20 predictive steps. Therefore, for an equivalent recursive feasibility, a barrier constraint requires a shorter constraint horizon compared with a pointwise constraint. When $\lambda = 0.02$, the barrier constraint becomes too conservative. As seen from Figure 9.12, in some state trajectories, the car even stops when the remaining distance is still greater than zero.

9.2.5   Classification of Constrained RL/ADP Methods

There are three kinds of optimization approaches that can deal with constrained OCPs: (1) transfer it into unconstrained OCPs using the penalty function method; (2) solve it using the Lagrange multiplier method; and (3) solve it using the feasible descent direction (FDD) technique. In the penalty function method, a penalty term is added to the cost function to penalize the states that cause constraint violation. In the Lagrange multiplier method, duality theory is used to determine the lower bound of primal problems, which will become an optimal solution according to Slater's condition. In the FDD method, the updating direction that is both feasible and descent must be found, and the original constrained OCP is divided into a series of locally convex optimization problems. Like their unconstrained counterparts, constrained RL/ADP algorithms are composed of two basic families: direct methods (e.g., constrained policy gradient) and indirect methods (e.g., constrained policy iteration, constrained value iteration). In the direct RL/ADP algorithm, the original OCP is viewed as a constrained optimization problem, and its optimum is found by stochastic optimization. In the indirect RL/ADP algorithm, the solution of a constrained Bellman equation, which is generally both sufficient and necessary in terms of optimality, is calculated as the optimal policy. Table 9.2 lists the three kinds of constrained optimization and their suitability for direct and indirect methods.

Table 9.2 List of RL/ADPs for constrained OCPs

	Direct methods			Indirect methods		
	Penalty	Lagrange	FDD	Penalty	Lagrange	FDD
Finite horizon	●	●	●	○	○	○
Infinite horizon	●#	●	●	●	●#	●#

● Corresponding RL/ADP exists;

○ Corresponding RL/ADP does not exist;

Constrained RL/ADP that is introduced in this chapter.

9.2.5.1   Direct RL/ADP for constrained OCPs

Without loss of generality, we take the full-horizon pointwise constraint to demonstrate how to formulate a standard constrained OCP. The basis of direct RL/ADP methods is to solve the associated constrained optimization problem. The most popular algorithm attributes to the first-order optimization, such as stochastic gradient descent (SGD). A standard constrained OCP with infinite horizon is formulated in the following form:

$$\min_{u} V(x) = \sum_{i=0}^{\infty} l(x_{t+i}, u_{t+i}),$$

subject to

$$x_{t+i+1} = f(x_{t+i}, u_{t+i}), \qquad (9\text{-}11)$$
$$x \overset{\text{def}}{=} x_t$$

with the full-horizon pointwise constraint:

$$h(x_{t+i}) \leq 0, i = 1, 2, \dots, \infty,$$

where ∞ represents an infinite-horizon problem. In the problem definition above, the virtual-time subscript $x_{i|t}$ is replaced with the real-time subscript x_{t+i} so as to keep consistency with traditional notations. Please always remember that in the RL/ADP community, constrained OCPs are actually defined in the virtual-time domain rather than the real-time domain. The domain separation perspective can help us to understand the infeasibility phenomenon in a more convenient manner.

To perform the first-order optimization, direct RL/ADP needs to calculate the constrained version of policy gradient. In addition to adding a penalty function, the Lagrange and FDD methods are also suitable for building constrained RL/ADP algorithms. The primal-dual optimization (PDO) algorithm, in which the dual ascent technique is used for constrained policy updates, is a pioneering study of the Lagrange multiplier method [9-12]. Trust-region policy optimization (TRPO) can be viewed as a constrained model-free RL algorithm that uses a second-order oval constraint to confine the length of policy updates [9-39]. The local linearization allows TRPO to use a conjugant gradient optimizer to compute gradients more efficiently. The success of TRPO inspired the constrained policy gradient (CPO) algorithm (Achiam et al., 2017) [9-2]. The CPO algorithm can guarantee that every policy update would not violate the state constraints of a certain expected form.

9.2.5.2   Indirect RL/ADP for constrained OCPs

The indirect RL/ADP method does not directly optimize constrained OCPs. Instead, it takes the solution of constrained Bellman equation as the optimal policy. The study of indirect RL/ADP has a longer history than its direct counterpart. The penalty technique dates back to the 1970s with risk-sensitive MDPs, which regard the constraint violation as a risk in the design of reward systems [9-19]. In the penalty function technique, the constrained Bellman equation can degenerate to an unconstrained version, which generates a series of risk-sensitive TD-based algorithms, including risk-sensitive TD(0) and risk-sensitive Q-learning [9-13]. To apply the Lagrange or FDD method into reality, it is required to utilize a constrained actor and an unconstrained critic for solving the constrained Bellman equation. The constrained actor needs to build a constrained optimization problem, and both Lagrange and FDD methods are suitable to guarantee recursive satisfaction of the original state constraints. For example, Duan et al. (2019) proposed a constrained model-based ADP algorithm with deep neural networks [9-16], in which constrained policy iteration is integrated with pointwise state constraints. In this study, a trust-region retrieval mechanism is adopted to eliminate the potential infeasibility issue. Here, we will take (9-11) as an example to demonstrate how to build the constrained Bellman equation. First, let us formally clarify the definition of feasible policy:

- Definition 9.3 (feasible policy): For a given region $X \subseteq X_{Edls}$, a policy is said to be feasible in this region if for any initial state $x_0 \in X$, its following states still stay in the same region; i.e., $x_1, x_2, \cdots, x_\infty \in X$.

Whenever a policy is said to be feasible, its working region must be specified. Of course, the perfect working region is the largest endlessly feasible region, denoted as X_{Edls}, and it must originate from the deployment of full-horizon pointwise constraints. Unfortunately, the largest EFR is not easily accessible, and some of its special subsets, denoted as X, can also serve as feasible working regions. In short, a feasible policy is either defined in the region X_{Edls} or in one of its special subsets X. The term "special" means that the subset X itself must be an invariant set, and any state trajectory starting from X will always stay in this region. We have mentioned that the property of recursive feasibility is discussed in the real-time domain. Obviously, this property must be described in a specific feasible region. The policy iteration algorithm, especially in its model-free version, needs to compute a sequence of intermediate policies, and these policies should be recursively feasible to ensure safe environment exploration. The design of constrained indirect RL/ADP algorithms relies on the so-called constrained Bellman equation:

$$V^*(x) = \min_u \{l(x,u) + V^*(x')\},$$

subject to

$$x' = f(x,u), \tag{9-12}$$

with one-step constraint:

$$x' \in X_{Edls}.$$

This constrained Bellman equation comes from the problem definition of infinite-horizon constrained OCP in (9-11). In the Bellman equation, there is only a one-step constraint left, i.e., the hard constraint of the next state. To be compatible with recursive feasibility, a special working region $x' \in X$ must be chosen as the field of definition for the next state, where X is an EFR. Naturally, the best choice is the largest EFR and it provides the most explicit information about all feasible states. Therefore, in (9-12), we should select the largest EFR, i.e., $x' \in X_{Edls}$, to maximize the working region of the final policy. One might ask why the original state constraint $x' \in X_{Cstr}$, i.e., $h(x') \le 0$, is not selected as the one-step constraint. Obviously, this selection does not need to pre-calculate any endlessly feasible region, and its algorithm design will become much easier. In fact, $x' \in X_{Cstr}$ is also a reasonable option in the design of constrained Bellman equation because it allows to implicitly search for a constrained policy. A brief explanation comes from the working mechanism of Bellman equation. In essence, its one-step constraint states the field of definition in which constrained optimization is mathematically meaningful. The key of ensuring meaningfulness is that its state-value function is finite in the field of definition (shown in Figure 9.14(a)). Therefore, choosing $x' \in X_{Cstr}$ is not a bad option due to the self-maintenance ability of policy feasibility. As a result of the sweeping mechanism in a reinforcement learner, any state point that leads to subsequent constraint violation will be excluded, and those feasible state points will be kept. Eventually, the feasible region will be identified as more information about constrained dynamics is discovered. One shortcoming of selecting $x' \in X_{Cstr}$ as the one-step constraint is that the true field of definition is not explicitly given, and a large amount of tedious computation must be required to identify its accurate EFR.

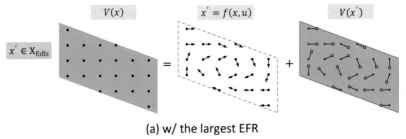

(a) w/ the largest EFR

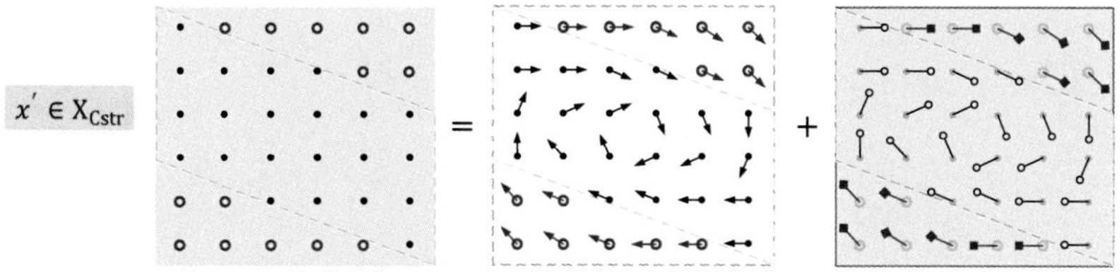

(b) w/ the original constraint

Figure 9.14 Working mechanism of constrained Bellman equation

The next question is how to design an iterative algorithm that computes the solution of constrained Bellman equation. The first option is constrained policy iteration. The key to its algorithm design is how to deal with the one-step constraint. One common understanding is that any constraint in the Bellman equation can be regarded as a certain part of environment dynamics. Therefore, if no violation occurs, constrained policy iteration has no difference from its unconstrained version. Therefore, inside an EFR, the solution of constrained Bellman equation is optimal to its primal OCP. Outside this region, its solution becomes meaningless because this type of policy will lead to constraint violation at some future time. Obviously, available state constraints should have the Markov property, i.e., each constraint only depends on the current state as well as its last state and action. Otherwise, there is no way to view it as a part of the Markovian environment.

Now, let us formally discuss how to design the constrained policy iteration algorithm. As usual, this type of algorithm includes two alternating steps: (1) policy evaluation, i.e., critic update, and (2) policy improvement, i.e., actor update. In the former step, no constraint needs to be considered since the current policy is already feasible. The implementation of this step will not cause any constraint violation. This is why PEV is unconstrained. Accordingly, its self-consistent condition is the same as its unconstrained version:

$$V^k(x) = l(x, \pi_k) + V^k(x').$$ (9-13)

In contrast to unconstrained PEVs, PIMs must be constrained so as to search for a better yet feasible policy. Otherwise, the self-consistent condition at the next iteration will not exist because its state-value function cannot be meaningful in its working field. Therefore, we need to formulate a constrained optimization problem in the PIM step:

$$\pi_{k+1}(x) = \arg\min_u \{l(x, u) + V^k(x')\},$$

with

$$x' = f(x, u),$$
$$x' \in X_{Edls}.$$ (9-14)

By restricting the next state by $x' \in X_{Edls}$, each constrained PIM outputs a better yet feasible policy and its corresponding state-value function will keep meaningful in this region. Hence, (9-14) aims to build a recursively meaningful optimization problem. To successfully repeat (9-13) and (9-14), we need to ensure that π_k is recursively feasible in the region X_{Edls}. The key is to prove that such a constrained PIM problem must have a feasible solution; that is, π_{k+1} always exists in the largest EFR X_{Edls}. Here, we assume that the initial guess π_0 is feasible in X_{Edls}. Under this assumption, it is easy to prove that $\pi_k(x), k = 1,2, \dots, \infty$ holds the property of recursive feasibility.

- Lemma 9.2 (recursive feasibility): Policy π_{k+1} is feasible in X_{Edls} if its previous policy π_k is feasible in X_{Edls}.

Proof:

When π_k is feasible, $V^k(x)$ is finite-valued in X_{Edls}. Due to the finiteness of $V^k(x)$, a meaningful optimization is built at each constrained PIM. If there is a nontrivial solution, π_{k+1} is exactly a feasible policy. Otherwise, one can at least select $\pi_{k+1} = \pi_k$ as a trivial solution. Therefore, there always exists a next policy π_{k+1}, which is feasible in X_{Edls}.

∎

In addition to recursive feasibility, convergence is also critical to a constrained RL/ADP algorithm. Convergence means that this algorithm can gradually shift its intermediate policy to the optimum even if the hard state constraint is enforced. Inside an EFR, its mathematical proof is not very difficult. In other words, one can easily examine whether $V^k(x)$ is monotonically decreasing with respect to the main iteration. As usual, the proof is divided into two steps: (1) value decreasing property, and (2) converge to the optimum. Considering the characteristics of recursive feasibility, their proofs are almost identical to that of their unconstrained versions.

Furthermore, it is worth mentioning that the same concepts of recursive feasibility and endlessly feasible region (EFR) have been illustrated in some MPC literature. Specifically, the book by Borrelli et al. (2017) has indicated that feasibility at the initial time does not necessarily imply feasibility for all future times [9-7]. Therefore, it is desirable to design an optimization-based receding horizon controller to guarantee feasibility for all future times, referring to a property called persistent feasibility. This book also defines a term named control invariant set for systems subject to external inputs, i.e., the set of initial states for which a controller exists such that the system constraints are never violated. Note that the definition of EFR is almost similar to that of forward invariant set, which has been discussed in some control barrier literature. A set is called forward invariant for a dynamical system if every solution to the system from this set stays in it [9-4].

9.3   Penalty Function Methods

Here, we take the direct version of model-based ADP as an example to demonstrate how to apply penalty function methods. This kind of methods are easy to understand and have been widely used in static optimization [9-8]. The basic idea is to convert a constrained optimization problem into an unconstrained one by implementing a higher cost for any constraint violation. Optimizing a penalized criterion will encourage the learned policy to satisfy the state constraints as much as possible. Usually, penalty function methods can be classified into two major subclasses: (1) additive smoothing penalty and (2) absorbing state penalty. In the former, a weak penalty term is added to the original criterion, and in the latter, it is replaced with a new penalty constant. The searching behaviors of these two subclasses are illustrated in Figure 9.15. In the former subclass, the learned policy is slowly dragged into an admissible space as more evidence about constraint violation is accumulated. In the latter, the one-shot ability is obtained to identify admissible actions since the accumulative return changes more quickly.

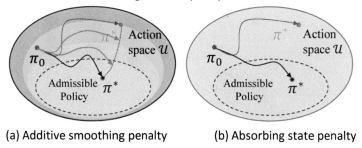

(a) Additive smoothing penalty   (b) Absorbing state penalty

Figure 9.15 Policy updates with penalty functions

9.3.1 Additive Smoothing Penalty

In the method of additive smoothing penalty, the penalty term is added into every utility function, thus spreading the violation penalty to the overall objective function. Both exterior and interior penalties can be seen in the RL community. The exterior penalty is more popular due to its ability to relax the hardness of state constraints. Constraint relaxation is helpful to avoid potential infeasibility issues. Taking infinite-horizon pointwise constraint as an example, the overall objective function is built with a group of penalized utility functions:

$$J_{\text{mod}} = \sum_{i=0}^{\infty} l_{\text{mod}}(x_{t+i}, u_{t+i}),$$

with

$$l_{\text{mod}}(x, u) = l(x, u) + \rho \cdot \phi(x),$$

where the penalty term $\phi(\cdot)$ is a function that is correlated with state constraints and $\rho > 0$ is called the penalty coefficient. There are various popular strategies of designing the penalty term $\phi(\cdot)$. One strategy is called the rectified linear penalty:

$$\phi(x) = \begin{cases} 0 & h(x) \leq 0 \\ h(x) & \text{otherwise} \end{cases}.$$

This rectified function gives a linear penalty to the unsatisfactory constraint. Other penalty strategies include one-shot constant penalty, exponential risk penalty, and rectified quadratic penalty [9-13][9-19][9-23]. The nature of additive smoothing penalty is to slightly increase the original cost function when the state constraint is not satisfied. This mechanism is more likely to learn a policy at a local minimum. One special advantage of this method is that the penalized value function is a good indicator of whether a state is feasible or not. One can examine the feasibility of a policy by observing the magnitude of its penalized state-value function.

9.3.2 Absorbing State Penalty

Unlike additive smoothing penalty, the method of absorbing state penalty replaces the original cost with a completely new penalty constant when the state constraint is violated. This idea has been adopted in many RL tasks because of its high exploration efficiency in constrained environments. The utility function with absorbing state penalty is

$$l_{\mathrm{Mod}}(x,u) = \begin{cases} l_{\mathrm{absorb}} & h(x) > 0 \\ l(x,u) & h(x) \leq 0 \end{cases},$$

where l_{absorb} is a sufficiently large penalty constant. Once the state reaches the upper bound of the state constraint, the environment exploration stops at once, and a high negative reward is returned. As a result of high negative reward, exploration is often terminated earlier than before. Hence, more trials can be completed in the same amount of time, which means that more experience is collected and better training efficiency is obtained. In nature, absorbing state penalty is similar to additive smoothing penalty since both of them pursue admissible policy updates. The difference is that additive smoothing penalty needs to balance the penalty term and original reward more carefully. If the penalty coefficient is poorly adjusted, the reinforcement learner may easily fail and stay in an unwanted local optimum. In comparison, absorbing state penalty is more like a one-shot strategy. Its early-stopping behavior has the ability to strengthen the penalization on potential constraint violation. In addition, if the penalty constant l_{absorb} is sufficiently large, the global optimum is often achievable.

9.4   Lagrange Multiplier Methods

Here, we take the constrained PIM step as an example to demonstrate how to deploy the Lagrange multiplier methods. In theory, each constrained PIM better obeys the one-step constraint from the largest endlessly feasible region, i.e., $x' \in X_{\mathrm{Edls}}$. However, X_{Edls} is actually unknown in each constrained PIM. Therefore, this unknown region needs to be replaced with accessible regional knowledge. One popular method is to use a one-step pointwise constraint, i.e., $h(x') \leq 0$, as the replacement of original unknown constraint. This kind of replacement introduces a relaxed confinement to the next state in each constrained PIM. Considering the sweeping mechanism in the whole state space, a feasible policy can still be found as the final output of the reinforcement learner. The rationality of this replacement will be further analyzed in the design of the actor-critic-scenery (ACS) learning architecture.

The Lagrange duality is a fundamental theory of the Lagrange multiplier methods. It is used to convert the constrained PIM problem into a Lagrange duality problem, in which at least a lower bound of the primal problem must be proven (if the actor loss function is minimized) [9-8]. In particular, their solutions are identical under the strong duality condition. The strong duality condition holds if both the actor loss function and the state constraint are convex and Slater's condition is satisfied. In this situation, the optimal duality gap is zero, and the optimal solutions of the primal and dual problems are the same. The strong duality condition also explains where the complementary slackness condition comes from. This condition provides a useful feasibility indicator in the Lagrange multiplier methods.

9.4.1   Dual Ascent Method

The original problem is a primal problem; in contrast, the optimization of the Lagrange function is known as the dual problem. In general, the dual problem is easier to solve than the primal problem because it is a concave maximization even if its primal problem is not convex. In the dual ascent method, the dual problem is solved by alternately performing maximization and minimization. These alternating operations finally converge to a saddle point if the strong duality condition is satisfied. In the constrained PIM, the actor loss function with a one-step pointwise constraint is converted into a Lagrange function:

$$L(\theta, \lambda) \stackrel{\text{def}}{=} l(x, u) + V(x') + \lambda h(x'), \tag{9-15}$$

where $L(\theta, \lambda) \in \mathbb{R}$ is called the Lagrange function, $h(\cdot) \in \mathbb{R}$ is the scalar constraint function, θ is the policy parameter (i.e., the primal variable), and $\lambda \in \mathbb{R}$ is the Lagrange multiplier (i.e., the dual variable). In constrained PIM, the Lagrange multiplier is not a fixed scalar; instead, it is a function of the state. Hence, its parameter updates need to sweep the whole state space. The dual problem is essentially a maximin optimization:

$$\max_{\lambda \geq 0} \min_{\theta} \{ L(\theta, \lambda) \}. \tag{9-16}$$

The dual ascent updates will converge to the saddle point of the maximin problem if the strong convexity condition holds. The primitive dual ascent algorithm is

$$\theta \leftarrow \arg\min_{\theta} L(\theta, \lambda),$$
$$\lambda \leftarrow \lambda + \alpha_\lambda \cdot \nabla_\lambda L(\theta, \lambda). \tag{9-17}$$

Here, λ must be no less than zero. It is easy to see that (9-17) must still solve a minimization problem. Replacing it with a gradient descent formula, the standard dual ascent algorithm is obtained:

$$\theta \leftarrow \theta - \alpha_\theta \cdot \nabla_\theta L(\theta, \lambda),$$
$$\lambda \leftarrow \lambda + \alpha_\lambda \cdot \nabla_\lambda L(\theta, \lambda),$$

where $\alpha_\theta > 0$ and $\alpha_\lambda > 0$ are the step sizes. The gradients above are easily derived:

$$\nabla_\theta L(\theta, \lambda) = \frac{\partial u^{\mathrm{T}}}{\partial \theta} \frac{\partial L(\theta, \lambda)}{\partial u}, \qquad \nabla_\lambda L(\theta, \lambda) = h(x').$$

In practice, one should select $\alpha_\theta \gg \alpha_\lambda$ to approximate the argmin operator to the greatest extent. The constrained policy iteration algorithm consists of two cyclic steps, including unconstrained PEV and constrained PIM. In the dual ascent technique, a new inner loop is built inside each constrained PIM. Even though there is no theoretical proof, this type of double-loop iteration often works very well. A deeper look into its working mechanism reveals that the Lagrange multiplier is essentially a special penalty coefficient, which has the self-adjustment ability so that any potential constraint violation can be avoided. When the constraint is unsatisfactory, the Lagrange multiplier will automatically increase to penalize the Lagrange function more strongly. Automatic penalization can be utilized to depress the value of the constraint function and keep it below zero. One can also integrate the fixed penalty coefficient and self-adjustable Lagrange multiplier together to build a faster and more stable constrained RL algorithm [9-34]. It is easy to observe that the Lagrange multiplier itself is a good feasibility indicator. One can check the feasibility of a given state simply by the value of its Lagrange multiplier. If this value is larger than a given threshold, one can roughly make a decision that its associated state is infeasible.

9.4.2   Geometric Interpretation

A feasible policy must work in an endlessly feasible region. Finding an optimal yet feasible policy is the central task of constrained RL/ADP algorithms. Based on prior experience, searching for an optimal policy is already laborious. The demand on feasibility introduces a lot of new challenges to constrained optimization. The use of an inadequate learning algorithm may lead to very bad results because constrained PIM has obviously different solvability inside and outside the feasible region. Hence, in a useful RL/ADP algorithm, the chosen constrained optimization technique, as well as how it performs constrained updates and its feasibility property, must be deeply understood.

Let us graphically demonstrate how feasibility evolves in the Lagrange multiplier method. The Lagrange multiplier in the constrained PIM is not a scalar. Instead, it is a function that is defined in the whole state space. We choose two special state points to demonstrate the behaviors of dual ascent updates. The first state point is from an endlessly feasible region, but its initial policy is temporarily infeasible due to a bad guess. The second state point is from the infeasible region, in which no feasible initial policy exists. The automatic adjustments of their Lagrange multipliers show very different behaviors, whose geometric interpretations are shown in Figure 9.16 and Figure 9.17, respectively. Let us define a new space that contains the values of constraint function and actor loss function:

$$G = \{(h(x), l(x,u) + V(x')) \mid x \in X\},$$

where G is the space of pair $\{h, l + V\}$. The constraint function h and actor loss function $l + V$ are regarded as the two variables in the new space G. In Figure 9.16, the horizontal axis is the value of constraint function h, and the vertical axis is the value of actor loss function $l + V$. Now, the Lagrange dual problem can be rewritten as

$$\max_{\lambda \geq 0} \min_{(h, l+V) \in G} \{(l + V) + \lambda h\}. \tag{9-18}$$

Figure 9.16 shows the geometric explanation of how the Lagrange multiplier works in a feasible region. All the policies starting from a given feasible state form a policy set, which is represented as the crescent shape in the G-plane. Even though the state is in a feasible region, its associated policy can be either feasible or infeasible. The subset of feasible policies is on the left side of the G-plane, where the constraint function is less than zero, and the subset of infeasible policies is on the right side of the G-plane, where the constraint function is greater than zero. The red star represents the constrained optimum $(h^*, l + V^*)$, which is also the optimal policy to be learned. The dashed line represents the contour of the Lagrange dual function, and each line is a graphical description of the linear equation $(l + V) + \lambda h = $ const. The value of the Lagrange multiplier determines the slope of each contour line.

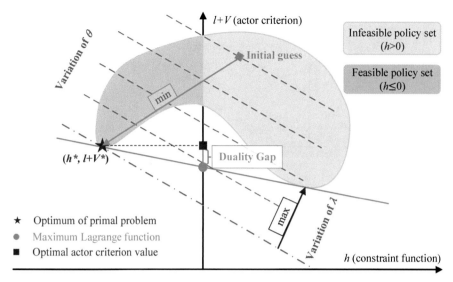

Figure 9.16 Lagrange duality in the feasible region

In Figure 9.16, the initial policy, which is located on the right side of the $\mathcal{G}$-plane, is a bad guess. Let us rethink the working principle of dual ascent updates. In fact, the maximin problem in (9-18) contains two variational operations: an inner minimization with a fixed Lagrange multiplier and an outer maximization through the variation of the Lagrange multiplier. When using the inner minimization operation, a search is performed to find the smallest Lagrange function that descends along the perpendicular direction of the linear contour. This search continues until the lowest tangent line with the crescent shape is found. When using the outer maximization operation, a search is performed to find the largest actor loss function. Meanwhile, the tangent property with the crescent shape needs to be maintained. The tangent line does not allow entering the infeasible policy set because it contradicts the feasibility guarantee. By maintaining the tangency property, an admissible policy is eventually found after dual ascent updates. The optimization degree of the actor loss function is represented by the intersection of the tangent line with the vertical axis. The green circle is the optimal solution of the Lagrange dual problem, and the red rectangle is the true constrained optimum of the primal problem. Their distance is called the duality gap. When the policy set is nonconvex, a duality gap exists. When this set is convex, the duality gap becomes zero, and the optimal solutions of the dual and primal problems are the same. Obviously, the Lagrange multiplier cannot be less than zero, and it is at least equal to zero when the duality gap diminishes.

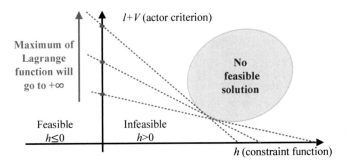

Figure 9.17 Dual ascent in the infeasible region

Figure 9.17 illustrates the behavior of the Lagrange multiplier in the infeasible region. The primal problem has no feasible solution at a given state point, and therefore, its admissible policy set must lie on the right side of the $\mathcal{G}$-space. The outer maximization operation will drive the Lagrange multiplier to go to infinity since its actor loss function has no upper bound. This infinitely large Lagrange multiplier provides an effective indicator to examine whether a feasible policy can be found or not at the chosen state.

9.4.3   Augmented Lagrange Multiplier

Compared with normal Lagrange method, the augmented Lagrange method is more robust and has faster convergence behavior. The basic idea behind this new method is to augment the Lagrange function by adding a quadratic penalty. In theory, dual ascent updates can reach a constrained optimum, but their convergence is often unsatisfactory in some nonconvex problems. As previously discussed, the penalty function method has the ability of slightly changing the constrained policy gradients, thus enabling the improvement of the convergence quality. Generally, the augmented Lagrange multiplier can be seen as a combination of Lagrange duality and penalty function. The augmented Lagrange function is defined as

$$L_{\text{Aug}}(\theta, \lambda, \zeta) = l(x, u) + V(x') + \lambda h(x') + \frac{\rho}{2} \|h(x') + \zeta\|_2^2,$$

where $\rho > 0$ is a penalty coefficient. A slack variable $\zeta \in \mathbb{R}$ is introduced to convert the inequality constraint to an equality constraint:

$$h(x') + \zeta = 0,$$

where $\zeta \geq 0$ is a slack variable. This equality constraint is then converted to a quadratic penalty term. The introduction of a quadratic penalty term can strengthen the convexity of the augmented Lagrange function if the penalty coefficient ρ is large enough. The use of convexity enhancement can accelerate convergence in nonconvex tasks. Considering the new slack variable, the optimization of the augmented Lagrange function has three updating formulas:

$$\theta^{\text{new}} \leftarrow \arg\min_{\theta} L_{\text{Aug}}(\theta, \lambda, \zeta),$$

$$\zeta^{\text{new}} \leftarrow \arg\min_{\zeta} L_{\text{Aug}}(\theta, \lambda, \zeta),$$

$$\lambda^{\text{new}} \leftarrow \lambda + \alpha_{\lambda} \cdot \nabla_{\lambda} L_{\text{Aug}}(\theta, \lambda, \zeta).$$

The geometric interpretation of the augmented Lagrange multiplier method is very similar to the dual ascent version, where the scalar Lagrange multiplier λ can be interpreted as an adaptive penalty to avoid potential constraint violation. The main difference between these two methods lies in the fact that the augmented Lagrange multiplier introduces an additional quadratic penalty, which results in stronger convexity to stabilize the optimization process.

9.5 Feasible Descent Direction Methods

Both penalty and Lagrange methods focus on how to modify the original criterion, either the overall objective function or actor loss function. In large-scale optimization problems, relatively slow convergence is their common shortcoming due to improper penalty selection or unstable dual updates. As another popular constrained optimization family, feasible descent direction (FDD) methods seek to find an updating direction that is both descent and feasible, which is able to equip RL/ADP with a few new advantages. In this family, multi-stage optimization is broken into a sequence of simple convex programming problems, and the parameterized policy is updated along the feasible descent direction that originates from each simple convex problem.

In the field of static optimization, the study of FDD dates back to Zoutendijk's feasible descent direction in the 1960s, which first linearizes the primal problem and then bounds the updating length by a box constraint [9-44]. Later, its convergence performance was improved by introducing active constraints and an augmented objective function [9-35][9-40]. The FDD method with a linearized cost function is called sequential linear programming (SLP). A quadratic approximation version of this method was presented by Polak in the 1970s [9-36]. Later, in the 1990s, quadratic approximation of the objective function began to merge with the technique of constraint linearization, which has become today's sequential quadratic programming (SQP) (West and Polak) [9-43]. To date, SQP has been one of the most powerful algorithms in the field of constrained optimization. When introducing FDD into the RL/ADP community, how to integrate it with constrained OCP is a much more complicated task than that of static optimization. The main difference is that static optimization only addresses a fixed state point, while RL/ADP needs to search for a feasible policy in the whole state space.

9.5.1 Feasible Direction and Descent Direction

Similar to the Lagrange multiplier method, we still take the constrained PIM as an example to demonstrate how to deploy FDD. The constrained policy iteration algorithm, which can either be a model-free or model-based algorithm, consists of a sequence of constrained PIM problems. Each constrained PIM problem is individually solved with FDD to output a better and more feasible policy. The largest EFR is still replaced with $h(x') \leq 0$ to avoid the use of unknown feasible region. With FDD, the update formula inside each constrained PIM is

$$\theta \leftarrow \theta + \Delta\theta, \tag{9-19}$$

where $\Delta\theta$ is the parameter increment. The parameter increment must be both feasible and descent, satisfying two inequality conditions:

- Descent direction condition: $\qquad$ (9-20)

$$J_{\text{Actor}}(\theta + \Delta\theta) \leq J_{\text{Actor}}(\theta),$$

• Feasible direction condition:

$$h(x'(\theta + \Delta\theta)) \leq 0,$$

where $J_{\text{Actor}}(\theta) = l(x, u) + V(x')$ is the actor loss function and $h(x')$ is the one-step pointwise constraint. Here, $x' = f(x, u)$ is the environment model, representing the one-step transition from the current state x to the next state x'. The notation $x'(\theta)$ emphasizes that the action u is dominated by a parameterized policy $\pi(\theta)$. A two-dimensional example is shown in Figure 9.18. The blue triangular-like area is a set of admissible policies. The orange sector indicates the set of descent directions, which always decreases the value of actor loss function. The green sector indicates the set of feasible directions, in which the next state is not allowed to move outside the constraint boundary. The FDD is in their coincident sector, heading toward the right-bottom direction.

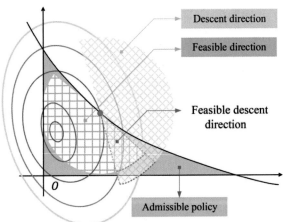

Descent direction

Feasible direction

Feasible descent direction

0

Admissible policy

Figure 9.18 Illustration of the feasible descent direction

In most tasks, analytically calculating FDD is often an intractable task. The optimization problem needs to be converted into a series of linear programming (LP) or quadratic programming (QP) subproblems for more efficient computation. In contrast to static optimization, sequentially solving LP or QP is not always necessary in each step of the constrained PIM. This is because there is already an iteration cycle between PEV and PIM, and the sequential computation of LQ or QP can be merged into this main cycle. In other words, instead of repeated conversions such as SLP or SQP in static optimization, each constrained PIM can be converted to LP/QP only once in RL/ADP. Of course, in each constrained PIM, LP or QP can still be converted and solved multiple times, which is beneficial for obtaining a more accurate search at each step.

How to examine feasibility is an emerging challenge in FDD-based optimization. An intuitive idea is to check whether the feasible direction condition in (9-20) has a solution. The satisfaction of this condition is one evidence for feasibility determination. Obviously, a given state point is infeasible if the feasible direction condition has no solution. However, after LP or QP conversion, the linearization of this condition will lose the ability to examine feasibility. More reliable evidence about the solvability of constrained PIM needs to be collected as it is repeatedly solved at the given state point.

9.5.1.1   Transformation into linear programming

The FDD condition of constrained PIM is converted into a constrained linear programming (LP) problem through the local linearization of both actor loss function (i.e., descent direction condition) and state constraint (i.e., feasible direction condition). Therefore, at each main iteration, finding FDD for the constrained PIM becomes a constrained LP problem:

$$\min_{\Delta\theta} g^{\mathrm{T}}\Delta\theta,$$

subject to   (9-21)

$$h(x')|_\theta + \nabla_\theta h(x')^{\mathrm{T}}\Delta\theta \leq 0,$$

where $g^{\mathrm{T}}\Delta\theta$ is the first-order Taylor approximation of the actor loss function:

$$g = \frac{\partial J_{\mathrm{Actor}}}{\partial\theta} = \frac{\partial\{l(x,u) + V(x')\}}{\partial\theta}.$$

The new cost function is actually based on the linear approximation of $J_{\mathrm{Actor}}(\theta + \Delta\theta) - J_{\mathrm{Actor}}(\theta)$. The minimization outputs the steepest descent direction around the current policy parameter. For the constrained PIM, its associated LP always has a feasible solution. One can at least select $\Delta\theta = 0$ to ensure a feasible solution. When $\Delta\theta = 0$, the constrained PIM often stops, and the RL agent has reached its optimum. The inequality constraint is the linear approximation of the second part of FDD condition. After converting the FDD condition to an LP problem, the new constraint condition loses the ability to examine feasibility because of linear approximation.

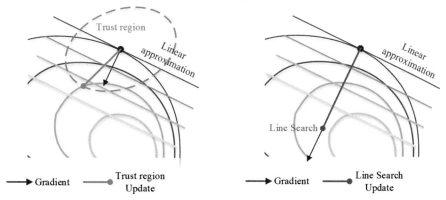

Figure 9.19 Trust region and line search

The updating length of the policy parameter cannot be determined by constrained LP optimization. Determining a proper updating length is helpful to avoid a fast policy change and to stabilize the LP-based update. A small constant updating length is a mundane choice. Two advanced strategies with automatic adjustment ability are the trust region and line search strategies, as shown in Figure 9.19. The trust region is actually an inequality constraint that restricts the range of the exploration zone. A typical trust region is in a quadratic form:

$$\Delta\theta^T H \Delta\theta \leq \delta,$$

where $H > 0$ is a weighting matrix. Constrained policy optimization (CPO) is actually a kind of model-free algorithm with a linearized actor loss function and a trust-region constraint [9-2][9-42]. In the line search strategy, an optimal length that can minimize the original actor loss function is found along the updating direction from LP-based FDD. A variety of line search methods can be integrated into the LP-based PIM, including the Armijo rule, limited line search and diminishing length.

9.5.1.2   Transformation into quadratic programming

Compared to first-order optimization, second-order derivative information has the potential to speed up the convergence of constrained RL/ADP algorithms. To use the second-order information, each constrained PIM is converted into a constrained quadratic programming (QP) problem. The quadratic cost function originates from the second-order Taylor approximation of the descent direction condition, and its constraint maintains the linear approximation of the feasible direction condition. In this case, at each main iteration, calculating FDD for the constrained PIM becomes a constrained QP problem:

$$\min_{\Delta\theta} \frac{1}{2}\Delta\theta^T F \Delta\theta + g^T \Delta\theta,$$

subject to

(9-22)

$$h(x')|_\theta + \nabla_\theta h(x')^T \Delta\theta \leq 0,$$

where g is the gradient of the actor loss function and F is its Hessian matrix. The Hessian matrix of the actor loss function is

$$F = \frac{\partial^2 J_{Actor}}{\partial\theta^2} = \frac{\partial^2\{l(x,u) + V(x')\}}{\partial\theta^2}.$$

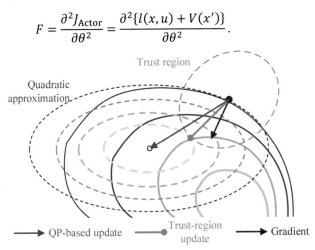

Figure 9.20 Trust region in quadratic programming

The quadratic cost function is based on the second-order Taylor approximation of $J_{Actor}(\theta + \Delta\theta) - J_{Actor}(\theta)$. The minimization of (9-22) can reach the optimum of the quadratic cost function if no constraint is enforced. This implies that $\Delta\theta$ contains information on both the descent direction and updating length. If $\Delta\theta = 0$ is found to be one solution of (9-22), $\theta + \Delta\theta$ satisfies the stationary condition of optimality. One obvious drawback of QP is that the quadric approximation might have large errors when departing far from the operating point. This drawback can be compensated for by a trust-region constraint, which bounds the updating length to avoid fast policy movement (Figure 9.20). The feasibility examining ability is also lost in QP-based FDD since it shares the same linearized constraint as LP-based FDD.

9.5.2   Constrained LP Optimization

Constrained LP optimization is employed to solve the constrained PIM after transforming it into a linear programming problem such as the one in (9-21). During iteration, the FDD condition is often linearized only once in each constrained PIM. Sequential linearization is performed in synchronization with the main iteration in RL/ADP. One benefit of performing a single conversion at each PIM is that the overall training efficiency is often better than that of multiple conversions. In deep learning, the first-order optimization algorithm is more suitable for training RL/ADP with deep neural networks than high-order optimization. A well-known example for training neural networks is the Adam (adaptive moment estimation) [9-21]. Adam is a standard first-order stochastic optimization and can be used to update large-scale network weights more efficiently. To better understand how to solve FDD in reinforcement learning, two practical constrained LP algorithms are introduced: (1) Lagrange-based LP algorithm and (2) gradient projection algorithm. The details of these LP-based algorithms are listed in Table 9.3.

Table 9.3 Instances of RL with first-order optimization

Algorithm	Guidelines	
	Updating direction	Updating length
Lagrange-based LP	Lagrange duality	Trust region
Gradient projection	Gradient projection	Constant/Diminishing

9.5.2.1   Lagrange-based LP Method

The Lagrange-based LP method is featured by the Lagrange duality and a trust-region constraint. The Lagrange duality technique has a few benefits, such as low sensitivity to bad initial guess and fast convergence in a local region. The trust-region constraint can limit the updating length to avoid potential training instability. The KL divergence between $\pi(\theta)$ and $\pi(\theta^{\mathrm{new}})$ is selected to construct the trust-region constraint:

$$D_{\mathrm{KL}}(\pi(\theta), \pi(\theta^{\mathrm{new}})) \approx \frac{1}{2}\Delta\theta^{\mathrm{T}}H\Delta\theta \leq \delta,$$

where δ is a hyperparameter for the trust region and H is the Hessian matrix from the KL divergence. The Lagrange function of (9-21) is defined as

$$\max_{\mu,\lambda\geq 0} \min_{\Delta\theta} g^{\mathrm{T}}\Delta\theta + \lambda(h(x')|_\theta + \nabla_\theta h(x')^{\mathrm{T}}\Delta\theta) + \mu\left(\frac{1}{2}\Delta\theta^{\mathrm{T}}H\Delta\theta - \delta\right),$$

where $\lambda \geq 0$ is the Lagrange multiplier for the linearized constraint and $\mu \geq 0$ is the Lagrange multiplier for the trust region. One main advantage of Lagrange-based LP is that its optimal solution $\Delta\theta^*$ can be analytically derived:

$$\Delta\theta^* = -\frac{H^{-1}(g + \lambda^*\nabla_\theta h)}{\mu^*},$$

where λ^*, μ^* are the optimal Lagrange multipliers. As discussed before, traditional Lagrange-based optimization requires inserting a computationally inefficient inner loop. The Lagrange-based LP method subtly avoids the loop nesting issue because its optimum can be analytically computed in one shot.

9.5.2.2   Gradient projection

Gradient projection is an effective technique to solve high-dimensional LPs. It utilizes a positive definite matrix to project an unconstrained gradient into a feasible direction. As shown in Figure 9.21(a), the gradient projection task is built to satisfy the feasible direction condition in (9-20) through a proper projection matrix P_T. The gradient projection is executed individually after normal policy gradient is calculated. Therefore, this method can be easily integrated with many existing RL/ADP algorithms.

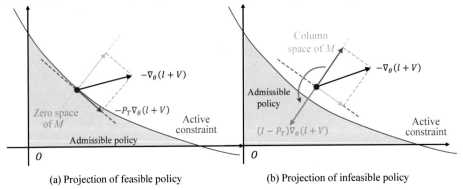

(a) Projection of feasible policy   (b) Projection of infeasible policy

Figure 9.21 Gradient projection

One classic gradient projection method is called Rosen gradient projection, in which the gradient is projected to the tangent plane of the active constraints, i.e., zero space of constrained gradient [9-38]. The projection matrix is formulated as

$$P_T = I - M^{\mathrm{T}}(MM^{\mathrm{T}})^{-1}M,$$

where $M = \nabla_\theta h^{\mathrm{T}}$ is the gradient vector of active constraints. This type of projection matrix P_T must be positive definite, and its projection can be used to find an FDD along the tangent direction of the active constraints. Note that the dimension of the projection matrix is fixed even though its inner vector M depends on the number of active constraints. Generally, gradient projection is more suitable for large-scale problems with sparse linear constraints. One may ask whether the Rosen gradient projection can start from an infeasible policy. As demonstrated in Figure 9.21(b), P_T can yield a gradient projection that moves toward the feasible region. This type of matrix $(I - P_T)$ can project any policy gradient to the normal line of the active constraint, which is the fastest direction to enter the feasible region. Therefore, this method can be used with either a feasible policy or an infeasible arbitrary guess.

9.5.3   Constrained QP Optimization

Constrained QP optimization is used to solve constrained PIM after its conversion into a quadratic programming problem such as the one in (9-22). The Hessian information of the actor loss function has the potential to accelerate a policy search. For low-dimensional tasks, two kinds of constrained QP algorithms have been introduced: (1) the active-set QP algorithm and (2) the quasi-Newton algorithm. When extending to high-dimensional tasks, the high computational burden of the Hessian matrix is one of its major challenges. Recently, a few studies have attempted to address this issue with approximate Hessian techniques [9-28]. However, accurate Hessian estimation is significantly impaired in large-scale policy searches, especially when the sparsity of the Hessian matrix is lost. The loss of sparsity makes the inverse operation computationally inefficient and even impossible in neural network-based policy and value functions. Therefore, constrained QP optimization is not very popular among today's RL community.

9.5.3.1   Active-set QP

In active-set QP, an inner iteration loop is inserted into the constrained PIM. At each PIM step, a working set of active constraints is first realized, and then, this set is iteratively refined through Lagrange duality optimization. Hence, this process consists of at least two steps: (1) solving an equality-constrained QP; (2) refining the working set by adding and removing constraints.

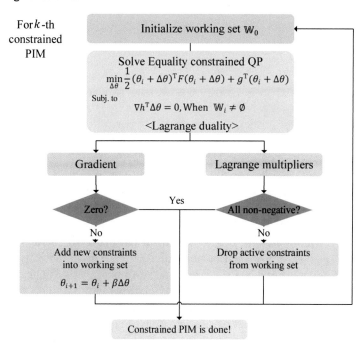

Figure 9.22 Demonstration of the active-set QP algorithm

In the first step, the Lagrange multiplier method, which outputs both the gradient direction and the Lagrange multiplier, is implemented. In the second step, the working set is refined according to the following empirical rules: (1) if the gradient is nonzero, the closest constraint along the gradient direction is added; (2) if there exists a negative Lagrange multiplier, its corresponding constraint is removed. The optimal solution is found when the gradient is zero and all Lagrange multipliers are nonnegative. It is easy to see that active-set QP can only calculate an equality-constrained QP since all the inequalities in their working set must be classified into active or inactive equalities. This property makes its computation very fast for low-dimensional problems with sparse constraints [9-33].

9.5.3.2   Quasi-Newton method

The exact solution of the Hessian matrix is often computationally expensive. Unlike the Newton method, the quasi-Newton method uses an approximated Hessian estimation for a more efficient calculation. More importantly, its resulting second-order optimizer has a more robust learning behavior, thus having the potential to stabilize reinforcement learners. In the constrained PIM, its true Hessian matrix is $\nabla^2 J_{\text{Actor}}$. We use the first-order Taylor approximation of J_{Actor} to build an equality condition for the Hessian estimation:

$$\hat{F}\Delta\theta = \nabla J_{\text{Actor}}|_{\theta+\Delta\theta} - \nabla J_{\text{Actor}}|_{\theta},$$

where $\hat{F}$ is the Hessian matrix to be estimated. In a few recursive estimation algorithms, the error of two successive Hessian matrices is enforced to hold the low-rank property. Other regularity conditions, such as symmetry and strictly positive definiteness, are often needed to maintain necessary matrix properties. Popular regularity conditions include the symmetric rank-one (SR1) rule [9-14] and the Broyden-Fletcher-Goldfarb-Shanno (BFGS) rule [9-10].

9.6   Actor-Critic-Scenery (ACS) Architecture

As stated in previous chapters, existing actor-critic (AC) methods contain two cyclic elements. One element is called the "actor", whose core is a parameterized policy, and the other is called the "critic", whose core is a parameterized value function. The architecture that alternately updates these two elements is very popular in today's RL community because of it benefits in stability enhancement and efficiency improvement. Under this architecture, typical RL algorithms include advantage actor-critic (A2C), asynchronous advantage actor-critic (A3C), soft actor-critic (SAC), and distributional soft actor-critic (DSAC). Based on the names of these AC algorithms, we can see their intrinsic connection with this two-element architecture.

Imagine this actor-critic architecture as a playing field with a child ("actor") and his parent ("critic"). The child learns from playing but is supervised by his parent. The child wants to explore all the possible options, such as sliding down a slide and swinging on a swing. The parent observes the child and criticizes or compliments him based on his actions. The parent's decisions must be environmentally dependent. For example, if the child wants to swim in a pool, swimming deserves far less praise than sliding or swinging. The reason is easy to understand. A swimming pool is a more dangerous scene for a child than many other playing fields. For the child, the sliding and swinging fields are endlessly feasible regions, while the swimming pool is an infeasible region. To let the child safely learn new skills, the parent needs to recognize which region is feasible and which region is infeasible.

This kind of region-based supervision inspires us to add a new element into the existing AC architecture. The added element is named "scenery", whose purpose is to calculate the feasible working region of the trained policy. The new scenery update, together with traditional actor and critic updates, are able to build more advanced learning algorithms, which have a three-element architecture, namely, actor-critic-scenery (ACS). In this ACS architecture, actor and critic still maintain the same functionality, i.e., policy improvement and policy evaluation. The newly added scenery element seeks to identify an endlessly feasible region (EFR). The functionality of this element is referred to as "region identification". In summary, the ACS architecture has three basic functional modules: policy improvement (PIM), policy evaluation (PEV), and region identification (RID).

9.6.1   Two Types of ACS Architectures

In the ACS architecture, designing the scenery element requires two key steps. First, a quantitative description of the EFR should be provided. Usually, the feasible region is described as the subzero level set of a feasibility function. Then, a scenery optimization problem should be built according to how to examine feasibility. The resulting scenery update is to search for the perfect feasibility function, whose region is equal to the largest EFR. The definitions of feasibility functions are very different depending on their feasibility examination mechanism. Classified by the definition of feasibility function, there are two kinds of scenery elements: one is called reachability-based scenery, and the other is called solvability-based scenery. The reachability-based scenery can be viewed as a special critic that is related to risk evaluation, and the definition of its feasibility function originates from the Hamilton-Jacobi reachability analysis. The associated feasibility examination is based on how to evaluate the risk level of each given state. As shown in Figure 9.23 (a), this kind of scenery element stays in parallel with the critic element since risk evaluation in the scenery is similar to policy evaluation in the critic. Different from the reachability-based scenery, the solvability-based scenery more resembles an adversarial player against the actor element, and the definition of its feasibility function depends on whether each local constrained PIM has a feasible solution. The associated feasibility examination depends on the penalty degree of potential constraint violation. As demonstrated in Figure 9.23 (b), this kind of scenery element simultaneously updates itself with the actor element because they actually solve the same optimization problem. In fact, these two kinds of scenery elements can work in one ACS architecture. For example, they are used in positions that are parallel to the critic and actor, respectively. However, its complex learning structure may easily result in unpredictable policy behavior. In most tasks, we usually only retain one scenery for better theoretical completeness and algorithm conciseness.

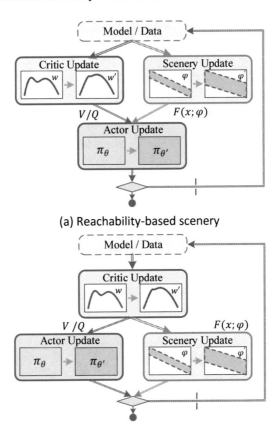

(a) Reachability-based scenery

(b) Solvability-based scenery

Figure 9.23 Actor-critic-scenery architecture

How to mathematically quantify the endlessly feasible region (EFR) with a feasibility function is the first key question in building a useful ACS algorithm. The EFR is defined as the subzero level set of a feasibility function:

$$X = \{x | F(x) \leq 0\},$$

where X represents an EFR, and $F(x)$ is the feasibility function. The feasibility function that corresponds to the largest EFR is referred to as "perfect". An optimal policy must be feasible in the region that is governed by the perfect feasibility function. The subzero level set of the perfect feasibility function is

$$X_{\text{Edls}} = \{x | F^*(x) \leq 0\},$$

where X_{Edls} is the largest EFR and $F^*(x)$ is the perfect feasibility function. In a reachability-based scenery, the definition of the feasibility function is motivated by the Hamilton-Jacobi reachability analysis. Accordingly, this function is also referred to as the reachable feasibility function. It describes the most dangerous constraint value starting from a given state. A larger reachable value means that the state is more dangerous in the future and a constraint violation is more likely to occur. One benefit of this definition is that its perfect version holds an optimality condition, which is the basis for building the scenery optimization problem. In a solvability-based scenery, the definition of the feasibility function depends on whether each local constrained optimization in the actor element is solvable or not. Due to its connection to problem solvability, this function is also referred to as the solvable feasibility function. A typical design of this function is based on the complementary slackness condition in constrained optimization. The value of complementary slackness is an effective indicator for regional feasibility. This value remains zero if there is a feasible solution; otherwise, it becomes a positive real value.

9.6.2   Reachability-based Scenery

The Hamilton-Jacobi (HJ) reachability analysis is a scalable safety verification method for constrained dynamic systems [9-30][9-5]. Its popularity is based on several well-known reasons, including its high compatibility with nonlinear dynamics, formal treatment of bounded disturbances, and well-developed numerical tools. Given an arbitrary constraint function, the basis of HJ reachability analysis is to compute a reachable set. This set is composed of dangerous states and is actually a representation of an infeasible region [9-20]. The complement of the reachable set represents an endlessly feasible region. Classic HJ reachability analysis only considers checking the risk of constrained dynamic systems, and it has no connection with how to find an optimal policy. As an extension to reinforcement learning, the ACS architecture is used to consider two entangled requirements: safety guarantee and optimality pursuit [9-45]. In reachability-based ACS, searching for an optimal policy is equally important to identifying an endlessly feasible region. Even though policy search is the first priority, an endlessly feasible region must be simultaneously identified to ensure the usefulness of the learned policy.

9.6.2.1   Reachable feasibility function

The theory of HJ reachability analysis inspires us to design a reachable version of feasibility function. One benefit of reachable feasibility function is its high compatibility with general nonlinear dynamics. Let us consider a deterministic dynamic system in the discrete-time domain. The system dynamics are assumed to be fully controllable, i.e., there exists a policy that is able to move the state from one point to any other point. Here, we use the notation $x_{t+i}^{\pi}, i = 0,1,2,\cdots,\infty$ to describe the state trajectory starting from an arbitrary initial state $x_t = x$ under the policy π. The reachable feasibility function $F^{\pi}(x)$ is defined as

$$F^{\pi}(x) \overset{\text{def}}{=} \max_i h(x_{t+i}^{\pi}), i \in \{0,1,2,\cdots,\infty\}, \tag{9-23}$$

where $h(\cdot)$ is the original constraint function. Here, the superscript in $F^{\pi}(x)$ represents a feasibility function that has the policy π as one of its arguments. Figure 9.24 shows the relationship between $h(x)$ and $F(x)$. The constraint $h(x) \leq 0$ means that the current state is safe. As its extension to the whole temporal domain, $F(x)$ does not describe the risk level of the current state, but the worst case in the future. It is the maximizer that attempts to find the most dangerous constraint point along the whole state trajectory. When $F^{\pi}(x) \leq 0$, the policy π is always safe in the infinite horizon. When $F^{\pi}(x) > 0$, a constraint violation is doomed to happen at some instance in the future.

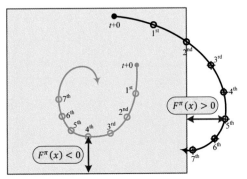

Figure 9.24 Definition of the reachable feasibility function

With the ability to describe the risk level in the future, this feasibility function provides directional guidance to search for a less risky policy. Obviously, the smaller the reachable value is, the safer the policy will become. Among all the possible policies, there should exist one special policy that has the smallest reachable value. The outcome that corresponds to this policy is the perfect feasibility function:

$$F^*(x) \stackrel{\text{def}}{=} \min_{\pi} \max_{i} h(x_{t+i}^{\pi}), i \in \{0,1,2,\cdots,\infty\}. \tag{9-24}$$

It is observed that the perfect feasibility function contains two opposite optimizers. The inner maximizer attempts to find the most dangerous constraint point, while the outer minimizer seeks to find the safest policy. Here, the safest policy is defined as the policy that has the lowest reachable value. Interestingly, the outer minimizer in (9-24) is an unconstrained optimization, which naturally builds an optimization problem for region identification. Like any multistage optimization, it holds an optimality condition from Bellman's principle of optimality.

9.6.2.2   Risky Bellman equation

The reachable feasibility function has an intrinsic recursive structure, and it has a self-consistency condition similar to that of a value function. Thus, the associated scenery optimization problem holds an optimality condition, which is named as "risky Bellman equation". In a general multistage optimization, Bellman's principle tells us that a local path from the global optimal trajectory remains optimal regardless of where it starts and where it ends. A similar conclusion is found for the risky Bellman equation, which states that a part of the safest trajectory always remains the safest.

- Theorem 9.3: Given the definition in (9-24), the perfect reachability function satisfies the following risky Bellman equation:

$$F^*(x) = \min_{u} \max\{h(x), F^*(x')\}. \tag{9-25}$$

Proof:

$$F^*(x_t) \stackrel{\text{def}}{=} \min_{\{u_t,u_{t+1},\cdots\}} \max_{i}\{h(x_{t+i})\}, i \in \{0,1,2,\cdots,\infty\}$$

$$= \min_{\{u_t,u_{t+1},\cdots\}} \max\left\{h(x_t), \max_{i} h(x_{t+i})\right\}, i \in \{1,2,\cdots,\infty\}$$

$$= \min_{u_t} \max \left\{ h(x_t), \min_{\{u_{t+1}, u_{t+2}, \cdots\}} \max_i h(x_{t+i}) \right\}$$

$$= \min_{u_t} \max\{h(x_t), F^*(x_{t+1})\}.$$

∎

Figure 9.25 illustrates how to derive the risky Bellman equation. The core of this derivation is the division of the outer minimization into two separable stages. The first stage contains the constraint function at the current time, and the second stage is composed of the remaining feasibility function starting from the next state. According to Bellman's principle of optimality, $F^*(x)$ is equal to the smaller value of the two stages.

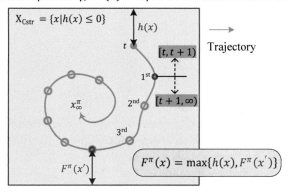

Figure 9.25 Explanation of how to derive risky Bellman equation

As previously discussed, a Bellman equation can always reduce to a self-consistency condition if no optimality is pursued. This self-consistency condition is a built-in property of a separable multistage control problem. Considering the structural similarity of feasibility function and value function, the risky version of self-consistency condition emerges after removing the outer minimizer:

$$F^\pi(x) = \max\{h(x), F^\pi(x')\}.$$

Obviously, the risky version of self-consistency condition has two possible solutions: one is $F^\pi(x) = F^\pi(x')$, and the other is $F^\pi(x) = h(x)$. The risky Bellman equation plays a central role in region identification, which is similar to that of the Bellman equation in policy search. One can simply use tabular dynamic programming (DP) to calculate the perfect feasibility function if a low-dimensional dynamic model is available. When facing high-dimensional tasks with continuous state spaces, the feasibility function has better be generalized, and a parameterized scenery problem is formulated for more efficient computation.

9.6.2.3   Model-based reachable ACS algorithm

The goal of an ACS algorithm is to output a policy that is both optimal and feasible. It needs to fulfill two tasks: the first task is to find the optimal policy, and the second task is to identify the largest EFR. The feasibility function is parameterized for the task of region identification (RID):

$$F(x; \varphi) \cong F(x),$$

where $\varphi \in \mathbb{R}^{l_\varphi}$ is the scenery parameter and l_φ is the dimension of the scenery parameter. A sufficient condition for the largest EFR is built using the risky Bellman equation. This condition, together with the conventional Bellman equation, are utilized to finish both policy search and region identification. Without using the conventional Bellman equation, the risky Bellman equation itself is not sufficient to determine a safe policy and its working region. There is one short explanation why the scenery update must work with the critic update, i.e., regarding the reachability-based scenery as a special reward system with the risk evaluation ability. The critic element and scenery element build two separated evaluation systems: one is the policy evaluation system, which outputs the level of policy optimality, and the other is the safety evaluation system, which outputs the risk level of constraint violation. Under this perspective, the design of scenery loss function is very close to that of critic loss function. Here, we choose the squared Bellman error from the risky Bellman equation as the scenery loss function:

$$J_{\text{Scenery}} = \frac{1}{2}\left(\max\{h(x), \min_u F(x'; \varphi)\} - F(x; \varphi)\right)^2, \tag{9-26}$$

where J_{Scenery} is the scenery loss function. The risky Bellman error has two terms, in which the first term provides a target value for updating the feasibility function, and it contains an inner minimizer and an outer maximizer. Upon a deep inspection, it is found that this scenery loss function follows a similar idea as parameterized Q-learning (e.g., deep Q-network), in which value iteration is utilized to update the feasibility function and no explicit policy is involved. One benefit is that it avoids the usage of an additional risk-related policy, which can decrease the structural complexity of the overall ACS algorithm. Similar to deep Q-network, this design is only suitable for a discrete action space or a control affine model, which means that the inner minimizer has an explicit solution in the target calculation. Otherwise, the inner minimization problem must be solved with a certain numerical optimization method, and it will introduce additional computational complexity.

Even though both maximizer and minimizer are contained in the scenery loss function, its scenery updates can be rather simple when only semi-gradients are considered. The derivation of semi-gradients only addresses the second term in the Bellman error and leaves the complicated first term unchanged. Thus, semi-gradient computation does not need to go through complex maximization and minimization. The semi-gradient of the scenery loss function is given as

$$\nabla_\varphi J_{\text{Scenery}} = -\left(F^{\text{target}}(x) - F(x; \varphi)\right)\frac{\partial F(x; \varphi)}{\partial \varphi},$$

$$F^{\text{target}}(x) \stackrel{\text{def}}{=} \max\{h(x), \min_u F(x'; \varphi)\}$$

Thus far, learning a feasibility function is very similar to learning a value function. As demonstrated in Figure 9.23 (a), the scenery update works in parallel with the critic update at each ACS iteration. The ACS learner can either be a model-driven or data-driven learner depending on whether a dynamic model or a set of data is accessible. If all aspects are designed properly, fast convergence to the largest EFR will occur when the model is accurate enough or the environment is sufficiently explored.

Let us discuss how to modify the constrained PIM to fit the algorithm design. After a few scenery updates, the parameterized function $F(x; \varphi)$ emerges and becomes known by the reinforcement learner. Conventionally, constrained PIM is formulated to obey the one-step constraint, i.e., $x' \in X_{\text{Edls}}$ (see (9-12) for more details). However, there is no pre-known knowledge about the largest EFR. The actor update requires a small but critical change so as to build a practical ACS algorithm. A known feasible region is used to replace the largest EFR in each constrained PIM. The best candidate is the EFR indicated by each known feasibility function. The rationality behind is that in a constrained Bellman equation, the one-step constraint is not just a state confinement. It is also a field of definition for meaningful optimization. Any constraint replacement is acceptable as long as it yields a meaningful optimization problem. Therefore, the constrained PIM is modified to be

$$J_{\text{Actor}} = l(x, u) + V(x')$$

subject to   (9-27)

$$F(x'; \varphi) \leq 0 .$$

One advantage of this modification is its excellent compatibility with various constrained optimization methods. Methods such as penalty function, Lagrange duality and FDD can be seamlessly applied to solve this new constrained PIM. In this section, an exterior penalty function is taken to illustrate how to design a reachable ACS algorithm. In the model-based setting, a rectified linear penalty is utilized to solve each modified constrained PIM (9-27), in which a linear penalty is posed when the next state moves to the outside of the state constraint. The penalty strength on constraint violation depends on how to choose the penalty coefficient. The larger the penalty coefficient is, the stronger the penalty becomes. However, a too strong penalty may reduce the possibility of the learned policy reaching its true optimum.

Algorithm 9-1: Model-based ACS with reachability-based scenery

---

Hyperparameters: critic learning rate α, actor learning rate β, scenery learning rate ζ, maximum batch size M, penalty coefficient ρ

Initialization: state-value function $V(x; w)$, policy function $\pi(x; \theta)$, feasibility function $F(x; \varphi)$

Repeat

       (1)   Use environment model

       Initialize memory buffer $\mathcal{D} \leftarrow \emptyset$

       Repeat M times environment reset

            $x \sim d_{\text{init}}(x)$

            $u = \pi(x; \theta)$

            $x' = f(x, u)$

            Compute $\partial V / \partial w, \partial l / \partial u, \partial V / \partial x, \partial f^{\mathsf{T}} / \partial u,$

                    $\partial h / \partial x, \partial \pi^{\mathsf{T}} / \partial \theta, \partial F / \partial \varphi, \partial F / \partial x$

$$D \leftarrow D \cup \left\{ \left(l, V, \frac{\partial V}{\partial w}, \frac{\partial l}{\partial u}, \frac{\partial V}{\partial x}, \frac{\partial f^{\mathrm{T}}}{\partial u}, \frac{\partial h}{\partial x}, \frac{\partial \pi^{\mathrm{T}}}{\partial \theta}, \frac{\partial F}{\partial \varphi}, \frac{\partial F}{\partial x} \right) \right\}$$

End

(2) Critic update

$$\nabla_w J_{\text{Critic}} \leftarrow -\frac{1}{M} \sum_{D} \left(l(x, u) + V(x'; w) - V(x; w) \right) \frac{\partial V(x; w)}{\partial w}$$

$$w \leftarrow w - \alpha \cdot \nabla_w J_{\text{Critic}}$$

(3) Scenery update

$$\nabla_\varphi J_{\text{Scenery}} \leftarrow -\frac{1}{M} \sum_{D} \left\{ \max \left\{ h(x), \min_u F(x'; \varphi) \right\} - F(x; \varphi) \right\} \frac{\partial F(x; \varphi)}{\partial \varphi}$$

$$\varphi \leftarrow \varphi - \zeta \cdot \nabla_\varphi J_{\text{Scenery}}$$

(4) Actor update

$$\nabla_\theta J_{\text{Actor}} \leftarrow \frac{1}{M} \sum_{D} \frac{\partial \pi^{\mathrm{T}}(x; \theta)}{\partial \theta} \left(\frac{\partial l}{\partial u} + \frac{\partial f^{\mathrm{T}}}{\partial u} \frac{\partial V(x')}{\partial x'} + \rho \frac{\partial f^{\mathrm{T}}}{\partial u} \frac{\partial \max\{F(x'), 0\}}{\partial x'} \right)$$

$$\theta \leftarrow \theta - \beta \cdot \nabla_\theta J_{\text{Actor}}$$

End

In summary, the reachability-based scenery quantifies the risk level of a policy from the perspective of the most dangerous future state. The risky Bellman equation is derived from the definition of reachable feasibility function, which provides directional guidance to search for a safe policy. The reachability-based scenery can learn the largest EFR, while other methods, such as handcrafted control barrier function (CBF), offer no guarantee that their invariance set will be the largest one. Although not explicitly discussed, a reachable ACS algorithm often has an excellent ability to resist model mismatch and external disturbances. Despite these advantages, this algorithm has an intrinsic limit when facing general nonlinear systems. The scenery loss function contains an inner minimization problem with respect to actions. If the action space is not discrete or the model is not input-affine, numerical optimization must be employed to search for the safest action. Therefore, a more generic reachable ACS algorithm should parameterize two conflicting policies: one is the original policy to pursue control optimality, and the other is a new risk-evaluation policy to pursue the best safety. In this situation, the algorithm complexity will significantly increase, and its training process may easily lose stability due to the overly complex algorithm structure.

9.6.3   Solvability-based Scenery

Ensuring reachability and examining solvability are two basic ideas to design scenery elements. The former is to evaluate the risk level by searching for the most dangerous future state, and the latter is to identify the potential constraint violation by checking whether the constrained PIM has a feasible solution. In the reachability-based ACS, the design of scenery loss function has no connection with the critic loss function or actor loss function. In contrast, in the solvability-based ACS, its scenery element and actor element share the same optimization problem. Their parameter updates act as two adversarial players that compete against each other. The actor seeks to learn a better policy, and the scenery aims to pursue a stricter constraint satisfaction. This adversarial optimization also implies that feasibility guarantee must come at the cost of sacrificing a certain level of policy optimality. Often, their competition will eventually reach a Nash equilibrium, i.e., an optimal policy and its perfect feasibility function. Existing studies in this field have mainly focused on how to search for an optimal policy under the constraint confinement, but do not consider how large its feasible region is [9-13][9-16]. As discussed before, a useful policy must work in a known feasible region. Therefore, accurately computing the endlessly feasible region is equally important to searching for an optimal policy. This section will introduce the definition of solvable feasibility function and discuss how to design a solvability-based ACS trainer to identify the largest feasible region [9-26].

9.6.3.1   Basis of solvability-based scenery

The central task of feasibility examination is to check whether a feasible solution exists or not in each constrained PIM. This is not an easy task since most constrained optimization problems in RL/ADP are highly nonlinear and even nonconvex. One must rely on various numerical optimization algorithms to search for their optimal solution. As a result, how to check solvability varies a lot depending on the working mechanism of each optimization method. For instance, the penalty strength is a feasibility indicator in the penalty function method, while the Lagrange multiplier can serve as an indicator in the Lagrange multiplier method. Therefore, the definition of solvable feasibility function depends on what kind of optimization method is utilized.

Adding a penalty term may be the simplest approach to solve constrained PIM. A rectified penalty term can be inserted into the actor loss function in an attempt to avoid constraint violation. The penalty term is often the product of a fixed penalty coefficient and the constraint function. Through this penalty term, the constrained PIM is transferred to an unconstrained optimization, which is associated with both actor and scenery. A simple choice for solvable feasibility function is the penalty term itself because it has the property of monotonically increasing with the degree of constraint violation. The feasibility of a given region is examined by observing how much of a penalty is beyond an empirical threshold.

The Lagrange multiplier method is another popular optimization method. This method provides an intrinsic indicator to examine the regional feasibility, namely, the Lagrange multiplier and its complementary slackness. The Lagrange multiplier can be regarded as a special penalty coefficient that has the ability to adjust itself to match the penalization strength. When a constraint violation is about to occur, the Lagrange multiplier automatically increases to penalize those states that move outside the boundary. The automatic adjustment makes the Lagrange multiplier itself a good solvability indicator, but its judgment must require an empirical threshold. An advanced choice of solvability indicator is the complementary slackness in the Karush-Kuhn-Tucker (KKT) condition. The complementary slackness is the product of the Lagrange multiplier and constraint function, which is an analogy of the aforementioned penalty term. This variable becomes zero in the following cases: (1) the Lagrange multiplier is equal to zero, or (2) the state constraint is equal to zero. The former means that an optimal solution is found, and no constraint is active. The latter means that exact constraint satisfaction is achieved and the corresponding constraint becomes active. If none of them occur, the complementary slackness will become a positive value, which means that no feasible solution exists.

9.6.3.2   Solvable feasibility function

What kind of constraint is needed to build a computationally tractable constrained PIM? The previous strategy was to replace $x' \in X_{Edls}$ with $F(x'; \varphi) \leq 0$. This replacement does not work very well in solvability-based scenery updates. The reason attributes to the mechanism that the actor and scenery share the same optimization problem. The adversarial feasibility identifier may easily cause self-excited instability in its competition with the policy searcher. Therefore, in the design of solvability-based scenery, one needs to rely on more consistent constraints, for example, one-step pointwise constraint and one-step barrier constraint. Either of them provides a more relaxed but fixed constraint to stabilize the feasibility identifier. In addition, these new constraints are still posed on the next state, which only rolls out forward one step and does not add much computational burden. Compared to one-step pointwise function, one-step barrier constraint is more sensitive to the constraint violation. As a result, the feasible region can be somewhat enlarged with only a single inequality. Here, for the sake of narrative conciseness, we use the one-step pointwise constraint $h(x') \leq 0$ to illustrate how to redesign the constrained PIM.

Let us formally introduce the definition of solvable feasibility function. The Lagrange multiplier is a function of the state vector rather than a scalar variable. The function approximation technique is often utilized to parameterize the Lagrange multiplier. The Lagrange function for parameterized constrained PIM is built as follows:

$$L(\theta, \varphi) = l(x, u) + V(x') + \lambda(x; \varphi)h(x').$$

where θ is the policy parameter, and φ is the scenery parameter. As previously discussed, this loss function is shared to perform both actor updates and scenery updates. The actor updates seek to search for the best policy, while the scenery updates aim to find the perfect feasibility function. Hence, solving this constrained PIM is equivalent to finding the optimal variables of the following dual problem:

$$\max_{\varphi} \min_{\theta} \{L(\theta, \varphi)\}. \tag{9-28}$$

In this dual problem, the Lagrange multiplier must be no less than zero. This nonnegative requirement can be satisfied by some parameterization tricks, for example, using a nonnegative activation function in the output layer of a neural network. If (9-28) is solvable in a given region X, its Lagrange multiplier must satisfy the complementary slackness condition:

$$\lambda(x; \varphi)h(x') = 0, \forall x \in X. \tag{9-29}$$

The complementary slackness is zero only when there exists a feasible solution. If there exists no feasible solution, severe constraint violation is unavoidable regardless of what kind of policy is chosen. In this situation, the Lagrange multiplier will automatically go to infinity, and the complementary slackness becomes greater than zero. Therefore, a more reasonable solvable feasibility function is defined as

$$F(x; \varphi) \overset{\text{def}}{=} \lambda(x; \varphi)h(x').$$

(9-30)

In this definition, φ is the parameter of both Lagrange multiplier and feasibility function. To examine the feasibility of a given state point, the value of feasibility function $F(x; \varphi)$ can simply be checked. The state point is feasible when $F(x; \varphi)$ is zero; otherwise, it is infeasible. The condition of feasibility examination is stated in Table 9.4.

Table 9.4 How to examine feasibility

Region	Solvable feasibility function
x is in EFR	$F(x; \varphi) = 0$
x is not in EFR	$F(x; \varphi) > 0$

A common approach to solve the dual problem is the so-called dual ascent algorithm, in which one-step gradient descent for the actor update and the one-step gradient ascent for the scenery update are alternately executed. The dual ascent algorithm is expressed as

$$\theta^{\text{new}} \leftarrow \theta - \alpha_\theta \cdot \nabla_\theta L(\theta, \varphi),$$
$$\varphi^{\text{new}} \leftarrow \varphi + \alpha_\varphi \cdot \nabla_\varphi L(\theta, \varphi),$$

where $\alpha_\theta > 0$ and $\alpha_\lambda > 0$ are the step sizes. Their respective gradients are

$$\nabla_\theta L(\theta, \varphi) = \frac{\partial u^T}{\partial \theta} \left(\frac{\partial l}{\partial u} + \frac{\partial f^T}{\partial u} \frac{\partial V(x')}{\partial x'} + \lambda \frac{\partial f^T}{\partial u} \frac{\partial h(x')}{\partial x'} \right),$$

$$\nabla_\varphi L(\theta, \varphi) = h(x') \frac{\partial \lambda}{\partial \varphi}.$$

In the Lagrange multiplier method, the zero condition of complementary slackness provides a natural threshold to distinguish feasible states from infeasible states. Since the threshold is equal to zero in theory, an effective solvability indicator is naturally built without having difficulty to specify its threshold. A small empirical threshold can be chosen to distinguish feasible states from infeasible states. This empirical threshold is a hyperparameter that represents how much error is tolerated in the feasibility examination. The same analysis also applies to penalty function methods, in which an improper threshold setting on the penalty level may lead to a large regional discrepancy. One challenge of complementary slackness condition is that its associated ACS algorithm actually has two adversarial players from dual ascent updates. Easy divergence may occur if some key hyperparameters are not properly chosen, especially those coefficients that adjust the actor and scenery learning rates. A few empirical tricks can be introduced to eliminate the issue of easy divergence, for example, assigning lower learning rates to avoid fast changes, adding target networks to stabilize updates, and updating parameters less frequently.

9.6.3.3   Monotonic extension of feasible region

The solvability-based scenery shares the same loss function with the actor. The dual ascent iteration algorithm alternately searches for a better policy and a larger feasible region. One might be interested in why such an adversarial update can enlarge the area of the feasible region instead of just keeping it or even shrinking it. Formally, this characteristic is expressed as monotonic region extension property, which plays a central role in fulfilling the ultimate goal of ACS. The region extension property can be proven by the dual ascent mechanism in a perfect ACS algorithm. Here, perfectness means that there is no numerical error in the function approximation and minimax optimization, and the environment model is also accurately known.

- Theorem 9.4: Given the Lagrange multiplier method, each newly learned EFR X_{k+1} is at least a superset of its previous EFR X_k; i.e.,

$$X_k \subseteq X_{k+1}.$$

Proof:

At the k-th step, we have a feasible policy π_k in the EFR X_k. The EFR is defined as the zero level set of the solvable feasibility function: $X_k = \{x | F_k(x) = 0\}$. It is known that $F_k(x) = \lambda_k(x)h(x'(\pi_k)) = 0$ when $x \in X_k$. Moreover, the state-value function $V^{\pi_k}(x), x \in X_k$, is finite because π_k is a feasible policy. Using the Lagrange multiplier method, let us define the actor loss function as the Lagrange function:

$$J_{Actor}(\pi, \lambda) \overset{\text{def}}{=} l(x, \pi) + V(x'(\pi)) + \lambda(x)h(x'(\pi))$$

The actor loss function forms a meaningful optimization problem because its three terms are all finite functions. At the $(k + 1)$-th step, its actor and scenery updates are based on the following minimax optimization:

$$[\pi_{k+1}, \lambda_{k+1}] = \arg \max_\lambda \min_\pi \{J_{Actor}(\pi_k, \lambda_k)\}, x \in X_k.$$

Here, $\pi_{k+1} = \pi_k$ and $\lambda_{k+1} = \lambda_k$ can be selected as one potential solution. The complementary slackness condition gives the following equality:

$$F_{k+1}(x) = \lambda_{k+1}(x)h(x'(\pi_{k+1})) = 0, x \in X_k.$$

Since X_{k+1} is the zero level set of $F_{k+1}(x)$, X_k must be a subset of X_{k+1} according to the definition of complementary slackness condition. That is, the newly learned π_{k+1} is at least feasible in its last EFR X_k after it updates via the Lagrange multiplier method.

∎

In addition, we can further prove that when the ACS iteration stops, we have

$$X_{Edls} = X_\infty \overset{\text{def}}{=} \lim_{k \to \infty} X_k.$$

This conclusion can be proven by contradiction. Let us assume that there exists a subset of the largest EFR, i.e., $X_{sub} \subseteq X_{Edls}$; however, this subset has no intersection with the region X_∞, i.e., $X_{sub} \cap X_\infty = \emptyset$. For any state $x \in X_{sub}$, the environment exploration at this state point provides a new updating space for policy improvement. This is because its information as a feasible point has not been exploited before. Based on this analysis, the scenery update has not reached its optimum and the region extension can still continue. This introduces a contradiction into the previous assumption. Together with

Theorem 9.4, it is easily concluded that the learned EFR monotonically increases as the policy is gradually improved, and eventually, their respective optimum will be reached.

9.6.3.4   Model-based solvable ACS algorithms

Thus far, we have explained the mechanism of solvability-based scenery updates. Let us introduce how to design its model-based ACS algorithm. This ACS algorithm inherits the mechanism of dual ascent updates due to the shared loss function between actor and scenery. One of its key features is that scenery updates should run synchronously with actor updates. Because of its similarity to traditional AC architecture, the solvable ACS architecture can have various algorithm variants, including on-policy and off-policy, model-based and model-free, stochastic policy and deterministic policy. The pseudocode in Algorithm 9-2 shows an on-policy version in a model-based setting. This ACS algorithm has three parameterized functions: policy function, state-value function, and Lagrange multiplier function. The feasibility identifier is built through the complementary slackness condition.

Algorithm 9-2: Solvable ACS algorithm in model-based setting

Hyperparameters: critic learning rate α, actor learning rate β, scenery learning rate ζ, times of environment reset M

Initialization: state-value function $V(x; w)$, policy function $\pi(x; \theta)$, Lagrange multiplier function $\lambda(x; \varphi)$.

Repeat (indexed by k)

      (1) <u>Use environment model</u>

      Initialize memory buffer $\mathcal{D} \leftarrow \emptyset$

      Repeat M times environment reset

$$x \sim d_{\text{init}}(x)$$
$$u = \pi(x; \theta)$$
$$x' = f(x, u)$$

          Compute $\partial V / \partial w, \partial l / \partial u, \partial V / \partial x, \partial f^{\text{T}} / \partial u,$
$$\partial h / \partial x, \partial \pi^{\text{T}} / \partial \theta, \partial F / \partial \varphi$$
$$\mathcal{D} \leftarrow \mathcal{D} \cup \left\{ \left(l, V, \frac{\partial V}{\partial w}, \frac{\partial l}{\partial u}, \frac{\partial V}{\partial x}, \frac{\partial f^{\text{T}}}{\partial u}, \frac{\partial h}{\partial x}, \frac{\partial \pi^{\text{T}}}{\partial \theta}, \frac{\partial F}{\partial \varphi} \right) \right\}$$

      End

      (2) <u>Critic update</u>

$$\nabla_w J_{\text{Critic}} \leftarrow -\frac{1}{M} \sum_{\mathcal{D}} (l(x, u) + V(x'; w) - V(x; w)) \frac{\partial V(x; w)}{\partial w}$$

$$w \leftarrow w - \alpha \cdot \nabla_w J_{\text{Critic}}$$

      (3) <u>Actor update</u>

$$\nabla_\theta J_{\text{Actor}} \leftarrow \frac{1}{M} \sum_{\mathcal{D}} \frac{\partial \pi^{\text{T}}(x; \theta)}{\partial \theta} \left(\frac{\partial l}{\partial u} + \frac{\partial f^{\text{T}}}{\partial u} \frac{\partial V(x')}{\partial x'} + \lambda(x) \frac{\partial f^{\text{T}}}{\partial u} \frac{\partial h(x')}{\partial x'} \right)$$

$$\theta \leftarrow \theta - \beta \cdot \nabla_\theta J_{\text{Actor}}$$

(4) <u>Scenery update</u>

$$\nabla_\varphi J_{\text{Scenery}} \leftarrow \frac{1}{M} \sum_{\mathcal{D}} h(x') \frac{\partial \lambda(x; \varphi)}{\partial \varphi}$$

$$\varphi \leftarrow \varphi + \zeta \cdot \nabla_\varphi J_{\text{Scenery}}$$

End

∎

Let us briefly summarize the characteristics of this solvable ACS algorithm. The joint loss function of actor and scenery is the Lagrange function, in which optimality pursuit and safety guarantee are two competing sides of the maximin optimization. Instead of generalizing the whole feasibility function, only the Lagrange multiplier is parameterized, which is a part of complementary slackness. The dual ascending method is used to simultaneously update actor and scenery. In general, its scenery updates and actor updates are more computationally efficient than those in a reachable ACS algorithm. When policy improvement and region extension finally stop, the algorithm iteration will reach the saddle point of the two adversarial players, which are the optimal policy and its perfect feasibility function.

9.7   Safety Considerations in the Real World

Strictly guaranteeing safety is an indispensable task in controlling real-world systems. Here, safety guarantee is defined as the ability to avoid fatal violation of hard state constraints. The hard constraint is a concept that contrasts the soft constraint. A constraint is said to be hard when it cannot be violated at any circumstance. The physical meaning of constraint violation varies with different systems. For example, collision avoidance in autonomous driving means that inter-vehicle distances must be bigger than zero and conflict avoidance in multi-arm robot control means that arms should have no spatial contact. By equipping an OCP with hard state constraint, safety guarantee is equivalent to solving this task to find its constrained policy. The resulting policy is said to be safe because it never violates any hard state constraint when starting from a known endlessly feasible region (EFR). Hence, the safe policy and its corresponding EFR must be simultaneously solved from constrained OCP. This section will discuss how to strictly consider the issue of safety guarantee in real-world control tasks. Two basic training modes are proposed for safe policy search, and their corresponding safety-critical ACS algorithms are designed in both model-free and model-based settings. In reality, a perfect safe policy is rarely accessible due to function approximation errors, unknown modeling uncertainties, state measurement noises, etc. In many control tasks, the safety shield mechanism is also introduced as the last protection for unsafe actions.

9.7.1   Two Basic Modes to Train Safe Policy

The safe policy can be trained from either a virtual environment or a real environment. The virtual environment is actually a high-fidelity simulation model. Safety is a mandatory requirement in the real environment, but it is not quite necessary in a virtual environment. This is because constraint violation in a computer-simulated model is not a big deal, and one can always restart the simulation without any property loss or fatalities. For example, safe exploration is not critical in a virtual race car game that run in computers. However, in a real autonomous car, any constraint violation can be destructive if there is no safe policy to interact with the real traffic and road environment.

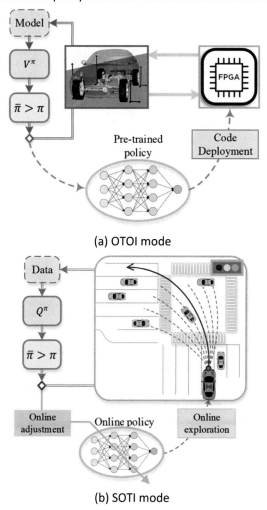

(a) OTOI mode

(b) SOTI mode

Figure 9.26 Two modes of training safe policies

Generally, there are two basic training modes to search for a safe policy. One mode is called offline training and online implementation (OTOI). As shown in Figure 9.26(a), it considers RL/ADP as an offline policy trainer that works in the virtual environment. The learned policy is then implemented as a real-world controller. Similar to MPC, the trained policy from the OTOI mode can perform receding horizon control and moreover it can achieve very high computational efficiency as an online controller. The reason is easy to understand: such a policy like tabular function or neural network is more computationally efficient than an online optimizer in MPC. The other mode is called simultaneous online training and implementation (SOTI). In this mode, RL/ADP is utilized to learn a series of safe policies through online interaction with the real environment. Since a series of online interactions are involved, each intermediate policy must be safe in a known working region in order to deploy it into reality. As illustrated in Figure 9.26(b), an autonomous car under this training mode can run safely on the real road while continuously improving its self-driving policy. The SOTI mode actually provides a basic training mechanism to build the self-evolving ability in a reinforcement learner.

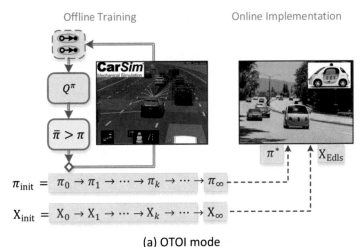

(a) OTOI mode

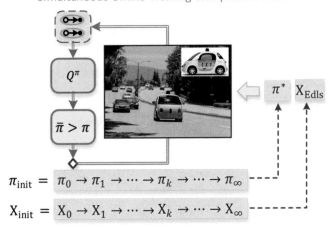

(b) SOTI mode

Figure 9.27 Safety consideration in two training modes

How to choose a proper training mode depends on whether the environment model on hand is perfect or not. A perfect model means that it has no model error with respect to the real environment. When a perfect model is available, the OTOI mode can be selected to train a safe policy. In reality, a perfect model can never be built due to those issues like structural uncertainties, parametric errors and unknown external disturbances. In many control tasks, the high-fidelity simulation model can be roughly regarded as a perfect model. In the case of imperfect model, the SOTI mode must be utilized to explore the real environment. The collected experience from the real environment interaction is used as supplementary information to the model imperfectness. In this situation, a mixed ACS trainer should be designed to blend the information from environment model and collected experience, in which some parts of environment dynamics come from the imperfect model and the other parts come from the real-world exploration. It is easy to observe that the real-world exploration is helpful to collect more accurate environment information, but it requires each intermediate policy to be safe enough. One additional observation is that such a safety-critical ACS learner must assume to have a pre-known model, even though the model may have some errors. This is because sufficient environment exploration always needs to extend from known regions to unknown regions. Without a pre-known model to guide the exploration direction, a random trial-and-error sampler in unknown regions will inevitably lead to unpredictable risk behaviors. In contrast, a pre-known model is able to avoid completely aimless exploration, and thus help the mixed trainer to continue its safe interaction with the real environment.

Table 9.5 Two training modes for safety-critical ACS

	OTOI	SOTI
Perfect model	Only the final policy needs to be safe	--
Imperfect model	--	Each intermediate policy must be safe for real environment interaction

As shown in Figure 9.26(a), OTOI obtains a safe policy at the final step. In the virtual environment, any constraint violation is simply a penalty signal to push the policy update toward a safer direction. The aforementioned model-based ACS algorithms can be easily used if the perfect model is differentiable. Compared with OTOI, SOTI is more suitable to handle the case of model imperfection. It provides a self-evolutionary learning mechanism to output a better and better policy by continuously interacting with the real environment. As shown in Figure 9.26(b), the environment interaction is a safety-critical task, which requires each intermediate policy to be safe enough; that is, each learned policy must work in a known endlessly feasible region.

9.7.2 Model-free ACS with Final Safe Guarantee

One can utilize the OTOI mode to train a safe policy when the model is perfect. We have known that there are two kinds of environment models, i.e., simplified analytical model and high-fidelity simulation model. The former model is differentiable but often less inaccurate. In terms of perfectness, we are more likely to discuss the high-fidelity simulation model. This kind of model, including its reward signal and risk signal, is usually nondifferentiable, but its accuracy is much higher than the former. In the OTOI mode, model-free ACS is a more reasonable algorithm choice, where one needs to compute sample-based gradients rather than model-based gradients. Each sample to be collected includes the following three elements: one-step transition triple, reward signal, and risk signal. Learning a risk evaluation function (i.e., accumulative return of risk signals) is similar to that of an action-value function (i.e., that of reward signals). To be compatible with action-value function, the risk evaluation function must be a function of both state and action. A new evaluation function is introduced to measure the next-step risk level:

$$H(x, u; \eta) \cong h(x'),$$

where $H(\cdot, \cdot)$ is the risk evaluation function and η is the risk parameter. Here, the risk evaluation function inherently mimics the risk level of the next state. The squared error of real and predicted risk signals is minimized:

$$\min_{\eta} J_{\text{Risk}} = \left(c' - H(x, u; \eta)\right)^2,$$

where c' is the next-step risk signal. The actor loss function is accordingly modified by replacing $h(x')$ in the model-based setting with $H(x, u; \eta)$ in the model-free setting. The Lagrange multiplier method is then utilized to design the dual ascent algorithm. In the OTOI mode, safety guarantee is only required in the final step and does not necessarily hold for each intermediate policy. Therefore, this model-free algorithm is allowed to fully explore every corner of the virtual environment and dig out sufficient information about rewards and risks.

Algorithm 9-3: Model-free ACS algorithm in the OTOI mode

Hyperparameters: critic learning rate α_{Critic}, risk learning rate α_{Risk}, actor learning rate β, scenery learning rate ζ, maximum batch size B.

Initialization: action-value function $Q(x, u; w)$, risk evaluation function $H(x, u; \eta)$, policy function $\pi(x; \theta)$, Lagrange multiplier function $\lambda(x; \varphi)$.

Repeat (indexed by k)

    (1) Collect samples

    $\mathcal{D} \leftarrow \emptyset$

    $x_0 \sim d_{\text{init}}(x)$

    For i in $0, 1, 2, \ldots, B - 1$ or until termination

        Apply $u_i = \pi(x_i; \theta)$ and then observe x_{i+1}, c_{i+1} and r_i

        $\mathcal{D} \leftarrow \mathcal{D} \cup \{(x_i, u_i, r_i, x_{i+1}, c_{i+1})\}$

    End

    (2) Critic update

$$\nabla_w J_{\text{Critic}} \leftarrow -\frac{1}{|\mathcal{D}|} \sum_{\mathcal{D}} (r + \gamma Q(x', u'; w) - Q(x, u; w)) \nabla_w Q(x, u; w)$$

$$\nabla_\eta J_{\text{Risk}} \leftarrow -\frac{1}{|\mathcal{D}|} \sum_{\mathcal{D}} (c' - H(x, u; \eta)) \nabla_\eta H(x, u; \eta)$$

    $w \leftarrow w - \alpha_{\text{Critic}} \cdot \nabla_w J_{\text{Critic}}$

    $\eta \leftarrow \eta - \alpha_{\text{Risk}} \cdot \nabla_\eta J_{\text{Risk}}$

    (3) Actor update

$$\nabla_\theta J_{\text{Actor}} \leftarrow \frac{1}{|\mathcal{D}|} \sum_{\mathcal{D}} \nabla_\theta \pi(x; \theta) (\nabla_u Q(x, u) + \lambda(x) \nabla_u H(x, u))$$

    $\theta \leftarrow \theta - \beta \cdot \nabla_\theta J_{\text{Actor}}$

---

(4) Scenery update

$$\nabla_\varphi J_{\text{Scenery}} \leftarrow \frac{1}{|\mathcal{D}|} \sum_{\mathcal{D}} H(x, u) \nabla_\varphi \lambda(x; \varphi)$$

$$\varphi \leftarrow \varphi + \zeta \cdot \nabla_\varphi J_{\text{Scenery}}$$

End

■

The pseudocode in Algorithm 9-3 describes the model-free version of safety-critical ACS algorithm, in which the Lagrange multiplier is learned to perform feasibility identification. Unlike its model-based counterpart, the model-free ACS algorithm requires learning an additional risk evaluation function, which is a natural replacement for accumulative risk signals. Moreover, this risk evaluation function should choose a parametric structure like the action-value function, in which both action and state are independent variables. The utilization of the action variable is to embed the environment information, which is required in model-free actor and scenery updates. Note that this ACS algorithm needs a perfect virtual environment because its sampler has to meet all the constraint violations before learning how to avoid them. Luckily, interacting with a virtual environment is not a safety-critical task, and the simulation can be reset easily whenever an unsafe exploration behavior occurs. Moreover, the time and cost consumption of virtual interaction is much less than that of real-world interaction; and the former is both faster and cheaper than the latter. Thus, a large amount of virtual exploration can be performed to collect sufficient information about the environment dynamics, reward signals and risk signals.

9.7.3   Mixed ACS with Safe Environment Interaction

Compared with the OTOI mode, safety guarantee in the SOTI mode is a much more challenging and complex task. The reason lies in its continuing interaction with the real-world environment, whose purpose is to collect samples and compensate for the model imperfectness. Fully exploring the real-world environment is definitely a very dangerous behavior, especially in a region that has never been visited before. In the SOTI mode, we do not assume that a perfect model is available. Instead, the pre-known model has some uncertainties due to its structural or parametric errors, or neglected high-order dynamics. The importance of the pre-known model, even though it is not perfect, is that temporarily unknown regions need the guidance of a priori model information to avoid aimless and dangerous exploration. Without loss of generality, an imperfect model with additive uncertainty is considered:

$$x_{t+1} = f(x_t, u_t) + \delta_t, \tag{9-31}$$

where δ is an additive uncertainty. The model uncertainty is bounded with a known threshold, for example, $\|\delta_t\| \le 1$. Such an uncertain model cannot serve as the exclusive source to provide accurate information to safe ACS learners. Some parts of the environment dynamics must be explored to compensate for the deficiency of this model. Other parts rely on the model information to avoid aimless and dangerous exploration. The blended utilization of virtual model and explored data is the central idea of mixed ACS learners. The term "mixed" refers to the two sources, i.e., model and data that are used to train the safe policy. Since this learner works in a mixed mode, it has an ability to predict future risky behaviors and avoid their occurrence before actually meeting them. Without the support of an imperfect model, an arbitrary trial-and-error sampler has to aimlessly explore unknown regions, and dangerous behavior is unavoidable during the interaction with a real-world environment.

9.7.3.1   Mechanism of safe environment interactions

In a mixed ACS algorithm with safety guarantee, each main iteration must simultaneously learn a safe policy and its corresponding feasible region. The safe policy at each cycle is denoted as π_k, and its working region is denoted as X_k. In the SOTI mode, π_k must be feasible in its working region X_k to ensure safe interaction in the next step, and accordingly, its environment exploration at the $(k + 1)$-th step is limited to the known local region X_k. To recursively guarantee safety, the mixed learner must output a new safe pair (π_{k+1}, X_{k+1}), and these two components are still compatible with each other; that is, π_{k+1} is also feasible in X_{k+1}. When the next-step exploration is retained in the local region X_{k+1}, safe interaction is recursively maintained with the policy π_{k+1}.

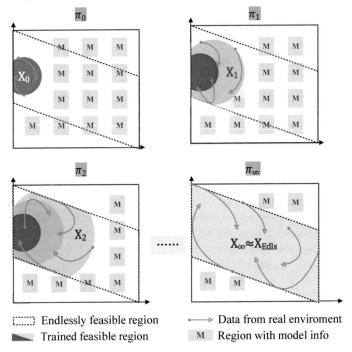

| ┈┈┈ Endlessly feasible region | ●——▶ Data from real enviroment |
| ◥ Trained feasible region | M   Region with model info |

Figure 9.28 Mixed information with safe exploration

The ability of safe exploration is ensured by the safe pair from the previous iteration. This interaction mechanism, however, has a defect that the environment exploration is limited to each known local region. Local data are not sufficient to train a new policy in the whole state space. To overcome this limitation, the data information in the local region and the model information in the whole space must be simultaneously utilized; that is how the mechanism of mixed ACS learners is conceived. The mixed ACS learner executes the regional policy improvement with local explored data and switches to rely on the model information outside of this local region. The explored data are accurate (if sufficiently explored) and can be used to build a perfect optimal policy in the local region. In the outside of this region, the model information is imperfect, but a certain prediction ability can be provided to avoid aimless exploration. The mechanism of safe interaction is demonstrated in Figure 9.28, in which online environment exploration is performed with each intermediate policy. Note that this online exploration is limited to the known EFR from the last step.

A practical ACS trainer must parameterize its key functions, including policy function, value function and feasibility function, to reduce the search space. Their generalization with parameterized functions has an additional benefit due to their continuity properties. The function continuity can extend the accurate environment information from each local region to its neighboring area, which is beneficial in accelerating the process of feasibility examination. If function continuity is properly chosen, the size of learned feasible region will gradually increase and eventually reach the largest EFR. As a Chinese proverb says, ideal is very plump, but reality is skinny. Mostly, the process of regional extension will stop at a subset of the largest EFR because one can never safely explore the environment dynamics of the infeasible region. If there is no accurate information about the infeasible region, there is no way to exactly determine its boundary with the feasible region.

9.7.3.2 Mixed ACS with regional weighted gradients

In summary, a safety-critical ACS algorithm must contain two key components: (1) a safe sampler for online exploration, and (2) a mixed learner with policy improvement and region extension abilities. In the former, the real-world environment is safely explored in each local region. In the latter, a new safe policy and its endlessly feasible region are learned for next round of safe exploration. Let us demonstrate how to design this mixed ACS algorithm with safety guarantee ability (Figure 9.29). With local real data and an imperfect model, the mixed ACS learner needs to simultaneously output a new policy and its local EFR. Both policy improvement and region extension are performed after safely exploring the real-world environment.

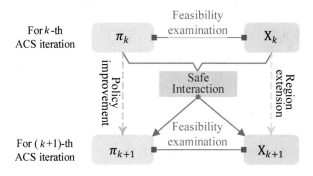

Figure 9.29 Iterative structure of mixed ACS

One main challenge of mixed ACS comes from how to deal with the imperfect model, which has uncertain knowledge about the environment dynamics. The strategy is to build a robust learner whose purpose is to search for a suboptimal policy and a subperfect region in the worst case. The worst-case solution is able to conservatively maintain the ability of safe exploration, even if its policy optimality is somewhat sacrificed. Following this perspective, additive uncertainty is regarded as an adversarial player, whose purpose is to search for the worst-case uncertainty. The Lagrange dual problem (9-16) is converted into its robust optimization version:

$$\max_{\varphi} \min_{\theta} \max_{\delta} \{L(\theta, \varphi, \delta)\}$$

with   (9-32)

$$L(\theta, \varphi, \delta) = Q(x, u) + \lambda(x; \varphi)h(x').$$

where θ is the policy parameter and φ is the scenery parameter. The robust version of the Lagrange dual problem has three unknown variables. Among them, the adversarial uncertainty is denoted as $\delta(x)$, whose maximization operation is placed at the innermost position. The maximization of the adversarial uncertainty seeks to find the worst-case disturbance. Meanwhile, the middle minimizer aims to output the worst-case policy, and the outer maximizer attempts to identify the worst-case feasible region. To simultaneously utilize the information of locally explored data and imperfect virtual models, mixed deterministic gradients are designed to update the three key elements: critic, actor, and scenery. Their gradients are in a mixed structure, and each gradient contains two subparts: one is referred to as the data-driven gradient ∇J^D from local exploration data, and the other is called the model-based gradient ∇J^M from the imperfect virtual model. The mixed deterministic gradient is the weighted sum of two subparts:

$$\kappa = \text{Area}(X_k)/\text{Area}(\mathcal{X})$$
$$\nabla J_{\#}^{\text{Mix}} = \kappa \nabla J_{\#}^D + (1 - \kappa)\nabla J_{\#}^M \qquad (9\text{-}33)$$
$$\# = \text{Actor}, \text{Critic}, \text{Scenery},$$

where κ is the area ratio, $\text{Area}(\cdot)$ is an area measure function, and $\nabla J_{\#}^{\text{Mix}}$ represents the mixed gradient. In fact, calculating the area of a working region is rather difficult because there is no access to its explicit mathematical description. The area ratio can be approximately calculated by counting the number of random samples that are in the subzero level set of the feasibility function.

The mixed version of safety-critical ACS algorithm is shown in Algorithm 9-4, in which a solvable feasibility function is utilized to quantify the local feasible region. Assume that X_0 is chosen as a subset of X_{Edls} and the initial policy guess π_0 is safe in X_0. In Algorithm 9-4, each element, including critic, actor and scenery, is updated multiple times instead of only once. A single update of each element is not sufficient to find the safe pair because early termination may cause large numerical error. A mature ACS learner should sufficiently reach the saddle point of each Lagrange dual problem. As a result, it has a capability of outputting the safe pair at each main iteration.

Algorithm 9-4: Mixed ACS algorithm in SOTI mode

Hyperparameters: critic learning rate α, actor learning rate β, scenery learning rate ζ, uncertainty learning rate ξ, maximum batch size B, critic updating frequency n_c, and actor updating frequency n_a

Initialization: action-value function $Q(x, u; w)$, policy function $\pi(x; \theta)$, adversarial uncertainty function $\delta(x; \psi)$, Lagrange multiplier function $\lambda(x; \varphi)$

//Note that initial policy guess π_0 must be safe in X_0.

Repeat (indexed by k)

    <u>(1) Collect data in the local region</u>

    $\mathcal{D}_D \leftarrow \emptyset, \mathcal{D}_M \leftarrow \emptyset$

    $x_0 \sim d_{\text{init}}(x)$

    If $x_0 \in X_k$

        //Safe environment interaction

        For i in $0, 1, 2, \dots, B - 1$ or until termination

            Apply $u = \pi_k$ in the real-world environment, observe x' and r

            $\mathcal{D}_D \leftarrow \mathcal{D}_D \cup \{(x, u, r, x')\}$

        End

    Else

        //Use imperfect model

        For i in $0, 1, 2, \dots, B - 1$ or until termination

            $u = \pi(x; \theta), \delta = \delta(x; \psi)$

            $x' = f(x, u) + \delta$

            Compute $r, h, \partial Q/\partial u, \partial h/\partial x, \partial \pi^{\mathrm{T}}/\partial \theta, \partial f^{\mathrm{T}}/\partial u, \partial \lambda/\partial \varphi$

$$\mathcal{D}_M \leftarrow \mathcal{D}_M \cup \left\{ \left(r, h, Q, \frac{\partial Q}{\partial u}, \frac{\partial h}{\partial x}, \frac{\partial \pi^{\mathrm{T}}}{\partial \theta}, \frac{\partial f^{\mathrm{T}}}{\partial u}, \frac{\partial \lambda}{\partial \varphi} \right) \right\}$$

        End

    End

    $\kappa = \text{Area}(X_k)/\text{Area}(\mathcal{X})$

    <u>(2) Mixed critic update</u>

    Repeat n_c times

$$\nabla_w J_{\text{Critic}}^{D} \leftarrow -\frac{1}{|\mathcal{D}_D|} \sum_{\mathcal{D}_D} (r + \gamma Q(x', u') - Q(x, u)) \nabla_w Q(x, u; w)$$

$$\nabla_w J_{\text{Critic}}^{M} \leftarrow -\frac{1}{|\mathcal{D}_M|} \sum_{\mathcal{D}_M} (r + \gamma Q(x', u') - Q(x, u)) \nabla_w Q(x, u; w)$$

$$w \leftarrow w - \alpha \left(\kappa \nabla_w J_{\text{Critic}}^{D} + (1 - \kappa) \nabla_w J_{\text{Critic}}^{M} \right)$$

    End

    <u>(3) Robust actor and scenery updates</u>

    Repeat n_a times

        //Actor policy parameter

$$\nabla_\theta J^{\mathrm{D}}_{\mathrm{Actor}} \leftarrow \frac{1}{|\mathcal{D}_{\mathrm{D}}|} \sum_{\mathcal{D}_{\mathrm{D}}} \nabla_\theta \pi(x;\theta) \left(\nabla_u Q(x,u) + \lambda(x) \frac{\partial f^{\mathrm{T}}}{\partial u} \frac{\partial h(x')}{\partial x'} \right)$$

$$\nabla_\theta J^{\mathrm{M}}_{\mathrm{Actor}} \leftarrow \frac{1}{|\mathcal{D}_{\mathrm{M}}|} \sum_{\mathcal{D}_{\mathrm{M}}} \nabla_\theta \pi(x;\theta) \left(\nabla_u Q(x,u) + \lambda(x) \frac{\partial f^{\mathrm{T}}}{\partial u} \frac{\partial h(x')}{\partial x'} \right)$$

$$\theta \leftarrow \theta - \beta \left(\kappa \nabla_\theta J^{\mathrm{D}}_{\mathrm{Actor}} + (1-\kappa) \nabla_\theta J^{\mathrm{M}}_{\mathrm{Actor}} \right)$$

//Adversarial uncertainty parameter & model-based update

$$\nabla_\varphi J^{\mathrm{M}}_{\mathrm{Adv}} \leftarrow \frac{1}{|\mathcal{D}_{\mathrm{M}}|} \sum_{\mathcal{D}_{\mathrm{M}}} \nabla_\psi \delta(x;\psi) \frac{\partial h(x')}{\partial x'} \lambda(x)$$

$$\psi \leftarrow \psi + \xi J^{\mathrm{M}}_{\mathrm{Adv}}$$

//Scenery parameter

$$\nabla_\varphi J^{\mathrm{D}}_{\mathrm{Scenery}} \leftarrow \frac{1}{|\mathcal{D}_{\mathrm{D}}|} \sum_{\mathcal{D}_{\mathrm{D}}} h(x') \nabla_\varphi \lambda(x;\varphi)$$

$$\nabla_\varphi J^{\mathrm{M}}_{\mathrm{Scenery}} \leftarrow \frac{1}{|\mathcal{D}_{\mathrm{M}}|} \sum_{\mathcal{D}_{\mathrm{M}}} h(x') \frac{\partial \lambda(x;\varphi)}{\partial \varphi}$$

$$\varphi \leftarrow \varphi + \zeta \left(\kappa \nabla_\varphi J^{\mathrm{D}}_{\mathrm{Scenery}} + (1-\kappa) \nabla_\varphi J^{\mathrm{M}}_{\mathrm{Scenery}} \right)$$

End

(5) Update policy and feasible region

$$\pi_{k+1} = \pi(x;\theta)$$
$$\delta_{k+1} = \delta(x;\psi)$$
$$X_{k+1} = \{x | \lambda(x;\varphi) h(x') \leq \varepsilon\}$$

End

∎

9.7.4   Safety Shield Mechanism in Implementation

In previous sections, how to train a safe policy and identify its working region has been discussed in both OTOI and SOTI modes. The trained policy may still be unsafe for several noticeable reasons, such as insufficient exploration, function approximation errors, and early termination of algorithm iterations. In addition, some optimal policies may be trained without any consideration of safe constraints. The safety shield technique can serve as the last protection in a wide range of safety-critical control tasks, in which no dangerous state is allowed to be visited during action implementation. The mechanism of safety shield is to continuously monitor the agent's behavior in the stage of action implementation and compute a safe action to replace any unsafe policy. One common safety shield method is to project the unsafe action to a new position that makes its next state satisfy the constraint. Usually, the nearest projection is performed by formulating a minimax optimization problem:

$$u_{\mathrm{Safe}} = \arg\min_u \max_{\|\delta\|\leq 1} \|u - \pi(x)\|^2, \tag{9-34}$$

subject to

$$x' = f(x, u) + \delta,$$
$$h(x') \le 0,$$

where u_{Safe} is the new action after safe projection, f is the certain part of the environment model, and $\|\delta\| \le 1$ is its bounded uncertainty. The deployment of a safety shield must rely on a model to find a feasible action. In the literature, many studies assumed that a perfect model was available. When the model is perfect, the minimax optimization is reduced to a minimization problem, in which the bounded uncertainty is precisely equal to zero. This assumption is definitely not realistic, and a perfect model is often inaccessible. Moreover, when a perfect model is available, one can always call a model-based ACS algorithm and train a safe policy under the OTOI mode. In this situation, safety shield is no longer necessary. Therefore, the standard safety shield should calculate a robust safe action with an uncertain model like the one in (9-34). The corresponding minimax optimization may sacrifice the optimality of projected actions but is able to strictly maintain safety even in the worst-case situation.

Figure 9.30 demonstrates the concept of safe projection in the state and action spaces. Whenever a potential risk is monitored, an action is projected into an allowable region, and the next state is forced to return to the constrained region. This type of projection can guarantee the safety of one-step state transition but neglects the recursive feasibility of the following states. In practice, it may not quickly drag the state away from the infeasible region. One remedy is to use a more conservative safety shield; for example, posing one-step barrier constraint into the projection problem. As previously discussed, a barrier constraint can more strongly drag the future states into the feasible region.

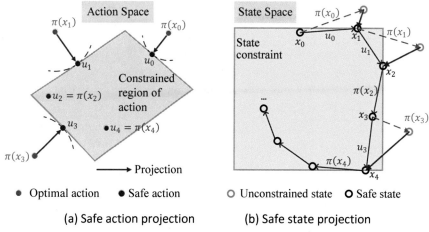

(a) Safe action projection   (b) Safe state projection

Figure 9.30 Working mechanism of safety shield

The safety shield method has the potential to prevent incorrect actions by an immature agent. However, it is a big mistake to believe that this method can always guarantee safety. One counterexample is that its projection operation may have no feasible solution due to overly tight shield constraints. The cause is similar to that of infeasibility phenomenon in constrained OCPs. If entering the infeasible region due to some occasional reasons, there is no way for the agent to correct unsafe actions and recover it from sliding into the status of constraint violation. Therefore, a reasonable application of safety shield should carefully consider the following two requirements: (1) guaranteed correctness, i.e., the safety specification is satisfied as much as possible during online execution, and (2) minimal interference, i.e., shielding may restrict exploration and lead to suboptimal policies.

9.8   References

[9-1]   Abu-Khalaf M, Lewis FL (2005) Nearly optimal control laws for nonlinear systems with saturating actuators using a neural network HJB approach. Automatica 41: 779-791

[9-2]   Achiam J, Held D, Tamar A, Abbeel P (2017) Constrained policy optimization. ICML, Sydney, Australia

[9-3]   Agrawal A, Sreenath K (2017) Discrete control barrier functions for safety-critical control of discrete systems with application to bipedal robot navigation. RSS, Massachusetts, USA

[9-4]   Ames AD, Coogan S, Egerstedt M et al (2019) Control barrier functions: Theory and applications. ECC, Naples, Italy

[9-5]   Bansal S, Chen M, Herbert S, Tomlin C (2017) Hamilton-Jacobi reachability: A brief overview and recent advances. IEEE CDC, Melbourne, Australia

[9-6]   Bertsekas D (1997) Nonlinear programming. Athena Scientific, Belmont

[9-7]   Borrelli F, Bemporad A, Morari M (2017) Predictive control for linear and hybrid systems. Cambridge university press, Cambridge

[9-8]   Boyd S, Boyd S, Vandenberghe L (2004) Convex optimization. Cambridge university press, Cambridge

[9-9]   Boyd S, Parikh N, Chu E (2011) Distributed optimization and statistical learning via the alternating direction method of multipliers. Now Publishers, Inc., Boston.

[9-10]   Byrd R, Lu P, Nocedal J et al (1995) A limited memory algorithm for bound constrained optimization. SIAM Journal on Scientific Computing 16(5):1190-1208

[9-11]   Chmielewski D, Manousiouthakis V (1996) On constrained infinite-time linear quadratic optimal control. Systems Control Letters 29(3):121-129

[9-12]   Chow Y, Ghavamzadeh M, Janson L et al (2017) Risk-constrained reinforcement learning with percentile risk criteria. J of Machine Learning Research, 18(1): 6070-6120

[9-13]   Gaskett C (2003) Reinforcement learning in circumstances beyond its control. ICCIM, Vienna, Austria

[9-14]   Conn A, Gould N, Toint P (1991) Convergence of quasi-Newton matrices generated by the symmetric rank one update. Math Programming 50(1):177-195

[9-15]   Dong L, Zhong X, Sun C et al (2016) Event-triggered adaptive dynamic programming for continuous-time systems with control constraints. IEEE Trans Neural Networks & Learn Syst 28(8): 1941-1952

[9-16] Duan J, Liu Z, Li SE et al (2022) Adaptive dynamic programming for nonaffine nonlinear optimal control problem with state constraints. Neurocomputing 484: 128–141

[9-17] Fisac J, Chen M, Tomlin C, Sastry S (2015). Reach-avoid problems with time-varying dynamics, targets and constraints. HSCC, New York, USA

[9-18] Guan Y, Ren Y, Ma H et al. Learn collision-free self-driving skills at urban intersections with model-based reinforcement learning. IEEE ITSC, Indianapolis, USA

[9-19] Howard R, Matheson J (1972) Risk-sensitive Markov decision processes. Management Science 18(7): 356-369

[9-20] Hsu K, Rubies-Royo V, Tomlin C, Fisac J (2021). Safety and liveness guarantees through reach-avoid reinforcement learning. RSS, Virtual Conference

[9-21] Kingma D, Ba J (2014) Adam: A method for stochastic optimization. ICLR, San Diego, USA

[9-22] Kong H, He F, Song X et al (2013) Exponential-condition-based barrier certificate generation for safety verification of hybrid systems. CAV, Saint Petersburg, Russia

[9-23] Levine S, Popovic Z, Koltun V (2010) Feature construction for inverse reinforcement learning. NeurIPS, Vancouver, Canada

[9-24] Lyshevski S (1998) Optimal control of nonlinear continuous-time systems: design of bounded controllers via generalized nonquadratic functionals. ACC, Philadelphia, USA

[9-25] Ma H, Chen J, Li SE et al (2021) Model-based constrained reinforcement learning using generalized control barrier function. IEEE/RSJ IROS, Prague, Czech Republic

[9-26] Ma H, Liu C, Li SE et al (2022) Joint synthesis of safety certificate and safe control policy using constrained reinforcement learning. L4DC, Stanford, USA

[9-27] Ma H, Zhang X, LI SE et al (2021) Feasibility enhancement of constrained receding horizon control using generalized control barrier function. IEEE ICPS, Victoria, Canada

[9-28] Martens J (2016) Second-order optimization for neural networks. Dissertation, University of Toronto

[9-29] Mayne D, Rawlings J, Rao C et al (2000) Constrained model predictive control: Stability and optimality. Automatica 36(6): 789-814

[9-30] Mitchell I, Bayen A, Tomlin C (2005) A time-dependent Hamilton-Jacobi formulation of reachable sets for continuous dynamic games. IEEE Trans Automatic Control, 50: 947–957

[9-31] Li SE, Jia Z, Li K, Cheng B (2015) Fast online computation of a model predictive controller and Its application to fuel economy–oriented adaptive cruise control. IEEE Trans Intelligent Transp Syst 16(3): 1199-1209

[9-32]   Nguyen Q, Sreenath K (2016) Exponential control barrier functions for enforcing high relative-degree safety-critical constraints. ACC, Boston, MA, USA

[9-33]   Nocedal J, Wright S (2006) Numerical optimization. Springer, New York

[9-34]   Peng B, Mu Y, Duan J et al (2021) Separated proportional-integral Lagrangian for chance constrained reinforcement learning. IEEE IVS, Nagoya, Japan

[9-35]   Pironneau O, Polak E (1972) On the rate of convergence of certain methods of centers. Math Programming 2(1): 230-257

[9-36]   Polak E (1971) Computational methods in optimization: a unified approach. Academic press, New York

[9-37]   Rawlings J, Muske K (1993) The stability of constrained receding horizon control. IEEE Trans Automatic Control 38(10): 1512-1516

[9-38]   Rosen J (1960) The gradient projection method for nonlinear programming. Part I. Linear constraints. J of Society for Industrial & Applied Math, 8(1): 181-217

[9-39]   Schulman J, Levine S, Abbeel P et al (2015) Trust region policy optimization. NeurIPS, Montréal, Canada

[9-40]   Topkis D, Veinott J, Arthur F (1967) On the convergence of some feasible direction algorithms for nonlinear programming. SIAM Journal on Control, 5(2): 268-279

[9-41]   Uchibe E, Doya K (2007) Constrained reinforcement learning from intrinsic and extrinsic rewards. IEEE CDL, London, UK

[9-42]   Wen L, Duan J, Li SE et al (2020) Safe Reinforcement learning for autonomous vehicles through parallel constrained policy optimization. IEEE ITSC, Rhodes, Greece

[9-43]   West E, Polak E (1992) A generalized quadratic programming-based phase I-phase II method for inequality-constrained optimization. Applied Math Optimization 26: 223–252

[9-44]   Wolfe P (1972) On the convergence of gradient methods under constraint. IBM Journal Res & Dev, 16: 407–411

[9-45]   Yu D, Ma H, Li SE et al (2022) Reachability constrained reinforcement learning. ICML, Baltimore, USA

Chapter 10. Deep Reinforcement Learning

> There is no law
>
> except the law that there is no law.
>
> -- John Wheeler (1911-2008)

Similar to humans, RL agents use interactive learning to successfully obtain satisfactory decision strategies. In many examples, it is desirable to learn directly from measurements of raw video or image data without any hand-engineered features or domain heuristics. This kind of task requires the so-called deep reinforcement learning (DRL), which is an in-depth combination of artificial neural network (ANN) and reinforcement learning (RL). The "deep" portion of DRL refers to the multiple layers of neural network that replicate the structure of a human brain. Its application often requires large amounts of interactive data and significant computing power. In recent years, computing power has dramatically increased, which has enabled successful DRL applications. Many researchers became aware of the importance of DRL after a series of well-publicized defeats of Go grandmasters by DeepMind's AlphaGo. In addition to successfully playing Go, DRL has achieved human-level performance in several competitive video games like Atari and Starcraft.

Early attempts to combine RL with ANNs often failed due to several issues, such as non-iid sequential data, easy divergence, overestimation, and sample inefficiency. The first successful combination attempt is attributed to the deep Q-network (DQN) (Mnih et al., 2015) from Google DeepMind [10-16]. In this work, Q-learning was implemented with a convolutional neural network, which learns an estimate for the optimal Q-function in the continuous state space. The utilization of DQN introduces two critical tricks, i.e., experience replay and target network, to stabilize the learning process. Surprisingly, human-level performance can be achieved using DQN in almost half of all Atari games, which has led to an increasing number of DRL studies that have ultimately initiated many more robust algorithms. In double DQN (Hasselt et al., 2016), the idea behind double Q-functions is incorporated to alleviate value overestimation. Two Q-functions are learned to decouple action selection and action evaluation [10-31]. As a well-known variant of DQN, the deep deterministic policy gradient (DDPG) (Lillicrap et al., 2016) is used to concurrently learn a Q-function and a deterministic policy for continuous control tasks [10-14]. The overestimation of Q-values is a common problem in DDPG that leads to a brittle policy, even though some tricks, such as double Q-functions, have been added. To address this issue, Fujimoto et al. (2018) introduced twin delayed DDPG (TD3) with a few newly proposed tricks, including bounded double Q-functions and delayed policy update [10-4]. The soft actor-critic (SAC) algorithm (Haarnoja et al., 2019) builds a bridge between stochastic exploration and deterministic policy search with reliance on the maximum entropy RL framework. Here, 'soft' means that the conventional reward is regularized with a policy entropy term as an exploration and exploitation tradeoff [10-6]. The distributional soft actor-critic (DSAC) (Duan et al., 2020) is an enhanced version of SAC that further combines the distributional return to conquer the overestimation issue and improve learning efficiency [10-3].

© The Author(s), under exclusive license to Springer Nature Singapore Pte Ltd. 2023, Corrected Publication 2024
S. E. Li, *Reinforcement Learning for Sequential Decision and Optimal Control*,
https://doi.org/10.1007/978-981-19-7784-8_10

Training instability represents a thrashing behavior in DRL and is often accompanied by severely low efficiency. One common strategy to eliminate training instability is to avoid changing the policy too much at each step. It is difficult to manually determine how much the policy step size should be chosen because a too small step size may lead to very slow convergence and a too big step size may result in the risk of training instability. Trust region policy optimization (TRPO) (Schulman et al., 2015) addresses this kind of trade-off by formulating a trust-region constraint, which measures how far the new and old policies are from each other [10-24]. The benefit of TRPO is that it can guarantee a monotonic improvement for policy optimization at each iteration. Proximal policy optimization (PPO) (Schulman et al., 2017) is an enhanced version of TRPO that relies on the first-order clipping technique to avoid solving a second-order optimization problem in TRPO [10-25]. When using this technique, it remarkably increases the computing efficiency while still maintaining monotonic policy updates. Since on-policy RL suffers from poor efficiency, the parallel exploration technique was introduced into DRL in the asynchronous advantage actor-critic (A3C) algorithm (Mnih et al., 2016). In A3C, the main agent spawns multiple slave agents to execute actions in the parallel environments [10-15]. Additionally, entropy regularization is often utilized in A3C to improve the exploration ability by discouraging premature convergence to suboptimal policies.

10.1 Introduction to Artificial Neural Network

Artificial neural networks (ANNs) are the foundation of deep learning. An artificial network is referred to as a "neural" network because its structure mimics the working mechanism of neural networks in the human brain. ANNs are composed of individual units that act in serial or in parallel. Each unit resembles a neuron because it receives inputs from other units and subsequently fires itself after reaching a certain threshold. Generally, ANNs can be analyzed from a hierarchical structure, starting from neurons, accumulating to layers, and finally stacking to a network. A layer is a group of neurons organized in the width direction, in which neurons are connected to extract useful features. A network is composed of different kinds of layers connected along the depth direction. The more layers a network has, the deeper it becomes. Usually, one can call a network "deep" if it has more than five layers.

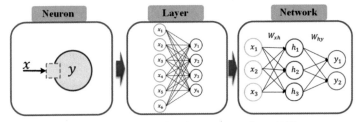

Figure 10.1 Hierarchical structure of an ANN

The most considerable merit of ANNs is their excellent approximation ability. It has been proven that even a feedforward network with a single hidden layer is the universal approximation of continuous functions on a compact subset under some mild assumptions. This statement is the well-known universal approximation theorem. In 1989, Cybenko proved the simplest version of the universal approximation theorem using logistic activation functions [10-2]. Further studies have demonstrated that most ANNs have the universal approximation ability as long as the activation function is not a linear polynomial.

10.1.1   Neurons

An artificial neuron is a mathematical model that is used to mimic a biological neuron. While an actual neuron receives electrical signals from other neurons, artificial neurons represent electrical signals as numerical values. At the synapses, electrical signals are modulated in various amounts according to their importance. An actual neuron fires an output signal only when the total strength of the electrical signal exceeds a certain threshold. In an artificial neuron, this phenomenon is represented by calculating the weighted sum of inputs and applying an activation function to determine its output. As in a human brain, the output of each artificial neuron is sent to other neurons in the network. A typical artificial neuron consists of a linear affine function and a nonlinear activation function. Given the input vector $x \in \mathbb{R}^n$, a neuron sends out its output $y \in \mathbb{R}$ with the following equation:

$$y = \sigma(z),$$
$$z = w^T x + b.$$

(10-1)

where $z \in \mathbb{R}$ is called the activation signal. Both weight $w \in \mathbb{R}^n$ and bias $b \in \mathbb{R}$ are trainable parameters, and $\sigma(\cdot)$ is the activation function. Without causing much confusion, we do not distinguish "trainable parameters" and "weights" in the following context. A simple neuron is illustrated in Figure 10.2.

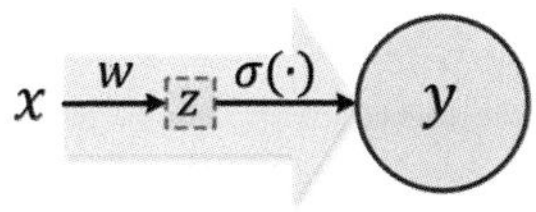

Figure 10.2 A simple neuron

Neurons differ from each other in terms of their activation functions. The selection of activation function is critical to effective feature extraction. Three kinds of activation functions are prevalent in deep learning (Figure 10.3): (1) logistic function; (2) hyperbolic tangent (tanh) function; and (3) rectified linear unit (ReLU) function. Both logistic and tanh functions belong to the sigmoid classification, which has a characteristic S-shaped curve. The logistic function is suitable for binary classifiers because its output is in the range of (0,1). The tanh function is the zero normalization of the logistic function. Since the output of the tanh function is symmetric about zero, there is usually less training oscillation when taking it as the activation function [10-13]. However, both logistic and tanh functions suffer from the vanishing gradient problem. The slopes of the logistic and tanh functions are close to zero when the input is far away from zero. The gradient becomes even worse during backpropagation in deep networks, which further results in training failure. The ReLU function is an effective solution to avoid the vanishing gradient problem because it outputs the same value for all positive inputs and zero for all negative inputs. In addition, due to its ability to output zeros, the ReLU function has an additional property called representational sparsity, i.e., some neurons can automatically become silent if they are not necessary.

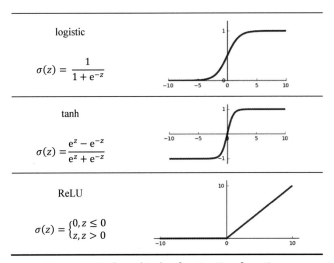

Figure 10.3 Three kinds of activation functions

10.1.2   Layers

A layer is the organization of neurons in the width direction. The input layer receives external data, and the output layer produces the ultimate result. The layers between the input and output layers are called hidden layers, and each hidden layer only connects to its adjacent two layers. A neural network may contain one or more hidden layers. It is generally believed that the more hidden layers there are, the better approximation ability the network has. Common hidden layers include fully connected layers, convolutional layers, and recurrent layers. In some neural networks, e.g., DenseNet, a layer may connect to every other layer [10-9]. Each layer in DenseNet receives additional inputs from distant preceding layers and passes on their own feature maps to subsequent layers. Therefore, each layer in this kind of network can receive collective knowledge from all preceding layers, and higher accuracy are achieved with fewer parameters.

10.1.2.1   Fully Connected Layer

In a fully connected layer, as illustrated in Figure 10.4 (left), neurons are connected with all the neurons in their previous or successive layers. In this layer, the activation signals of all the neurons can be computed via a linear affine function, which contains a weighting matrix multiplication and a bias offset.

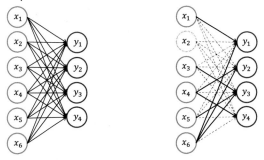

Figure 10.4 Fully connected layer (left: fully connected layer; right: dropout layer)

The overly large number of trainable weights is the major shortcoming of fully connected layers. In addition to a large computational burden, this shortcoming makes a fully connected network prone to overfitting, thus resulting in poor generalization. Dropout is a popular regularization technique to reduce overfitting [10-28]. The term "dropout" refers to manually or automatically removing some connections between some neurons. The dropout technique is helpful to prevent overfitting and maintain the preferred model accuracy.

10.1.2.2 Convolutional Layer

The convolutional layer, which is a special variant of the fully connected layer, utilizes a mathematical operation called convolution. The convolution operation is introduced to reduce the number of trainable weights, allowing the neural network to become much deeper. This kind of operation replaces trainable weights with learnable kernels, and each convolution operation only works with a restricted subarea of the previous layer, i.e., the so-called receptive field. Take one-dimensional convolution operation as an example:

$$y = \sigma(w * x),$$

$$y_i = \sigma(w^T \bar{x}_i) = \sigma\left(\sum_{j=1}^{N} w_j \cdot x_{i+j-1}\right), i = 1, 2, \cdots, m, \quad (10\text{-}2)$$

where $x = [x_1, x_2, \cdots, x_n]^T \in \mathbb{R}^n$ is the input, $w = [w_1, \cdots, w_N]^T$ is the convolutional kernel, $*$ is the convolution operation, $y = [y_1, y_2, \cdots, y_m]^T \in \mathbb{R}^m$ is the output and $\bar{x}_i = [x_i, x_{i+1}, \cdots, x_{i+N-1}]^T$ is the receptive field with length N. As illustrated in Figure 10.5, a convolutional layer is equivalent to a fully connected layer with two special properties: a) local connectivity: the narrow width of receptive fields enforces sparse local connectivity between some neurons of adjacent layers; and b) weight sharing: each receptive field is enforced to use the same weights, i.e., the same convolutional kernels are shared. These two properties allow a convolutional layer to keep space invariant characteristics, which is particularly important in large-dimensional data processing, such as images and videos.

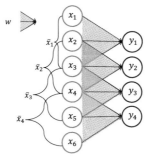

Figure 10.5 Convolutional layer

Convolution involving one-dimensional input is referred to as a 1D convolution. A 2D convolution is the extension of one-dimensional input to two-dimensional input. The 3D convolution involves three-dimensional inputs, whose dimensions include width, height and depth. As illustrated in Figure 10.6 (a), a 2D kernel of size 2 × 2 (2 wide, 2 high) strides across a 2D input of size 4 × 4 (4 wide, 4 high), moving forward one element at each step. Each receptive field performs elementwise multiplication and elementwise summation, whose output is a scalar value. Therefore, the final output is a 3 × 3 feature map. A 3D kernel can be used for a multichannel input, where the channel is defined in the depth dimension. The number of input channels is equal to the thickness of kernels C and that of the output channels is equal to the number of kernels K (shown in Figure 10.6 (b)).

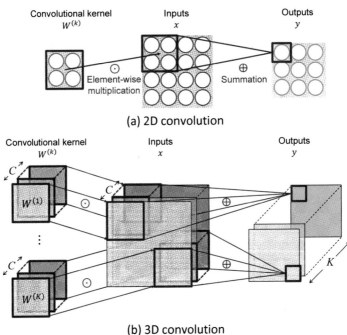

(a) 2D convolution

(b) 3D convolution

Figure 10.6 2D and 3D convolution

A convolutional layer records the precise position of every input element, and it may become too sensitive to small changes in the input. One approach to address this sensitivity issue is downsampling. Pooling is a widely used downsampling technique. In pooling, the output is partitioned into a set of non-overlapping rectangles and their outputs are calculated with minimization, maximization or average operation. The pooling technique makes an ANN more robust to changes in the input element position, which is referred to as local translation invariance.

10.1.2.3   Recurrent Layer

The recurrent layer involves neural connections with a temporal feedback mechanism. As illustrated in Figure 10.7, the output of the last time step is fed as the input of the current time step. The feedback mechanism contains a hidden state (memory) at each step; otherwise, a static algebraic loop inevitably occurs. This hidden state carries pertinent prior input information, which allows temporal dynamic behaviors to be exhibited:

$$y_t = \sigma\big(W_{xy}x_t + W_{yy}y_{t-1}\big), \tag{10-3}$$

where W_{xy} is the weight matrix from the current input x_t to the current output y_t, and W_{yy} is the weight matrix from the previous output y_{t-1} to the current output y_t. Once unfolded in the temporal domain (Figure 10.11), a recurrent layer can be seen as a feedforward network, in which all the hidden layers share the same weights.

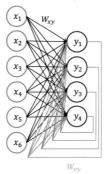

Figure 10.7 Recurrent layer

Training a recurrent network is prone to become very problematic because the network level gradients can grow or shrink more quickly in backpropagation. After only a few time steps, the gradients of some layers may explode to infinity or vanish to zero [10-18]. One common idea to alleviate this issue is to create additional paths to prevent exponential operations on gradients during backpropagation. This idea leads to two popular variants of recurrent layers (Figure 10.8): (1) long short-term memory (LSTM) (Hochreiter & Schmidhuber, 1997) [10-8] and (2) the gated recurrent unit (GRU) (Cho et al., 2014) [10-1].

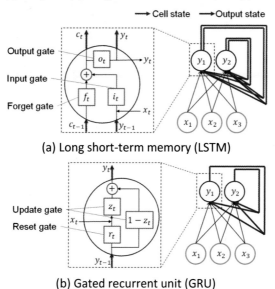

(a) Long short-term memory (LSTM)

(b) Gated recurrent unit (GRU)

Figure 10.8 Two recurrent layer variants

The LSTM has a unique self-recurrent unit called a cell (a memory part), which is composed of an input gate, a forget gate, and an output gate. The cell remembers a value over the temporal dimension, and the three gates regulate the information flow into and out of the cell. The gates are often activated with logistic functions, where 1 refers to "open" and 0 refers to "close". The connection weights determine how the gates operate. In the GRU, the output gate is removed from the cell in LSTM, thus reducing the number of trainable weights. In the GRU, the hidden state is controlled by only two gates. A reset gate controls how much of the previous state is remembered. An update gate controls how much of the new state is simply a copy of the old state.

10.1.3   Typical Neural Networks

Typically, an ANN model is constructed by stacking many different layers together. Its structure can be described with a directed graph, which is a mathematical description of the layer organization (e.g., Figure 10.9). The function of ANNs varies significantly with their structures and compositions. For example, models without recurrent connections form directed acyclic graphs, which are known as feedforward networks. A large class of feedforward networks are convolutional neural networks. Models that allow self-connections between neurons in the same layer, e.g., the recurrent layer, form directed cyclic graphs, which are known as recurrent networks.

To date, diversified ANN models have been proposed in various engineering scenarios. Convolutional neural networks (CNNs) and recurrent neural networks (RNNs) are the two mainstream deep learning models. CNNs are models with convolutional layers that specialize in processing a grid of values with spatial information. As one of the early representative models, LeNet was proposed by LeCun and Bottou in 1998 [10-12]. A standard LeNet consists of two convolutional layers, a flattening convolutional layer, two fully connected layers, and a softmax output layer. Since then, various convolutional networks have been proposed, including AlexNet (Krizhevsky et al., 2012) [10-11], GoogLeNet (Szegedy et al., 2015) [10-29], VGGNet (Simonyan et al., 2015) [10-27], ResNet (He et al., 2016) [10-7], and MobileNet (Sandler et al., 2018) [10-22]. The convolution operations in these networks can be used to encode high-dimensional information, making CNNs quite suitable for image processing problems. Moreover, CNNs can be used to automatically extract spatial features without any human supervision.

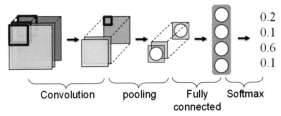

Convolution   pooling   Fully   Softmax
connected

Figure 10.9 A typical model of a feedforward network

Table 10.1 Some representative CNN models

Number of convolutional layers	Size of kernels	Number of kernels	Number of channels	Sliding stride

LeNet	2	5	20,50	1,20	1
AlexNet	5	3,5,11	96~384	3~256	1,4
GoogLeNet	21	1,3,5,7	16~384	3~832	1,2
VGG-16	13	3	64~512	3~512	1
ResNet-151	151	1,3,7	64~2048	3~2048	1,2
MobileNet	27	1,3	32~1024	3~1024	1,2

RNNs, which are models with recurrent layers, specialize in processing a sequence of values with temporal information. For example, stacked RNN, which was introduced in 2013 by Graves et al., is a recurrent model in which several recurrent layers are simply stacked in a depth-wise manner [10-5]. Other well-known RNN models include bidirectional RNNs (Schuster and Paliwal, 1997) [10-26], structural RNNs (Jain et al., 2016) [10-10], and recursive NNs (Pollack, 1990) [10-19]. Basically, in RNNs, the output of recurrent layers is retained and then fed back into the model. Therefore, remembering temporal information and sequential encoding relationships are the distinctive capabilities of RNNs, making them more suitable for time series problems such as natural language processing (NLP).

10.2 Training of Artificial Neural Network

Artificial neural networks (ANNs) can be used to perform automatic feature extraction without human intervention. Deep neural networks can be distinguished from shallow versions in terms of their impressive depth. Early NNs, which were composed of at most one or two hidden layers, were considered to be very shallow. More than three hidden layers (without considering input and output layers) is often qualified as "deep" learning. The training of ANNs is often formulated as a gradient-based optimization problem. The goal of training a model is to find the best weights that minimize a loss function that measures the model error. One common optimization method is called the stochastic gradient descent (SGD) algorithm, whose cornerstone is the backpropagation (BP) technique.

10.2.1 Loss Function

The loss function is a measure of the output error of an ANN model. The training of ANNs is similar to finding the best estimate of trainable weights from labeled data. As the number of samples increases in the training dataset, the accuracy of weight estimation will gradually improve until a specific performance limit is reached. Cross entropy and mean squared error are two popular loss functions in deep learning.

10.2.1.1 Cross Entropy

Cross entropy is a measure of the error between two probability distributions. One probability distribution is from the model prediction, and the other is from labeled training data. Cross entropy builds upon the basis of information entropy and calculates the average number of bits required to transmit an event from one distribution p to another distribution p^{target}:

$$J \overset{\text{def}}{=} \mathcal{H}(p^{\text{target}}, p) = -\sum_i p_i^{\text{target}} \log(p_i), \tag{10-4}$$

where p is the predicted distribution of an ANN model, p^{target} is the target distribution of training data, and $\mathcal{H}(\cdot, \cdot)$ is the cross entropy function. Cross entropy is often used in classification problems when outputs are interpreted as membership probabilities in an indicated class.

10.2.1.2 Mean Squared Error

The mean squared error (MSE) is the average of the squared difference between the predicted and target outputs:

$$J \overset{\text{def}}{=} \text{MSE}(y^{\text{target}}, y) = \frac{1}{n} \sum_{i=1}^{n} \left(y_i^{\text{target}} - y_i\right)^2, \tag{10-5}$$

where y is the predicted output of an ANN model and y^{target} is the target output of the training data. The MSE loss function is widely used for regression problems. Nevertheless, MSE is equivalent to cross entropy under some special conditions, for example, when the target obeys the Gaussian distribution.

10.2.1.3 Bias-Variance Trade-Off

Bias and variance are two important statistical measures of training performance. High bias means that a model does not "fit" well on the training set, and the training error is large. High variance means that the model is sensitive to noise, i.e., a small fluctuation in the input will cause a large error in the output. In this situation, the model cannot be used to accurately predict the validation set even if its training error is small. Hence, high variance often suggests that the validation error far exceeds the training error. The overfitting issue occurs when a model is very good at learning the training set but cannot accurately generalize to the validation set. This is also called the generalization issue. The underfitting issue occurs when a model cannot generate accurate predictions on the training set or the validation set.

In deep learning, overfitting is more common than underfitting because one ANN's superiority is its ability to fit large-size data. In general, increasing model complexity is helpful to reduce the model bias, but it also increases the risk of severe overfitting (i.e., poor generalization). Regularization is a popular technique that is used to improve the generalization ability. It discourages adopting complex models by penalizing the model size in the loss function:

$$J_{\text{Reg}}(w) = J(w) + \rho \Omega(w), \tag{10-6}$$

where $\rho \in [0, \infty)$ is the regularization parameter, and $\Omega(w)$ represents the penalty on the model complexity. Different choices for the penalty function Ω, e.g., L1 and L2 regularization, can result in very different training results [10-17]. In the L1 regularization, overfitting is alleviated by shrinking some unimportant parameters toward zero. This might make some desirable features obsolete. In the L2 regularization, overfitting is alleviated by forcing weights to be as small as possible but not making them exactly zero. Interestingly, even though equipped with the ability to fit large-scale data, modern deep learning suffers from insufficient data. One central task in deep learning is how to build standard and reliable databases for the whole community.

10.2.2 Training Algorithms

The training process is often formulated into a stochastic optimization problem, in which the first-order optimization is the most popular strategy in today's deep learning community. The first-order optimization strategy involves adjusting trainable weights along the gradient direction until a local minimum of the loss function is reached. Of course, finding the global minimum is the ultimate goal in deep learning. The backpropagation (BP) algorithm plays a central role in computing the gradients of all trainable weights in feedforward networks. This technique is renamed as backpropagation through time (BPTT) in the training of recurrent networks.

10.2.2.1 Backpropagation

The backpropagation (BP) algorithm allows the error information from the loss function to flow backward through layers. BP is a recursive gradient-calculating algorithm that uses the chain rule with a specific order of operations. Figure 10.10 compares two paths in a fully connected network: value forward-propagation and gradient backpropagation. In the former, the input information is received and the outputs of each neurons are calculated. In the latter, the gradients of all the weights in the network are calculated.

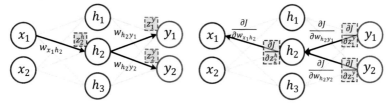

(a) Value forward-propagation   (b) Gradient backpropagation

Figure 10.10 Two types of propagation in NN

To illustrate the backpropagation process, we take the one-hidden-layer fully connected network as an example (Figure 10.10). The weight gradient between the hidden neuron $h_j (j = 1,2,3)$ and the output neuron $y_m (m = 1,2)$ is

$$\frac{\partial J}{\partial w_{h_j y_m}} = \frac{\partial J}{\partial z_m^y} \frac{\partial z_m^y}{\partial w_{h_j y_m}} = \frac{\partial J}{\partial y_m} \sigma'(z_m^y) h_j, \qquad (10\text{-}7)$$

where $w_{h_j y_m}$ is the weight from neuron h_j to y_m, $\sigma'(\cdot) = \partial \sigma(z)/\partial z$ is the derivative of the activation function, and z_m^y is the activation signal of output neuron y_m. Since the loss function J is the explicit function of the final output y_m and $\sigma'(\cdot)$ has an explicit expression, the gradient $\partial J/\partial z_m^y$ is easy to calculate. Then, we use the chain rule to calculate the weight gradient between the input neuron $x_i (i = 1,2)$ and the hidden neuron $h_j (j = 1,2,3)$:

$$\frac{\partial J}{\partial w_{x_i h_j}} = \frac{\partial J}{\partial z_j^h} \frac{\partial z_j^h}{\partial w_{x_i h_j}} = \frac{\partial J}{\partial h_j} \sigma'(z_j^h) x_i = \left(\sum_m w_{h_j y_m} \frac{\partial J}{\partial z_m^y} \right) \sigma'(z_j^h) x_i, \qquad (10\text{-}8)$$

where z_j^h is the activation signal of hidden neuron h_j. Note that $\partial J / \partial h_j$ is the weighted sum of gradients from the output layer $\partial J / \partial z_m^y$, which has been precomputed and stored in (10-7). Through the backpropagation process, the computational burden can be significantly reduced. Specifically, the concurrent calculation of $\partial J / \partial z_j^h$ can also be stored and reused once there are more than two hidden layers in the NN. In engineering practice, BP is often misunderstood as the whole learning algorithm for neural networks. Actually, it only refers to how to compute the gradient of each trainable weight. To train the model, one needs an additional updating formula from stochastic optimization. This formula is called stochastic gradient descent (SGD) [10-21]:

$$w \leftarrow w - \eta \mathbb{E}_{\mathcal{D}} \left\{ \frac{\partial J}{\partial w} \right\},$$

$$\frac{\partial J}{\partial w} = \left[\underbrace{\frac{\partial J}{\partial w_{x_i h_j}} \quad \frac{\partial J}{\partial w_{h_j y_m}}}_{12=2\times3+3\times2} \right]^{\mathrm{T}} \in \mathbb{R}^{12},$$

(10-9)

where η is the learning rate, w represents the whole network weights and $\mathcal{D}$ is the utilized small sample set. The neural network in Figure 10.10 has 12 trainable weights. Note that the bias parameter is also trainable but omitted for narrative brevity. In deep learning, large-scale training data are necessary for accurate prediction and good generalization, making a training process computationally expensive. SGD partially alleviates the issue of heavy computational burden, as its gradient is only estimated based on a small batch of samples (i.e., a mini-batch). The mini-batch is randomly picked from the training set to treat all samples equally. Sometimes, mini-batch training may result in a large variance in the updating direction. If one uses a large learning rate, SGD might generate apparent oscillation behavior. The momentum technique is a remedy to smooth out the oscillating updates by partially reusing the past gradients [10-20]. Momentum-based SGD introduces some damping characteristics into the updating dynamics, thus helping to stabilize convergence.

10.2.2.2 Backpropagation Through Time

Once unfolded in the temporal domain, an RNN model can be seen as a deep feedforward network in which all the hidden layers share the same trainable weights. The RNN model cannot be unfolded by infinite steps. Instead, one needs to stop at a specific temporal depth to match the length of the training data. This method is called backpropagation through time (BPTT) [10-33]. Therefore, training an RNN model is approximately equivalent to training an unfolded feedforward network with a finite number of hidden layers with shared weights.

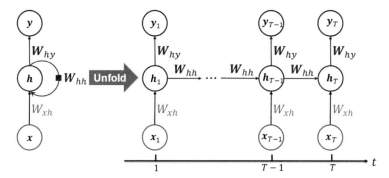

Figure 10.11 Temporal unfolding of an RNN

As shown in Figure 10.11, RNN with a single hidden layer is expanded along the time axis. The number of unfolded layers is the same as the input sequence length T. The hidden layer at each time step can be seen as a fully connected layer, and the state from the previous time step (i.e., the output of the previous layer in the unfolded network) is taken as an input at the subsequent time step. The gradient of each layer is then calculated using the backpropagation technique. In an unfolded RNN model, the gradient backpropagation involves a cascading exponential multiplication, thus resulting in a stronger vanishing or exploding gradient issue. Fortunately, LSTM and GRU can be used to create additional direct paths for gradients through hidden layers, with which the vanishing or exploding gradient is effectively reduced [10-18].

10.3 Challenges of Deep Reinforcement Learning

Compared with other approximate functions, such as polynomials and radial basis functions, ANNs provide a much stronger approximation ability for both policy and value function. The combination of RL and an ANN makes it possible to learn directly from the measurements of raw video or image data in complex and high-dimensional tasks. Moreover, in deep reinforcement learning (DRL), one can start with a blank candidate network without any hand-engineered features or domain heuristics. Eventually, the outcome of DRL is often better than that of professional humans.

Previously, it was believed that DRL was a natural product of combining RL and ANN, and its design was a trivial task. In practice, DRL is fundamentally complicated because it inherits a few serious challenges from both RL and ANN. Some challenges, including non-iid sequential data, easy divergence, overestimation, and sample inefficiency yield particularly destructive outcomes if they are not well treated. Since 2015, a few empirical but useful tricks have been proposed to address these prominent issues, which build the basis of various advanced DRL algorithms. Figure 10.12 lists 11 popular tricks, as well as what kind of challenges each trick mainly addresses. These tricks include experience replay (ExR), parallel exploration (PEx), separated target network (STN), delayed policy update (DPU), constrained policy update (CPU), clipped actor criterion (CAC), double Q-functions (DQF), bounded double Q-functions (BDQ), distributional return function (DRF), entropy regularization (EnR), and soft value function (SVF).

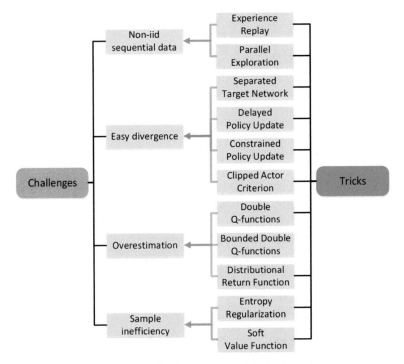

Figure 10.12 Challenges and tricks in deep RL

10.3.1   Challenge: Non-iid Sequential Data

For most statistical training algorithms, it is assumed that the data samples are independent and identically distributed (iid). This assumption no longer holds when the samples are sequentially generated from a dynamic environment. Instead, each state-action pair is correlated both temporally and spatially to its neighboring pairs. Due to local exploration and limited interaction time, an RL agent with deep neural network may suffer from large distribution bias due to non-iid sequential data. At each update cycle, the collected samples may not match the true data distribution of the policy that is currently used. This imperfect data distribution leads to a very inaccurate value estimation, which will cause a large discrepancy in the policy updating direction and eventually lead to a poor policy.

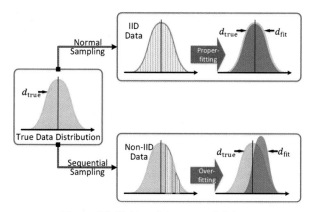

Figure 10.13 Non-iid sequential data

10.3.1.1 Trick: Experience Replay (ExR)

Experience replay is widely used in off-policy RL to address the issue of non-iid data [10-16]. An optimal action-value function always satisfies the Bellman equation for all state-action pairs, regardless of the behavior policy. This inherent advantage of Q-learning and its variants allows the reuse of historical samples without any explicit IS transformation. Therefore, the collected transition sample (s_t, a_t, s_{t+1}) can first be stored in a replay buffer and then the Q-function can be updated by randomly sampling a mini-batch of samples from the stored buffer. Random sampling over previous experience is used to not only remove temporal correlations in samples but also to average the training data distribution. Therefore, this approach can be used to effectively handle non-iid data issues.

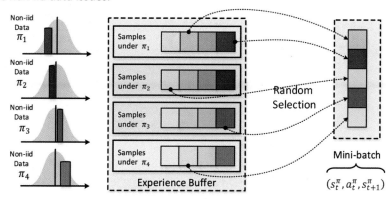

Figure 10.14 Experience replay

In addition, experience replay can be used to increase the sample efficiency. The sample efficiency is defined as how many new samples are needed before a specific policy performance is reached. Obviously, experience replay helps to accelerate policy updates but does not add new samples. In regular experience replay, historical data are equally treated and sampled. It is better to more frequently select high-value samples and less frequently select low-value samples. Prioritized experience replay can be used to more efficiently learn by preferentially searching more important samples [10-23]. The importance of a sample is defined as how much useful information is stored inside that sample. Therefore, it is a good choice to use the magnitude of TD error as an index of sample importance. The TD error indicates how far a value is from its bootstrapping estimate, and a large TD error means that the value function is not well fitted in the associated state region.

10.3.1.2 Trick: Parallel Exploration (PEx)

Another trick to handle non-iid sequential data is parallel exploration, which has enabled a large spectrum of distributed RL algorithms [10-15]. Unlike experience replay, parallel exploration is suitable for both on-policy and off-policy RL algorithms. With parallel exploration, multiple samplers are engaged to explore their corresponding environment instances. The parallel samplers simultaneously generate a large number of data samples to address the occurrence of non-iid data. The distribution of collected data will be closer to the true distribution since multiple samplers are more likely to cover all parts of environment dynamics. Moreover, under the off-policy setting, samplers can explicitly utilize different behavior policies to maximize the diversity of environment exploration.

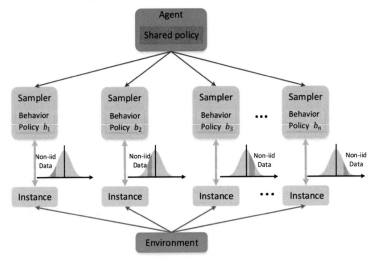

Figure 10.15 Parallel exploration

10.3.2 Challenge: Easy Divergence

An RL agent may easily become unstable or even diverge when its value function and policy are represented with deep neural networks. This phenomenon is manifested by oscillating target values and unstable policy updates. In an actor-critic algorithm, the leading cause of this phenomenon is the self-excitation mechanism between coupling critic and actor updates. Due to the tight connection between actors and critics, poor value estimation may result in very wrong policy update direction, and in return, inadequate policy improvements will significantly deteriorate the critic quality.

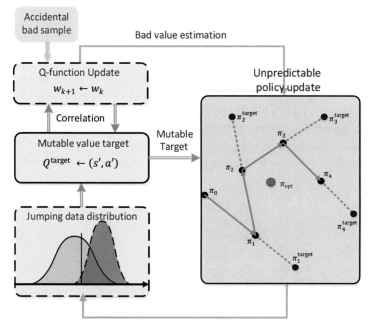

Figure 10.16 Easy divergence in deep RL

Poor value estimation may be attributed to several reasons, including accidentally poor samples, large approximation errors and insufficient critic updates. Take the accidentally poor samples as an example (Figure 10.16). An accidentally poor sample may come from an explorative policy, erroneous measurement or environmental stochasticity. Its occurrence may deteriorate the quality of the Q-value estimation. A low-quality Q-value function provides a highly uncertain direction for policy improvement and eventually leads to unpredictable policy updates. In return, environment exploration with the wrong policy will lead to an irregularly jumping data distribution and further mutable critic updating targets. The feedback loop will eventually deteriorate effective Q-value updates. In addition, this negative self-excitation behavior may be enlarged by an large actor learning step. The overlarge learning step may drag the policy to a completely new position. To reduce easy divergence, a few tricks, including maintaining a steady value target, shrinking value estimation error, and controlling fast policy updates, need to be deployed to lessen the self-excitation of this feedback loop.

10.3.2.1 Trick: Separated Target Network (STN)

A separated target network can be used to reduce the tight correlation between the Q-value and the target value. Maintaining a target network and updating it periodically is an effective way to address the issue of easy divergence. With this trick, the target network is updated less frequently than the online Q-network. This periodic updating mechanism allows the target network to keep fixed for a period between two successive updates of the online Q-network. The updating formula of the target value is

$$Q_{\bar{w}}^{\text{target}} = r + \gamma Q_{\bar{w}}(s', a'),$$

(10-10)

either a hard update

$$\bar{w} \leftarrow w, \text{every } n \text{ steps,}$$

or a soft update

$$\bar{w} \leftarrow \rho w + (1 - \rho)\bar{w},$$

where $Q_{\bar{w}}^{\text{target}}$ is the target value, $Q_{\bar{w}}(s, a)$ is the target network, $\bar{w}$ denotes the weight of the target network, and $\rho \geq 0$ is the updating rate of the target network. Specifically, $\rho = 1$ means that the weight of the target network is directly copied from the online Q-network. Easy divergence is alleviated if the target network is updated at a low frequency. However, a low updating frequency slows the training speed. Therefore, it is important to choose proper updating rate to balance between training stability and training speed.

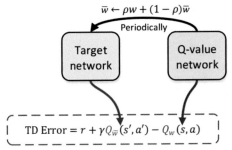

Figure 10.17 Separated target network

10.3.2.2 Trick: Delayed Policy Update (DPU)

Divergence can still occur when a policy is searched due to an imprecise value function. The simultaneous updating of value and policy functions is a major reason for an imperfect policy evaluation. The critic often needs high-frequency updates to avoid insufficient critic updates. A straightforward solution is to subtly delay policy updates (DPU), i.e., start each policy update after a certain number of value updates [10-4]. The fundamental idea behind this solution is to wait until the value approximation error becomes sufficiently small, and hence high-quality policy updates become possible. This simple trick is helpful to reduce the volatility that typically arises in actor-critic RLs.

10.3.2.3 Trick: Constrained Policy Update (CPU)

Another trick to alleviate easy divergence is to avoid changing the policy parameter too quickly. One solution is to constrain the distance of two adjacent policies, i.e., constrained policy update (CPU). Generally, a large update length allows a fast learning speed but may result in a massive change in policy weights. The weight change that is too fast is detrimental to stability. In contrast, a small update step size can effectively stabilize the policy updates, which is helpful in finding a satisfactory final policy. The CPU trick is implemented by bounding a certain distance measure of two adjacent policies:

$$\|\pi_{k+1} - \pi_k\| \leq \delta_\pi, \tag{10-11}$$

where $\|\cdot\|$ is the distance measure of two policies and δ_π is the upper bound. Specifically, the distance measure can be any vector norm, cross entropy, KL divergence, etc.

10.3.2.4 Trick: Clipped Actor Criterion (CAC)

In addition to confining two adjacent policies, the increment of the actor criterion value can also be limited. A common technique is the clipped actor criterion (CAC), in which the incentive of a new policy is penalized when its actor cost function moves farther from the old policy. The CAC trick limits the fast improvement of the actor criterion by

$$0 \leq J_{\text{Actor}}(\pi_{k+1}) - J_{\text{Actor}}(\pi_k) \leq \delta_J, \tag{10-12}$$

where $\delta_J > 0$ is the upper bound of the actor increment. The criterion value change must be maintained to no less than zero to achieve monotonic policy improvement.

10.3.3 Challenge: Overestimation

The overestimation issue states that DRL may quickly learn unrealistically high Q-values at some state-action pairs, resulting in a distorted Q-value estimation and destructive policies. Q-value updates always induce an overly optimistic estimation with a maximum operator, irrespective of whether their original errors are positive or negative [10-3]. In fact, optimistic value estimation itself is not a severe problem. If all values are uniformly higher, relative action preferences are still preserved, and it is not expected that the resulting policy will be any worse. However, Q-value overestimation is often not uniform and it may concentrate on some specific state-action pairs. This will lead to a "better" policy that is very realistic. We use the following update rule to demonstrate how the overestimation issue occurs:

$$w' \leftarrow w + \alpha \left(Q_w^{\text{target}}(s, a) - Q_w(s, a) \right) \nabla Q_w(s, a),$$

where $\alpha > 0$ is the learning rate and Q_w^{target} is the target value. In practice, $Q_w(s, a)$ and $\nabla Q_w(s, a)$ inevitably contain random errors, which are caused by state measurement noise or imperfect function approximation. Denoting the true weight of the Q-network as $\hat{w}$, we have the following error condition:

$$Q_w(s, a) = Q_{\hat{w}}(s, a) + \epsilon(s, a),$$

where ϵ is the random error, which is a function of the state-action pair. Let $\hat{w}'$ represent the next true weight from the true weight $\hat{w}$ at the current step. It can be proven that the estimation bias after the Q-value update is

$$\Delta(s, a) = \alpha \gamma \cdot \delta(s, a) \cdot \|\nabla Q_{\hat{w}'}(s, a)\|_2^2,$$

$$\delta(s, a) = \mathbb{E}_{s' \sim \mathcal{P}} \left\{ \mathbb{E}_\epsilon \left\{ \max_{a'} Q_w(s', a') \right\} - \max_{a'} \mathbb{E}_\epsilon \{ Q_w(s', a') \} \right\}, \tag{10-13}$$

where $\Delta(s, a)$ is the estimation bias of the updated Q-value and $\delta(s, a)$ is the target error between the estimated target and the true target. The target error is caused by the random error of the Q-value at the current step. Due to the maximum operator, $\delta(s, a)$ is not equal to zero, and it must be nonnegative, i.e., $\delta(s, a) \geq 0$. Therefore, an upward bias always exists, and it is not uniform since both target errors and Q-value gradients at different state-action pairs are very unpredictable. It may be expected that the upward bias at each update is insignificant. In reality, the overestimation error often accumulates rather quickly through bootstrapping, and it eventually results in poor policy updates.

10.3.3.1 Trick: Double Q-Functions (DQF)

The overestimation issue can be reduced to some extent by double Q-functions (DQF). The idea behind DQF is to decouple the maximum operation in the target calculation into two separable steps, i.e., action selection and action evaluation [10-30]. This disentangled design is fulfilled by maintaining two Q-functions, and each Q-function is used as a target to correct the other. The optimal action is first selected by one Q-function, the value of which is then evaluated by the other Q-function. Hence, in DQF, two individual Q-networks, $Q_{w_1}(s, a)$ and $Q_{w_2}(s, a)$, must be trained. For the first Q-network Q_{w_1}, the optimal action a_1^* is the maximum action value at state s':

$$a_1^* = \arg\max_{a'} Q_{w_1}(s', a'). \tag{10-14}$$

Instead of evaluating its actions with its own Q-network Q_{w_1}, DQF uses the other Q-network for action evaluation, i.e., $Q_{w_2}(s', a_1^*)$. In this trick, Q_{w_2} serves as an evaluation function, while a_1^* is the greedy action from Q_{w_1}. This type of selection and evaluation separation allows the creation of a new update target for the first Q-network:

$$Q_{w_1}(s, a) \leftarrow Q_{w_1}(s, a) + \alpha \left(r + \gamma Q_{w_2}(s', a_1^*) - Q_{w_1}(s, a) \right). \tag{10-15}$$

The update process of the second Q-network is symmetrical to that of the first Q-network. To maintain the separable behavior of action selection and action evaluation, the two Q-networks need to be sufficiently independent of each other. A popular strategy to achieve this goal is to respectively update Q_{w_1} and Q_{w_2} with two different datasets, which come from parallel samplers or are separately sampled from a replay buffer. In this situation, one can regard each Q-function as an unbiased estimate of the other. Accordingly, their cross evaluation can correct each other to obtain a more accurate value estimation. In addition, actions can come from either of the two Q-networks to interact with the environment.

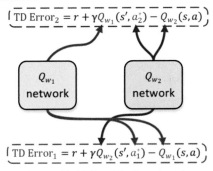

Figure 10.18 Double Q-networks

Let us explain why DQF can be used to alleviate the overestimation issue. We assume that Q_{w_1} and Q_{w_2} are two unbiased estimates, i.e., $\mathbb{E}\{Q_{w_1}(s, a)\} = \mathbb{E}\{Q_{w_2}(s, a)\} = q^\pi(s, a)$. Define the set of actions that maximize the true value as

$$\mathcal{M} \overset{\text{def}}{=} \left\{ a_{\text{opt}} \,|\, a_{\text{opt}} = \arg\max_{a'} q^\pi(s', a') \right\}.$$

Then, we have

$$\mathbb{E}\{Q_{w_2}(s', a_1^*)\}$$

$$= \Pr\{a_1^* \in \mathcal{M}\}\mathbb{E}\{Q_{w_2}(s', a_1^*)\big|a_1^* \in \mathcal{M}\} + \Pr\{a_1^* \notin \mathcal{M}\}\mathbb{E}\{Q_{w_2}(s', a_1^*)\big|a_1^* \notin \mathcal{M}\}$$

$$= \Pr\{a_1^* \in \mathcal{M}\}\max_{a'} q^\pi(s', a') + \Pr\{a_1^* \notin \mathcal{M}\}\mathbb{E}\{Q_{w_2}(s', a_1^*)\big|a_1^* \notin \mathcal{M}\} \qquad \text{(10-16)}$$

$$\leq \Pr\{a_1^* \in \mathcal{M}\}\max_{a'} q^\pi(s', a') + \Pr\{a_1^* \notin \mathcal{M}\}\max_{a'} q^\pi(s', a')$$

$$= \max_{a'} q^\pi(s', a').$$

The "less than" inequality in (10-16) means that there is a high probability that underestimation may occur when greedy actions are evaluated with another Q-function. Due to the usage of the DQF trick, an underestimated target value, which has the potential to compensate for the upward Q-value estimation, is created. Therefore, this compensation mechanism is able to somewhat eliminate the overestimation issue. Note that DQF can blend into the STN trick in an elegant manner. The target network provides a natural candidate for the second Q-function, which is employed to evaluate those actions selected by the online Q-network [10-31].

10.3.3.2   Trick: Bounded Double Q-functions (BDQ)

The estimates of the two Q-functions are not exactly independent of each other since each of them creates an update target for the other. Therefore, the DQF trick may not perfectly create underestimated target values for compensation, and a certain level of overestimation could still exist in some local regions of the state-action space. The bounded double Q-functions (BDQ) trick is an enhanced version of DQF that is used to actively generate underestimated target values. One can choose the minimum between the two estimates to calculate target values:

$$Q_{w_1}^{\text{target}}(s, a) = r + \gamma \min_{i=1,2} Q_{w_i}(s', a_1^*),$$

$$Q_{w_2}^{\text{target}}(s, a) = r + \gamma \min_{i=1,2} Q_{w_i}(s', a_2^*). \qquad \text{(10-17)}$$

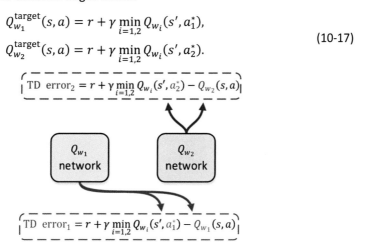

Figure 10.19 Bounded double Q-functions

With the BDQ trick, the underestimated target occurs if either of two Q-value estimates is lower than the true value. Thus, when utilizing this approach, there is a high possibility to depress the occurrence of any additional overestimation.

10.3.3.3 Trick: Distributional Return Function (DRF)

As previously discussed, in both DQF and BDQ, two independent Q-functions are utilized to learn a more accurate Q-value. It is a natural idea to consider learning an infinite number of Q-functions to maximize the accuracy of value estimates. However, the training efficiency will primarily suffer from the increasing number of parameterized functions. Instead of handling multiple Q-values, the distribution of state-action returns can be learned, as demonstrated in Figure 10.20. By introducing the distributional function, a single distributional return function can be learned to mimic the behavior of infinite Q-functions [10-3].

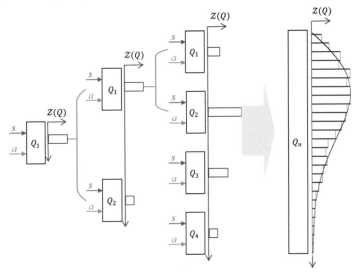

Figure 10.20 Concept of the distributional return function

More specifically, the state-action return is defined as

$$Z(s,a) = r_{ss'}^a + \gamma G_{t+1}, \tag{10-18}$$

where G is the long-term return under the current policy π. The long-term return is usually a random variable due to the randomness in the state transition $\mathcal{P}$, reward function r and policy π. Thus, the state-action return is still a random variable, and it is clear that

$$
\begin{aligned}
q^\pi(s,a) &= \mathbb{E}_{s'\sim\mathcal{P}}\{r_{ss'}^a + \gamma v^\pi(s')\} \\
&= \mathbb{E}_{s'\sim\mathcal{P}}\{r_{ss'}^a + \gamma\mathbb{E}_\pi\{G_{t+1}|s'\}\} \\
&= \mathbb{E}_\pi\{r_{ss'}^a + \gamma G_{t+1}\} \\
&= \mathbb{E}_\pi\{Z(s,a)\} \\
&= \mathbb{E}_{Z\sim Z^\pi}\{Z(s,a)\}.
\end{aligned}
\tag{10-19}
$$

In other words, the action-value function is equivalent to the expected state-action return under policy π. Furthermore, the state-action return must obey a specific distribution $Z^\pi(\cdot\,|s,a)$:

$$Z(s,a) \sim Z^\pi(\cdot \,|\, s, a).$$

As demonstrated in Figure 10.20, the distributional return function is an extension of the double Q-function. The trick of double Q-functions can easily be extended to multiple Q-functions. The collection of an infinite number of Q-functions naturally becomes the distributional state-action return function $Z^\pi(\cdot \,|\, s, a)$. It has been proven that this type of distributional design can be used to effectively depress the overestimation issue [10-3]. Let us take Q-learning with a Gaussian distribution as an example. Suppose that the mean and standard deviation of the state-action return are two independent parameters $q(s, a)$ and $\sigma(s, a)$. Then, it can be induced that the overestimation bias of distributional Q-learning has the following property:

$$\bar{\Delta}(s, a) = \frac{\Delta(s, a)}{\sigma(s, a)^2} , \qquad (10\text{-}20)$$

where $\Delta(s, a)$ is the overestimation error in traditional RL in (10-13) and $\bar{\Delta}(s, a)$ is the overestimation error with the distributional return. The overestimation error will decrease at a quadratic rate with increasing standard deviation $\sigma(s, a)$. The standard deviation $\sigma(s, a)$ tends to be larger in areas with inaccurate value estimates and uncertain future states. Thus, learning a distributional return can effectively reduce overestimation errors that are caused by large approximation errors and environmental randomness.

10.3.4   Challenge: Sample Inefficiency

The sample efficiency describes how many samples are required to reach a specific policy performance. Here, a "sample" is defined as a transition triple (s_t, a_t, s_{t+1}) that is newly generated through environment interactions. The reuse of old samples does not change the level of sample efficiency. The sample efficiency depends on two aspects: exploration efficiency and exploitation efficiency. The former describes whether an RL agent can fully cover the overall environment with fewer samples, and the latter describes how to learn a specific region more effectively with the collected samples. The difference in exploration efficiency and exploitation efficiency is demonstrated in Figure 10.21.

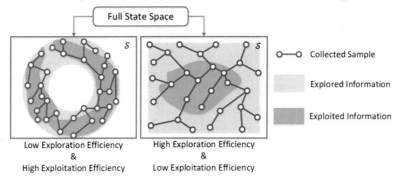

Figure 10.21 Sample inefficiency

Exploring environment is notoriously expensive in complex tasks with a large state-action space. Thousands or even millions of samples may be required to accurately represent environment dynamics. Higher exploration efficiency needs more random agent initialization behaviors, for example, explorative starts. A better approach is to actively drive the agent to explore the unexplored region. Entropy regularization and soft value function are two popular active tricks to enhance sample efficiency. Both of them come at the cost of heavy storage burden, but they do not affect the exploitation efficiency. The experience replay is an effective trick to increase the exploitation ability because it reuses old samples to exact their hidden information more sufficiently. Regular experience replay equally emphasizes all the samples, regardless of their significance on information extraction. An improved version of this technique is called prioritized experience replay, which can reuse important samples more frequently than less important ones.

10.3.4.1 Trick: Entropy Regularization (EnR)

Entropy regularization is a classical solution to increase the exploration efficiency. The entropy of a policy is a measure of how random the agent's behavior is. The higher the policy entropy is, the more stochastically the agent chooses its actions. Entropy regularization can be used to augment the conventional RL criterion to encourage exploration by preventing the agent from being too decisive [10-34]. In RL, the overall objective function with entropy regularization is to maximize

$$J(\theta) = \mathbb{E}_{\pi_\theta}\{v^{\pi_\theta}(s) + \alpha\mathcal{H}(\pi_\theta(\cdot\,|s))\}, \tag{10-21}$$

where $v^{\pi_\theta}(\cdot)$ is the state-value function, $\mathcal{H}(\pi_\theta(\cdot\,|s))$ is the policy entropy, and α is the temperature parameter, which controls the strength of entropy regularization. This kind of overall objective function has several advantages. The resulting stochastic policy is incentivized to explore more widely, capturing multiple modes of near-optimal behavior while discouraging premature convergence to suboptimal policies. Since the agent will explore more regions in the state space, maximum entropy policies become more robust to model mismatch and estimation errors.

10.3.4.2 Trick: Soft Value Function (SVF)

Another solution to increase sample efficiency is to add a policy entropy to each reward signal, yielding the so-called maximum entropy RL. When using this RL framework, a substantial improvement is observed for both exploration ability and robustness guarantees [10-6]. The augmented reward signal with policy entropy is defined as

$$r_{\text{aug}}(s, a, s') = r(s, a, s') + \alpha\mathcal{H}(\pi(\cdot\,|s)). \tag{10-22}$$

The corresponding value function is called the soft value function:

$$v^\pi(s) = \mathbb{E}_\pi\left\{\sum_{i=0}^{\infty}\gamma^i\left(r_{t+i} + \alpha\mathcal{H}(\pi(\cdot\,|s_{t+i}))\right)\middle|\, s_t = s\right\},$$

$$q^\pi(s, a) = \mathbb{E}_\pi\left\{\sum_{i=0}^{\infty}\gamma^i r_{t+i} + \alpha\sum_{i=1}^{\infty}\gamma^i\mathcal{H}(\pi(\cdot\,|s_{t+i}))\middle|\, s_t = s, a_t = a\right\}. \tag{10-23}$$

Unlike entropy regularization, a more thorough use of policy entropy occurs when the soft value function is utilized. By adding this function to each reward signal, $q^\pi(s, a)$ is changed to include the entropy bonuses from every time step. Thus, soft state values and soft action values naturally have the following self-consistency relation:

$$v^\pi(s) = \mathbb{E}_{a \sim \pi}\{q^\pi(s,a)\} + \alpha \mathcal{H}\big(\pi(\cdot \,|s)\big). \tag{10-24}$$

In the less-explored regions, the conventional reward is much smaller than the weighted entropy, thus the variance of the policy distribution will be increased to enlarge the entropy term and facilitate the exploration. In the well-explored regions where the optimal actions have been well-studied, the policy prefers to apply a more deterministic action to exploit the maximum reward. This mechanism results in a more active exploration ability, i.e., exploring familiar regions less frequently and unfamiliar regions more frequently. Therefore, an RL agent is more likely motivated to collect samples in unexplored regions, which will result in better sample efficiency.

10.4 Deep RL Algorithms

Deep neural networks are responsible for recent AI breakthroughs, and they can be combined with reinforcement learning to create something astounding, such as DeepMind's AlphaGo. Deep reinforcement learning (DRL) algorithms can start from a blank policy candidate and achieve superhuman performance in many complex tasks, including Atari games, StarCraft and Chinese Go. Mainstream DRL algorithms include Deep Q-Network (DQN), Dueling DQN, Double DQN (DDQN), Trust Region Policy Optimization (TRPO), Proximal Policy Optimization (PPO), Asynchronous Advantage Actor-Critic (A3C), Deep Deterministic Policy Gradient (DDPG), Twin Delayed DDPG (TD3), Soft Actor-Critic (SAC), Distributional SAC (DSAC), etc. These algorithms are proposed with one or several of the abovementioned tricks to alleviate one or some challenges. The mainstream DRL algorithms are listed in Table 8.1.

Table 10.2 Tricks of core DRL algorithms

Alg.	ExR	PEx	STN	DPU	CPU	CAC	DQF	BDQ	DRF	EnR	SVF	π
DQN	★		★									off
Dueling DQN	•		•									off
DDQN	•		•				★					off
TRPO					★							on
PPO						★						on
A3C		★								★		on
DDPG	•		•				•					off
TD3	•		•	•			•	★				off
SAC	•		•				•	•			★	off

DSAC	●	●	●	●		●	★	●	off

★ Formally proposed for the first time

● Inherited tricks from previous DRLs

ExR: Experience Replay	DQF: Double Q-Functions
PEx: Parallel Exploration	BDQ: Bounded Double Q-functions
STN: Separated Target Network	DRF: Distributional Return Function
DPU: Delayed Policy Update	EnR: Entropy Regularization
CPU: Constrained Policy Update	SVF: Soft Value Function
CAC: Clipped Actor Criterion	

10.4.1 Deep Q-Network

Equipping Q-learning with function approximation is a natural extension from tabular forms to parametric forms. The deep Q-network (DQN) was the first DRL algorithm that successfully used deep neural networks. When DQN was first proposed in 2013, there was a focus on replacing the tabular Q-function with neural networks for continuous state space. A naïve DQN design often suffers from occasional policy divergence. An advanced version of DQN was later proposed in 2015 that addressed the instability issue by embedding two empirical tricks: experience replay and separated target network [10-16]. This advanced DQN algorithm contains 2 Q-networks. Only one of them needs backpropagation for its weight updates, and the other slowly copies weights from the former to avoid fast varying targets.

- **Trick 1: Experience Replay.** Experience replay is used to remove correlations in the transition sequence and smooth changes in the data distribution.
- **Trick 2: Separated Target Network.** A target network is embedded to reduce correlations between Q-values and target values. Its weight is periodically copied from the online Q-network.

More specifically, in DQN, the agent's experience $e_t = (s_t, a_t, r_t, s_{t+1})$ at each time step t are stored in a replay buffer $\mathcal{D}_{\text{replay}} = \{e_0, \cdots, e_t, \cdots, e_N\}$. To reduce the computational burden and smooth the training process, each update of DQN is applied on a mini-batch of samples that are uniformly drawn from the replay buffer. Essentially, DQN is used to minimize the mean squared Bellman error:

$$J(w) = \mathbb{E}_{s,a,s' \sim \mathcal{D}_{\text{Replay}}} \left\{ \left(r + \gamma \max_{a'} Q_{\bar{w}}(s', a') - Q_w(s, a) \right)^2 \right\}, \qquad (10\text{-}25)$$

where $\mathcal{D}_{\text{Replay}}$ is the replay buffer. DQN is the first successful DRL algorithm, which can learn a desirable policy directly from high-dimensional sensory measurements, such as those from images or videos. Tests on Atari 2600 games have shown that DQN surpasses most previous RL algorithms and achieves a level of performance that is comparable to that of a professional human player.

The experience replay trick is the key to become success when using DQN. An RL agent often becomes unstable if learning from consecutive samples as they sequentially occur in the temporal domain. Drawing random samples from replay memory is able to break this correlation. One might ask what kind of DRL algorithm can use experience replay. The answer is from a special updating mechanism accompanied with Q-learning. In essence, Q-learning is an off-policy algorithm that does not explicitly contain any importance sampling (IS) ratio. This property provides high flexibility to use historical data without limitation. In fact, any off-policy RL algorithm, as long as it does not rely on the IS transformation, e.g., Q-learning and expected SARSA, can seamlessly utilize history data samples. Therefore, most action-value-based DRL algorithms can use experience replay, but it must be limited to one-step TD. In contrast, state-value-based DRL algorithms has not possibility to use historical data.

10.4.2 Dueling DQN

The main objective of DQN is to approximate the optimal Q-function by deep neural network. In standard DQN, training a Q-network can be done by minimizing the mean squared Bellman error. Dueling DQN performs similar training task by utilizing an innovative dueling Q-network, which better suits for Q-function approximation [10-32]. In dueling DQN, the stream of Q-network is split into two separate streams, of which one represents the state-value function (i.e., V-function) and the other represents the advantage function (i.e., A-function). The learned V-function indicates which states are or are not valuable without having to learn the effect for each action. This is very useful in state evaluation where its actions do not affect the environment in any relevant way. The learned A-function represents how beneficial it is to take an action over others. In a particular state, A-function provides pure guidance for greedy action selection. Since dueling DQN only differs from DQN in the architecture of the Q-network, they contain the same number of networks and adopt the same deep tricks.

- **Trick 1: Experience Replay.**
- **Trick 2: Separated Target Network.**

Specifically, dueling DQN consists of two separated streams that represent the V-function and the A-function while sharing a common feature extraction module. The two streams are combined via a special aggregating layer to produce an approximation of Q-function.

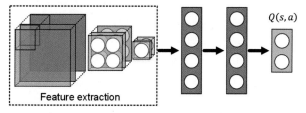

(a) Single-stream Q-network (i.e., normal network)

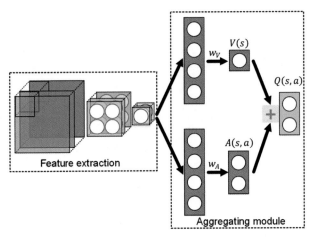

(b) Two-stream Q-network (i.e., dueling network)

Figure 10.22 DQN and dueling DQN network architectures

The aggregating layer requires a thoughtful design because it combines the two streams of fully connected layers to output the Q-value estimate. Using the definition of advantage function, we might be tempted to construct the aggregating operation as follows:

$$Q(s, a; w_V, w_A) = V(s; w_V) + A(s, a; w_A), \tag{10-26}$$

where w_V and w_A are the parameters of the two streams. However, (10-26) is unidentifiable in the sense that given any Q-value, we cannot uniquely recover its V-value and A-value. One may find that the same Q-value is maintained if we simultaneously perform two operations, i.e., adding a constant to V-value and subtracting the same constant from A-value. The lack of identifiability is mirrored by poor practical performance when (10-26) is used directly. To address this issue, the aggregating operation is changed to take a greedy search term:

$$Q(s, a; w_V, w_A) = V(s; w_V) + A(s, a; w_A) - \max_{\hat{a}} A(s, \hat{a}; w_A). \tag{10-27}$$

One might question that why the new equality allows identifiable estimation. Because the target policy π is greedy, we have:

$$\max_a A^\pi(s, a) = \mathbb{E}_{a \sim \pi}\{A^\pi(s, a)\} = \mathbb{E}_{a \sim \pi}\{V^\pi(s) - Q^\pi(s, a)\} = 0.$$

Therefore, (10-27) is equivalent to (10-26). Further, let us consider the following property of greedy actions:

$$a^* = \arg\max_{\hat{a}} Q(s, \hat{a}; w_V, w_A) = \arg\max_{\hat{a}} A(s, \hat{a}; w_A).$$

The equality above provides an additional condition that the Q-value of greedy action must be the same as the V-value, i.e., $Q(s, a^*; w_V, w_A) = V(s; w_V)$. With this condition, the parameters of V-function and A-function are uniquely identifiable.

The advantage of the dueling architecture lies partly in its ability to efficiently learn the V-function. In the dueling architecture, every update of Q-values will result in an update of the V-value stream. This frequent updating of the V-value stream allows for a better approximation of the V-function, thus benefiting the Q-value estimation. However, in the single-stream architecture of standard DQN, the update of a certain Q-value cannot fertilize the estimation of the other Q-values. Furthermore, the differences among Q-values for a given state are often much smaller than the magnitude of the Q-values. A small amount of noise in the Q-value updates could reorder the actions and thus cause an abrupt switch in the nearly greedy policy. The dueling architecture, with its separate advantage stream, is able to expose this negative effect and accordingly becomes more robust to noisy measurements. Experiments have shown that dueling DQN outperforms the state-of-the-art models on Atari 2600 games.

10.4.3   Double DQN

The aforementioned DQN algorithm uses the same Q-network for action selection and action evaluation. As a result, overoptimistic Q-values, which leads to the overestimation issue, may be learned due to the maximization operator. As an enhanced version of DQN, double DQN (DDQN) neatly blends the DQF trick into DQN to alleviate the occurrence of overestimation issues [10-31]. The DDQN algorithm utilizes two separated Q-networks: one for action selection, and the other for action evaluation. Since DDQN shares the same architecture as DQN, it also contains 2 Q-networks. One of these networks needs backpropagation, and the other serves as a target, whose weight is copied from the first network.

- **Trick 1: Experience Replay.**
- **Trick 2: Separated Target Network.**
- **Trick 3: Double Q-Functions.** The target network in DQN provides a natural candidate for action evaluation so that there is no need to add another online Q-network. More precisely, the target network evaluates actions that are selected from the online Q-network.

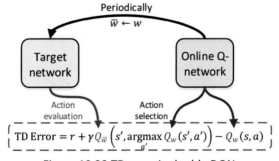

Figure 10.23 TD error in double DQN

Different from DQN, an optimal action a^* is selected in DDQN by the online Q-network rather than the target network itself. This design can eliminate the overestimation issue in high-dimensional function approximation. In DDQN, two Q-networks, including an online network and its target network, are simultaneously stored. The former selects the most valuable action a^*:

$$a^* = \arg\max_{a'} Q_w(s', a').$$

(10-28)

The latter is applied to evaluate the selected action. The target value with action evaluation becomes

$$Q_{\bar{w}}^{\text{target}} = r + \gamma Q_{\bar{w}}(s', a^*).$$

(10-29)

Finally, the mean squared Bellman error is minimized:

$$J(w) = \mathbb{E}_{s,a,s' \sim \mathcal{D}_{\text{Replay}}} \left\{ \left(r + \gamma Q_{\bar{w}}(s', a^*) - Q_w(s, a) \right)^2 \right\}.$$

(10-30)

One special merit of DDQN is that it shares the same architecture as DQN. Therefore, an extra network is not required to support the utilization of the DQF trick, and hence, no additional computational resources are needed. In practice, more accurate value estimates are achieved when using DDQN and better policy are often found compared with DQN. This ability has been demonstrated by the successful application of DDQN in Atari games.

10.4.4   Trust Region Policy Optimization

When facing complex problems, naïve policy gradients often have trouble in scaling to high-dimensional neural networks. The challenge may attribute to the high sensitivity in training deep neural networks. A slight difference in the parameter space can create a large discrepancy in the policy performance. A large step size becomes rather dangerous to stable policy updates in this situation. Trust region policy optimization (TRPO) is a natural policy gradient method that can be used to avoid fast policy changes. TRPO borrows insights from the mechanism of minorize-maximization (MM) optimization, which can provide theoretical guarantees for monotonic improvements. The basis of this approach is to reformulate a constrained optimization problem with the trust region constraint about the KL divergence of the new and old policies [10-24]. Finally, TRPO has 1 V-network and 1 policy network. Both of them require backpropagation for weight updates.

- **Trick 1: Constrained Policy Update.** TRPO avoids the large policy change by constraining the policy distance before and after each update. The trust region constraint comes from the KL divergence of the two adjacent policies.

The derivation of natural policy gradients depends on the minorize-maximization (MM) mechanism, whose basis is to maximize the surrogate function, i.e., a lower bound of the primal objective function. The monotonic improvement is guaranteed by the sandwich inequality property. The penalty term in the surrogate function is then regarded as the trust region constraint. Eventually, the following surrogate advantage is maximized with the KL divergence constraint:

$$\theta \leftarrow \arg\max_{\theta} \mathbb{E}_{s \sim d_{\pi_{\text{old}}}, a \sim \pi_{\text{old}}} \left\{ \frac{\pi_{\theta}(a|s)}{\pi_{\text{old}}(a|s)} A^{\pi_{\text{old}}}(s, a) \right\}$$

(10-31)

subject to

$$\bar{D}_{\text{KL}}(\pi_{\text{old}}, \pi_{\theta}) \le \delta_{\pi},$$

where $\bar{D}_{\text{KL}}(\pi_{\text{old}}, \pi_{\theta})$ is an average KL divergence between the old policy π_{old} and the learned policy π_{θ}:

$$\bar{D}_{\mathrm{KL}}(\pi_{\mathrm{old}}, \pi_\theta) = \mathbb{E}_{s \sim d_{\pi_{\mathrm{old}}}}\{D_{\mathrm{KL}}(\pi_{\mathrm{old}}(\cdot \mid s), \pi_\theta(\cdot \mid s))\}.$$

The advantage function is equivalent to

$$A^{\pi_{\mathrm{old}}}(s, a) = r + \gamma V^{\pi_{\mathrm{old}}}(s') - V^{\pi_{\mathrm{old}}}(s). \tag{10-32}$$

In practice, TRPO is a very effective DRL algorithm with few hyperparameter tunings. The monotonic improvement ability of TRPO allows it to work robustly on a wide variety of complex continuous control tasks. However, TRPO is rooted on on-policy design, and experience replay cannot be included. Despite a few successes, TRPO may be easily trapped in a local optimum because of insufficient exploration in on-policy settings.

10.4.5   Proximal Policy Optimization

Proximal policy optimization (PPO) is a refined successor of TRPO. The MM optimization is used to enhance the training stability with a monotonic policy improvement. Its computational cost, however, is expensive because it addresses the inverse Hessian of the average KL divergence. In fact, TRPO does not truly calculate the inverse of the Hessian matrix. Instead, it uses a conjugate gradient algorithm to solve linear equations. However, it is still computationally expensive. The central idea of PPO is to replace the trust region constraint with a clipped actor criterion. This replacement limits the largest performance improvement, without stepping a policy too far to accidentally cause collapse [10-25]. In its algorithm design, PPO shares the same architecture as TRPO, containing 1 V-network and 1 policy network. Both of these networks require backpropagation for weight updates.

- **Trick 1: Clipped Actor Criterion.** PPO relies on clipping the actor criterion and the IS ratio to remove incentives for the new policy to move farther away from the old policy.

In PPO, the surrogate function is modified with a clipped technique. Its core idea is to saturate the IS ratio to discourage excessively large policy updates. Specifically, the update rule of PPO is

$$\theta \leftarrow \arg\max_\theta \mathbb{E}_{s \sim d_{\pi_{\mathrm{old}}}, a \sim \pi_{\mathrm{old}}} \left\{ \min \left(\rho_{t:t} A^{\pi_{\mathrm{old}}}(s, a), \rho_{\mathrm{clip}} A^{\pi_{\mathrm{old}}}(s, a) \right) \right\},$$
$$\rho_{\mathrm{clip}} \stackrel{\mathrm{def}}{=} \mathrm{clip}(\rho_{t:t}, 1 - \epsilon, 1 + \epsilon), \tag{10-33}$$

where $\rho_{t:t} = \pi_\theta(a \mid s)/\pi_{\mathrm{old}}(a \mid s)$ is the IS ratio, $A^\pi(s, a)$ is the advantage function, ρ_{clip} is the clipped IS ratio, and ϵ is the clipped bound. The first term inside the minimization operator is equal to the surrogate advantage in TRPO. The second term modifies the surrogate advantage by clipping overly large or overly small IS ratios. Fortunately, the minimum of the clipped and unclipped criteria is still a pessimistic bound of the original objective function. Therefore, PPO is a computationally efficient first-order optimization algorithm that retains the monotonic improvement property.

10.4.6   Asynchronous Advantage Actor-Critic

The asynchronous advantage actor-critic (A3C) is rooted in the on-policy AC architecture [10-15]. It is the first algorithm to utilize the well-known asynchronous parallel framework, which enables a large spectrum of distributed DRL algorithms. Both the sampler and learner can be designed with a parallel mechanism. Each local learner needs backpropagation to compute gradients. In addition to multiple local learners, a typical

A3C algorithm must have a global learner, which contains 1 global V-network and 1 global policy network. The global networks do not need backpropagation, and their gradients come from the newest result of local learners.

- **Trick 1: Parallel Exploration.** The parallel design applies to both sampler and learner. In A3C, each local learner contains 1 local V-network and 1 local policy network. The local learner periodically copies parameters from the global networks to synchronize the newest learning results. The global network updates at once whenever a local learner finishes its own gradient computation.
- **Trick 2: Entropy regularization.** The conventional RL criterion has been augmented by the entropy regularization to encourage exploration.

Even though it seemingly belongs to off-policy design, A3C is actually an on-policy DRL algorithm because its critic and actor gradients are computed only with data samples from the current policy. Therefore, in A3C, historical data is not reused and it mainly relies on parallel exploration to maintain satisfactory training stability. Usually, its actor update is strengthened with an entropy regularization term to increase the exploration ability:

$$\nabla_\theta J(\theta) = \mathbb{E}_\pi\{\nabla_\theta \log \pi_\theta (a|s) A_w(s,a) + \alpha \nabla_\theta \mathcal{H}(\pi_\theta(\cdot|s))\}, \tag{10-34}$$

where $A_w(s,a)$ is an advantage function. In addition, practical A3C algorithms often use n-step TD to estimate the advantage function. Due to its on-policy behavior, this design is beneficial to speed up convergence without adding much algorithm complexity. Thus, its advantage function is equal to

$$A_w(s_t, a_t) = \sum_{i=0}^{n-1} \gamma^i r_{t+i} + \gamma^n V_w(s_{t+n}) - V_w(s_t). \tag{10-35}$$

The advantage of A3C mainly comes from its parallel mechanism (including both parallel samplers and parallel learners), which updates a common network in an asynchronous manner. This approach is suitable for a server with multicore processors, and each processor can run an individual learner. Since parallel exploration can generate a large amount of data simultaneously, A3C achieves satisfactory performance similar to those of off-policy RLs even though it cannot explicitly use experience replay.

10.4.7 Deep Deterministic Policy Gradient

DQN and DDQN are suitable only for discrete action spaces because of their intrinsic greedy search in the target value. The greedy search eliminates explicitly training a parameterized policy but is limited to discrete action spaces. The deep deterministic policy gradient (DDPG) is an extension of DQN from the discrete action space to the continuous action space. Inheriting the off-policy specialty from Q-learning, DDPG can use the history samples to learn both the Q-function and policy [10-14]. Unlike DQN, DDPG has 2 Q-networks and 2 policy networks, where 1 Q-network (called the online Q-network) and 1 policy network (called the online policy network) need backpropagation for weight updates. The other two networks do not need computationally expensive backpropagation, and their parameters are periodically copied or slowly updated from their corresponding online networks.

- **Trick 1: Experience Replay.**

- **Trick 2: Separated Target Network.** In DDPG, both the value network and policy network have their corresponding targets. The target networks are always slowly updated by

$$\bar{w} \leftarrow \rho_w w + (1 - \rho_w)\bar{w},$$
$$\bar{\theta} \leftarrow \rho_\theta \theta + (1 - \rho_\theta)\bar{\theta}, \qquad (10\text{-}36)$$

where $\rho_w, \rho_\theta > 0$ are their learning rates.

- **Trick 3: Double Q-Functions.** When calculating the target Q-value, DDPG uses the target policy $\pi_{\bar{\theta}}(s')$ to approximate the most valuable action, i.e., $\arg\max_{a'} Q_{\bar{w}}(s', a')$. Thus, action selection is decoupled from action evaluation, which has the same effect as DQF.

In DDPG, both critic update and actor update are formulated as stochastic optimization problems. The critic loss function in DDPG is

$$J(w) = \mathbb{E}_{s,a,s' \sim \mathcal{D}_{\text{Replay}}} \left\{ \left(r + \gamma Q_{\bar{w}}(s', \pi_{\bar{\theta}}(s')) - Q_w(s,a) \right)^2 \right\}. \qquad (10\text{-}37)$$

DDPG must contain a policy improvement step because it has an explicit policy function. Hence, the deterministic policy $\pi_\theta(s)$ is learned by maximizing the Q-function:

$$\max_\theta \mathbb{E}_{s \sim \mathcal{D}_{\text{Replay}}} \left\{ Q_w(s, \pi_\theta(s)) \right\}. \qquad (10\text{-}38)$$

Therefore, simple gradient ascent search for actor updates can be performed. In DDPG, the policy to be learned is deterministic, and it might not try enough to explore the environment. To increase exploration ability, zero-mean random noise can be added to deterministic actions and initial states can be sampled from a given state distribution at the beginning of environment interaction.

10.4.8 Twin Delayed DDPG

One disadvantage of DDPG is that the learned Q-network dramatically overestimates true action values, which makes its corresponding policy very brittle. In some extreme examples, DDPG can become very sensitive to its hyperparameters. The common solution is to add two more tricks into DDPG, which yields the so-called twin delayed DDPG (TD3) [10-4]. In the TD3 algorithm, there are 4 Q-networks and 2 policy networks, of which 2 online Q-networks and 1 policy network need backpropagation for weight updates. Unfortunately, such a TD3 algorithm has a much higher computational burden than DDPG due to the increased amount of backpropagation computation.

- **Trick 1: Experience Replay.**
- **Trick 2: Separated Target Network.**
- **Trick 3: Double Q-Functions.** It is implemented simultaneously in Trick 4.
- **Trick 4: Bounded Double Q-functions.** In TD3, 2 Q-functions are learned and the smaller Q-value is used to evaluate the target value as

$$Q_{\bar{w}}^{\text{target}} = r + \gamma \min_{i=1,2} Q_{\bar{w}_i}(s', \pi_{\bar{\theta}}(s')). \qquad (10\text{-}39)$$

Both Q-functions are updated with the same target. The formulations of the two critics are

$$J(w_1) = \mathbb{E}_{s,a,s'\sim\mathcal{D}_{\text{Replay}}}\left\{\left(Q_{\bar{w}}^{\text{target}} - Q_{w_1}(s,a)\right)^2\right\},$$

$$J(w_2) = \mathbb{E}_{s,a,s'\sim\mathcal{D}_{\text{Replay}}}\left\{\left(Q_{\bar{w}}^{\text{target}} - Q_{w_2}(s,a)\right)^2\right\}. \tag{10-40}$$

- **Trick 5: Delayed Policy Update.** In TD3, the policy update is delayed, i.e., the weight of the policy network is changed less frequently, to help reduce the volatility that generally arises in DDPG, where frequent policy updates may lead to a large oscillation of the target.

In addition, a unique trick called target policy smoothing is implemented in TD3 to avoid risky policy divergence. Usually, clipped noise is added to the target action to make it difficult to exploit erroneous Q-values. Thus, the target action becomes

$$a^* = \text{clip}\left(\pi_{\bar{\theta}}(s') + \text{clip}(\epsilon, -\epsilon_{\text{bnd}}, \epsilon_{\text{bnd}}), a_{\text{Low}}, a_{\text{High}}\right),$$

$$\epsilon \sim \mathcal{N}(0, \sigma). \tag{10-41}$$

where ϵ_{bnd} is the bound of action noise, and a_{Low} and a_{High} are the lower bound and upper bound of actions, respectively. Essentially, target policy smoothing is a kind of special policy regularization that avoids a particular failure mode in which the Q-function estimate develops an incorrect sharp peak for some actions. Overestimation is limited in TD3 by taking the minimum value between a pair of Q-functions and delaying policy updates to reduce the error arising in each update. Thus, significant improvements to the policy performance is achieved for complex control tasks.

10.4.9   Soft Actor-Critic

The soft actor-critic (SAC) forms a bridge between stochastic policy optimization and deterministic policy optimization. In SAC, the reparameterization trick is used to derive the expected gradient of a stochastic policy. Although SAC does not directly originate from TD3, it contains a few similar core tricks, including bounded double Q-functions and target policy smoothing. In essence, in SAC, a stochastic policy is trained to balance the expected return and accumulated policy entropy. The increase in the accumulated policy entropy results in more exploration, and increasing the expected return results in higher learning performance [10-6]. In total, SAC contains 4 soft Q-networks and 1 stochastic policy network, of which 2 online Q-networks and 1 policy network need backpropagation.

- **Trick 1: Experience Replay.**
- **Trick 2: Separated Target Network.** SAC concurrently learns a stochastic policy and two Q-networks. Both Q-networks have related target networks, while the policy does not.
- **Trick 3: Double Q-Functions.**
- **Trick 4: Bounded Double Q-functions.**
- **Trick 5: Soft Value Function.** In SAC, the maximum entropy RL framework, where each reward signal is augmented with policy entropy, is utilized.

To update the two Q-networks, SAC uses the same target value as in TD3:

$$J_{\text{Critic}}(w_i) = \mathbb{E}_{s,a,s'\sim\mathcal{D}_{\text{Replay}}}\left\{\left(r + \gamma V_{\bar{w}}(s') - Q_{w_i}(s,a)\right)^2\right\}, \forall i = 1,2, \tag{10-42}$$

where $V_{\bar{w}}(\cdot)$ is an implicit parameterized V-function, which can be calculated via (10-24). With the support of bounded double Q-functions, (10-24) is reformulated as

$$V_{\bar{w}}(s') = \mathbb{E}_{a' \sim \pi_\theta} \left\{ \min_{i=1,2} Q_{\bar{w}_i}(s', a') - \alpha \log \pi_\theta(a'|s') \right\}. \tag{10-43}$$

The policy parameter is learned by maximizing the expectation of V-values. Due to the self-consistency condition between V and Q, the corresponding actor loss function is

$$J_{\text{Actor}}(\theta) = \mathbb{E}_{s \sim \mathcal{D}_{\text{Replay}}, a \sim \pi_\theta} \{ Q^{\pi_\theta}(s, a) - \alpha \log \pi_\theta(a|s) \}. \tag{10-44}$$

Here, the reparameterization trick is applied to calculate the expected gradient of actor loss function $J_{\text{Actor}}(\theta)$, which has been tested to have a low variance estimate. It is critical to choose the best temperature parameter, which controls the strength of entropy regularization. In SAC, this hyperparameter can be automatically updated by formulating an optimization problem, where the policy entropy is enforced to be no less than a lower bound:

$$\max_{\pi_0, \cdots, \pi_N} \mathbb{E}_{s_{t+i} \sim d_{\pi_i}, a_{t+i} \sim \pi_i} \left\{ \sum_{i=0}^{N} r(s_{t+i}, a_{t+i}) \right\}, \tag{10-45}$$

subject to

$$\mathcal{H}(\pi_i(\cdot|s_t)) \geq \mathcal{H}_{\text{des}},$$

where $\mathcal{H}_{\text{des}}$ is the lower bound of the desired entropy. Its purpose is to find a stochastic policy with the maximal expected return that satisfies a minimum entropy constraint. This optimization problem (10-45) can be iteratively solved by dual gradient descent, in which the temperature parameter is updated gradually to obtain a balance between the expected return and the accumulated policy entropy. In short, SAC provides an off-policy RL algorithm with high sample efficiency because it retains the self-propelled exploration ability. In practice, SAC often outperforms many DRL algorithms, including TD3 and PPO, without any environment-specific hyperparameter adjustments.

10.4.10 Distributional SAC

Distributional soft actor-critic (DSAC), which is an enhanced version of SAC that is used to integrate the distributional return function into the maximum entropy RL framework, has the ability to vastly improve the estimation accuracy of Q-values. Unlike previous distributional RLs, which learn only a discrete return distribution, DSAC can be used to directly learn a continuous return distribution. Additionally, a clipping technique is implemented to truncate the difference between the target and current return distributions to prevent the gradient explosion issue [10-3]. In total, DSAC concurrently learns a parameterized stochastic policy and a single return distribution. It contains 2 distributional Q-networks and 2 policy networks, of which 1 online Q-network and 1 policy network need backpropagation.

- **Trick 1: Experience Replay.**
- **Trick 2: Parallel Exploration.** Actors explore environment instances in parallel, and multiple learners asynchronously update a few shared networks.

- **Trick 3: Separated Target Network.** Both policy and return distribution networks have their own targets.
- **Trick 4: Delayed Policy Update.** In DSAC, both policy and temperature updates are delayed. Their updating frequencies are both lower than those of critic updates to achieve more accurate value estimation.
- **Trick 5: Double Q-Functions.**
- **Trick 6: Soft Value Function.** In DSAC, each reward signal is augmented with a policy entropy term.
- **Trick 7: Distributional Return Function.** Instead of learning multiple independent Q-networks, DSAC learns a single distributional return function to avoid Q-value overestimation.

In DSAC, instead of an expected return, a return distribution is learned as the critic to enhance its robustness against environment stochasticity and approximation errors. Specifically, the return distribution is searched by minimizing the KL divergence between the current return distribution and its target distribution:

$$J_{\text{Critic}}(w) = \mathbb{E}_{s,a\sim\mathcal{D}_{\text{Replay}}}\left\{D_{\text{KL}}\left(Z_{\overline{w}}^{\text{target}}(\cdot\,|s,a), Z_w(\cdot\,|s,a)\right)\right\}, \tag{10-46}$$

where $Z^\pi(\cdot\,|s,a)$ is the return distribution and $Z_{\overline{w}}^{\text{target}}(\cdot\,|s,a)$ is the target distribution. The target of the state-action return obeys

$$Z_{\overline{w}}^{\text{target}}(s,a)\sim Z_{\overline{w}}^{\text{target}}(\cdot\,|s,a),$$
$$Z_{\overline{w}}^{\text{target}}(s,a) = r + \gamma Z_{\overline{w}}(s', \pi_{\overline{\theta}}(a'|s')),$$

where $Z(s,a)$ is the state-action return. Note that $Z(s,a)$ has a different definition than $Q(s,a)$, which is an expectation of the state-action return. Similar to DQN and DDPG, the target network is updated at a much slower frequency to reduce the potential target oscillation. The gradient of such a distributional critic is derived by approximating the KL divergence with the sample average:

$$\nabla J_{\text{Critic}} = -\mathbb{E}_{s,a,s'\sim\mathcal{D}_{\text{Replay}},a'\sim\pi_{\overline{\theta}}}\left\{\nabla_w \log Z_w\left(Z_{\overline{w}}^{\text{target}}(s,a)\big|s,a\right)\right\}. \tag{10-47}$$

The actor of the DSAC must output a stochastic policy. The return distribution offers a more prosperous evaluation than traditionally used expected returns, such as Q-functions or V-functions. Here, the actor loss function to be maximized is derived as

$$J_{\text{Actor}} = \mathbb{E}_{s\sim\mathcal{D}_{\text{Replay}},a\sim\pi_\theta}\left\{\text{Perc}\left(Z_w(\cdot\,|s,a)\right) - \alpha \log \pi_\theta(a|s)\right\}, \tag{10-48}$$

where $\text{Perc}(Z)$ is a certain percentile of the return distribution. It naturally reduces to standard Q-functions when choosing the 50% quantile for the Gaussian distribution. To date, DSAC has shown significant benefits in training stability, critic accuracy, risk awareness, and policy performance. The use of return distribution can considerably improve the estimation accuracy of Q-values, and it outperforms many state-of-the-art RL algorithms by a large margin.

10.5 References

[10-1] Cho K, Van Merriënboer B, Gulcehre C et al (2014) Learning phrase representations using RNN encoder-decoder for statistical machine translation. EMNLP, Doha, Qatar

[10-2] Cybenko G (1989) Approximation by superpositions of a sigmoidal function. Math Control Signals Syst 2(4): 303-314

[10-3] Duan J, Guan Y, Li SE et al (2021) Distributional soft actor-critic: off-policy reinforcement learning for addressing value estimation errors. IEEE Trans Neural Network & Learn Syst. doi: 10.1109/TNNLS.2021.3082568

[10-4] Fujimoto S, Hoof H, Meger D (2018) Addressing function approximation error in actor-critic methods. ICML, Stockholm, Sweden

[10-5] Graves A, Mohamed A, Hinton G (2013) Speech recognition with deep recurrent neural networks. IEEE ASSP, Vancouver, Canada

[10-6] Haarnoja T, Zhou A, Abbeel P et al (2018) Soft actor-critic: off-policy maximum entropy deep reinforcement learning with a stochastic actor. ICML, Stockholm, Sweden

[10-7] He K, Zhang X, Ren S et al (2016) Deep residual learning for image recognition. CVPR, Las Vegas, USA

[10-8] Hochreiter S, Schmidhuber J (1997) Long short-term memory. Neural Computation 9(8): 1735-1780

[10-9] Huang G, Liu Z, Van Der Maaten L et al (2017) Densely connected convolutional network. CVPR, Honolulu, USA

[10-10] Jain A, Zamir AR, Savarese S et al (2016) Structural-RNN: deep learning on spatio-temporal graphs. CVPR, Las Vegas, USA

[10-11] Krizhevsky A, Sutskever I, Hinton G (2012) Imagenet classification with deep convolutional neural networks. NeurIPS, Lake Tahoe, USA

[10-12] Lecun Y, Bottou L (1998) Gradient-based learning applied to document recognition. Proceedings of IEEE 86(11): 2278-2324

[10-13] LeCun Y, Bottou L, Orr G et al (2012) Efficient BackProp. In: Neural networks: tricks of the trade. Springer, Heidelberg

[10-14] Lillicrap T, Hunt J, Pritzel A et al (2016) Continuous control with deep reinforcement learning. ICLR, San Juan, Puerto Rico

[10-15] Mnih V, Badia A, Mirza M et al (2016) Asynchronous methods for deep reinforcement learning. ICML, New York, USA

[10-16] Mnih V, Kavukcuoglu K, Silver D et al (2015) Human-level control through deep reinforcement learning. Nature 518 (7540): 529-533

[10-17] Ng AY (2004) Feature selection, L1 vs. L2 regularization, and rotational invariance. ICML, Banff, Canada

[10-18] Pascanu R, Mikolov T, Bengio Y (2013) On the difficulty of training recurrent neural networks. ICML, Atlanta, USA

[10-19] Pollack JB (1990) Recursive distributed representations. Artificial Intelligence 46(1-2):77-105

[10-20] Qian N (1999) On the momentum term in gradient descent learning algorithms. Neural Networks 12(1): 145-151

[10-21] Robbins H, Monro S (1951) A stochastic approximation method. Ann Math Stat 22(3): 400-407

[10-22] Sandler M, Howard A, Zhu M et al (2018) MobileNetV2: inverted residuals and linear bottlenecks. CVPR, Salt Lake City, USA

[10-23] Schaul T, Quan J, Antonoglou I et al (2016) Prioritized experience replay. ICLR, San Juan, Puerto Rico

[10-24] Schulman J, Levine S, Abbeel P et al (2015) Trust region policy optimization. ICML, Lille, France

[10-25] Schulman J, Wolski F, Dhariwal P et al (2017) Proximal policy optimization algorithms. https://arxiv.org/abs/1707.06347

[10-26] Schuster M, Paliwal KK (1997) Bidirectional recurrent neural networks. IEEE Trans Signal Process 45(11): 2673-2681

[10-27] Simonyan K, Zisserman A (2015) Very deep convolutional networks for large-scale image recognition. ICLR, San Diego, USA

[10-28] Srivastava N, Hinton G, Krizhevsky A et al (2014) Dropout: a simple way to prevent neural networks from overfitting. J Mach Learn Res 15(1): 1929-1958

[10-29] Szegedy C, Liu W, Jia Y et al (2015) Going deeper with convolutions. CVPR, Boston, USA

[10-30] Van Hasselt H (2010) Double Q-learning. NeurIPS, Vancouver, Canda

[10-31] Van Hasselt H, Guez A, Silver D (2016) Deep reinforcement learning with double Q-learning. AAAI, Phoenix, USA

[10-32] Wang Z, Schaul T, Hessel M et al (2016) Dueling network architectures for deep reinforcement learning. ICML, New York, USA

[10-33] Werbos P (1990) Backpropagation through time: what it does and how to do it. Proceedings of IEEE 78(10): 1550-1560

[10-34] Williams RJ, Peng J (1991) Function optimization using connectionist reinforcement learning algorithms. Connect Sci 3(3): 241-268

Chapter 11. Miscellaneous Topics

> Two things are infinite:
>
> the universe and human stupidity;
>
> and I'm not sure about the universe.
>
> -- Albert Einstein (1879-1955)

In practice, RL is comparatively more challenging for a beginner to use than supervised learning. The challenges of using RL are mainly related to (1) how to interact with the environment more efficiently and (2) how to learn an optimal policy with a certain amount of data. Studies on the former challenge include on-policy/off-policy, stochastic exploration, sparse reward enhancement, and offline learning, while those of the latter include mixed representation, minimax optimization, and long-term credit assignment. In this chapter, we will explore a few recent RL advances that have been used to solve some major challenges, such as how to deal with adversarial uncertainties, how to handle partial observability, how to learn with fewer samples, how to learn rewards from experts, how to solve multi-agent games, and how to learn from offline data. The state-of-the-art RL frameworks, libraries and simulation platforms are also briefly described to support the R&D of more advanced RL theories and algorithms.

A model mismatch can easily result in the poor performance of a control policy. Robust RL seeks to search for an optimal policy that can deal with the severe model uncertainty. Model uncertainty can be considered an adversarial agent of the control policy. From this perspective, robust RL is formulated as a minimax optimization problem. The control policy and its adversarial policy are regarded as the two parties in a two-player zero-sum game, where the former attempts to improve the performance and the latter deteriorates it. The optimality condition of this problem is the Hamilton-Jacobi-Isaacs (HJI) equation, which plays a central role in building robust RL algorithms.

In addition to the model mismatch issue, real-world sensors always contain noise. Noisy measurements introduce an additional stochastic process to the Markovian environment dynamics. These types of problems are modeled as partially observable Markov decision processes (POMDPs). One common way to solving POMDP is to utilize the separation principle to build a two-step training algorithm. The key is to construct an implicit state representation called "belief state" to replace each state measurement. The belief state, which can be calculated by Bayesian estimation, is a posterior distribution of the real state. By taking the belief state as the policy input, POMDP can be naturally divided into two separate elements, i.e., a belief state estimator and an optimal static policy.

Traditionally, model-free RL suffers from poor generalization, i.e., a massive number of samples are needed to cover different scenarios. Typically, model-free RL is only designed for a restricted task domain and cannot be adapted to unknown scenarios. In meta-RL, which is one way to increase sample efficiency, learned knowledge is transferred from a set of training tasks to unseen but related tasks. The key to meta-RL is to specify what kinds of transferable knowledge should be learned from known tasks.

© The Author(s), under exclusive license to Springer Nature Singapore Pte Ltd. 2023,
Corrected Publication 2024
S. E. Li, *Reinforcement Learning for Sequential Decision and Optimal Control*,
https://doi.org/10.1007/978-981-19-7784-8_11

Three types of transferable knowledge have been employed in meta-RL, including transferable experience, transferable policy and transferable loss function.

Currently, multi-agent RL is attracting wide attention because it provides a natural way to address multi-agent problems. Compared to a single-agent system, a multi-agent system introduces many new challenges, such as (1) dimensional explosion, in which the joint state/action space grows quickly with the number of agents and (2) proneness to easy divergence, i.e., when simultaneously training multiple agents, the optimal policy of each agent is strongly entangled with those of other agents. Worse still, the joint dynamics of other agents are actually nonstationary, which easily causes unstable learning processes. Thus, accurate prediction of other agents' behaviors becomes vital for stabilizing the policy update. In the stochastic multi-agent game, other agents' behaviors are often considered in the design of transition function and value function.

The aim of inverse RL is to learn the reward from expert demonstration as well as the policy to produce behavior that maximizes the accumulation of the predefined reward function. The main challenge of inverse RL is that the expert demonstration may correspond to many different reward designs. Thus, it is necessary to find a unique reward function that is able to differentiate between desired and undesired policies. One way to find a unique reward function is called maximum margin inverse RL, in which the best reward function should maximize the difference between the expert policy and all other policies. Another solution is called maximum entropy inverse RL, in which the best reward function should maximize the probability of observing the same expert trajectories.

Offline RL is another rapidly developing field that enables an offline training paradigm from precollected data. Most RL algorithms require a large number of online interactions with the environment to collect experience. This becomes a major obstacle in many real-world tasks, as collecting data via online interactions can be expensive or dangerous. Furthermore, building a high-fidelity simulation model can be costly or even unrealistic for complex systems. Fortunately, many real-world systems often have a sufficient amount of precollected data. Different from traditional RL, offline RL trains an agent by using the fixed data without ever interacting with the environment. However, because offline policy and value function are trained under the data distribution and evaluated on a different data distribution induced by the learned policy, offline RL faces an inherent issue called distributional shift, making it unstable and hard to train.

11.1 Robust RL with Bounded Uncertainty

Model errors and external disturbances are commonly observed in a feedback control system, resulting in a strong need for a robustly designed control policy. The aim of robust control, which has received much attention since the 1980s, is to design a robust feedback system. System analysis and controller synthesis based on the H-infinity norm have made great progress. The aim of H-infinity control is to restrict a certain norm that describes the energy gain from the uncertainty level to the desired objective. Usually, this method is limited to linear uncertain systems with a quadratic utility function. In RL, the environment dynamics can be affine nonlinear, and the utility function can be nonquadratic. A standard robust OCP is formulated with a nonlinear affine model:

$$\dot{x} = f(x, u) + g(x, w), \tag{11-1}$$

where w is the unknown disturbance. The dynamic model involves a nominal part f and an uncertain part g. The infinite-horizon cost functional is defined as an integral of the utility function:

$$J\big(x(t), u(\tau), w(\tau)\big)\big|_{\tau=[t,\infty)} = \int_t^\infty l(x, u, w) \, d\tau, \tag{11-2}$$

where t is the starting time, $l(x, u, w)$ is the utility function, and $J(\cdot, \cdot, \cdot)$ is the performance index. The performance index is a function of the initial state $x(t)$, control sequence $u(\tau)|_{\tau=[t,\infty)}$, and disturbance sequence $w(\tau)|_{\tau=[t,\infty)}$. This problem may be computationally challenging when the state and action spaces are large or when $g(x, w)$ does not benefit from having a tractable structure. When u and w can be written as stationary Markov policies, the state-value function is simply a function of the initial state $x(t)$:

$$V\big(x(t)\big) = \int_t^\infty l(x, u, w)|_{u=u(x), w=w(x)} \, d\tau. \tag{11-3}$$

Here, the control policy is $u = u(x)$, and the adversarial policy is $w = w(x)$. From the differential game perspective, robust RL can be formulated as a dynamic minimax optimization problem, where u and w are regarded as its two adversarial players. One player is the control policy, and the other is the disturbance policy. When the state-value function is finite-valued, a partial differential equation (PDE) can be derived by applying Leibniz's formula with the boundary condition $V(0) = 0$:

$$H\left(x, u, w, \frac{\partial V(x)}{\partial x}\right) = l(x, u, w) + \frac{\partial V(x)}{\partial x^\mathrm{T}}\big(f(x, u) + g(x, w)\big) = 0, \tag{11-4}$$

where $H(x, u, w, \partial V(x)/\partial x)$ is called the Hamiltonian. The optimal control policy $u^*(x)$ and the worst-case disturbance policy $w^*(x)$ naturally constitute a saddle point (u^*, w^*) of the differential game. This Nash equilibrium can be solved by the associated Hamilton-Jacobi-Isaacs (HJI) equation [11-2][11-59]. The HJI equation is a nonlinear partial differential equation, and its analytical solution is very difficult to find, especially for those nonlinear systems. As a numerical method, robust RL employs an iterative procedure similar to that of policy iteration, which consists of two alternating steps: (1) policy evaluation (PEV), and (2) policy improvement (PIM). In the PEV step, the solution of the self-consistency condition is calculated for given control and disturbance policies while a better control policy and a worse adversarial policy are needed in the PIM step.

11.1.1   H-infinity Control and Zero-Sum Game

One might be interested in the inherent connection between H-infinity control and robust RL. Both of them aim to provide a feedback controller with a certain level of robustness. Classic H-infinity control is a frequency-domain method. As most H-infinity control methods employ transfer function matrices to design optimization targets, they are restricted to linear time-invariant systems [11-25][11-60][11-44]. Consider a linear uncertain system

$$\dot{x} = (A + \Delta A)x + (B + \Delta B)u, \tag{11-5}$$

where (A, B) is the nominal model and ΔA and ΔB are the model errors. The disturbance $w = \Delta A \cdot x + \Delta B \cdot u$ reflects the model uncertainty. Define the objective output as

$$z^{\mathrm{T}} = \left[\left(\sqrt{Q} x \right)^{\mathrm{T}} \quad \left(\sqrt{R} u \right)^{\mathrm{T}} \right],$$

where $Q = Q^{\mathrm{T}} > 0$ and $R = R^{\mathrm{T}} > 0$ are positive definite weighting matrices. The disturbance can be rewritten as a linear function of the objective output:

$$w = \left[\Delta A \cdot \sqrt{Q^{-1}} \quad \Delta B \cdot \sqrt{R^{-1}} \right] \begin{bmatrix} \sqrt{Q} & 0 \\ 0 & \sqrt{R} \end{bmatrix} \begin{bmatrix} x \\ u \end{bmatrix} = \Delta \cdot z, \qquad (11\text{-}6)$$

where $\Delta = \left[\Delta A \cdot \sqrt{Q^{-1}} \quad \Delta B \cdot \sqrt{R^{-1}} \right]$ is the bounded perturbation. It is known from the small gain theorem that the closed-loop system is robustly stable if $\|T_{zw}\|_\infty \|\Delta\|_\infty < 1$, where T_{zw} is the closed-loop transfer function from the disturbance w to the objective output z [11-25][11-67]. The perturbation is unknown but has the property that $\|\Delta\|_\infty$ is bounded by a constant $1/\gamma$. Thus, this condition can be transformed into $\|T_{zw}\|_\infty \le \gamma$, i.e.,

$$\|T_{zw}\|_\infty^2 = \sup_w \frac{\|z\|_2^2}{\|w\|_2^2} = \sup_w \frac{\int_t^\infty z^{\mathrm{T}} z \, d\tau}{\int_t^\infty w^{\mathrm{T}} w \, d\tau} \le \gamma^2, \qquad (11\text{-}7)$$

in which $\|T_{zw}\|_\infty^2$ is the L_2-gain of the closed-loop system. The transformed condition reflects the ability to suppress the unknown disturbance lower than a given disturbance attenuation level.

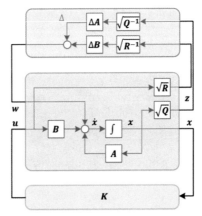

Figure 11.1 Standard formulation of H-infinity control

Figure 11.1 shows the standard formulation of H-infinity control, which contains two feedback loops: one is the main control loop, and the other is the disturbance attenuation loop. Traditionally, this type of robust control problem is reformulated into a convex optimization problem in the form of linear matrix inequalities (LMIs). The advent of LMI optimization has significantly influenced the research direction in robust control. A large number of efforts have been devoted in developing efficient LMI solvers for convex optimization problems from robust control. While the reduction of a robust control problem to an LMI problem provides a solution, it is also well recognized that in many practical applications, the resulting LMI problems are so large as to go beyond the limits of currently available LMI solvers. One new remedy is to convert H-infinity control into a minimax optimization problem. The first step of this conversion is to find a suboptimal controller such that the following integral is always less than zero:

$$\sup_{w} \int_{t}^{\infty} (z^{T}z - \gamma^{2}w^{T}w)\, d\tau \leq 0.$$

Its utility function naturally becomes

$$l(x, u, w) = z^{T}z - \gamma^{2}w^{T}w = x^{T}Qx + u^{T}Ru - \gamma^{2}w^{T}w. \tag{11-8}$$

Therefore, a feasible solution is to find a control policy to minimize the supremum of the above integral performance, in which a two-player differential game is finally constructed as

$$V^{*}(x) = J(x, u^{*}, w^{*}) = \min_{u} \max_{w} J(x, u, w), \tag{11-9}$$

where $V^{*}(x)$ is denoted as the optimal value function. This differential game has two players, namely, a control policy $u = u(x)$ and its adversarial policy $w = w(x)$. The control policy seeks to minimize the cost, and the adversarial policy seeks to maximize it. Since the benefit for the control policy is a loss for the adversarial policy, the minimax problem becomes a typical zero-sum game, which is the standard problem formulation of robust RL. The relationship between H-infinity control and the zero-sum game is shown in Figure 11.2. This relationship also hints us how to solve the general robust control problem from the perspective of a minimax optimization [11-54].

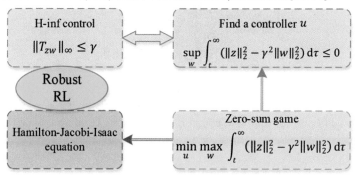

Figure 11.2 H-infinity control vs. zero-sum game

11.1.2   Linear Version of Robust RL

Robust RL for linear systems is of significance in both theoretical analysis and engineering practice. Basar et al. (2008) have shown that the existence of its solution is always guaranteed as long as the given disturbance attenuation level γ is large enough [11-9]. Zhang et al. (2020) found that convergence is greatly impacted by poor initialization in linear systems, and some robust admissible initializations were proposed to stabilize the learning process [11-65]. For a linear system with quadratic cost, the minimization and maximization steps are exchangeable in the minimax Hamiltonian (11-12). Therefore, it is convenient to apply polynomial bases, which are quadratic with respect to the current-time state, to approximate the value function [11-4][11-5]. One way to address nonlinear environments is to view the weak nonlinearity as the approximate error of linearized dynamics, although the stability region could shrink if a linear policy is applied [11-65].

Here, we revisit the H-infinity control from the perspective of a zero-sum game (i.e., a minimax optimization for dynamic systems). For a linear system (11-5), the self-consistency condition (11-4) is simplified as

$$H\left(x, u, w, \frac{\partial V(x)}{\partial x}\right)$$

$$= x^T Q x + u^T R u - \gamma^2 w^T w + \frac{\partial V(x)}{\partial x^T}(Ax + Bu + w) = 0, \tag{11-10}$$

with the boundary condition $V(0) = 0$. The zero-sum game (11-9) has a unique solution if the saddle point (u^*, w^*) exists:

$$\min_{u} \max_{w} J(x(t), u, w) = \max_{w} \min_{u} J(x(t), u, w). \tag{11-11}$$

This is the Nash condition, where two players stay at the Nash equilibrium (u^*, w^*) and neither of them has an incentive to unilaterally change. As shown in Figure 11.3, there is no benefit for the control policy u^* if it individually changes itself because any change will increase the performance index. The same is true for the adversarial policy w^* but with a reverse effect for the performance index.

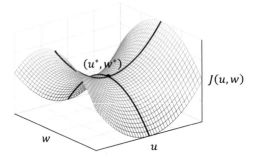

Figure 11.3 Saddle point of zero-sum game

A necessary condition for the Nash condition is the Isaacs condition, which can be seen as an extension of the Pontryagin maximum principle [11-42], i.e.,

$$\min_{u} \max_{w} H\left(x, u, w, \frac{\partial V(x)}{\partial x}\right) = \max_{w} \min_{u} H\left(x, u, w, \frac{\partial V(x)}{\partial x}\right). \tag{11-12}$$

The Isaacs condition can also constitute a sufficient condition for the Nash condition if the dynamic system is zero-state observable. The advantage of the Isaacs condition for linear systems is that its minimization and maximization operations are exchangeable. Hence, given the optimal value function, the stationary point (u^*, w^*) can be calculated via stationarity conditions, $\partial H/\partial u = 0$ and $\partial H/\partial w = 0$, i.e.,

$$u^* = -\frac{1}{2}R^{-1}B^T\frac{\partial V^*(x)}{\partial x},$$

$$w^* = \frac{1}{2\gamma^2}\frac{\partial V^*(x)}{\partial x}. \tag{11-13}$$

Note that $\partial^2 H/\partial u^2 = 2R > 0$ and $\partial^2 H/\partial w^2 = -2\gamma^2 < 0$; the Hamiltonian attains a minimum at u^* and a maximum at w^*. Thus, the stationary point (u^*, w^*) is actually a saddle point. Applying the saddle point (u^*, w^*) and the optimal value function $V^*(x)$ to the Hamiltonian (11-10), we have the following condition:

$$H\left(x, u^*, w^*, \frac{\partial V^*(x)}{\partial x}\right) = 0, V^*(0) = 0. \tag{11-14}$$

This formula is called the Hamilton-Jacobi-Isaacs (HJI) equation. It can be proven that if the HJI equation has a solution, the optimal controller derived from (11-13) not only ensures a disturbance attenuation level γ but also robustly stabilizes the linear system. For linear systems, the HJI equation can be reduced into an algebraic Riccati equation, in which the optimal value function maintains a quadratic form:

$$V^*(x) = x^\mathrm{T} P x, \tag{11-15}$$

where $P = P^\mathrm{T} > 0$ is called Riccati matrix. By using two stationarity conditions, the saddle point (u^*, w^*) can be calculated and substituted back into the HJI equation. The continuous-time algebraic Riccati equation for robust linear control is

$$A^\mathrm{T} P + P A - P\left(B R^{-1} B^\mathrm{T} - \frac{1}{\gamma^2} I\right) P + Q = 0. \tag{11-16}$$

The study of continuous-time algebraic Riccati equation has a very long history and lots of mature computational methods have been developed in the past decades. Even though (11-16) is also called Riccati equation, its detailed structure is slightly different from those Riccati equations from linear quadratic controller, Kalman filter and their combination, i.e., linear quadratic Gaussian controller. Fortunately, all of them can call the same numerical methods to calculate the solution of Riccati equation.

11.1.3   Nonlinear Version of Robust RL

Although a nonlinear system does not have the concept of a transfer function, the L_2-gain of the closed-loop system can be employed as an objective design [11-55]. The abovementioned procedure for linear systems is also applicable for nonlinear systems. The Isaacs condition (11-12) and the stationarity conditions still hold. Given an optimal value function, its optimal control policy u^* and worst-case adversarial policy w^* are

$$u^* = -\frac{1}{2} R^{-1} \frac{\partial f(x, u)}{\partial u}\bigg|_{u=u^*} \frac{\partial V^*(x)}{\partial x},$$

$$w^* = \frac{1}{2\gamma^2} \frac{\partial g(x, w)}{\partial w}\bigg|_{w=w^*} \frac{\partial V^*(x)}{\partial x}. \tag{11-17}$$

The specific forms of u^* and w^* are attributed to the separated structure in an affine dynamic model. This is also the reason why we want to consider an affine nonlinear system in robust RL. The associated HJI equation is then formulated as

$$\min_u \max_w \left[l(x, u, w) + \frac{\partial V^*(x)}{\partial x^\mathrm{T}}(f(x, u) + g(x, w))\right] = 0. \tag{11-18}$$

Unlike the HJB equation, the utility function in the HJI equation (11-18) is non-positive definite due to the existence of a disturbance item, which brings a remarkable challenge to its numerical computation. The policy iteration algorithm has been developed to iteratively solve this HJI equation [11-32]. A better control policy and a worse adversarial policy are obtained by alternatingly minimizing and maximizing the "weak" Hamiltonian. The policy iteration algorithm for robust RL is illustrated in Figure 11.4. Depending on how a control policy and its adversarial policy are updated, robust policy iteration algorithms can have two kinds of iteration loops: (1) simultaneous iteration loop and (2) asynchronous iteration loop. These two loops are shown in Figure 11.4 (a) and (b), respectively. In the first iteration loop, given a value function, the two policies are updated in parallel in the same PIM step [11-43]. In the second iteration loop, the adversarial policy is updated in an inner loop, while the control policy is updated in an outer loop [11-3].

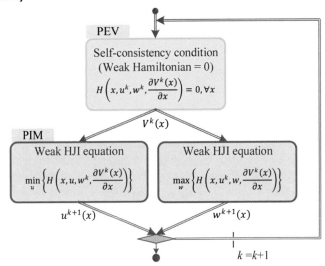

(a) Simultaneous iteration loop

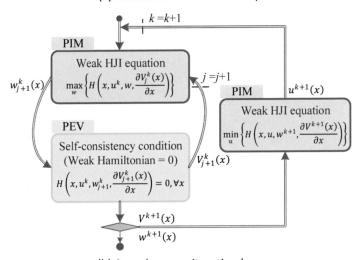

(b) Asynchronous iteration loop

Figure 11.4 Policy iteration to solve the HJI equation

In addition to the quadratic form in (11-8), the utility function $l(x, u, w)$ can be designed directly from real-world performance requirements. For example, to equip an autonomous car with the eco-driving ability, the utility function can include the term of fuel consumption rate, which is a nonlinear function of engine torque and engine speed. A saturated penalty term can be included when the error bound of the fuel actuator is viewed as an adversarial policy. Its final utility function will not be in the form of quadratic function. In most industrial applications, actual nonlinear dynamics cannot be described as accurate analytical models (let along in the form of affine models) but can be approximated by deep neural networks or high-fidelity simulation codes. In these cases, their plant uncertainties and approximation errors can be integrated into an unknown but bounded disturbance. The robust RL algorithm, which owns high robustness to resist various uncertainties, is capable of obtaining an optimal control policy in the worst case of the integrated disturbance [11-60].

11.2 Partially Observable MDP

The issues of imperfect environment modeling and measurements limit the application of RL. For example, in autonomous driving, a vehicle and its surrounding traffic environment always have some uncertainties, such as vehicle mass error, road slope disturbance, and the random behaviors of road users. Moreover, environment sensors may be noisy or even biased, making it challenging to find an optimal driving policy in an uncertain environment. Briefly, there are two kinds of uncertainties: (1) process uncertainty [11-51] and (2) observation uncertainty [11-7]. The process uncertainty refers to unmodelled high-order dynamics, parametric or structural errors, and unknown external disturbances. The observation uncertainty refers to the sensing imperfectness in measuring the state variables [11-18][11-30]. These uncertainties can be described as stochastic noises, which may obey some special but unknown distribution, such as a Gaussian distribution, uniform distribution, or beta distribution. One may be interested in how to handle uncertainties that arise from inaccurate actuators. In an inaccurate actuator, their received commands cannot be executed perfectly, but these uncertainties can be embedded into the process uncertainty by taking the control command as the action.

11.2.1 Problem Description of POMDP

Previous RL algorithms assume that the environment state is fully observable at every time step. This assumption does not always hold in reality; for example, noisy measurements may be primarily different from the real state. Such a sequential decision process can be formalized as a partially observable Markov decision process (POMDP), in which the underlying states cannot be perfectly observed [11-56]. In this section, how to search for an optimal policy with imperfect observations is described. A general discrete-time stochastic system with imperfect observations contains two equations: (1) a transition equation (11-19), which describes the environment dynamics, and (2) an observation equation (11-20), which represents the sensor characteristics.

$$x_{t+1} = f(x_t, u_t) + \xi_t, \qquad (11-19)$$

$$y_t = g(x_t) + \zeta_t,\tag{11-20}$$

where $x_t \in \mathbb{R}^n$ is the state, $u_t \in \mathbb{R}^m$ is the action, $f(\cdot,\cdot)$ is the deterministic part of the environment dynamics, $\xi_t \in \mathbb{R}^n$ is the process noise, $y_t \in \mathbb{R}^l$ is the measurement, $g(\cdot)$ is the observation function, and $\zeta_t \in \mathbb{R}^l$ is the observation noise. Here, we assume that both $\{\xi_0, \xi_1, \cdots, \xi_\infty\}$ and $\{\zeta_0, \zeta_1, \cdots, \zeta_\infty\}$ are independent and identically distributed (iid) and that they are independent of the initial state x_0. Therefore, the next state x_{t+1} only depends on the triple (x_t, u_t, ξ_t), which means that the environment dynamics have the Markov property. However, the current state x_t cannot be determined by only observation y_t due to the observation uncertainty. In POMDP, since the true state is not observable, the policy cannot be defined as a function of the current state. A reasonable alternative is to select a policy based on the set of historical information h_t, which contains all the known observations and actions from the initial time to the current time:

$$h_t \overset{\text{def}}{=} \{y_{1:t}, u_{0:t-1}\}.\tag{11-21}$$

The corresponding policy relying on historical information is

$$u_t = \pi(h_t).\tag{11-22}$$

The diagram of POMDP is shown in Figure 11.5. The goal of POMDP is to find a reasonable policy that minimizes the expectation of the cumulative cost under both process uncertainty and observation uncertainty. A discounted optimal control problem for POMDP is defined as

$$\min_{u=\pi(h)} V_h^\pi(h_t) = \mathbb{E}_\pi \left\{ \sum_{i=t}^\infty \gamma^{i-t} l(x_i, u_i) \,|\, h_t \right\},\tag{11-23}$$

where $V_h^\pi(\cdot)$ is the expectation of the cumulative cost, i.e., the state-value function, and $l(x_t, u_t) \geq 0$ is the utility function. The subscript "h" represents the fact that its argument is the historical information.

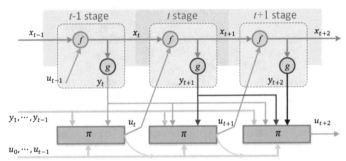

Figure 11.5 Partially observable MDP

The POMDP (11-23) cannot be optimized based on only its current observation because the observation sequence does not have the Markov property. We need to find a way to reformulate POMDP into a Markovian problem. Intuitively, if the historical information can be regarded as a "state", it is possible to update this "state" iteratively:

$$h_{t+1} = \text{Update}\{h_t, u_t, y_{t+1}\}.\tag{11-24}$$

From this perspective, the "state" that owns all the historical information approximately holds the Markov property:

$$p(h_{t+1}|h_t) = p(h_{t+1}|h_1, h_2, ..., h_t).$$ (11-25)

One intuitive explanation is that all past measurements and actions are already embedded inside the current historical information and that new information can be added in a step-by-step manner. Hence, the self-consistency condition and its corresponding Bellman equation can be derived as

$$
\begin{aligned}
V_h^\pi(h_t) \\
&= \mathbb{E}_\pi\{l_t|h_t\} + \mathbb{E}_\pi\{\textstyle\sum_{i=1}^\infty \gamma^i l(x_{t+i}, u_{t+i})|h_t\} \\
&= \mathbb{E}_\pi\{l_t|h_t\} + \gamma \sum_{h_{t+1}} p(h_{t+1}|h_t) \, \mathbb{E}_\pi\{\textstyle\sum_{i=1}^\infty \gamma^{i-1} l(x_{t+i}, u_{t+i})|h_{t+1}\} \\
&= \mathbb{E}_\pi\{l_t|h_t\} + \gamma \sum_{h_{t+1}} p(h_{t+1}|h_t) V_h^\pi(h_{t+1}).
\end{aligned}
$$

$$V_h^*(h_t) = \min_{u_t} \left\{ \mathbb{E}\{l_t|h_t\} + \gamma \sum_{h_{t+1}} p(h_{t+1}|h_t, u_t) V_h^*(h_{t+1}) \right\}.$$ (11-26)

The major challenge of taking the historical information as the "state" is that the length of historical information h_t is not fixed and it cannot be directly sent into the policy function with fixed input dimension. A straightforward idea is to encode the historical information h_t through a recurrent function, whose output is enforced to be a fixed-length latent representation. The most common recurrent function is the so-called recurrent neural network (RNN). In this case (see Figure 11.6), one needs to simultaneously address two learning problems by using recurrent functions: (i) representation learning, that is to learn the representation of the real state (which is encoded by hidden states in the recurrent function) from the historical sequential information, and (ii) policy learning, that is to minimize the expected cumulative cost with the learned representation. Clearly, these two learning processes are deeply coupled; and an unstable learning process may occur if these processes are not properly designed. In theory, simultaneously learning the recurrent function and its state representation will eventually result in a perfect policy that minimizes the cost function in POMDP. However, it was observed that POMDP has a representation learning bottleneck. A considerable portion of the learning activities must be spent to acquire good representations of the historical data, which has a negative impact on training stability and efficiency. Therefore, training a recurrent policy stably and efficiently is still an open question in the RL community.

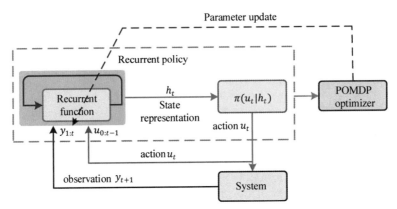

Figure 11.6 One-step approach with recurrent policy

11.2.2   Belief State and Separation Principle

POMDP is an extension of Markov decision process in which the system state cannot be directly observed. Because its noisy observation does not have the Markov property, the policy should take historical observations and actions into consideration. Obviously, the length of history observations keeps growing with the time and might even go to infinity. This poses a major challenge to solving POMDP tasks, which is also known as the curse of history. To deal with this problem, the belief state, which is the probability distribution of the current state conditioned on the history, is often used to replace the infinitely long history as the input of the policy. With this idea, a large class of POMDP algorithms follows a two-step approach, in which the first step (i.e., representation learning) is to estimate the belief state, and the second step (i.e., policy learning) is to solve for the optimal policy taking the belief state as input. The optimality guarantee of their combination relies on the so-called separation principle [11-33]. A famous example about separation principle is linear quadratic Gaussian (LQG) control, which decouples the closed-loop optimal controller into a Kalman filter and a linear quadratic regulator. Here, we will revisit the proof of separation principle in a general perspective, in which LQG is just a special case in linear and Gaussian problems.

Let us start the discussion from the definition of belief state. The belief state is a sufficient statistic for the historical information h_t, which can be used to bridge the gap between representation learning and policy learning. Mathematically, a statistic is called "sufficient" for a probability distribution if the collected samples from this distribution give no additional information than the statistic. For example, the sample mean and variance are a group of sufficient statistic for a process with normal distribution. If the variance is already known, the sample mean itself can serve as a sufficient statistic of this normal distribution.

- Definition 11.4 (Sufficient statistic): A function of historical data b_t is called sufficient statistic if the following condition holds:

$$p(x_t|h_t) = p(x_t|b_t), \tag{11-27}$$

where b_t is a certain distributional function of historical data h_t, for example, the Gaussian function. Obviously, sufficient statistic is an adequate representation of the historical information and can replace the historical information h_t as the input to the policy or value function. Note that the sufficient statistic is not unique, but one can find a physically meaningful and structurally concise statistic in POMDP. One common choice is the probability distribution of the current state conditioned on the historical information. This type of sufficient statistic is referred to as the belief state [11-57]. The sample mean is a typical belief state for a process with normal distribution.

- Definition 11.5 (Belief state): The belief state is a sufficient statistic in which b_t is reduced to a pure distribution of the current state x_t:

$$b_t(x) = p(x_t = x|h_t). \tag{11-28}$$

- Theorem 11-1: The belief state has the Markov property.

Proof:

It is easy to know that

$$b_t(x) = p(x_t = x|h_{t-1}, u_{t-1}, y_t). \tag{11-29}$$

According to Bayes' rule, one can derive a recursive formula of belief state:

$$
\begin{aligned}
b_t(x) &= \frac{p(x_t = x, y_t|h_{t-1}, u_{t-1})p(h_{t-1}, u_{t-1})}{p(y_t|h_{t-1}, u_{t-1})p(h_{t-1}, u_{t-1})} \\
&= \frac{\sum_{x_{t-1}} p(x_{t-1}|h_{t-1}, u_{t-1})Z}{p(y_t|h_{t-1}, u_{t-1})} \\
&= \frac{\sum_{x_{t-1}} p(x_{t-1}|h_{t-1})p(x_t = x|x_{t-1}, u_{t-1})p(y_t|x_t = x)}{p(y_t|h_{t-1}, u_{t-1})} \\
&= \frac{\sum_{x_{t-1}} b_{t-1}(x_{t-1})p(x_t = x|x_{t-1}, u_{t-1})p(y_t|x_t = x)}{\sum_x \sum_{x_{t-1}} b_{t-1}(x_{t-1})p(x_t = x|x_{t-1}, u_{t-1})p(y_t|x_t = x)}.
\end{aligned}
\tag{11-30}
$$

Here, the notation Z is defined for simplicity:

$$Z \overset{\text{def}}{=} p(x_t = x|h_{t-1}, x_{t-1}, u_{t-1})p(y_t|h_{t-1}, x_{t-1}, x_t = x).$$

Therefore, the belief state can be updated by the following recursive formula:

$$b_t = \text{Update}(u_{t-1}, b_{t-1}, y_t). \tag{11-31}$$

We can regard (11-31) as the new environment dynamics, in which b_t is the state and $\{u_{t-1}, y_t\}$ is the action. The probability of the belief state conditioned on the previous belief state and historical information has the following equality:

$$
\begin{aligned}
&p(b_t|b_{t-1}, u_{t-1}, y_t) \\
&= \mathbb{I}\big(b_t = \text{Update}(b_{t-1}, u_{t-1}, y_t)\big) \\
&= p(b_t|b_{t-1}, b_{t-2}, \dots, b_0, u_{t-1}, y_t),
\end{aligned}
\tag{11-32}
$$

where $\mathbb{I}(\cdot)$ is the discrete Dirac function. Obviously, the belief state is Markovian.

∎

The proof of the Markov property, especially the derivation in (11-30), implies an important fact, i.e., the update formula of the belief state is essentially equivalent to a Bayesian estimator. The Bayesian estimate is the best balance of the state information from noisy observations and uncertain dynamics. Similar to a Bayesian filter, the belief state is a combination of the newly received information (i.e., likelihood function) and the pre-known probability distribution (i.e., prior probability). In POMDP, the belief state can be updated through any computationally efficient Bayesian algorithm, for example, Kalman filter, extended Kalman filter, and variational inference. The belief state, together with its Bayesian estimator, form the cornerstone of representation learning in a two-step POMDP algorithm. In addition, the belief state has the following properties [11-50]:

(1) The expected reward conditioned on h_t is equal to that conditioned on b_t:

$$\mathbb{E}\{l_t|h_t, u_t\} = \mathbb{E}\{l_t|b_t, u_t\}. \tag{11-33}$$

(2) The probability of y_{t+1} conditioned on h_t is equal to that conditioned on b_t:

$$p(y_{t+1}|h_t, u_t) = p(y_{t+1}|b_t, u_t). \tag{11-34}$$

In addition to the Markov property, one important advantage of belief state is its fixed dimensionality. For example, if we know the belief state has a Gaussian structure, one can only use two parameters, its mean and variance, to represent this belief state. By substituting the belief state into a hidden Markov process, a new environment is obtained with both Markov property and fixed-dimensional state representation. Such a new environment holds a more concise Bellman equation, which is computationally manageable by traditional RL algorithms. Another potential benefit of belief state is the reduction of high computational complexity. When facing a large-scale problem, one can select a fairly low-dimensional belief state as the policy input.

In fact, the optimality of a two-step approach depends on the existence of separation principle, which states that the optimal controller of POMDP can be separated into an optimal state estimator and an optimal deterministic controller. The separation principle of simple LQG systems has been long studied, which states that the solution of an LQG system includes a Kalman filter and a linear quadratic regulator. In this simple case, the regulator does not take the whole Gaussian distribution as the belief state but its mean. This result is still optimal because if we replace the regulator input with the true state, the expectation of state-value function remains unchanged in the LQG case. However, in nonlinear stochastic systems, similar result does not hold again because a single-valued state estimate may not be sufficient to represent the belief state, which should be a complex non-Gaussian distribution.

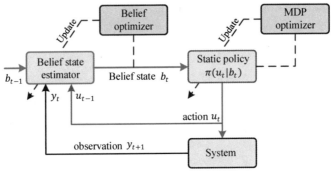

Figure 11.7 Two-step approach with belief state

Here, we will prove the separation principle for general POMDPs. In a two-step algorithm, POMDP can be divided into two subproblems: one is how to estimate the belief state, and the other is how to train a policy with the belief state [11-15]. On this basis, the solution of POMDP can be broken into two separate parts, including a belief state estimator and an optimal static policy, as shown in Figure 11.7. The former is actually a Bayesian estimator while the latter is an optimal policy for the MDP with belief state. This two-step approach yields an optimal solution that is exactly equivalent to that of the primal POMDP. The second subproblem actually formulates an MDP with the belief state

$$\min_{u=\pi(b)} V_b^\pi(b_t) = \mathbb{E}\left\{\sum_{i=t}^\infty \gamma^{i-t} l(x_i, u_i) \,|b_t\right\}, \tag{11-35}$$

where $V_b^\pi(\cdot)$ is the state-value function with the belief state and $\pi(b)$ is a new policy that takes the belief state as its exclusive input. The beauty of separation principle lies in the equivalence between the primal POMDP (11-23) and belief-state-based MDP (11-35). Their equivalence means that the same optimal action can be obtained from either original POMDP or belief-state-based MDP.

● Theorem 11-2 (Separation principle): For POMDP, the optimal value function given historical information h_t is equal to that given belief state b_t.

Proof:

Without loss of generality, we assume a one-to-one mapping from the historical information to the belief state. Define two transition probability functions:

$$\begin{aligned}\mathcal{P}_{t+1}^h &= p(h_{t+1}|h_t, u_t),\\ \mathcal{P}_{t+1}^b &= p(b_{t+1}|b_t, u_t).\end{aligned} \tag{11-36}$$

Since the belief state is a sufficient statistic of the historical information, we have the following equality:

$$\sum_{h_{t+1}} \mathcal{P}_{t+1}^h \mathbb{E}\{l_{t+1}|h_{t+1}\} = \sum_{b_{t+1}} \mathcal{P}_{t+1}^b \mathbb{E}\{l_{t+1}|b_{t+1}\}. \tag{11-37}$$

The equivalence condition allows the reuse of the belief state-based environment dynamics in the derivation of Bellman equation. The Bellman function with historical information can be written as

$$\begin{aligned}V_h^*&(h_t)\\ &= \min_{u_t}\left\{\mathbb{E}\{l_t|h_t\} + \gamma \sum \mathcal{P}_{t+1}^h V_h^*(h_{t+1})\right\}\\ &= \min_{u_t}\left\{\mathbb{E}\{l_t|b_t\} + \gamma \sum \mathcal{P}_{t+1}^h V_h^*(h_{t+1})\right\}\\ &= \min_{u_t, u_{t+1}}\left\{\mathbb{E}\{l_t|b_t\} + \gamma \sum \mathcal{P}_{t+1}^h \left\{\mathbb{E}\{l_{t+1}|h_{t+1}\} + \gamma \sum \mathcal{P}_{t+2}^h V_h^*(h_{t+2})\right\}\right\}\\ &= \min_{u_t, u_{t+1}}\left\{\mathbb{E}\{l_t|b_t\} + \gamma \sum \mathcal{P}_{t+1}^b \left\{\mathbb{E}\{l_{t+1}|b_{t+1}\} + \gamma \sum \mathcal{P}_{t+2}^h V_h^*(h_{t+2})\right\}\right\}.\end{aligned} \tag{11-38}$$

By rolling the right-hand side of (11-38) forward until infinity, one has

$$V_h^*(h_t) = \min_{u_{t+i},i=0,1,\cdots,\infty} \left\{ \textstyle\sum_{i=0}^{\infty} \gamma^i \left(\prod_{j=0}^{i} \sum \mathcal{P}_{t+j}^b \right) \mathbb{E}\{l_{t+i}|b_{t+i}\} \right\}$$

$$= V_b^*(b_t), \tag{11-39}$$

in which $\sum \mathcal{P}_t^b = 1$ is needed to match the derivation.

∎

11.2.3 Linear Quadratic Gaussian Control

In this section, the separation principle for LQG control problem is discussed in details [11-24]. LQG is one of the most fundamental optimal control problems, which concerns linear systems driven by additive white Gaussian noise. The output measurements are assumed to be corrupted by Gaussian noise. The optimal policy for such a stochastic system can be separated into a Kalman filter and a linear quadratic regulator. This principle was first proposed by Wonham and Zachrisson in 1968 [11-62][11-64]. It has been demonstrated that the separation principle also holds in some nonlinear controllers, even though the system dynamics must be linearized [11-11]. In some exceptional cases, this principle even holds when the process and observation noises are non-Gaussian martingales [11-8][11-24][11-46]. Consider a linear time-invariant stochastic system with Gaussian noise:

$$\begin{aligned} x_{t+1} &= Ax_t + Bu_t + \xi_t, \\ y_t &= Cx_t + \zeta_t, \end{aligned} \tag{11-40}$$

where $A \in \mathbb{R}^{n \times n}$ is the transition matrix, $B \in \mathbb{R}^{n \times m}$ is the control matrix, $C \in \mathbb{R}^{l \times n}$ is the observation matrix, and ξ_t and ζ_t are both Gaussian noises with zero mean. Define the utility function $l(x, u)$ as the quadratic function of state x and action u:

$$J = V_h^\pi(h_t) = \mathbb{E}\left\{ \textstyle\sum_{i=t}^{\infty} \gamma^{i-t} \left(x_i^{\mathrm{T}} Q x_i + u_i^{\mathrm{T}} R u_i \right) \big| h_t \right\}. \tag{11-41}$$

In linear systems with Gaussian noise, the conditional mean of the state is equal to its minimum variance estimate $\hat{x}_t$, i.e.,

$$\mathbb{E}\{x_t|b_t\} = \hat{x}_t. \tag{11-42}$$

Hence, the recursive Kalman filter is a standard minimum variance estimator:

$$\begin{aligned} \hat{x}_{t+1} &= \mathrm{Kalman}(y_{t+1}, u_t, \hat{x}_t), \\ \hat{x}_{t+1} &= A\hat{x}_t + Bu_t + L\left(y_{t+1} - C(A\hat{x}_t + Bu_t) \right), \end{aligned} \tag{11-43}$$

where L is the steady-state Kalman gain. According to the property of the Kalman filter, the minimum variance estimate is a sufficient statistic for the real state. In addition, the optimal cost function of LQG can be divided into two parts, including the linear quadratic cost and the estimation cost. The policy can only be derived from the linear quadratic cost, which contains the action and the mean of the state distribution. Surprisingly, the variance of the state distribution appears only in the estimation cost, thus it does not need to be taken as the policy input. This is why only the mean estimate is considered as the policy input in LQG. Here, we introduce how to find an optimal policy based on the state estimate $\hat{x}_t$. This design follows the linear version of the separation principle.

- Lemma 11-3: If $\mathbb{E}\{\xi_t\} = 0$ and $\mathbb{E}\{\zeta_t\} = 0$, the optimal cost function J^* is a quadratic function of the current state x_t; i.e.,

$$J^*(x_t) = \mathbb{E}\{x_t^T P^* x_t | b_{t-1}\},\tag{11-44}$$

where $P^* = P^{*T} > 0$ is the solution of the algebraic Riccati equation [11-7][11-18].

■

- Theorem 11-4: For LQG, if $\mathbb{E}\{\xi_t\} = 0$ and $\mathbb{E}\{\zeta_t\} = 0$, its optimal policy is

$$u_t = -K_\gamma \mathbb{E}\{x_t | b_t\},\tag{11-45}$$

where K_γ is the discounted optimal feedback gain of deterministic LQR.

Proof:

According to Theorem 11-2, the minimization of the conditional quadratic cost given the historical information can be written as

$$V_h^*(h_t) = V_b^*(b_t) = \min_{u_t}\left\{\mathbb{E}_{x_t}\{l(x_t, u_t)|b_t\} + \gamma \mathbb{E}_{b_{t+1}}\{V^*(b_{t+1})\}\right\},$$

where $V_b^*(b_t)$ is the optimal value function with belief state. We take the derivative of the right-hand side of this equation and force the derivative to be zero:

$$\frac{\partial\left\{\sum p(x_t|h_t)(x_t^T Q x_t + u_t^T R u_t) + \gamma \mathbb{E}\{x_{t+1}^T P^* x_{t+1}|b_t\}\right\}}{\partial u_t} = 0,$$

$$\gamma\frac{\partial\mathbb{E}\{x_{t+1}^T P^* x_{t+1}|b_t\}}{\partial u_t} + 2R u_t = 0.\tag{11-46}$$

Defining $z_{t+1} = x_{t+1} - \mathbb{E}\{x_{t+1}|b_t\}$ as the error of the minimum variance estimation, one has

$$\mathbb{E}\{z_{t+1}^T P^* z_{t+1}|b_t\}$$
$$= \mathbb{E}\{x_{t+1}^T P^* x_{t+1}|b_t\} + \mathbb{E}\{x_{t+1}|b_t\}^T P^* \mathbb{E}\{x_{t+1}|b_t\} - 2\mathbb{E}\{x_{t+1}|b_t\}^T P^* \mathbb{E}\{x_{t+1}|b_t\}$$
$$= \mathbb{E}\{x_{t+1}^T P^* x_{t+1}|b_t\} - \mathbb{E}\{x_{t+1}|b_t\}^T P^* \mathbb{E}\{x_{t+1}|b_t\}$$
$$= 0.$$

Further, substituting the above equation into (11-46), we have

$$\gamma\frac{\partial\mathbb{E}\{Ax_t + Bu_t + \xi_t|b_t\}^T P^* \mathbb{E}\{Ax_t + Bu_t + \xi_t|b_t\}}{\partial u_t} + 2R u_t = 0.$$

Finally, the optimal control policy can be derived as

$$u_t = -K_\gamma \mathbb{E}\{x_t|b_t\},$$
$$K_\gamma = \gamma(\gamma B^T P^* B + R)^{-1}B^T P^* A.\tag{11-47}$$

■

Figure 11.8 shows the diagram of the separation principle in LQG control. The optimal policy for this linear stochastic system can be separated into a Kalman filter and a linear quadratic (LQ) regulator.

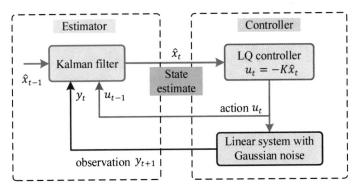

Figure 11.8 Separation principle in LQG control

In the classical setting, an LQG controller may be difficult to find the optimal solution when the dimension of system state is very large. The reduced-order LQG problem overcomes this challenge by limiting the number of controlled states into a fixed range. In addition, LQG control does not automatically ensure good robustness. The robust stability should be guaranteed by properly tuning the hyperparameters in the LQG controller. To promote good robustness, the feedback controller had better be designed and tested under the assumption that some of the system parameters are stochastic or even unknown.

11.3 Meta Reinforcement Learning

RL has achieved many impressive success, including gaining professional skills to play games (e.g., Atari and Chinese Go) and acquiring advanced manipulation and locomotion abilities in robot control. However, many of them come at the expense of high sample complexity. To master a skill, RL agents need to learn from millions of episodes per task. In contrast, humans and animals are capable of learning a new task after just a few attempts. Learning quickly, for example, recognizing new skills with past experience, is a hallmark of human intelligence. Our artificial agents should be able to do the same, learning and adapting quickly from only a few examples and continuing to adapt as more data become available. This kind of fast and flexible learning is rather challenging since the agent must integrate its prior experience with a small amount of new information while avoiding overfitting the new data.

Given this situation, meta-learning holds the promise of enabling learning systems to compile a diverse set of prior experiences and using the prior knowledge compiled to learn new skills or to rapidly and efficiently adapt to new environments. Meta-learning is referred to by many different names, e.g., learning to learn, multitask learning, few-shot learning, lifelong learning and transfer learning. In meta-RL, we consider a distribution $\mathcal{T}_i \sim p(\mathcal{T})$ over tasks, where each task $\mathcal{T}_i$ is a different MDP and $p(\mathcal{T})$ represents all the tasks. The reward functions and transitions vary across tasks. The aim in meta-RL is to train for transferable knowledge over a set of tasks from $p(\mathcal{T})$ to quickly adapt to a new task from $p(\mathcal{T})$ as efficiently as possible using only a small number of examples or trials. According to the specific type of learned transferable knowledge, such algorithms can be divided into three categories: transferable experience-based meta-RL, transferable policy-based meta-RL, and transferable loss function-based meta-RL. The framework of meta-RL is shown in Figure 11.9.

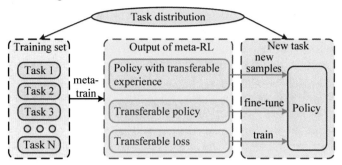

Figure 11.9 Framework of meta-RL

11.3.1   Transferable Experience

Transferable experience-based methods address meta-RL by training recurrent models on past states s, actions a, rewards r, and termination flags d and predicting new actions that maximize rewards. Memory is retained across several episodes of interaction. Unlike the recurrent policy in traditional RL, the policy under transferable experience-based methods entails the memory after each episode. At each time step, the agent learns transferable experience by taking all (s, a, r, d) as its policy input instead of only considering the state-action pairs. After the transferable experience is trained for a set of training tasks, the agent can rapidly adapt to a new task from $\rho(\mathcal{T})$ with fixed policy weights because the learned policy makes decisions based on past experience. Typical algorithms that belong to this category include RL^2 and Simple Neural Attentlve Learner (SNAIL) [11-17][11-49].

11.3.1.1   RL2

The RL^2 algorithm employs a recurrent neural network to approximate the optimal policy. During the training process, a set of tasks are drawn from $\rho(\mathcal{T})$. For each task, the meta-RL agent explores a series of episodes. If the episode has terminated, the agent sets the termination flag d_t to 1; otherwise this value defaults to 0. At each time step, the policy network receives the tuple (s_t, a_t, r_t, d_t) as input. At the end of an episode, the hidden state of the policy is preserved. Preservation occurs between different episodes but not between different tasks. The aim of this design is to maximize the accumulated discounted reward during a single task rather than a single episode. Since the underlying state transition model changes across tasks, the agent will learn different policies according to different transition trajectories during the meta-learning process. Furthermore, the internal state of the model is preserved across episodes so that it has the capacity to perform learning in its hidden activation signals. Hence, the agent is forced to integrate all the information it has received, including past actions, rewards, and termination flags and continually adapt its policy. After RL^2 is trained, the agent can rapidly adapt to a new task from $\rho(\mathcal{T})$ based on past experience with fixed policy weights.

11.3.1.2 SNAIL

The temporally linear dependency of traditional RNN architectures limits their capacity to perform sophisticated computation on a stream of inputs. Compared to traditional RNNs, temporal convolution (TC) layers offer more direct, high-bandwidth access to past information, allowing TC layers to perform more sophisticated computations over a fixed size temporal context. However, the dilation rates generally increase exponentially when scaling to long experience sequences. Therefore, the required number of layers scales logarithmically with the sequence length. In contrast, soft attention allows a model to pinpoint a specific piece of information from a potentially infinitely large context. Hence, SNAIL is constructed by combining TC and soft attention: the former aggregates information from experience, and the latter pinpoints specific pieces of information. By interleaving TC layers with causal attention layers, SNAIL can effectively encode long sequences of historical experience. By using attention within an end-to-end model, SNAIL can extract key information from the experience it gathers. As an additional benefit, SNAIL architectures are easier to train than traditional RNNs, such as LSTM or GRUs, and can be efficiently implemented to process an entire sequence in a single forward pass.

11.3.2 Transferable Policy

Over a set of tasks $\mathcal{T}_i \sim p(\mathcal{T})$, transferable policy-based meta-RL directly trains on a transferable policy with gradient descent that can quickly adapt to new tasks drawn from the task distribution $p(\mathcal{T})$. This method has the benefit of allowing for asymptotic performance, which is similar to learning from scratch, since adaptation is performed using gradient descent while also enabling acceleration of meta-training. Typical algorithms that belong to this category include pretraining, model-agnostic meta-learning (MAML) and model-agnostic exploration with structured noise (MAESN) [11-21][11-26].

11.3.2.1 Pretraining method

The most straightforward way to learn a transferable policy is to first pretrain policy π_θ on a set of tasks $\mathcal{T}_i \sim p(\mathcal{T})$ and then fine-tune the policy with gradient descent to fit new tasks. The policy is updated by maximizing the expected value function for all $\mathcal{T}_i$:

$$\theta^* = \arg\max_{\theta} \mathbb{E}_{\mathcal{T}_i \sim p(\mathcal{T})} \left\{ \mathbb{E}_{s \sim \mathcal{T}_i, \pi_\theta} \{ V_i^{\pi_\theta}(s) \} \right\}. \tag{11-48}$$

In fact, the pretraining method is worse than random initialization in some cases because by pretraining on all tasks, the transferable policy learns to output an average result for a particular input value. In some instances, the policy may learn very little information about the actual domain and instead learn about the range of the output space.

11.3.2.2 MAML

Unlike the pretraining method, MAML is designed to learn a transferable policy π_θ that has strong adaptability to new tasks drawn from $p(\mathcal{T})$. Essentially, MAML aims to find policy parameters that are sensitive to changes in the task during the training process such that small changes in the parameters will produce large improvements on the loss function of any new tasks drawn from $p(\mathcal{T})$ when altering the direction of the gradient of that loss. To achieve this goal, MAML divides the learning process into inner and outer loops. In the inner loop, an agent learns to solve a task, which is sampled from a particular distribution over a family of tasks $\mathcal{T}_i \sim p(\mathcal{T})$. The agent learns a post-update policy $\pi_{\theta_i'}$ by maximizing the value function $V_i^{\pi_\theta}$ of task $\mathcal{T}_i$:

$$\theta_i' = \theta + \alpha_\theta \nabla_\theta \mathbb{E}_{s \sim \mathcal{T}_i, \pi_\theta} \{V_i^{\pi_\theta}(s)\}. \qquad (11\text{-}49)$$

In the outer loop, the objective is to update the policy by maximizing the expected value function of the post-update policy $\pi_{\theta'}$ for all training tasks $\mathcal{T}_i \sim p(\mathcal{T})$ rather than that of the pre-update policy π_θ:

$$\theta^* = \arg\max_\theta \mathbb{E}_{\mathcal{T}_i \sim p(\mathcal{T})} \mathbb{E}_{s \sim \mathcal{T}_i, \pi_{\theta_i'}} \{V_i^{\pi_{\theta_i'}}(s)\}. \qquad (11\text{-}50)$$

The intuition behind this approach is that some internal representations are more transferable than others. MAML enables an agent to quickly acquire a policy for a new test task using few-shot learning with only a single gradient update, which means that only a small amount of experience is needed for the new task.

11.3.2.3 MAESN

The MAESN algorithm employs prior experience to initialize a policy and learn a latent exploration space to add structured stochasticity into a policy, producing exploration strategies informed by prior knowledge that are more effective than adding random action noise. The resulting policies can be written as $\pi_\theta(s, z)$, where $z \sim q_\omega(z)$ and $q_\omega(z)$ comprise the latent variable distribution with parameters ω. For example, $q_\omega(z)$ can be a diagonal Gaussian distribution of the form $q_\omega(z) = \mathcal{N}\{\mu, \sigma^2\}$ such that $\omega = \{\mu, \sigma^2\}$. The structured stochasticity of this form can provide more coherent exploration by sampling entire behaviors or goals rather than by simply relying on independent random actions. Note that the latent variable z is sampled only once per episode to provide temporally coherent stochasticity. Similar to MAML, the agent learns a post-update policy $\pi_{\theta_i'}$ by maximizing the value function $V_i^{\pi_\theta}$ of task $\mathcal{T}_i$:

$$\theta_i' = \theta + \alpha_\theta \nabla_\theta \mathbb{E}_{z_i \sim q_{\omega_i}, s \sim \mathcal{T}_i, \pi_\theta(s, z_i)} \{V_i^{\pi_\theta(s, z_i)}(s)\}. \qquad (11\text{-}51)$$

In addition, MAESN also learns a post-update latent variable distribution $q_{\omega_i'}$ for each task $\mathcal{T}_i$:

$$\omega_i' = \omega_i + \alpha_\omega \nabla_{\omega_i} \mathbb{E}_{z_i \sim q_{\omega_i}, s \sim \mathcal{T}_i, \pi_\theta(s, z_i)} \{V_i^{\pi_\theta(s, z_i)}(s)\}. \qquad (11\text{-}52)$$

MAESN aims to meta-train the policy parameters θ so that they can use the latent variables z to perform coherent exploration on a new task, and the behavior can be adapted as fast as possible. Therefore, the outer loop objective is to maximize the expected state-value under the post-update parameters for each task and the KL divergence between each task's pre-update latent distribution and the prior $p(z)$:

$$\{\theta^*, \omega_i^*\} = \arg\max_{\{\theta, \omega_i\}} \mathbb{E}_{\mathcal{T}_i \sim p(\mathcal{T})} \left\{ \mathbb{E}_{z_i' \sim q_{\omega_i'}, s \sim \mathcal{T}_i, \pi_{\theta_i'}(s, z_i')} \{V_i^{\pi_{\theta_i'}(s, z_i')}(s)\} \right.$$
$$\left. - D_{\mathrm{KL}}(q_{\omega_i'} \| p(z)) \right\}. \qquad (11\text{-}53)$$

The KL divergence term is added to encourage the post-update distributions to be close to the prior distribution.

11.3.3 Transferable Loss

Rather than explicitly encoding transferable knowledge through a learned behavioral policy, transferable loss-based meta-RL encodes it implicitly through a learned loss function. The end goal is that agents can use this transferable loss function to quickly learn a novel task. The Evolved Policy Gradients (EPG) algorithm is one of the most well-known algorithms in this category [11-28].

11.3.3.1 EPG

The EPG algorithm aims to learn a transferable loss function $L_\varphi(\pi_\theta, \tau)$ that is more efficient than the traditional objective function of policy optimization such that the agent can quickly learn a novel task using this learned loss function, where τ is a sampled sequence $(s_0, a_0, r_0, \cdots)$ of state, action and reward. Similar to MAML, EPG consists of two optimization loops. Given a sampled training task $\mathcal{T}_i \sim p(\mathcal{T})$, the inner loop optimization problem is to learn a new policy $\pi_{\theta'_i}$ by minimizing the transferable loss $L_\varphi(\pi_\theta, \tau)$ provided by the outer loop:

$$\theta'_i = \theta - \alpha_\theta \nabla_\theta \mathbb{E}_{\tau \sim \mathcal{T}_i, \pi_\theta}\{L_\varphi(\pi_\theta, \tau)\}. \tag{11-54}$$

The goal of the outer loop is to solve for φ^* such that policy π_{θ^*} trained with surrogate loss function $L_{\varphi^*}(\pi_\theta, \tau)$ achieves the highest expected accumulated reward:

$$\varphi^* = \arg\max_\varphi \mathbb{E}_{\mathcal{T}_i \sim p(\mathcal{T})} \mathbb{E}_{s \sim \mathcal{T}_i, \pi_{\theta'_i}} \left\{ V_i^{\pi_{\theta'_i}}(s) \right\}. \tag{11-55}$$

The accumulated reward or value function of policy $\pi_{\theta'_i}$ cannot be represented as an explicit function of the loss function L_φ. Hence, in practical applications, the EPG algorithm relies on evolution strategies to optimize the loss function.

In summary, transferable experience-based agents make decisions on new task $\mathcal{T}_i \sim p(\mathcal{T})$ based on their transferred experience learned from other but related tasks drawn from the task distribution $p(\mathcal{T})$. These methods do not explore new tasks or fine-tune their policies based on new experiences. In other words, these methods aim to learn the entire "learning algorithm" using a recurrent model, so this method is also called recurrent model-based meta-RL. While this process allows quick adaption via a single forward pass of the recurrent models, it limits the asymptotic performance compared to learning from scratch since the learned policy generally does not correspond to a convergent iterative optimization procedure and is not guaranteed to keep improving. However, the transferable policy or loss-based meta-RL directly trains for some transferable functions over $\mathcal{T}_i \sim p(\mathcal{T})$ that can quickly adapt with gradient descent for new tasks. In some research, these two methods are also called gradient-based meta-RL, in which the asymptotic performance is similar to that by learning from scratch since adaptation is performed using gradient descent while also enabling acceleration from meta-training.

11.4 Multi-agent Reinforcement Learning

A multi-agent RL system (also called a distributed RL system) is defined as multiple autonomous agents operating in a shared environment [11-66]. Each agent needs to consider interactions with the environment and other agents to optimize its long-term reward, resulting in more complex behaviors. An independent learner directly applies the single-agent RL algorithm to a multi-agent system and treats other agents as a part of the environment. This approach violates the assumption of the Markov property since the environment is no longer stationary [11-27]. If other agents adapt or learn based on their previous interactions, the independent learner will fail.

Multi-agent RL algorithms can be divided into different categories based on different learning objects, including fully cooperative tasks, fully competitive tasks, and multi-agent RL with hybrid rewards. Typical algorithms for fully cooperative tasks are distributed Q-learning [11-41] and frequency maximum Q-value [11-14]. Fully competitive tasks are generally designed as two-player zero-sum games, and the most well-known RL algorithm for these tasks is minimax Q-learning. For multi-agent RL with hybrid rewards, each agent is self-interested, and their rewards may or may not conflict with those of others. However, these tasks are challenging and not well understood. Some typical algorithms for hybrid rewards include Nash Q-learning [11-29], friend-or-foe [11-48] and mean field regime [11-12].

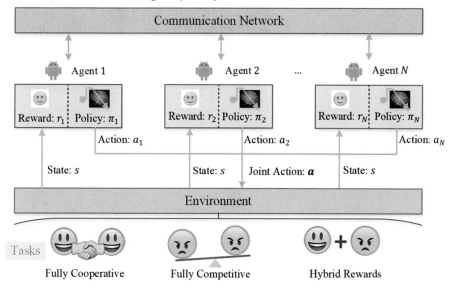

Figure 11.10 Framework of multi-agent RL tasks

The multi-agent RL system can be categorized from another viewpoint. Considering the partial observation situation, each agent can only obtain the partial observation s_i rather than the whole state information s, from the environment. Thus, the MARL methodologies can be categorized into three types based on the level of information that training and execution procedures can access: (a) fully decentralized in both training and execution, (b) fully centralized in both training and execution, and (c) centralized training with decentralized execution (CTDE). In the setting of fully decentralized MARL, each agent can only choose an action based on local information and can use this information to update its value function. In contrast, an agent in a fully centralized setting has access to other agents' actions and observations, thus making decisions based on the integrated information. The above two methods have their own disadvantages: a fully decentralized setting ignores the nature of the MARL system, while a fully centralized setting acquires stringent assumptions that are hard to achieve in real life. CTDE can combine those two settings while still performing well. During the training procedure of CTDE, agents have access to the complete information to update their value functions. However, the agent can only use local information to make decisions during the execution procedure. This allows the agent to use global knowledge to train its policy but to act in a decentralized manner.

Table 11.1 Methodology of multi-agent RL

Method	Policy	Value function	Typical algorithm
Fully decentralized	$\pi_i(a_i\|s_i)$	$Q_i(s_i, a_i)$	Single-agent RL algorithm
Fully centralized	$\pi_i(a_i\|s)$	$Q_i(s, a)$	Nash Q-learning
CTDE	$\pi_i(a_i\|s_i)$	$Q_i(s, a)$	MADDPG, VDN

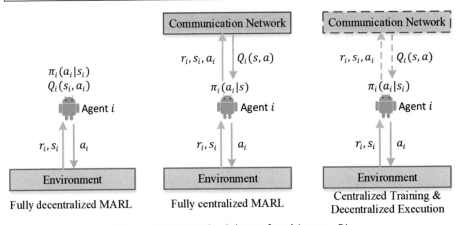

Figure 11.11 Methodology of multi-agent RL

11.4.1 Stochastic Multi-Agent Games

The generalization of single-agent RL to multi-agent RL can follow the framework of a stochastic multi-agent game. A stochastic multi-agent game, which is an extension of an MDP, is denoted as a tuple $\{S, \mathcal{A}, \mathcal{P}, \mathcal{R}\}$, where S is the shared state space, $\mathcal{A} = \mathcal{A}_1 \times \dots \times \mathcal{A}_N$ is the joint action space, $\mathcal{A}_i$ is the action space available to agent i, $\mathcal{P}$ is the transition probability, $\mathcal{R}$ is the joint reward function, r_i is the reward function for agent i, and N denotes the number of agents. Given a learning agent i and using shorthand notation $-i = \{1, 2, \dots, N\}\backslash\{i\}$ for the rest of the opponents, the value function depending on joint action $a = (a_i, a_{-i})$ and joint policy $\pi(a|s) = \prod_i \pi_i(a_i|s)$ is derived as follows:

$$V_i^\pi(s) = \sum_{a \in \mathcal{A}} \pi \sum_{s' \in \mathcal{S}} \mathcal{P}\big(r_i + \gamma V_i^\pi(s')\big), i = 1,2,\cdots,N, \tag{11-56}$$

where

$$\pi = \pi(a|s),$$
$$\mathcal{P} = \mathcal{P}(s'|s, a) = \mathcal{P}(s'|s, a_i, a_{-i}),$$
$$r_i = r_i(s, a, s') = r_i(s, a_i, a_{-i}, s').$$

Then, we can derive the optimal policy for agent i, which depends on other agents' policies:

$$\pi_i^*(a|s, \pi_{-i}) = \arg\max_{\pi_i} V_i^{(\pi_i, \pi_{-i})}(s)$$
$$= \arg\max_{\pi_i} \sum_{a \in \mathcal{A}} \pi_i(a_i|s)\pi_{-i}(a_{-i}|s) \sum_{s' \in \mathcal{S}} \mathcal{P}\left(r_i + \gamma V_i^{(\pi_i, \pi_{-i})}(s')\right). \tag{11-57}$$

The opponents' joint policy $\pi_{-i}(a_{-i}|s)$ can be nonstationary, i.e., it can change over time, which makes it difficult to guarantee the convergence property of multi-agent RL. In this case, strong assumptions are needed to ensure convergence. For example, a saddle point or maximal Nash equilibrium is necessary for Nash Q-learning to converge. One popular convergence property for the stochastic game is the Nash equilibrium:

$$V_i^{(\pi_i^*, \pi_{-i}^*)}(s) \geq V_i^{(\pi_i, \pi_{-i}^*)}(s), \forall \pi_i. \tag{11-58}$$

The Nash equilibrium characterizes an equilibrium point π^* where none of the agents has the incentive to deviate. Under such equilibrium, policy π_i^* is the best response of π_{-i}^* for any agent i. As a standard convergence property, the Nash equilibrium always exists for discounted and average-reward stochastic games but may in general not be unique [11-20].

11.4.2 Fully Cooperative RL

The fully cooperative task constitutes a great portion of multi-agent RL and can be categorized into two settings based on rewards: same reward and average reward. In the first setting, all the agents usually share the same reward function, i.e., $r_1 = r_2 = \cdots = r_N$. Thus, the state-value function and action-value function are identical for all agents. If all the agents are coordinated as one decision-maker, single-agent algorithms, such as Q-learning updates, can be applied to multi-agent RL systems. In the average-reward setting, agents have different and potentially private reward functions. The goal is to optimize the long-term average reward, i.e., $\bar{r} = 1/N \sum r_i$. The average-reward setting allows more heterogeneity and includes the same-reward setting as a particular case. Under the average reward setting, more coordination is needed as the global value function cannot be estimated locally without knowing the other agents' reward functions.

(1) Distributed Q-learning

Homogeneous agents usually share the common reward function that aligns all agents' interests. A straightforward value-based algorithm applies standard Q-learning to the joint action space. Its distributed version assigns each agent i with a local policy π_i and a local Q-function $Q_i(s, a_i)$ [11-41]. A local Q-value is updated when it can lead to an increase, i.e.,

$$Q_i(s, a_i) \leftarrow \max\left\{Q_i(s, a_i), r(s, a_i) + \gamma \max_{a_i'} Q_i(s', a_i')\right\}. \tag{11-59}$$

The above formula ensures that the local Q-value always captures the maximum of the joint-action Q-values at all time steps:

$$Q_i(s, a_i) = \max_{a_{-i}} \hat{Q}_i(s, a). \tag{11-60}$$

However, its convergence property is not guaranteed. The combination of each agent's equilibrium policies may not constitute a global equilibrium policy if multiple equilibria exist. The policies of each individual agent may update towards different equilibrium, which contribute to suboptimal solutions.

(2) Frequency Maximum Q-value (FMQ)

The frequency maximum Q-value (FMQ) algorithm is a static game algorithm, which means that the state space is empty [11-36]. An agent should choose coordinated actions based on the frequency with which actions yielded good rewards in the past. The likelihood of good values is evaluated using the models of other agents estimated by the learner or the statistics of the values observed in the past. Agent i uses the Boltzmann exploration strategy. By plugging in modified Q-values $\tilde{Q}_i, i = 1, 2, \cdots, N$, one has

$$\tilde{Q}_i(a_i) = Q_i(a_i) + \lambda \frac{C_{\max}^i(a_i)}{C^i(a_i)} r_{\max}(a_i), \tag{11-61}$$

where $r_{\max}(a_i)$ is the maximum reward observed after action selection, $C_{\max}^i(a_i)$ represents the count of how many times the maximum reward has been observed, $C^i(a_i)$ is the count of how many times a_i has been taken, and λ is a weighting factor. The only additional complexity in frequency maximum Q-value learning compared to single-agent Q-learning comes from the storing and updating of these counters [11-14]. In this case, increasing the Q-values of actions that previously produced good rewards steers the agent toward cooperation. However, the algorithm only works for deterministic tasks, where variance in the rewards can only result from the other agents' actions [11-14].

11.4.3   Fully Competitive RL

The competitive task is generally designed as a zero-sum game, i.e., $\sum r_i(s, a, s') = 0$. Most algorithms focus on two-player games since the simplest 3-player matrix games are PPAD-complete (PPAD is shortened from Polynomial Parity Argument on Directed Graphs). Zero-sum games also serve as a minimax optimization framework for robust control. Uncertainty can be seen as a fictitious opponent who is always against the main agent. Thus, the Nash equilibrium yields a robust policy that optimizes the worst-case long-term return.

(1) Minimax Q-learning

Here, we consider the two-player zero-sum game, in which the main agent is represented by A and its opponent is represented by B. The value function satisfies $V_A^{(\pi_A, \pi_B)}(s) =$

$-V_B^{(\pi_A,\pi_B)}(s)$ since the sum of their rewards equals zero and the Nash equilibrium $\pi^* = (\pi_A^*, \pi_B^*)$ satisfies

$$V_A^{(\pi_A,\pi_B^*)}(s) \le V_A^{(\pi_A^*,\pi_B^*)}(s),$$

$$V_B^{(\pi_A^*,\pi_B)}(s) \le V_B^{(\pi_A^*,\pi_B^*)}(s). \tag{11-62}$$

Thus, according to the minimax theorem, the optimal state-value functions for agents are defined as

$$V_A^*(s) = \max_{\pi_A(\cdot|s)} \min_{a_B} \sum_{a_A} Q_A^*(s, a_A, a_B)\pi_A(a_A|s),$$

$$V_B^*(s) = \max_{\pi_B(\cdot|s)} \min_{a_A} \sum_{a_B} Q_B^*(s, a_A, a_B)\pi_B(a_B|s), \tag{11-63}$$

where $\pi_i(\cdot\,|s)$ represents all the possible policies of agent $i = \{A, B\}$ under state s and $Q_i^*(s, a_A, a_B)$ represents the expected return of agent i when both agents choose their actions and follow the Nash policy from then on. $V_i^*(s)$ can be obtained by solving a linear program based on the knowledge of $Q_i^*(s, a_A, a_B)$. In the minimax Q-learning algorithm [11-47], to propagate the values across state-action pairs, a temporal-difference rule similar to Q-learning is given as

$$Q_A(s, a_A, a_B) \leftarrow (1 - \alpha)Q_A(s, a_A, a_B) + \alpha\big(r_A(s, a_A, a_B) + \gamma V_A(s')\big),$$

$$Q_B(s, a_A, a_B) \leftarrow (1 - \alpha)Q_B(s, a_A, a_B) + \alpha\big(r_B(s, a_A, a_B) + \gamma V_B(s')\big). \tag{11-64}$$

If all the state-action pairs are infinitely visited, Q_i will converge to the optimal action-value function Q_i^*. Minimax Q-learning is truly opponent-independent because even if the minimax optimization has multiple solutions, any of them will achieve at least the minimax return regardless of the behavior of the opponent.

11.4.4   Multi-agent RL with Hybrid Rewards

Multi-agent RL with hybrid rewards imposes no restrictions on the goal and relationship between agents. The setting may simultaneously contain fully cooperative agents and competitive agents, i.e., two competitive teams with cooperative agents on each team. Hence, multi-agent RL with hybrid rewards is a general-sum game, where each agent is self-interested, and its reward may or may not conflict with those of the other agents. This makes finding the Nash equilibrium more challenging than other tasks. It has been proven that finding the Nash equilibrium in the simplest 2-player general-sum normal-form game is PPAD-complete. Under strong assumptions, several value-based methods extending Q-learning are guaranteed to find a Nash equilibrium, such as Nash Q-learning [11-29], friend-or-foe [11-48] and mean field regime [11-12].

(1) Nash Q-learning

The Nash Q-learning algorithm is agent-independent and shares a common structure with Q-learning. Policies and state values are computed with game-theoretic solvers [11-29]. Given the Q-values of all agents, the Q-value of agent i updates according to

$$Q_i(s, a) \leftarrow (1 - \alpha)Q_i(s, a) + \alpha(r_i(s, a) + \gamma\text{Nash}[Q_i(s')]), \tag{11-65}$$

where Nash$[Q(s')]$ is the expected return of the next state generated by the Nash equilibrium policy. The updates use the Q-tables of all agents. Therefore, the Q-table of each agent should be accessible to the other agents, which requires that all actions and rewards of all agents are measurable. The algorithm provably converges to the Nash equilibrium for all states if either 1) every stage of the game encountered by the agents during learning has a Nash equilibrium under which the expected return of all the agents is maximal or 2) every stage of the game has a Nash equilibrium, which is a saddle point, i.e., the learner will benefit from any other agent deviating from this equilibrium. This requirement is satisfied only in a small class of problems. In all other cases, some external mechanism for equilibrium selection is needed for convergence.

(2) Friend-or-foe Q-learning (FFQ)

Friend-or-foe is a hybrid reward algorithm where agent i categorizes the other agents into two groups: friend or foe [11-48]. The friends of agent i work together to maximize the expected return of agent i, while the foes try to minimize it. This type of learning can be seen as an extension of the two-player zero-sum game. In addition, in games with coordination or adversarial equilibrium, FFQ converges to the value that Nash-Q ought to obtain. The updating of the Q-value of agent i is as follows:

$$V_i(s) = \max_{\pi_A} \min_{a_B} \sum_{a_A} Q_i(s, a_A, a_B)\pi_A,$$

$$Q_i(s, a_A, a_B) \leftarrow (1 - \alpha)Q_i(s, a_A, a_B) + \alpha(r_i + \gamma V_i(s')),$$

(11-66)

with

$$a_A = \{a_1 \quad \cdots \quad a_{AN}\},$$
$$a_B = \{a_1 \quad \cdots \quad a_{BN}\},$$
$$\pi_A = \prod_{i=1}^{AN} \pi_i(a_i|s),$$

where a_A represents the action set of agent i and its friends, a_B represents the action set of the opponents, and A_N and B_N represent the number of agents of each group. Compared to Nash-Q learning, FFQ does not require estimating the Q-functions of opposing players and provides stronger guarantees due to the minimax theorem.

(3) Mean-field RL

The mean-field game is proposed to alleviate the scalability issue [11-12]. In this game, each agent i has a local state s_i and a local action a_i. The interaction among agents is captured by an aggregate effect μ_i, i.e., the mean field, which is a function of the empirical distribution of the local states and actions of all agents.

$$\mu_i = \frac{1}{N_i} \sum_j a_j, \quad a_j \sim \pi_j(\cdot \,|s_i, \mu_j),$$

$$\pi_i(a_i|s_i, \mu_i) = \frac{\exp(-\beta Q_i(s_i, a_i, \mu_i))}{\sum_{\tilde{a}_i \in \mathcal{A}_i} \exp(-\beta Q_i(s_i, \tilde{a}_i, \mu_i))},$$

(11-67)

where N_i is the number of agent i's neighbors, a_j is the action taken by agent j that belongs to the neighbor set, and π_i is the policy of agent i parametrized by μ_i. Thus, the agent in the mean-field game is faced with a time-varying MDP parametrized by μ instead of multiple agents. Its solution is a mean-field equilibrium, which is a sequence of pairs of policies and the mean field $\{\pi^*, \mu^*\}$ under the following two conditions: (a) π^* is the optimal policy for the time-varying MDP; and (b) μ^* is generated when all the agents follow the optimal policy π^*.

11.5 Inverse Reinforcement Learning

The problem of deriving a reward function from observed behavior is inverse reinforcement learning (IRL). IRL is also known as apprenticeship learning, learning by watching, imitation learning, or learning from demonstration. The goal of standard RL is to find an optimal policy, in which it is usually assumed that a reward function is already known. In some cases, the reward function is difficult to manually specify. For example, drivers typically make a tradeoff between different performances, such as driving safety, fuel economy, and ride comfort; however, it is hard to represent them with accurate mathematical formula. In practice, the reward function is often manually modified and tested until the desired behavior is obtained. The goal of inverse RL is to learn a reward function from expert experiences. This is not to mimic the external actions that experts conduct under each state but to acquire the internal reasoning mechanism. The expert is assumed to optimize an unknown reward function that can be expressed as a combination of known features. Even though it is difficult to recover the expert's true reward, it has been demonstrated that IRL often finds a policy that performs as well as the expert. There are two common IRL algorithms. One is called max-margin learning, which finds the reward function that maximizes the difference between the value functions of expert policy and other policies[11-1]. The other is called maximum entropy learning, which finds the reward function that maximizes the probability of observing the same expert trajectories [11-68].

11.5.1 Essence of Inverse RL

The basis of IRL is to minimize the feature expectation between the expert trajectories and the generated trajectories. It assumes that the reward function is expressed as a linear function of known features:

$$r_\psi(s, a) = \psi^{\mathrm{T}} f(s, a),\qquad(11\text{-}68)$$

where $f = [f_1, f_2, \cdots, f_n]^{\mathrm{T}} \in \mathbb{R}^n$ is the feature vector and $\psi \in \mathbb{R}^n$ is the reward parameter. Each feature f_i is a function that maps a state-action pair to a real value, capturing some relevant characteristics. Define τ as all the state-action pairs along the trajectory, i.e., $\tau = [s_0, a_0, s_1, a_1, s_2, a_2, \cdots, s_{T-1}, a_{T-1}]$. Define the accumulated value of the feature along a trajectory as $\mu(\tau)$, i.e.,

$$\mu(\tau) = \sum_{i=0}^{T-1} \gamma^i f(s_i, a_i).\qquad(11\text{-}69)$$

The learned reward of IRL should have the same feature expectation as the demonstration, i.e.,

$$\mathbb{E}_{\pi_\psi}\{\mu(\tau)\} = \mathbb{E}\{\mu(\tau^*)\}. \tag{11-70}$$

Here, π_ψ is the optimal policy for the parameterized reward function $r_\psi(s, a)$, and τ^* is the trajectory generated by the expert. The key question is how to calculate the expectation in (11-70). The right-hand side can be evaluated by only using the expert trajectories to approximate the expected value, while the left-hand side is much more difficult to evaluate. It requires solving the RL problem and then using the learned policy to generate samples via environment interactions. Given a series of expert trajectories, there might be a variety of potential reward functions $r_\psi(s, a)$ that match the feature expectation of the expert trajectories. Existing IRL algorithms always select the best function from these reward functions according to certain criteria, such as max-margin and max entropy, to reduce the ambiguity of multiple solutions.

11.5.2 Max-Margin Inverse RL

Max-margin IRL, also called apprenticeship learning, employs the maximum margin principle to achieve the most robust distinction between expert trajectories and trajectories generated by all other policies [11-1][11-52]. It aims to pick the reward function that, in some sense, draws the largest value margin between the expert policy and the optimal policy. The corresponding optimization problem can be expressed as

$$\max_\psi m,$$

subject to $\hspace{8cm}$ (11-71)

$$\psi^T \mathbb{E}\{\mu(\tau^*)\} \geq \max_\psi \psi^T \mathbb{E}_{\pi_\psi}\{\mu(\tau)\} + m,$$

where m is the value margin between the expert policy and the optimal policy. Like the support vector machine (SVM), this optimization is equivalent to finding the maximum margin hyperplane separating two sets of points. Therefore, the SVM trick can be used to formulate a quadratic programming problem:

$$\max_\psi \frac{1}{2}\|\psi\|^2,$$

subject to $\hspace{8cm}$ (11-72)

$$\psi^T \mathbb{E}\{\mu(\tau^*)\} \geq \max_\psi \psi^T \mathbb{E}_{\pi_\psi}\{\mu(\tau)\} + 1.$$

Therefore, we can use any quadratic programming solver to find the optimal reward function[11-1]. This method will terminate in a few iterations and guarantee that the searched policy will have similar performance comparable to that of the expert. Unfortunately, this method may suffer from overfitting because it only concentrates on the policy that matches the demonstrated trajectory and does not concern the suboptimal policy.

11.5.3 Max Entropy Inverse RL

Unlike max-margin inverse RL, the IRL proposed by Ziebart et al. entails selecting the reward function that maximizes the policy entropy, i.e., the maximum possibility of observing the same expert trajectories [11-68]. Following this principle, the distribution with maximum entropy describes the data best since it is the least biased distribution. The max entropy inverse RL solves a constrained optimization problem:

$$\max_{\psi} \mathcal{H}(\pi_{\psi}),$$

subject to   (11-73)

$$\mathbb{E}_{\pi_{\psi}}\{\mu(\tau)\} = \mathbb{E}\{\mu(\tau^*)\},$$

where $\mathcal{H}(\pi)$ is the policy entropy. It has been found that the constrained optimization problem (11-73) has the optimal solution that follows the form

$$p(\tau|\psi) = p(\tau)e^{\psi^{\mathrm{T}}\mu(\tau)} / \int p(\tau)e^{\psi^{\mathrm{T}}\mu(\tau)}\, d\tau,$$ (11-74)

where $p(\tau)$ is the probability density of the trajectory generated by actions selected from a uniform distribution. Therefore, (11-73) is equivalent to maximizing the likelihood of the observed data under the maximum entropy distribution derived in (11-74), i.e.,

$$\max_{\psi} L(\psi),$$ (11-75)

where $L(\psi)$ is the objective function, which is expressed as

$$L(\psi) = \sum_{i=1}^{N} \log p(\tau_i^*|\psi)$$

$$= \sum_{i=1}^{N} \log p(\tau_i^*) + \psi^{\mathrm{T}}\mu(\tau_i^*) - \log \int p(\tau)e^{\psi^{\mathrm{T}}\mu(\tau)}d\tau.$$ (11-76)

Here, $\{\tau_1^*, \tau_2^*, \dots, \tau_N^*\}$ are the trajectories generated by the expert. The gradient of $L(\psi)$ is equal to the difference between the empirical and expected feature values

$$\nabla L(\psi) = \mathbb{E}\{\mu(\tau^*)\} - \mathbb{E}_{\tau \sim p(\tau|\psi)}\{\mu(\tau)\}.$$ (11-77)

There is an intuitive explanation for this gradient: when the expected value $\mathbb{E}_{\tau \sim p(\tau|\psi)}\{\mu_i(\tau)\}$ is higher than $\mathbb{E}\{\mu_i(\tau^*)\}$, we should decrease the corresponding weight ψ_i, which in turn assigns a lower likelihood to any trajectories with a high value of $\mu_i(\tau)$. As a result, the expected value $\mathbb{E}_{\tau \sim p(\tau|\psi)}\{\mu_i(\tau)\}$ decreases. This method elegantly resolves the overfitting problem in maximum margin methods since it evaluates all the policies in a probabilistic view. In addition, it can provide a convex, computationally efficient procedure for optimization and maintain good performance guarantees.

11.6 Offline Reinforcement Learning

Although online RL algorithms have achieved a few successes in computer games and robotic control, they often require extensive interactions with the environment. This becomes a major obstacle for many real-world applications as collecting data with an imperfect policy via environment interactions can be expensive (e.g., industrial control) or dangerous (e.g., autonomous driving). Some real-world systems are overly complex or partially observable, and it may be impossible to build high-fidelity simulation models. Naïvely applying RL in imperfect simulation models often leads to a serious sim-to-real transfer issue. On the other hand, many real-world systems are designed to log sufficient data of historical states and actions. There is much practical value in deriving a data-driven offline RL paradigm to fully utilize the data information and learn optimized policies. The recently emerged offline RL (sometimes known as batch RL) provides a viable framework to address online interaction problems. Offline RL focuses on training RL policies from offline, static datasets without further interaction with the environment. In offline RL, the learning algorithm is provided with a static dataset of transitions $\mathcal{D} = \{(s_t, a_t, s_{t+1}, r_t)\}$, which is generated by one or multiple behavior policies $\pi_\beta(a|s)$. Offline RL aims to learn the best policy $\pi_\theta(a|s)$ using only this precollected dataset.

The most obvious challenge of offline RL is that the learning algorithm can only rely on a static dataset $\mathcal{D}$, without any possibility to explore and collect new information from unknown state-action pairs. The dataset $\mathcal{D}$ may only partially cover the entire state-action space, which creates a situation where the Bellman operator is no longer implemented with the true MDP but instead becomes a biased operator. The sampling error in the data as well as the function approximation error will become much worse in offline settings compared with standard RL settings. A more fundamental challenge is the distributional shift issue [11-38], which is the result of counterfactual reasoning in offline RL. The function approximators (including the policy and value functions) are trained under one data distribution but are evaluated on a potentially different distribution. This phenomenon is caused by the change in the state marginal distribution induced by the learned policy as well as the action of maximizing the expected return. For example, when the policy-induced data distribution largely deviates from the data distribution that is available, the policy can make counterfactual queries on unknown out-of-distribution (OOD) actions. Maximizing the Q-value under overly optimistic value estimates of OOD actions can cause severe overestimation, leading to nonrectifiable exploitation errors during training [11-38]. Due to the abovementioned reasons, off-policy RL algorithms typically fail to learn from only fixed offline data and often require a growing batch of online samples for good performance.

Recent offline RL methods have addressed the distributional shift issue in the following directions: 1) policy constraint offline RL methods attempt to constrain the learned policy to stay "close" to the behavior policy; 2) model-based offline RL algorithms adopt a pessimistic MDP framework, where the reward is penalized for OOD actions; 3) the value function regularization approach modifies the Q-function training objective to achieve conservatism; 4) uncertainty-based offline RL methods use the epistemic uncertainty of the Q-function to penalize the OOD actions; 5) in-sampling learning methods learn value function within data samples to avoid queries to OOD actions; and 6) goal-conditioned imitation learning methods convert the RL problem into a conditional, filtered or weighted imitation learning problem to remove the policy evaluation procedure that is easily impacted by distributional shift. A brief introduction of these methods is presented in the following content.

Offline RL is a rapidly developing area. Many open problems remain and are worth further exploration, including 1) designing data-efficient offline RL algorithms under small offline datasets with limited state-action space coverage; 2) finding better uncertainty and generalizability evaluation measures for models or value functions on unknown data samples; 3) exploring new offline RL learning schemes to combat distributional shifts while avoiding overly conservative policy learning; and 4) improving policy robustness with limited offline data and potential inaccuracies in the input data. Table 11.2 presents a summary of recent offline RL algorithms and provides a brief description of their key modifications, advantages, and disadvantages.

Table 11.2 Comparison of offline RL methods

Method	Idea	Advantage	Disadvantage
Policy constraint (BCQ, BEAR, TD3BC)	Constrain the learned policy using the behavior policy	Works well when behavioral data are easy to model; Straightforward modeling framework	Many algorithms needs to estimate behavior policy; Often tends to be over-conservative
Model-based (MOReL, MOPO)	Penalize reward using uncertainty	Works well on high-coverage/mixed datasets; Better generalizability	Heavily impacted by the learned dynamic model; Needs uncertainty estimation
Value regularization (CQL)	Modified Q-function training objective	Typically have good performance	Policy evaluation is generally more costly on continuous action space; sometimes can be over-conservative
Uncertainty-based (BEAR, MOPO, EDAC)	Penalize Q-function using uncertainty	Works well on high-coverage datasets; Typically have good performance	Needs uncertainty estimation
In-sample learning (IQL)	Modify policy evaluation scheme	Good performance; stable policy learning	Performance depends on the quality of the data
Goal-conditioned imitation learning (RvS)	Learn a conditioned policy	Bypasses the distributional shift issue during policy learning	Optimal policy is only guaranteed under some restrictive assumptions

11.6.1   Policy constraint offline RL

The basic idea of policy constraint methods is straightforward. Since evaluating the Q-values on OOD actions can cause errors in the computation of the Bellman update, what would happen if the actions fed into Q-function were restricted to be in-distribution of the training data $\mathcal{D}$? The answer is that errors in the Q-function should not accumulate, and the distributional shift will be avoided. A general optimization objective of policy constraint methods is presented as follows:

$$Q(s,a) \leftarrow r(s,a) + \gamma \mathbb{E}_{a' \sim \pi_\theta(a'|s')}\{Q(s',a')\},$$

$$\pi = \arg\max_\pi \mathbb{E}_{s \sim \mathcal{D}, a \sim \pi(a|s)}\{Q(s,a)\} \text{ or } \arg\max_\pi \mathbb{E}_{s \sim \mathcal{D}, a \sim \pi(a|s)}\{A(s,a)\} \quad (11\text{-}78)$$

subject to

$$D\big(\pi(a|s), b(a|s)\big) \leq \epsilon,$$

where $D(\cdot,\cdot)$ is a certain divergence metric. Policy constraint methods typically contain two components: 1) the learning of a behavior policy $b(a|s)$ from the training data $\mathcal{D}$, either explicitly using behavior cloning (BC) or by implicit modeling; and 2) a suitable divergence metric $D(\pi, b)$ that quantifies the "closeness" of the policy distribution and the behavioral distribution. Many model-free offline RL algorithms belong to policy-constraint methods. These algorithms can be broadly classified into two categories: 1) using explicit policy constraints, such as BEAR [11-38] and TD3BC [11-23]; 2) using implicit policy constraints, such as BCQ [11-22].

Explicit policy constraint methods use some divergence metrics to evaluate and restrain the deviation of policy from the behavioral data. For example, BEAR implements the support constraint using the maximum mean discrepancy (MMD) [11-38], which is a kernel-based metric that evaluates the divergence between the learned and behavior policy distributions based on finite action samples. Perhaps surprisingly, Fujimoto and Gu have showed that using the simple behavior cloning (BC) constraint sometimes can achieve reasonable performance in many tasks [11-23]. This forms a minimal offline RL algorithm called TD3BC, which uses a deterministic policy $\pi(s)$ and is given as follows:

$$\pi = \arg \max_{\pi} \mathbb{E}_{s,a \sim \mathcal{D}}\{\alpha Q(s, \pi(s)) - (\pi(s) - a)^2\}, \tag{11-79}$$

where α is a hyperparameter to balance between RL (maximize Q) and imitation (minimize the second BC term).

Implicit policy constraint methods do not use an explicit divergence metric to constrain policy distribution. As one of the earliest offline RL algorithms, BCQ enforces π_θ to be close to π_β by adding a small perturbation to π_β [11-22]. The learning of the target policy π_θ is calculated as

$$\pi_\theta(\cdot | s) = \arg \max_{a_i \sim b(\cdot|s)} Q\big(s, a_i + \xi_\theta(s, a_i)\big) \ \forall i, \tag{11-80}$$

where ξ_θ is the perturbation model with bounded output in $[-\xi_{\text{bnd}}, \xi_{\text{bnd}}]$ (ξ_{bnd} is a predefined hyperparameter), which is trained to maximize the Q-value through a deterministic policy gradient. Model-free policy constraint methods generally outperform behavior cloning methods. While performing well on single-modal datasets, policy constraint methods have been shown to have limited improvements on multimodal datasets or for some hard tasks due to overly strict behavioral constraints. Strictly matching the behavior distribution limits the performance of the learned policy. To surpass the performance of the behavior policy in data, it is reasonable to expect that the optimized policy has a different distribution than the data. As a result, previous distribution matching methods all tend to be over-conservative, and may have weak performance on some hard tasks.

11.6.2   Model-based Offline RL

Model-based offline RL methods take an alternative approach by adopting a pessimistic MDP framework, which incorporates a reward penalty based on the uncertainty of the dynamic model to handle the distributional shift issue [11-37]. The underlying assumption is that the model learned on state-action pairs that are far away from the data distribution will become less accurate, thus exhibiting larger uncertainty. Typical model-based offline RL algorithms include MOReL [11-37] and MOPO [11-63]. Their general reward penalization scheme can be summarized as follows:

$$\tilde{r}(s, a) = r(s, a) - \varphi(s, a), \tag{11-81}$$

where $\varphi(s, a)$ is some uncertainty measure and $\tilde{r}(s, a)$ is the conservative reward function used in offline RL. Different choices of uncertainty measures lead to different offline RL algorithms and potentially different model behaviors. MOReL learns an ensemble of multiple dynamic models $f(s, a; \phi_i), i = 1, 2, \ldots, N$ from data and uses the disagreements within the model ensemble as the uncertainty measure [11-37]. In general, the number of models N is chosen to be 2~5. During policy optimization, MOReL runs trajectory optimization or planning using the dynamic model but terminates the generated trajectories if the state-action pairs are unreliable, i.e., the disagreement within model ensembles is large. The equivalent reward penalization scheme is as follows:

$$\tilde{r}(s, a) = \begin{cases} -R_{\max} & \text{if } \max_{i,j} \|f(s, a; \phi_i) - f(s, a; \phi_j)\|_2 > \epsilon \\ r(s, a) & \text{otherwise} \end{cases} . \tag{11-82}$$

The MOPO algorithm learns an ensemble of dynamics models that output state-conditioned Gaussian distributions $(f(s'|s, a; \phi_i) = \mathcal{N}(\mu_i(s, a), \sigma_i^2(s, a)), i = 1, 2, \ldots N)$ [11-63]. A reward penalty is added on generated transitions with large variance from the learned dynamic models

$$\varphi(s, a) = \max_i \{\|\sigma_i(s, a)\|_F\} .$$

The learned dynamic model together with the modified conservative reward are used to generate new data, and a model-free off-policy RL algorithm is applied to learn the optimized policy. Model-based offline RL methods are generally less conservative and perform better than policy constraint methods when the dynamic model is learned well (e.g., on high-coverage or mixed datasets). Another advantage is that some degree of generalization is allowed due to the usage of the learned model. However, the performance of model-based offline RL methods is heavily impacted by the quality of the learned dynamics model. There are also some gaps in the literature on finding better epistemic uncertainty measures that can more accurately detect OOD state-action pairs.

11.6.3   Value Function Regularization

In addition to the policy constraint method, another approach to handling the Q-value overestimation issue on OOD samples is to regularize the value function by learning a conservative, underestimated value function. The most representative method is conservative Q-learning (CQL) [11-39]. CQL learns a Q-function such that the expected value of a policy derived from the learned Q-function underestimates the true policy value. The modified Q-function training objective in CQL is as follows:

$$Q_{\text{CQL}}^{\pi} = \arg\min_Q \mathbb{E}_{s,a,s'\sim\mathcal{D}} \left\{ \left(Q(s, a) - Q^{\text{target}}(s, a) \right)^2 \right\} + \alpha$$
$$\cdot \left(\mathbb{E}_{s\sim\mathcal{D}, a\sim v(a|s)} \{Q(s, a)\} - \mathbb{E}_{s,a\sim\mathcal{D}} \{Q(s, a)\} \right). \tag{11-83}$$

In addition to minimizing the standard TD error (the first term, in which $Q^{\text{target}}(s, a)$ is the target Q-value), CQL minimizes the Q-function under a chosen distribution $v(a|s)$

while maximizing it under the data distribution. The above objective leads to an underestimated Q-value, which effectively alleviates the overestimation issue. This treatment is shown to be equivalent to adding a "Q-function aware" penalty to the value function [11-39], and can be also perceived as an implicit KL divergence regularizer on a Boltzmann policy. CQL outperforms many previous policy constraint methods. It has strong performance on benchmark datasets with complex data distributions and hard control problems. A potential drawback of CQL is that one has to perform importance sampling for continuous action control tasks, which is somewhat complex. Value regularization methods tend to be less conservative than many policy constraints methods, as they do not constrain the learned policy π towards behavior policy b, thus often enjoy better performances.

11.6.4   Uncertainty-based Offline RL

Another branch of offline RL algorithms is the uncertainty-based methods. This approach focuses on estimating the epistemic uncertainty of Q-function and uses it to regularize the policy learning. The idea is that the uncertainty of the Q-values of OOD actions should be much larger than that of in-distribution actions and thus can be used as a value function penalty to address distributional shifts. Such methods require learning an uncertainty set or distribution over possible Q-functions from the dataset $\mathcal{D}$, denoted as $\mathcal{P}_{\mathcal{D}}(Q^\pi)$. The policy is then improved using a conservative estimate of the Q-function:

$$\pi' = \arg\max_\pi \mathbb{E}_{s \sim \mathcal{D}, a \sim \pi(a|s)}\{\mathbb{E}_{Q^\pi \sim \mathcal{P}_{\mathcal{D}}}\{Q^\pi(s,a)\} - \alpha\mathrm{Unc}(\mathcal{P}_{\mathcal{D}})\}, \qquad (11\text{-}84)$$

where $\mathrm{Unc}(\cdot)$ is some uncertainty metric. The most popular choice of the uncertainty metric is to use the disagreement of the ensemble of bootstrapped Q functions, which is used in BEAR [11-38], MOPO [11-63], and EDAC[11-6].

Uncertainty-based methods enjoy a tighter suboptimality bound as compared to policy constraint methods, and have exhibited strong performance enhancement, especially in EDAC. However, its core problem is still associated with the accuracy and reliability of uncertainty estimation. Currently, there is still no cheap and theoretically rigorous uncertain estimation method for nonlinear function approximators such as deep neural networks. Although the ensemble of bootstrapped Q-functions has shown promising results in benchmark tasks, it still may not be able to accurately capture the epistemic uncertainty. For example, under the case when there is very little diversity in an ensemble, the resulting uncertainty estimates can be inaccurate. Hence, such special treatment as diversified Q-ensemble is often needed in practical implementations [11-6]. Moreover, evaluating uncertainty through ensemble of bootstrapped Q-functions is very costly, as it needs to simultaneously learn lots of Q functions (2~500 for EDAC according to different tasks) in order to obtain the best performance.

11.6.5   In-sample Offline RL

Most of aforementioned offline RL methods use the actor-critic framework, which learns a policy π as well as a Q-function Q^π associated with the policy. However, this formalism inevitably faces the potential distributional shift during the policy evaluation step, when we compute the target values using the learned policy π during evaluating the Bellman error. In-sample learning methods take a very different approach. They learn the value function completely within data samples or simply use the value function $Q^b(s,a)$ with respect to behavior policy b in data instead of the optimized policy π. A representative in-sample learning method is implicit Q-learning (IQL) [11-40]. IQL learns an upper expectile value function for policy improvement, which has the following minimization objective for Q-function Q_w:

$$L_Q(w) = \mathbb{E}_{s,a,s'\sim\mathcal{D}}\left\{\left(r(s,a) + \gamma V_\psi(s') - Q_w(s,a)\right)^2\right\} \tag{11-85}$$

where $V_\psi(\cdot)$ is the upper expectile state-value function, learned completely using data samples with the following loss:

$$L_V(\psi) = \mathbb{E}_{s,a\sim\mathcal{D}}\left\{L_2^\tau\left(Q_{\hat{w}}(s,a) - V_\psi(s)\right)^2\right\} \tag{11-86}$$

where $L_2^\tau(u) = |\tau - \mathbb{I}(u < 0)|u^2$ is the expectile loss function with $\tau \in (0,1)$ as the given expectile, which can be used to penalize negative errors more than positive errors. Using above design, IQL avoids computing target Q-values using the learned policy π, thus achieves the same objective as in RL using one-step bootstrapping. Finally, the policy π_ϕ in IQL can be learned by maximizing the following advantage weighted regression (AWR) objective:

$$L_\pi(\phi) = \mathbb{E}_{s,a\sim\mathcal{D}}\left\{\exp\left(\beta\left(Q_{\hat{w}}(s,a) - V_\psi(s)\right)\right)\log\pi_\phi(a|s)\right\} \tag{11-87}$$

IQL is easy to implement and has strong performance in offline RL benchmark. Moreover, due to the avoidance of querying OOD actions, IQL also enjoys a stable learning process in many complex tasks.

11.6.6 Goal-conditioned Imitation Learning

The last branch of existing offline RL algorithms is the goal-conditioned imitation learning (GCIL) approach, which are summarized and discussed under the RvS framework [11-19]. These methods essentially adopt a conditional, filtered, or weighted imitation learning approach. GCIL learns an outcome-conditioned policy (conditioned either on a goal state, reward, or target return, etc.) by maximizing the following empirical negative log likelihood loss:

$$L(\pi) = -\sum_{\tau\sim\mathcal{D}}\sum_{1\leq t\leq H}\log\pi\left(a_t|s_t, o(\tau)\right), \tag{11-88}$$

where τ is a trajectory in the dataset, H is the horizon length, $o(\tau)$ is the designed outcome related to the current trajectory τ. Once learned the conditioned policy, we can set the desired outcome o at each step during inference to obtain actions for control.

As GCIL methods essentially solve the RL problem as a supervised learning problem, they avoid the inherent instabilities during the update procedure in RL. However, it has been shown in recent studies that GCIL requires a more strict set of assumptions than dynamic programming in order to obtain a near-optimal policy. Hence GCIL alone is unlikely to be a general solution for offline RL problems, but can be a viable solution to some specific situations, such as deterministic MDPs with high-quality behavior data.

11.7 Major RL Frameworks/Libraries

A unified RL framework (or library) serves to abstract the shared components in RL problems and algorithms. Proper RL frameworks can simplify the development process, help researchers focus on algorithm innovation, and improve the reusability and transferability of codes. Current popular RL frameworks include RLlib [11-45], OpenAI Baselines [11-16], GOPS [11-31], Tianshou [11-61]. Beyond some similarities, different requirements and corresponding approaches for implementation make these frameworks different from each other.

11.7.1 RLlib

RLlib is an open-source RL library that offers both high scalability and a unified API for a variety of applications [11-45]. It was developed by the UC Berkeley RISE Lab and the team at Anyscale Inc. RLlib is based on Ray, which provides a simple, universal API for building distributed applications. Therefore, RLlib is designed to support distributed sampling and updating, which is able to accelerate the training process. The module stack shown in Figure 11.12 indicates the generality of RLlib due to the abstraction for RL training. In addition to default training workflows, environments, preprocessors, neural network models, sample collection processes and RL algorithms can all be customized by users as long as the APIs of RLlib are maintained. Deep learning frameworks, such as TensorFlow, TensorFlow Eager and PyTorch, are supported in RLlib, which is convenient for custom development and application. Nearly all classical RL algorithms have been implemented according to their properties. The disadvantage of RLlib is its highly modularized and reusable code structure, making learning initially difficult. One must dive into the RLlib source code to locate places to modify or customize.

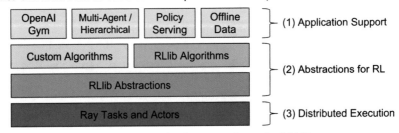

Figure 11.12 The framework stack of RLlib

11.7.2 OpenAI Baselines

OpenAI Baselines is a set of high-quality implementations of RL algorithms, offering a tool for comparing new algorithms with existing algorithms [11-16]. The OpenAI Baselines was first established in 2017 as one of the earliest RL training frameworks. After its publication, it was commonly used and provided researchers with algorithms, such as DQN, DDPG, A2C, ACER, TRPO, and PPO. Although OpenAI Baselines was published early, it is not quite popular due to the lack of documentation, unified modules and mixed code styles. Moreover, some early algorithms implemented by OpenAI Baselines sometimes had a few oscillations in training curves. Despite these flaws, the OpenAI Baselines framework still plays an important role because of its contributions to RL framework development. Stable Baselines (SB) and Stable Baselines3 (SB3) are two improved versions of OpenAI Baselines [11-53]. SB is still based on TensorFlow, featured with better training stability and code simplicity. The code readability of SB has been improved, and its structure was simplified after its framework was transferred to PyTorch. These two improved versions inherited the architecture of OpenAI Baselines such a way that they all do not support multi-GPU training. Thus, their training speeds are slower than those of distributed frameworks.

11.7.3 GOPS

General Optimal Problem Solver (GOPS) is a PyTorch-based lightweight RL framework that aims to apply RL algorithms to solve general OCPs [11-31]. It features a highly modularized design, and each application is broken down into several standard modules, such as the environment, algorithm, trainer, buffer, and sampler. Different modules can be flexibly combined through hyperparameter setups according to the users' needs. GOPS is not only compatible with Gym's standard environment but also supports some third-party interfaces, such as MATLAB/Simulink, CarSim, and SUMO. It has covered over a dozen mainstream RL/ADP algorithms, including infinite-horizon ADP, finite-horizon ADP, DQN, DDPG, TD3, SAC, TRPO, PPO, and DSAC. Furthermore, GOPS supports a variety of serial and parallel training methods, including on-policy serial trainers, off-policy serial trainers, synchronous parallel trainers, and asynchronous parallel trainers. Other special demands, for instance, hard state constraints and adversarial actions, are formally defined and supported in the GOPS framework.

11.7.4 Tianshou

Tianshou is an RL platform based on pure PyTorch, featuring a modularized framework and Python API for building the RL agent with fewer codes [11-61]. It supports more than 15 classic RL algorithms, including model-free, model-based and multi-agent algorithms, and provides the state-of-the-art benchmark of Gym's MuJoCo. Unit tests, which include the full agent training procedure for all of the implemented algorithms, comprise another feature of Tianshou. This feature guarantees good reproducibility and a high quality of training results. The modularized architecture of Tianshou makes it convenient for users to customize their environments, algorithms and training pipelines, which is helpful for agile development and quick verification. Despite all of its strengths, Tianshou only supports parallelized sampling, not parallelized training. Its training speed for large-scale and complicated problems may not be satisfactory.

11.8 Major RL Simulation Platforms

To facilitate the validation of RL algorithms, simulation platforms, such as OpenAI Gym [11-13], DeepMind Lab [11-10], StarCraft II Learning Environment [11-58], Microsoft Project Malmo [11-34], and Unity ML-Agent Toolkit [11-35], have been developed in recent years. These platforms provide various benchmarks or challenging tasks based mainly on existing game engines, which can serve as the environment of RL problem settings. Various classic and advanced RL algorithms have already been successfully trained and compared on these RL platforms. Major RL platforms will be described in the following content, addressing graphics rendering engines, algorithm compatibility, task complexity and extension flexibility.

11.8.1 OpenAI Gym

OpenAI Gym was first released in 2016 and is compatible with all RL agent structures and numerical computation libraries, such as TensorFlow and PyTorch [11-13]. The Gym library contains a collection of environments with a common interface. The environments are versioned to ensure that the results remain comparable and reproducible. The OpenAI Gym platform supports built-in environments, such as classic control (small-scale tasks), algorithmic computation, board games, and extensional environments, such as Atari games, Doom games, Box2D robots, and MuJoCo multibody dynamics. OpenAI Gym focuses on the episodic setting, where the RL agent's experience is broken down into a series of episodes. In each episode, the agent's initial state is randomly sampled from a predefined distribution, and the interaction proceeds until the environment reaches a terminal state. Typical environments in OpenAI Gym are shown in Figure 11.13.

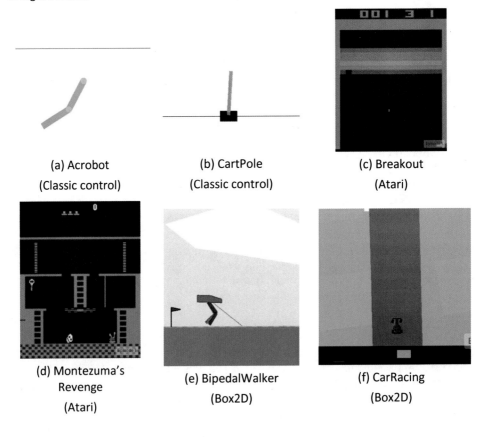

| (a) Acrobot | (b) CartPole | (c) Breakout |
| (Classic control) | (Classic control) | (Atari) |

| (d) Montezuma's Revenge | (e) BipedalWalker | (f) CarRacing |
| (Atari) | (Box2D) | (Box2D) |

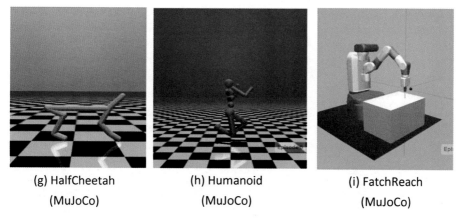

| (g) HalfCheetah | (h) Humanoid | (i) FatchReach |
| (MuJoCo) | (MuJoCo) | (MuJoCo) |

Figure 11.13 Typical environments in Open AI Gym

11.8.2 TORCS

The Open Racing Car Simulator (TORCS) is an open-source multi-platform car racing game (http://torcs.sourceforge.net/). It is written in C++ and can run on Linux, FreeBSD, OpenSolaris and Windows. TORCS allows the game player to drive in races against around 100 opponents simulated by the computer. It also enables multiple users to race against one another and allows pre-programmed AI or online reinforcement learner to automatically control a car. Its graphic features is developed with OpenGL, which contains lighting, smoke, skid marks and even glowing brake disks. The car dynamics is composed of sophisticated physical models, including damage model, collision detection model, tire and wheel model, aerodynamic model, etc. Thanks to its openness, modularity and extensibility, TORCS has been adopted as a base for many research projects, including automated computation of car setups and the application of AI techniques to autonomous driving.

Figure 11.14 Screenshots in TORCS

11.8.3 DeepMind Lab

Google's DeepMind Lab was released in 2016 as an agent-based AI research platform built upon the first-person 3D game engine Quake III Arena [11-10]. The interaction with this platform is lock-stepped according to a user-specified frame rate. The game is paused after an observation is provided until an RL agent provides the next action. At each step, the platform can provide reward signals, pixel-based observations and velocity information. The platform provides access to the raw RGB pixels rendered by the game engine from the player's first-person perspective. Available actions are looking around and moving in 3D space. RL agents can provide multiple simultaneous actions to control movement, looking and tagging. With the help of the DeepMind Lab API, researchers can define their environment variants and basic multi-agent interactions. DeepMind Lab provides a simple C-language API and ships with Python bindings. This platform includes an extensive-level API written in Lua to allow custom-level creation and mechanics. Typical tasks in DeepMind Lab are shown in Figure 11.15.

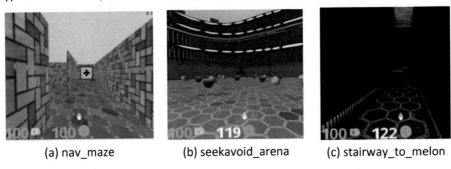

|(a) nav_maze|(b) seekavoid_arena|(c) stairway_to_melon|

Figure 11.15 Typical environments in the DeepMind Lab

11.8.4 StarCraft II Environment

The StarCraft II Learning Environment (SC2LE) is based on the game StarCraft II [11-58]. This platform provides the observation, action, and reward specifications for the StarCraft II domain. It poses a new challenge for RL, representing a more difficult class of problems than prior works and featuring multi-agent problems and imperfect information. It has a large action space involving the selection and control of hundreds of units. It has a large state space that needs to be observed solely from raw input feature planes. To communicate with the game engine, SC2LE provides a Python-based interface that is compatible with Windows, Mac and Linux systems. In addition to the main game maps, a suite of minigames focusing on different elements of StarCraft II gameplay is also available. For the main game maps, a dataset of game replay data from human expert players is available. The baseline neural network is trained from this dataset to predict game outcomes, and player actions are embedded in this RL platform. Snapshots of SC2LE are shown in Figure 11.16.

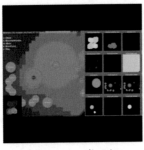 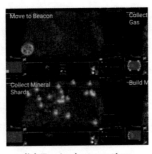 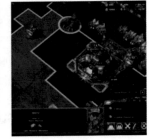

|(a) Map visualization|(b) Typical operation|(c) Agent demonstration|

Figure 11.16 Snapshots of SC2LE

11.8.5 Microsoft's Project Malmo

Microsoft's Project Malmo was developed based on the 3D exploration and building game Minecraft, which provides flexibility in defining scenarios and environment types [11-34]. The Project Malmo platform supports a range of experimentation needs and can support research in robotics, computer vision, and multi-agent systems. It provides a structured and dynamic environment where agents are coupled through a natural sensorimotor loop. The platform supports infinitely varied environments and missions, such as navigation, survival, and construction. Both AI-AI and human-AI interactions (and collaboration) are supported. Project Malmo also supports increasingly complex tasks to challenge current and future technologies. A low entry barrier is supported by providing different levels of abstractions for observations and actions. Several research projects have explored multi-agent communication, hierarchical control, and planning using this platform.

11.8.6 Unity ML-Agents Toolkit

The Unity ML-Agents Toolkit is a game development platform that consists of a game engine and graphical user interface called the Unity Editor, enabling rapid prototyping and development of games and simulation environments [11-35]. The Unity Editor provides various built-in components, including cameras, meshes, renders, and rigid bodies. It supports 2D, 3D and AR/VR experiences and is compatible with Windows, Mac and Linux systems. The platform provides a set of prebuilt indoor scenes and Python APIs to interact with those environments using a first-person agent. A set of baseline algorithms are embedded to validate new environments and support the development of novel algorithms. It is also possible to define custom components using C# scripts or external plugins. The underlying engine flexibility can help build simple grid world problems, complex strategy games, physics-based puzzles, and multi-agent competitive games. This platform enables the development of learning environments that include sensory and physical complexity, provides compelling cognitive challenges, and supports dynamic multi-agent interactions. Typical environments in the Unity ML-Agents Toolkit are shown in Figure 11.17.

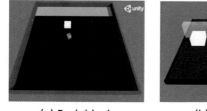

| (a) Push block | (b) Wall jump | (c) Soccer twos |

Figure 11.17 Typical environments in the Unity ML-Agents Toolkit

11.9 References

[11-1] Abbeel P, Ng AY (2004) Apprenticeship learning via inverse reinforcement learning. ICML, Banff, Canada

[11-2] Abu-Khalaf M, Lewis F, Huang J (2006) Policy iterations on the Hamilton–Jacobi–Isaacs equation for H-inf state feedback control with input saturation. IEEE Trans Automatic Control 51(12): 1989-1995

[11-3] Abu-Khalaf M, Lewis F, Huang J (2008) Neurodynamic programming and zero-sum games for constrained control systems. IEEE Trans Neural Networks & Learn Syst 19(7): 1243-1252

[11-4] Al-Tamimi A, Abu-Khalaf M, Lewis F (2007) Adaptive critic designs for discrete-time zero-sum games with application to H-infinity control. IEEE Trans Syst, Man, and Cyber, 37(1): 240-247

[11-5] Al-Tamimi A, Lewis F, Abu-Khalaf M (2007) Model-free Q-learning designs for linear discrete-time zero-sum games with application to H-infinity control. Automatica 43(3): 473-481

[11-6] An G, Moon S, Kim J et al (2021) Uncertainty-based offline reinforcement learning with diversified q-ensemble. NeurIPS, virtual, online

[11-7] Astrom K (2012) Introduction to stochastic control theory. Courier Corporation, Chicago

[11-8] Athans M (1971) The role and use of the stochastic linear-quadratic-Gaussian problem in control system design. IEEE Trans Automatic Control 16(6):529-552

[11-9] Başar T, Bernhard P (2008) H-infinity optimal control and related minimax design problems: a dynamic game approach. Springer Science & Business Media, New York

[11-10] Beattie C, Leibo J, Teplyashin D et al (2016) Deepmind lab. https://arxiv.org/abs/1612.03801

[11-11] Bensoussan A (2004) Stochastic control of partially observable systems. Cambridge University Press, Cambridge

[11-12] Bensoussan A, Frehse J, Yam P (2013) Mean field games and mean field type control theory. Springer, New York

[11-13] Brockman G, Cheung V, Pettersson L et al (2016) Openai gym. https://arxiv.org/abs/1606.01540

[11-14] Busoniu L, Babuska R, De Schutter B (2008) A comprehensive survey of multiagent reinforcement learning. IEEE Trans Syst, Man, and Cyber, 38(2): 156-172

[11-15] Chen X, Mu Y, Luo P et al (2022). Flow-based recurrent belief state learning for POMDPs. ICML, Baltimore, USA

[11-16] Dhariwal P, Hesse C, Klimov O et al (2017). OpenAI Baselines. Github repositories: https://github.com/openai/baselines

[11-17] Duan Y, Schulman J, Chen X et al (2016) RL 2: Fast reinforcement learning via slow reinforcement learning. https://arxiv.org/abs/1611.02779

[11-18] Duncan T, Varaiya P (1971) On the solutions of a stochastic control system. SIAM J on Control 9(3): 354-371

[11-19] Emmons S, Eysenbach B, Kostrikov I et al (2021) RvS: What is Essential for Offline RL via Supervised Learning? ICLR, Vienna, Austria

[11-20] Filar J, Vrieze K (2012) Competitive Markov decision processes. Springer Science & Business Media, Berlin Heidelberg, New York

[11-21] Finn C, Abbeel P, Levine S (2017) Model-agnostic meta-learning for fast adaptation of deep networks. ICML, Sydney, Australia

[11-22] Fujimoto S, Meger D, Precup D (2019) Off-policy deep reinforcement learning without exploration. ICML, California, USA

[11-23] Fujimoto S, Gu S (2021) A minimalist approach to offline reinforcement learning. NeurIPS, virtual, online

[11-24] Georgiou T, Lindquist A (2013) The separation principle in stochastic control. IEEE Trans Automatic Control 58(10): 2481-2494

[11-25] Green M, Limebeer D (1995) Linear robust control. Prentice Hall, Inc., New Jersey

[11-26] Gupta A, Mendonca R, Liu Y et al (2018) Meta-reinforcement learning of structured exploration strategies. NeurIPS, Montreal, Canada

[11-27] Hernandez-Leal P, Kartal B, Taylor M (2019) A survey and critique of multiagent deep reinforcement learning. Autonomous Agents and Multi-Agent Syst 33(6): 750-797

[11-28] Houthooft R, Chen R, Isola P et al (2018) Evolved policy gradients. NeurIPS, Montreal, Canada

[11-29] Hu J, Wellman MP (2003) Nash Q-learning for general-sum stochastic games. J of Machine Learning Res 4(Nov): 1039-1069

[11-30] Igl M, Zintgraf L, Le TA et al (2018) Deep variational reinforcement learning for POMDPs. ICML, Stockholm, Sweden

[11-31] Wang W, Zhang Y, Gao J et al (2023) GOPS: GOPS: A general optimal control problem solver for industrial control applications. Comm in Transportation Res: 1-25

[11-32] Jiang Z, Jiang Y (2013) Robust adaptive dynamic programming for linear and nonlinear systems: An overview. European J of Control 19(5): 417-425

[11-33] Jo N, Seo J (2000) Local separation principle for non-linear systems. International J of Control 73(4): 292-302

[11-34] Johnson M, Hofmann K, Hutton T et al (2016) The Malmo platform for artificial intelligence experimentation. IJCAI, New York, USA

[11-35] Juliani A, Berges V, Teng E et al (2018) Unity: A general platform for intelligent agents. https://arxiv.org/abs/1809.02627

[11-36] Kapetanakis S, Kudenko D (2004) Reinforcement learning of coordination in heterogeneous cooperative multi-agent systems. In: Adaptive Agents and Multi-Agent Systems II. Springer, Berlin, Heidelberg

[11-37] Kidambi R, Rajeswaran A, Netrapalli P et al (2020) Morel: Model-based offline reinforcement learning. NeurIPS, Vancouver, Canada

[11-38] Kumar A, Fu J, Soh M et al (2019) Stabilizing off-policy q-learning via bootstrapping error reduction. NeurIPS, Vancouver, Canada

[11-39] Kumar A, Zhou A, Tucker G et al (2020). Conservative q-learning for offline reinforcement learning. NeurIPS, Vancouver, Canada

[11-40] Kostrikov I, Nair A, Levine S (2021) Offline Reinforcement Learning with In-sample Q-Learning. ICLR, Vienna, Austria

[11-41] Lauer M, Riedmiller M (2000) An algorithm for distributed reinforcement learning in cooperative multi-agent systems. ICML, Stanford, USA

[11-42] Lewis F, Vrabie D, Syrmos V (2012) Optimal control. John Wiley & Sons, Inc., New York

[11-43] Li J, Li SE, Guan Y et al (2020) Ternary policy iteration algorithm for nonlinear robust control. https://arxiv.org/abs/2007.06810

[11-44] Li K, Li SE, Gao F et al (2020) Robust distributed consensus control of uncertain multiagents interacted by eigenvalue-bounded topologies. IEEE Internet of Things Journal 7(5): 3790-3798

[11-45] Liang E, Liaw R, Nishihara R et al (2018). RLlib: Abstractions for distributed reinforcement learning. ICML, Stockholm, Sweden

[11-46] Lindquist A (1973) On feedback control of linear stochastic systems. SIAM Journal on Control 11(2): 323-343

[11-47] Littman M (1994) Markov games as a framework for multi-agent reinforcement learning. Machine Learning Proceedings, New Jersey, USA

[11-48] Littman M (2001) Friend-or-foe Q-learning in general-sum games. ICML, Williams College, USA

[11-49] Mishra N, Rohaninejad M, Chen X et al (2017) A simple neural attentive meta-learner. ICLR, Toulon, France

[11-50] Movellan J (2009) Primer on POMDPs and infomax control. MPLab Tutorials, University of California San Diego, USA

[11-51] Mu Y, Li SE, Liu C et al (2020) Mixed reinforcement learning for efficient policy optimization in stochastic environments. IEEE ICCAS, Pusan, Korea

[11-52] Ng AY, Russell S (2000) Algorithms for inverse reinforcement learning. ICML, Stanford, USA

[11-53] Raffin A, Hill A, Ernestus M et al (2019). Stable Baselines3. Github repositories: https://github.com/DLR-RM/stable-baselines3

[11-54] Ren Y, Duan J, Li SE et al (2020) Improving generalization of reinforcement learning with minimax distributional soft actor-critic. IEEE ITSC, Rhodes, Greece

[11-55] Schaft A (1992) L2-gain analysis of nonlinear systems and nonlinear state feedback H-infinity control. IEEE Trans Automatic Control 37(6): 770-784

[11-56] Singh S, Jaakkola T, Jordan M (1994) Learning without state-estimation in partially observable Markovian decision processes. ICML, New Brunswick, USA

[11-57] Striebel C (1965) Sufficient statistics in the optimum control of stochastic systems. J of Math Analysis and Applications 12(3): 576-592

[11-58] Vinyals O, Ewalds T, Bartunov S et al (2017) Starcraft ii: A new challenge for reinforcement learning. https://arxiv.org/abs/1708.04782

[11-59] Wang D, He H, Liu D (2017) Adaptive critic nonlinear robust control: A survey. IEEE Trans Cybernetics 47(10): 3429-3451

[11-60] Wei Q, Liu D (2013) Adaptive dynamic programming for optimal tracking control of unknown nonlinear systems with application to coal gasification. IEEE Trans Automation Sci and Eng 11(4): 1020-1036

[11-61] Weng J, Chen H, Yan D et al (2021). Tianshou: a Highly Modularized Deep Reinforcement Learning Library. https://arxiv.org/abs/2107.14171

[11-62] Wonham W (1968) On the separation theorem of stochastic control. SIAM J on Control 6(2): 312-326

[11-63] Yu T, Thomas G, Yu L et al (2020) Mopo: Model-based offline policy optimization. NeurIPS, Vancouver, Canada

[11-64] Zachrisson L (1968) A proof of the separation theorem in control theory. Royal Institute of Technology, IOS Report No R-23

[11-65] Zhang K, Hu B, Basar T (2020) On the stability and convergence of robust adversarial reinforcement learning: A case study on linear quadratic systems. NeurIPS, Vancouver, Canada

[11-66] Zhang K, Yang Z, Başar T (2019) Multi-agent reinforcement learning: A selective overview of theories and algorithms. https://arxiv.org/abs/1911.10635

[11-67] Zhou K, Doyle J, Glover K (1996) Robust and optimal control. Prentice Hall, Inc., New Jersey

[11-68] Ziebart B, Maas A, Bagnell J et al (2008) Maximum entropy inverse reinforcement learning. AAAI, Chicago, USA

Correction to: Reinforcement Learning for Sequential Decision and Optimal Control

Shengbo Eben Li

Correction to:
S. E. Li, *Reinforcement Learning for Sequential Decision and Optimal Control*, https://doi.org/10.1007/978-981-19-7784-8

The original online versions of Chapters 1 to 11 of this book were inadvertently published with incorrect abstracts. The online version chapter abstracts have now been amended with this correction.

The updated version of the book can be found at
https://doi.org/10.1007/978-981-19-7784-8

Chapter 12. Index

© The Author(s), under exclusive license to Springer Nature Singapore Pte Ltd. 2023
S. E. Li, *Reinforcement Learning for Sequential Decision and Optimal Control*,
https://doi.org/10.1007/978-981-19-7784-8

Printed in the United States
by Baker & Taylor Publisher Services